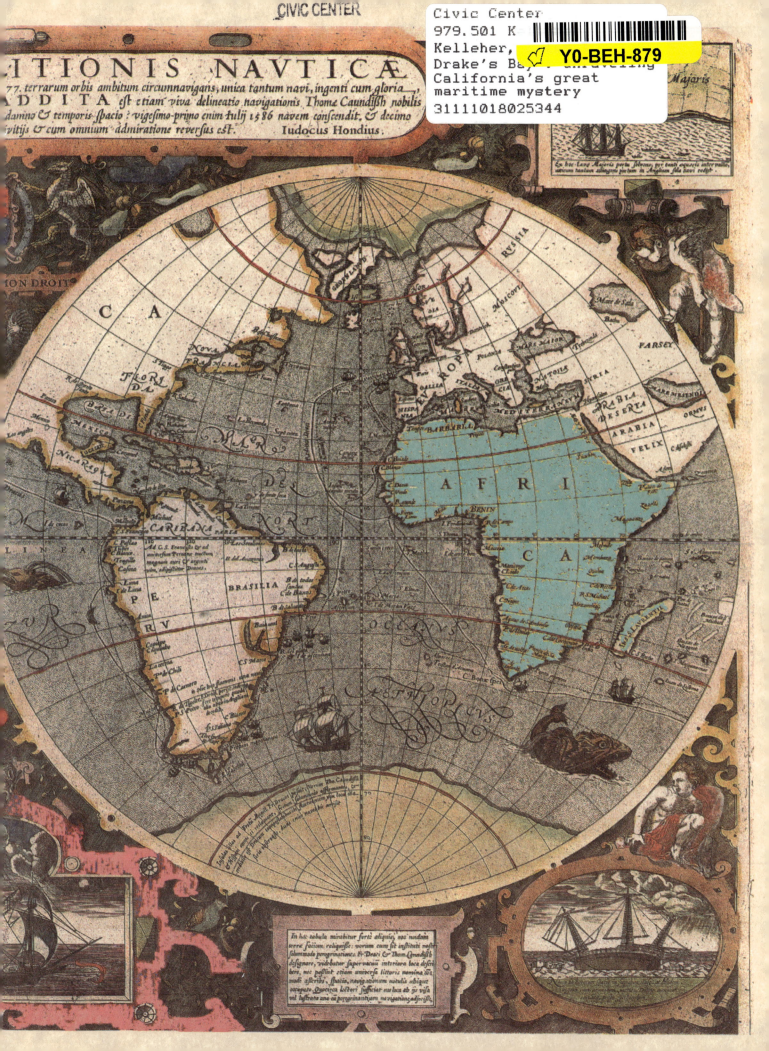

Drake's Bay

Drake's Bay

Unraveling California's Great Maritime Mystery

By

Brian T. Kelleher

Kelleher & Associates
P. O. Box 850
Cupertino, California 95014

Drake's Astrolabe: the instrument he used to make land-based latitude readings.
Courtesy of the National Maritime Museum, Greenwich, England

For information: Brian Kelleher, Kelleher & Associates
P.O. Box 850, Cupertino, California 95014
First U.S. Edition: 2,500 copies. ISBN: 0-9657609-0-1
Library of Congress Catalogue Card Number: 97-93385

Cover design and tint for world maps inside front and rear cover: Heidi Heath Garwood

Cover photo of Charles Lundgren's "Golden Hind"
courtesy of Unicover, Cheyenne, Wyoming; copyright 1983

Back cover:
Map of San Francisco Bay Area derived from IDENTIFICATION OF SIR FRANCIS DRAKE'S
ANCHORAGE ON THE COAST OF CALIFORNIA IN THE YEAR 1579 by George Davidson,
California Historical Society, San Francisco, 1890, Plate 1, Coast of New Albion
Interior:
Artwork for chapter headings is from Henry Wagner's epic
SIR FRANCIS DRAKE'S VOYAGE AROUND THE WORLD, San Francisco, 1926

Artwork for part headings is from John Robertson's
THE HARBOR OF ST. FRANCIS, San Francisco, 1926

Publication coordination, consultation and printing: Day Publishing, San Jose, CA

To Teri

Note on Punctuation

Spelling, capitalization and punctuation have been modernized throughout,
except in appendices A–C.

Contents

Illustrations

Maps

Tables

Preface

It was not entirely by chance that I embarked upon the quest to solve California's great maritime mystery. The writings of Jules Verne kindled a childhood fascination with shipwrecks and maritime mystery that had developed over the years into a hobby. In the summer of 1989, my friend Teri Fung and I took the Coast Highway north from San Francisco to explore Marin County's Point Reyes Peninsula. Teri was interested in walking the trails and beaches to enjoy the fresh salt air and scenery; I was anxious to get to the visitor's center to learn more about a sixteenth-century Spanish galleon wrecked in what is today called Drakes Bay. It was at the exhibits in the visitor's center that I learned that many, but not all, historians believed that the infamous Elizabethan seafarer, Sir Francis Drake, had beached the legendary *Golden Hind* at this bay in June 1579, leaving behind a large stone-walled fort that had never been discovered. My interest was immediately piqued.

Within a few weeks of the trip, I found a copy of Warren Hanna's *Lost Harbor* while browsing through one of San Francisco's many used bookstores. This book, published in 1979, summarizes the voluminous literature published over the last two centuries by the so-called anchorage sleuths: a host of historians, scientists, and laymen who have attempted to solve the anchorage mystery. After the first reading, I could not get the controversy about Drake's lost California harbor out of my mind. I kept Hanna's book on my bed stand and carried it along wherever I traveled, reexamining the contemporary accounts, reevaluating the theories and arguments, puzzling over the uncertainties and inconsistencies. In my free time I started visiting the candidate fort sites hoping I might find some stones that had been left unturned in the search. You will be surprised to see how many turned up, literally and figuratively.

It was in 1992 that I began in earnest the research leading to this book. The bulk of it was conducted in public libraries, mostly in Palo Alto where I kept the interlibrary loan staff scrambling for over three years. I suspect that never in their research-librarian careers had Roger Bonilla, Elnor Pahl, or Suzie Shark received such a constant and perplexing flow of requests. Thank you for your invaluable assistance. There were numerous trips to the Bancroft Library in Berkeley where I received much assistance from Richard Ogar and Cherise Sun. There were also trips to the Green Library at Stanford, the California Historical Society in San Francisco and the libraries and historical associations of Monterey, San Mateo, Marin, and Sonoma Counties. There was also a great deal of tramping, hiking, and boating along the shores of the Central California coast: my favorite part.

I found the research part of this work richly rewarding. The trails of evidence, as you will see, wind intricately through California's entire human saga, from the Stone Age to the Nuclear Age. It was like assembling a jigsaw puzzle cut from the pages of California's maritime history: sixteenth-century accounts and maps of Drake's voyage, the journals of early Spanish and English explorers, Russian accounts from their days in California, nineteenth-century California history books, old narratives, state archival records, antique nautical charts, ethnological and archaeological research papers, pioneer journals, old newspaper clippings, and the voluminous literature of the anchorage sleuths.

Although there are still some pieces missing, with those I have now, the picture that has emerged is clear enough to me. The big challenge I faced in researching and writing this book was making things clear to the reader.

I did have some help in this respect. In the summer of 1993, while I was vacationing in my hometown of Medfield, Massachusetts, my friend Wally Reynolds pointed out the home of Lisa Palson Priest, whom I remembered as the best writer in our high school class. It happened that Lisa and her colleague, Howard "Skip" MacMullen, had recently formed a partnership, Logos Editorial Associates, and they agreed to work as my editors. Their assistance covering general approach, organization, presentation, accuracy, grammar, syntax, and style is fully acknowledged. They also did their best to keep me out of trouble with the professional historians.

In addition, my close friends, Ron Floria and Teri Fung, offered constant encouragement and assistance when I set forth on the adventure of first-time writing. Teri was there when I discovered the fort site. Another close friend, Tony Meyer, accompanied me on many of the field trips. Ron, Tony, my friend Jim Nelson and my parents, Harry and Louise Kelleher, reviewed the draft manuscripts, providing excellent constructive criticism. Ron; Sharon Fournier, a business colleague degreed in literature; and especially my longtime friend from St. Petersburg, Florida, Debra DiGiacomandrea, a free-lance writer, provided much-needed assistance in proofing and correcting the text. Special thanks to Sylvia Barker Thalman of the Miwok Archaeological Preserve of Marin for providing input on Coast Miwok culture and to Ruth Burke and Nancy Conzett of Sonoma County for last minute input on points of local history.

Over the history of the anchorage debate, there has been no shortage of individuals who have convinced themselves that they had solved the riddle of Drake's lost California harbor. What nobody has been able to do so far, however, is present enough evidence to settle the matter conclusively. Have I done so? Follow the trails of evidence with me and decide for yourself. The solution I offer is this simple: Francis Drake on June 17, 1579, came to anchor in a California harbor that is located right where the most important contemporary accounts of the voyage place it; upon landing his goods and crew, he built a fortified encampment precisely where it is shown in a reasonably accurate sixteenth-century sketch of the harbor; and, without anyone realizing it, the stones of this walled encampment have been accidentally uncovered in modern times and are lying in open view near where they were placed over four centuries ago. I will show you how other anchorage detectives went astray and how, when properly interpreted, all the credible evidence leads to one inescapable and very surprising conclusion.

With the release of this book I am identifying what I believe to be the location of Francis Drake's 1579 California encampment, including the remnants of his stone fort: an important historical landmark that has eluded discovery for over four centuries. While I recognize this finding is speculative, I am confident it can be confirmed through the work of qualified archaeologists.

Fig. 1. Sir Francis Drake at age forty-three: an engraving attributed to Jodocus Hondius, circa 1590. *Courtesy The Bancroft Library, Berkeley, California.*

Fig. 2. The *Golden Hind* and her tender rounding Point Reyes Head: watercolor by Raymond Aker, 1971. *Courtesy of Raymond Aker, Palo Alto, California.*

Introduction

orning, June 17, 1579. A worried Francis Drake leaned over the port rail of a small three-masted bark and strained his eyes against the hazy gray air. He could see nothing of the land they were approaching, now some three miles distant, but looking back he could easily make out the *Golden Hind* following along at a safe distance. The wind was blowing hard northwest, his heading due east.

From his cross-staff readings, Drake believed that the ships were approaching thirty-nine degrees north latitude. According to the Spanish chart that had fallen into his hands, there should be a large bay surrounded by pine forests, the "Baia de los Pinos," situated in approximately this position. Here Drake hoped to find the harbor of refuge he and his beleaguered crew were so urgently seeking.

Their situation was becoming desperate. It had been more than sixty days since the Englishmen had last been at port. The *Golden Hind*, heavily laden with plundered Spanish treasure, was leaking badly. For the last ten days they had been sailing along a treacherous shoreline in violent winds and huge seas. Whenever the winds let up, blankets of dense fog rolled in, completely blinding them. The little they had seen of the coast was not promising: a steep, rocky wall fronted by murderous reefs and backed by barren plains and snowcapped hills. The crew was cold, exhausted and demoralized. Each league south brought them closer to enemy territory and certain doom.

Shadowy landforms began to emerge to the east—rugged black rocks upon which the sea was thundering. Standing within a mile of shore, Drake bore south and soon came upon a low sandy point projecting a mile seaward. Rounding the point with great expectations, he found the southern side composed of sandy cliffs, remarkably white, though interspersed with streaks of dull green. The land abreast was high, steep to the sea, and presented a rude and barren aspect with only a few scattered copses of forest trees. With no Baia de los Pinos in sight, Drake and his men filled the briny air with a torrent of curses for Spain and her inept pilots.

About seven o'clock that evening, just as the fog started to roll in again, ending the day's search, the bark's lookout sang out. They had fallen upon the entrance to a substantial bay lying under a fistlike headland of moderate height. Just inside was a small flat island, separated from the promontory by a channel, less than a mile in width, but clear of weed. The south side of the bay was already shrouded in mist.

The crew sprang to action! Roaring out orders, Drake took the bark on a small stretch to the southwest, tacked, and stood through the channel, lead constantly going, soundings regular from fifteen fathoms to four and a half, good bottom. They came to, took in sail, and dropped anchor a half mile off a sandy shore.

Fig. 3. The white-cliffed headland that lies towards the sea near Drake's former encampment. *Photograph by the author, 1994.*

Drake let out a deep sigh of relief while his eyes, ever alert for signs of opportunity or danger, quickly scanned the barren hills rolling back from the shoreline. Minutes after the *Golden Hind* hove to beside them, they were completely engulfed in fog.

Thus did the notorious English seafarer Sir Francis Drake become the first European to touch on Northern California's shores over four centuries ago.[1]

The Famous Voyage

Drake's California visit was but a brief sojourn in his spectacular 1577–80 circumnavigation of the globe, an expedition that rocked sixteenth-century Europe and influenced the course of modern European and American history. Under the guise of an innocent trading expedition to the Mediterranean, Drake left Plymouth, England, in December 1577. Bypassing the Strait of Gibraltar, he took his fleet of five ships across the Atlantic, slipped through the Strait of Magellan and entered the Pacific Ocean, then considered Spain's imperial lake. There the English were beset by a series of ferocious storms that drove them far enough south to discover that the Atlantic and Pacific Oceans were joined below the tip of South America. When the weather cleared two months later, Drake and his crew found themselves alone. The rest of the fleet was either abandoned en route, was lost, or was beating a retreat to England. Undaunted, Drake changed his ship's name from *Pelican* to *Golden Hind* and stormed up the west coast of South America, sacking Spanish ports and plundering her shipping at will. The climax of the voyage occurred off the coast of Ecuador when the English captured a Spanish galleon transporting an enormous horde of silver from Peru to Panama.

His vessel filled to the gunwales with booty, Drake now faced the problem of how to return to England. He weighed three options: go back the same way he came; cross the Pacific and round Africa; or sail north and wind his way back to the Atlantic through the mythical Straits of Anian. Choosing the latter, he struck boldly north into territory where no European had ever ventured. Frustrated by frigid temperatures, dense fog, and contrary winds, he finally abandoned this hopeless pursuit somewhere along the coast of what is now Washington State. Retreating south in search of a safe haven, on June 17, 1579, he fell upon a convenient harbor in present-day Northern California. During their thirty-six-day stay, the English established friendly relations with the natives and explored the hinterland. Before leaving, they set up a great post upon which Drake nailed a brass plate inscribed with a proclamation claiming the territory in the name of his queen—England's first formal claim to American soil. He named the country Nova Albion, Latin for "New England."

Departing California in late July, Drake crossed the Pacific. During a brief stop in the Spice Islands (Moluccas), he managed to strike a trade agreement with the sultan of Ternate, laying the foundation for Great Britain's enormously successful East India Company. After a series of further adventures in the East Indies (now Indonesia) and the Indian Ocean, the weary mariners rounded the Cape of Good Hope and in September 1580, after almost three years at sea, finally slipped back into Plymouth Harbor.

Despite some tense moments stemming from the international furor created by his raids on Spanish America, Queen Elizabeth I knighted Drake for his accomplishments shortly after his return.

The Anchorage Controversy

Most authorities agree that Drake's June 1579 landfall was in the vicinity of San Francisco, California. The question of exactly where he came to anchor, however, is one that has provoked heated controversy, alternately smoldering and erupting, for more than two centuries.

---

1. I have recreated this semifictional account of Drake's approach to his landfall from the journals of the next two English mariners who sailed these waters: Captain James Colnett in the *Argonaut*, 1790 (Colnett, *Journal*, pp. 173 and 174); and George Vancouver in the *Discovery*, 1792 (Vancouver, *Voyage of Discovery*, pp. 698–700). Many of the early navigators reported seeing snowcapped hills along the California coast including Drake, Captain James Cook, and Vancouver. It seems they were fooled by dense, low-lying fog and/or some very white sand dunes. Drake's own journal disappeared in the sixteenth century.

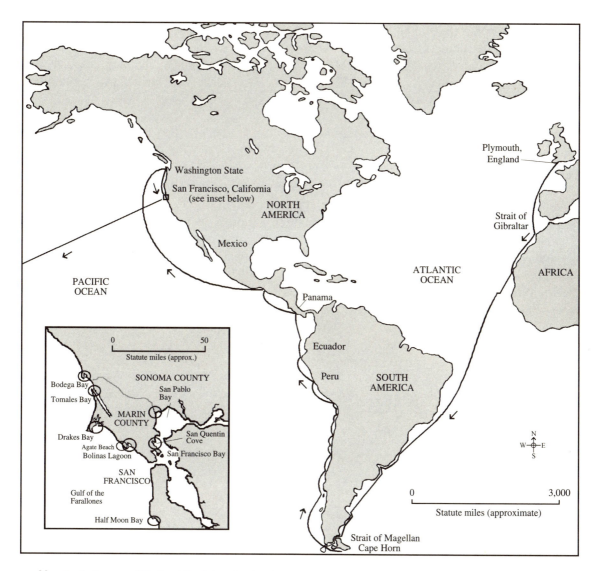

Map 1. Outline map of North and South America showing the outbound route of Drake's circumnavigation from Plymouth, England, in December 1577 to the California coast in June 1579. The *inset* shows eight candidate anchorage sites (*circled*) in the San Francisco area.

Historians, anthropologists, archaeologists, geographers, geologists, climatologists, sailors, navigators, cartographers, noted authors, politicians and a host of laymen from various walks of life have taken part in the fray. There have been five comprehensive books on the subject and hundreds of articles in newspapers, historical journals and magazines. There have been nautical reenactments, archaeological expeditions, and formal debates. Many California history books and Drake biographies contain a lively discussion of the subject, some including a special chapter devoted exclusively to the anchorage question. Despite all the attention, however, no one yet has been able to muster sufficient evidence to settle the matter conclusively. Some have become resigned to the opinion that there will never be a satisfactory solution to this puzzle.

Candidate Sites and Current Opinion

Since the late 1800s, consensus opinion has held that Drake anchored just north of San Francisco within the bay in Marin County that now bears his name. Thanks to the monumental efforts of the Drake Navigators Guild, a distinguished group of retired naval officers, mariners, and local citizenry who devoted more than four decades of intense research to the subject, many authorities now pinpoint the landfall at a small cove within Drakes Estero, a large estuary on the northern side of Drakes Bay. Sir Francis Drake Boulevard, Sir Francis Drake High School, a thirty-foot-tall statue, several monuments and plaques, and a host of hotels and tourist stops proudly commemorate Drake's historic visit to Marin County. In June 1994, the Marin County Board of Supervisors adopted a resolution officially declaring Drakes Estero as the landing site.

There are only three other locations that have received serious scholarly support as candidate anchorage sites in the past half century, all in Marin County: Bolinas Lagoon, a small harbor just south of Drakes Bay; Agate Beach at the north end of Bolinas Bay; and San Quentin Cove on the north shore of San Francisco Bay (see inset, map 1). The notion that Drake was the first European to enter San Francisco Bay has been popular since Americans first settled California in the Gold Rush days. Since then there have always been steadfast supporters of this theory.

Other San Francisco area sites that have had some support include Bodega Bay in Sonoma County; Tomales Bay, and San Pablo Bay, all in Marin County; and Half Moon Bay and Point Año Nuevo in San Mateo County. Arguments have also been raised in favor of Trinidad Bay in Mendocino County; Fan Shell Beach and the mouth of the Carmel River in Monterey County; and San Luis Obispo Bay and Santa Barbara Bay in Southern California. Two sites on the coast of Oregon that have received some attention are Nehalem Bay and Whale Cove.

The Mystery Unraveled?

The dense fog that has hung stubbornly over Drake's lost port for over four hundred years finally begins to lift. An eerie dawn breaks over a broad expanse of ice-still water. In the distance, shadowy images of the dark cliffs forming the perimeter of the cove appear on the opposite side of the harbor. Faint silhouettes of masts, spars, and rigging emerge from the gloom. . . .

What cove is that emerging through the mists ?

Fig. 4. The *Golden Hind* emerging through the mists of Drake's lost harbor. *Adapted from a photograph by the author, 1994.*

The goal of this book is to present not just the solution to the anchorage puzzle, but the complete story of one of America's greatest maritime mysteries. Part I, "Nova Albion," recounts the adventures of Drake's circumnavigation and the California sojourn derived from firsthand accounts. Part II, "Contemporary Evidence," lays out all the early maps and journals that bear evidence on the location of the California landfall. Part III, "The Great Debate," summarizes the two-hundred-year-old anchorage controversy. Part IV, "Assembling the Pieces," fits the many parts of this historic jigsaw puzzle into a surprisingly clear picture. Part V, "Lost Harbor," relates the fascinating human saga that took place over the bones of Drake's encampment, including the tragic fate of the Nova Albion Indians and the astounding saga of the hole in the head. Part VI, "Drake's American Legacy," bears on the historical significance of Drake's visit and Nova Albion proclamation and contains a biographical sketch highlighting the role Drake played in America's founding.

Along the way, I will endeavor to solve a number of other interesting Drake puzzles and perhaps, in the process, stir up a bit more controversy about the great Elizabethan mariner.

Drake's California Legacy

Was it by chance that the swashbuckling Sir Francis Drake, one of the most famous, heroic, and notorious seafarers of all time, played out the first chapter in California's spectacular modern age? Is it mere coincidence that Drake's treasure-laden *Golden Hind* was the first European vessel to touch on her golden shores?

Drake's importance in California history is well illustrated by the scale of the quadricentennial celebrations the state held to commemorate his 1579 visit. In Marin County there were exhibits, lectures, nine performances of the play *Drake in the Pacific*, and twelve performances of the pageant *Drake: King of Nova Albion*. Other celebrations were organized by the 100-member Sir Francis Drake Society and a special legislative body, the Sir Francis Drake Commission, and included religious services, pageants, parades, exhibitions, and an international conference.

The British consul general opened the Drake conference with an address at the University of California, Los Angeles (UCLA). The seven-day affair moved briefly to Berkeley and ended in San Francisco on June 17, 1979, with a banquet at the Sir Francis Drake Hotel and a special service at Grace Cathedral given by the lord bishop of London. Noteworthy ceremonies that took place in between included the unveiling of two plaques honoring Drake, one at Vista Point near the north end of the Golden Gate Bridge, the other on the shores of Drakes Bay. There was also a reenactment of Drake's landing at modern-day Drakes Bay. In addition to organizing and sponsoring the events, the Sir Francis Drake Commission was instrumental in the publication of three books: Warren Hanna's *Lost Harbor: The Controversy over Drake's California Anchorage*, which I mentioned in the preface; Norman Thrower's *Sir Francis Drake and the Famous Voyage, 1577–1580*, which includes nine scholarly papers presented at the Drake conference; and Benjamin Draper's *Drake Bibliography, 1569–1979*. All three authors were members of the commission.[2]

The next time you enter a public library, peruse the subject category for Sir Francis Drake. You will be amazed at the amount of literature this famous Elizabethan sea dog has inspired. The most recent Drake biographies by John Sugden (1990) and John Cummins (1995) have been critically acclaimed. There are more than a dozen others to choose from. Try Julian Corbett's, *Sir Francis Drake*, published 1890, if you can find it.

The mystique and quadricentennial celebrations notwithstanding, today's Californians are, for the most part, indifferent to the state's history and heritage. Many, like myself, are transplanted from other states and countries. With the stirring up of the lost harbor controversy, perhaps California's present generation will take a renewed interest in the Drake legacy.

---

2. Thrower, ed., *Sir Francis Drake*, pp. xiii – xvii.

Fig. 5. Sir Francis Drake gazes out to sea from Plymouth Hoe. This larger than life-size bronze statue was erected at Plymouth, England in 1884. It is a replica of an identical statue erected the year before that still stands in Tavistock, England, Drake's birthplace. The original was sculpted in Germany in 1883 by Joseph Edgar Boehm. *Courtesy of the Drake Navigators Guild, Palo Alto, California.*

PART I

Nova Albion

Fig. 6. "Inhabitant of Rumiantsev Bay": painting by Mikhail Tikhanov, 1818. *Courtesy of the Scientific Research Museum, Academy of Arts, St. Petersburg, Russia.*

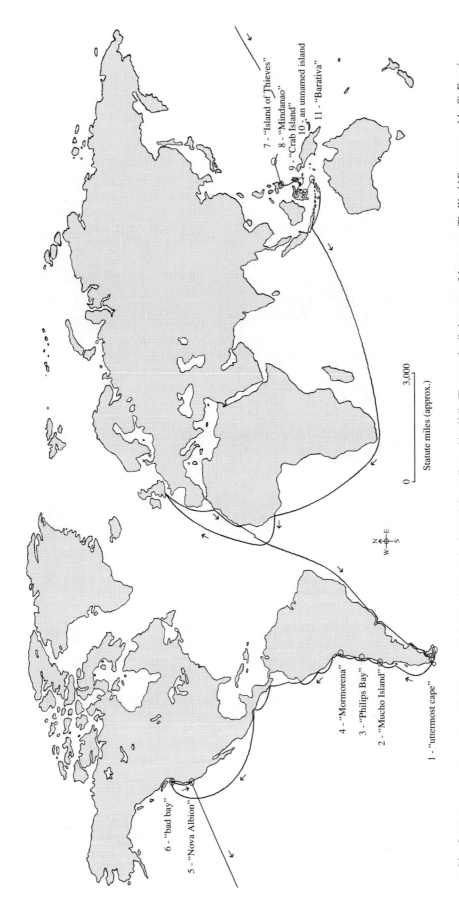

7 - "Island of Thieves"
8 - "Mindanao"
9 - "Crab Island"
10 - an unnamed island
11 - "Baratíva"

4 - "Mormorena"
3 - "Philips Bay"
2 - "Mucho Island"
1 - "uttermost cape"

6 - "bad bay"
5 - "Nova Albion"

N
W + E
S

0   3,000

Statute miles (approx.)

Map 2. Outline map of the world showing the route of Drake's circumnavigation and eleven disputed landfalls. The most detailed account of the voyage, *The World Encompassed*, by Sir Francis Drake (1628), reports fifty latitude readings, some thirty of which ostensibly pinpoint bays, harbors or islands where the English made anchor. Of these thirty, Drake's lost California harbor is one of nine where historians have rejected *The World Encompassed*'s latitude evidence and raised questions about the whereabouts of the anchorage. On the map above, I show the approximate locations of these nine disputed landfalls plus two others that I contend historians have traditionally misplaced ("Mucho Island" and "Philips Bay"). Putting aside California's lost harbor for the moment, in this chapter I will readily identify the locations of these ten other disputed circumnavigation landfalls based on the following simple premises: where the latitude readings of these ten other disputed circumnavigation landfalls were likely made on land, they are accurate to within fifteen minutes; where likely made aboard ship, they are accurate to within thirty minutes.

CHAPTER I
The Circumnavigation

To set the stage for our Nova Albion story, I provide here a brief account of Francis Drake's famous round-the-world voyage of 1577–80. By encompassing the globe, Drake commanded history's second circumnavigation. Ferdinand Magellan, who led the first circumnavigation in 1519–21, was killed by Philippine natives just halfway along his epic voyage, making Drake the first commander to survive the endeavor. Except where indicated, I have drawn the account from *The World Encompassed* by Sir Francis Drake, compiled, edited and published in 1628 by Drake's nephew, heir and namesake, Sir Francis Drake, baronet.

In the process of relating the key episodes of Drake's historic voyage, I will take the first step in unraveling California's great maritime mystery. I will readily identify all of Drake's other important circumnavigation landfalls that have come into question over the years by applying two simple premises, the rationale for which I will explain later on. The first premise is that the author of *The World Encompassed* drew from the original logbooks of the voyage to provide the navigational data found in the account. The second premise is that the navigational detail set forth in *The World Encompassed* is sufficiently accurate to identify Drake's landfalls. Appendix D contains figures and tables that summarize and evaluate all the latitude readings reported in this account. Appendix E has additional information on Drake's disputed anchorage on the Pacific Northwest Coast.

Fitting out the Fleet

There has been much written of the motives underlying Drake's circumnavigation. Some have branded him an unscrupulous freebooter and view the event as the dawn of England's age of piracy. Others see Drake as a great patriot, and view the event as the birth of British maritime greatness. This controversy is covered in chapters 38 and 39. In any event, Drake certainly recruited an impressive group of investors to help him finance the expedition. The syndicate included Her Majesty, Queen Elizabeth I; Sir Edward Clinton, the earl of Lincoln; Sir Christopher Hatton, the captain of the queen's guard; Sir Robert Dudley, the earl of Leicester; Sir Francis Walsingham, the secretary of state; and three of the English Crown's highest ranking naval officials and naval board members: Sir John Hawkins, and William and George Winter. They put together the following fleet manned by approximately 160 men and boys: the *Pelican*, 100 tons, owned and commanded by Drake; the *Elizabeth*, 80 tons, owned and commanded by John Winter, George Winter's son; the *Marigold*, 30 tons, captained by John Tomas; the *Swan*, 50 tons, owned by Drake and captained by John Chester; and the *Benedict*, 15 tons, owned by John Winter and captained by Tom Moone. Lashed to the decks, the ships also carried four small pinnaces, disassembled.

Down the Coast of Africa

The tensions between England and her former Iberian allies were running high when Drake's fleet departed Plymouth, England, in December 1577 and sailed south past the watchful eyes of Spain and Portugal. Elizabeth's uneasy three-year truce with King Philip of Spain was wearing thin. It had been just five years since broadsides had been exchanged in the English Channel, and just four since Drake returned from the Spanish Caribbean, his ships wallowing under the burden of the plunder he took at Panama. Rumors filled the air of both the English and Spanish courts: Elizabeth was to be assassinated; the Spanish army would attack from across the Channel; the dreaded *El Draque* (Spanish for "Drake") would sail again on the Caribbean.

Though the purported destination of the expedition was the Mediterranean Sea, Drake sailed his flagship right past the Strait of Gibraltar and continued south along the coast of Northwest Africa.

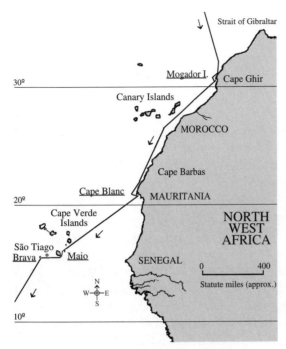

Map 3. Outline map of Northwest Africa showing Drake's four landfalls (<u>underlined</u>). *The World Encompassed* provides latitude readings for all four landfalls and there has been no dispute about any of the locations. The four readings are accurate to within fifteen minutes.

What some had known or suspected all along, now became evident to the entire company: this was not going to be an innocent trading expedition. After a week's stay at Mogador Island, Morocco, the English seized a half-dozen Spanish and Portuguese vessels between "Cape DeGuerre," (Cape Ghir) and Cape Barbas, Morocco, releasing them after taking their victuals. At "Cape Blanck," (Cape Blanc, Mauritania), they force-traded the *Benedict* for a slightly larger Portuguese ship they named the *Christopher* and took her Portuguese pilot. In the Cape Verde Islands, they landed at Maio and Brava and while off the coast of "Saint Iago" (São Tiago) seized a 100-ton Portuguese trader loaded with wine and merchandise. Putting the crew ashore on Brava, they took the ship, her cargo, her pilot Nuño da Silva and his valuable nautical charts. Drake renamed the prize the *Mary*, after his wife.[1]

Across the Atlantic and into the River of Plate

In the mid-Atlantic the English were astonished to observe fish that flew above the water's surface to escape their pursuers. The crews and officers must have been equally astonished when after approaching land, Drake directed the fleet south along the coasts of Brazil and Argentina, rather than heading north to plunder the Spanish Caribbean. They entered the River of Plate, carefully reconnoitered the surrounding area, and replenished their water supplies and victuals.

Storms scattered the ships as they continued south along the coast of Patagonia (southern Argentina). At "Cape Hope" (Cape Tres Puntas, Argentina), Drake put the Golden Hind into a large bay and by lighting signal fires on the shore, reassembled all the ships but the Swan and Mary. He then took the fleet into another bay a short distance to the south where they were soon joined by the *Swan*. During their two-week stay, the English befriended a tribe of Patagonian "giants" and having reprovisioned on the abundant seal population, named the locale "Seal Bay" (Port Desire, Argentina). To consolidate his forces, Drake had the crew scuttle the *Swan* for firewood, and in another bay a little farther south, abandoned the *Christopher*.

Port Saint Julian and the Doughty Affair

Continuing south, Drake took the fleet into a small harbor (Port Santa Cruz, Argentina) about one hundred miles north of the Strait of Magellan. A council was held, and it was to decided to make a final attempt to recover the *Mary* before entering the strait. Reversing course, they chanced upon the missing ship the very next day, finding her in a high state of distress. To refresh the men and refit the ships, Drake took the fleet into nearby Port St. Julian, the very harbor Magellan had used for the same purpose some sixty years before.

---

1. The Francis Fletcher narrative (Penzer, ed., *The World Encompassed*, p. 93) is the only contemporary account that mentions the taking of a Portuguese pilot at Cape Blanc. The pilot may have been one N. de Morena, whom, according to Spanish accounts, Drake later left for dead on the Pacific Northwest Coast. The fact that Drake carried two Portuguese pilots on the circumnavigation is confirmed by the Spanish ambassador, Bernardo de Mendoza, who in 1580 wrote that "Drake asserts that had it not been for two Portuguese pilots he took from one of the ships [actually two ships] he plundered on the coast of Brazil [actually Africa] on his way out, he could never have made the voyage." Cited by Wallis, "The Cartography of Drake's Voyage," in Thrower, ed., p. 131.

Fig. 7. Flying fish and their pursuers: bonito, dolphin and sea fowl. an illustration from the Francis Fletcher narrative, drawn circa 1591. This is one of sixteen drawings illustrating the only existing copy of a manuscript ostensibly authored by the expedition's minister, Francis Fletcher, circa 1591—the so-called Fletcher notes (Fletcher narrative) copied in 1677 by Londoner, John Conyers. It is likely that the original was drawn aboard the *Golden Hind* either by Francis Drake or his nephew John Drake. *By permission of the British Library, London.*

Their interlude at Port St. Julian turned out to be an evil one. On the day of arrival Drake lost two men in a skirmish with the Patagonian giants who were not as friendly as those at Seal Bay. "This bloody tragedy being ended, another more grievous ensued." On a small island within the harbor stood the grim gallows that Magellan had erected in putting down a mutiny, bones still scattered on the ground below it. Drake found himself compelled to carry out the same gruesome task on this "Island of Blood." Thomas Doughty, a gentleman Drake had befriended in Ireland and the expedition's highest-ranking land officer, was accused of fomenting mutiny, tried, found guilty, and beheaded.[2]

The *Mary* was too heavily damaged for repair, so they broke her up for firewood and spent several months trimming the three remaining ships. Between Doughty's execution, the hardships already endured, and the certain dangers that lay ahead, the company became so disheartened and quarrelsome that Drake was hard pressed to keep them from abandoning the venture. The Sunday before they were to leave port, he assembled the entire company and gave a long sermon, rallying his men, explaining the true motives of the voyage, and impressing upon them the urgency and importance it held for their queen and country:

> Thus it is, my masters, that we are very far from our country and friends; we are compassed in on every side with our enemies; wherefore we are not to make small reckoning of a man, for we can not have a [another] man if we would give for him ten thousand pounds. Wherefore we must have these mutinies

2. The quotations are from the Francis Fletcher narrative (Penzer, ed., *The World Encompassed*, pp. 124 and 127). There has been a great deal written about Doughty's execution, with Drake's detractors accusing him of murder. Drake was apparently vindicated of all such charges upon his return to England.

and discords that are grown among us redressed, for by the life of God it does even take my wits from me to think on it; here is such controversy between the sailors and the gentlemen, and such stomaching between the gentlemen and the sailors, that it does even make me mad to hear it. But, my masters, I must have it left, for I must have the gentlemen to haul and draw with the mariner, and the mariner with the gentlemen. What, let us show ourselves all to be of a company, and not let us give occasion to the enemy to rejoice at our decay and overthrow. . . .

. . . We have now set together by the ears three mighty princes, as first Her Majesty, [and then] the kings of Spain and Portugal, and if this voyage should have good success, we should not only be a scorning or a reproachful scoffing stoke unto our enemies, but also a great blot to our whole country forever, and what triumph would it be to Spain and Portugal, and again the like would never be attempted.[3]

Drake offered the *Marigold* to those who would not go on. There was not an Englishman among them, however, who could refuse the tantalizing alternative that Drake put forth: the opportunity to raze the immensely rich seaports along the unprotected and unsuspecting backside of Spanish America!

Through the Strait and into the Stormy Pacific

Drake changed his ship's name from *Pelican* to *Golden Hind* as they entered the strait with the *Elizabeth* and *Marigold*. A day's sail inside the mouth, they made a brief landfall on a "very fair and large" island that Drake named "Elizabeth Island" (Isla Isabel) after taking formal possession for his queen. They completed the transit in two weeks without mishap.[4]

Entering the Pacific, however, the tiny fleet was soon beset by a furious three-week storm that drove the English farther south than Europeans had ever ventured before. The *Marigold* foundered with all hands. When the storm subsided, they made their way back to an open bay a little north of where they had exited the strait. More storms battered them there, separating the ships. John Winter took the *Elizabeth* back into the strait and waited briefly for Drake to appear. When he did not, Winter turned back for England.

The storms chased the *Golden Hind* south again until Drake and his crew found themselves among the remote islands at the tip of the southern continent that he named the "Elizabeth Islands." Drake took the ship into "a haven of an island" (Hermite Island, Chile) and made an excursion to the "uttermost cape" of the islands (Cape Horn), where he went ashore, engraved the queen's name and the date on a stone, and claimed the territory for England. Drake was the first European to discover that the Atlantic and Pacific were joined south of Tierra del Fuego by today's Drake Passage:[5]

The uttermost cape or headland of all these islands [Cape Horn] stands near in 56 deg. without which there is no main nor island to be seen to the southwards: but that the Atlantic Ocean and the South Sea meet in a most large and free scope.[6]

The Mucho Island Incident

The weather had finally cleared during their short stay in the Elizabeth Islands and with the wind to his stern, Drake headed northwest in search of his two missing ships. Undaunted by their loss, the determined mariner sailed boldly up the coast of Chile. His first landfall, "Mucho Island"(Santa Maria Island, Chile), was nearly fatal. The large population of Indians who inhabited this island thought Drake and his men were the accursed Spanish who had recently driven them from their homelands. After first extending a friendly welcome to the unwary seamen, they set up an ambush and attacked the eleven-man shore party as they landed the next day. Drake and seven others were severely wounded, barely escaping with their lives. Two men were lost.

---

3. From the John Cooke narrative, in Vaux, ed., *The World Encompassed*, pp. 215–16.
4. Drake coined the name "Golden Hind" from the mythical animal resembling a deer that adorned the Hatton family crest. Sir Christopher Hatton was one of the principal investors in the voyage and Drake's friend.
5. The first quotation is from Nuño da Silva's second relation in Nuttall, *New Light on Drake*, p. 260; the excursion to Horn Island is described in the Francis Fletcher narrative (Penzer, ed., *The World Encompassed*, p. 135).
6. Sir Francis Drake and Francis Fletcher, *The World Encompassed*, p. 44 (here after cited as Drake, *The World Encompassed*).

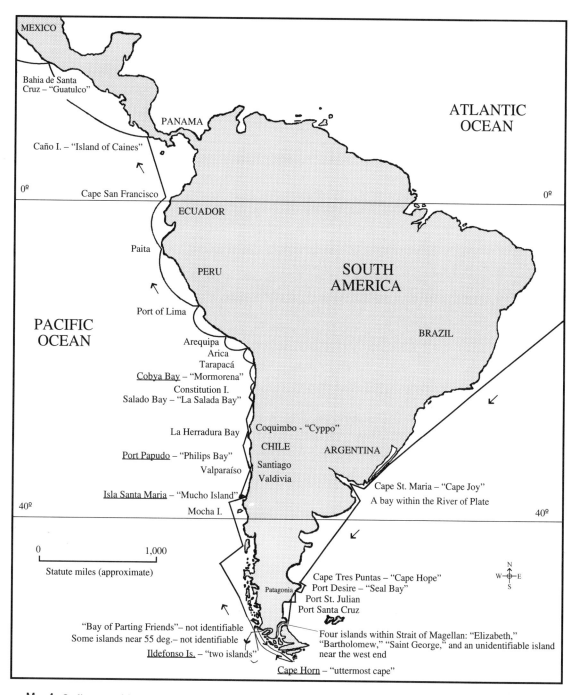

MEXICO

Bahia de Santa
Cruz – "Guatulco"

PANAMA

ATLANTIC
OCEAN

Caño I. – "Island of Caines"

0º   Cape San Francisco   0º

ECUADOR

Paita

PACIFIC
OCEAN

PERU

SOUTH
AMERICA

Port of Lima

BRAZIL

Arequipa
Arica
Tarapacá
Cobya Bay – "Mormorena"
Constitution I.
Salado Bay – "La Salada Bay"

La Herradura Bay   Coquimbo - "Cyppo"

CHILE   ARGENTINA

Port Papudo – "Philips Bay"

Valparaíso   Santiago
Valdivia

Isla Santa Maria – "Mucho Island"

40º   Mocha I.   40º

Cape St. Maria – "Cape Joy"
A bay within the River of Plate

0   1,000

Statute miles (approximate)

N
W—⊕—E
S

Cape Tres Puntas – "Cape Hope"
Patagonia   Port Desire – "Seal Bay"
Port St. Julian
Port Santa Cruz

"Bay of Parting Friends"– not identifiable
Some islands near 55 deg.– not identifiable
Ildefonso Is. – "two islands"

Four islands within Strait of Magellan: "Elizabeth,"
"Bartholomew," "Saint George," and an unidentifiable island
near the west end

Cape Horn – "uttermost cape"

Map 4. Outline map of Central and South America showing Drake's twenty-eight landfalls. *The World Encompassed* reports latitude readings for sixteen of the landfalls. I have underlined my solutions to five locations where potentially identifiable landfalls are in question. Of these five, there are three that have been historically disputed ("uttermost cape," "two islands," and "Mormorena") and there are two that I believe have traditionally been misplaced ("Mucho Island" and "Philips Bay"). *See appendix D.*

Our general notwithstanding he might have revenged this wrong with little hazard or danger; yet being desirous to preserve one of his own men alive, then to destroy one hundred of his enemies, committed the same to God, with this only punishment to them: that they did but know whom they had wronged; and that they had done this injury not to an enemy, but to an Englishman who would rather have been a patron to defend them, than any way an instrument of the least wrong . . .[7]

7. Ibid., pp. 48–49.

Razing the Coast of Chile and Peru

Upon exiting the Strait of Magellan, Drake had instructed the captains of the two other ships to rendezvous at thirty-two degrees south latitude in the event they were separated. Accordingly, Drake proceeded to the appointed position and anchored a short distance to the south in an inlet that he named "Philips Bay" (Port Papudo, Chile). He found no sign of the other ships, and his shore party was unsuccessful in efforts to secure water and provisions. They did, however, chance upon a Spanish-speaking Indian chief named Philip, who informed them that there was a ship at harbor not far south, and agreed to pilot them to it.

This nearby harbor turned out to be Valparaíso, the port of Santiago, Spain's southernmost mining outpost in South America. Departing with the ship's pilot, "Juan the Greek," Drake sailed north looking for an uninhabited harbor to refit the ship and reprovision. He found one at "La Salada Bay" (Salado Bay, Chile) after attempting a landing at the small harbor (La Herradura, Chile) just south of the town of "Cyppo" (Coquimbo, Chile). A month later, he resumed their course north intent on razing the rich seaports of Peru.[8]

They made a rich haul of plunder. Anchoring near the Indian town of "Mormorena" (Cobya Bay, Chile), they surprised a Spaniard driving a team of "Peruvian sheep" (llamas) and unburdened him of his sixteen bags of refined silver. At the town of Tarapacá, they happened upon a sleeping Spaniard and stole thirteen silver bars lying on the ground beside him. At the port of Arica they captured two ships, one laden with wine, the other with thirty-seven bars of silver and a chest of silver coins. At the port of Arequipa, they just missed taking five hundred bars of silver that the now-alarmed Spanish had frantically cast off the deck of a galleon just hours before they arrived. The English took the galleon and set her adrift along with the other two vessels they had taken at Arica. Outside the port of Lima, they again captured several ships and learned that a large treasure ship, the (nicknamed the *Caca Fuego*) had recently departed for Panama. Entering the port, they cut the cables or otherwise disabled the numerous ships they found after fruitlessly searching them for treasure. At Paita they found no treasure, but did pilfer some more wine and provisions. They learned that the *Caca Fuego* had left port just two days before.[9]

Capture of the Great Treasure Galleon

Now hot on the trail of the treasure galleon, off Cape San Francisco, Ecuador, they captured a small ship belonging to Benito Diaz Bravo. They took forty bars of silver, some gold, some large emeralds, a gold crucifix, and a good deal of provisions. After releasing the crew in the ship's boat, Drake set the vessel adrift without sails and only her owner aboard.

The high point of the voyage occurred the next day, March 1, 1579, when Drake finally caught up to the *Caca Fuego* and took her after a brief fight. The booty included 1,300 silver bars, 14 chests of silver coin, and some gold, jewelry and plate. He released the Spanish noble, Don Juan de Anton, his ship and his crew unharmed, giving them a few presents for their troubles.

Razing the Coast of New Spain

Needing a haven to refit his ship, Drake careened the *Golden Hind* on the coast of Central America near the "Island of Caines," (Caño Island, Costa Rica). Here they captured a tiny merchant ship that happened by, bound for Panama: Rodrigo Tello's bark. Along with some provisions, Drake took the ship's pilot who was familiar with the Manila trade, and his valuable charts. After careening the

8. Ibid. Drake, *The World Encompassed*, reports that Valparaíso "stands in 35. deg. 40. min." which is very likely a transcription error for 32. deg. 40. min.
9. In the case of Arica (18º 30′) that Drake, *The World Encompassed*, reports to "stand in 20. deg.," it is evident that the compiler assigned Arica the position that Drake recorded for his previous landfall at Tarapacá which indeed stands at about twenty degrees. This conclusion is reinforced by the fact that the compiler appears to have confused his source information at this point in the account, and reported the Tarapacá incident out of sequence.

Fig. 8. The *Golden Hind* capturing the *Caca Fuego*: reproduction of an engraving illustrating a book by Levinus Hulsius, Nürnberg, Germany, drawn circa 1600. Some view the event as the dawning of England's age of piracy. It can also be viewed as the birth of the British Empire. *Courtesy of the Rare Books and Manuscripts Division, The New York Public Library, Astor, Lenox and Tilden Foundations. KB 1598.*

Golden Hind, Drake left his last pinnace with Tello and his crew, and set sail with the bark and her pilot.

A few days out, they captured the ship of a wealthy Spanish official, Francisco de Zárate. It was en route from Acapulco to the Port of Lima laden with products of the Manila trade. They lightened her of some of her more valuable cargo including silks, porcelain and jewelry, as well as Zárate's African mistress, María. After an evening of lavish entertainment at the noble's expense, Drake released Zárate with his ship and crew. He also released Tello's pilot who had adamantly refused to assist Drake in any way.[10]

Drake's last stop on the coast of New Spain was at the port of "Guatulco," (Bahia de Santa Cruz, Mexico). They plundered the only ship in port and sacked the small town, taking a pot full of coins and a gold chain. Upon departing, Drake released Nuño da Silva and all his remaining prisoners except María, two other Africans, and, according to Spanish accounts, a Portuguese pilot named N. de Morena. Before leaving Guatulco, Drake attempted to explain his actions to one of the Spanish officials he had captured:[11]

> You will say this man who steals by day and prays by night in public is a devil, but what I do is because, just as when King Philip gives a written order to your viceroy, Don Martin Enriquez, in which he tells him what to do and how to govern, so the queen, my mistress, gives one to me to come to these parts. Therefore I am doing so, and if it is wrong she knows it and I am not to blame for anything, although it gives me pain to make his vassals pay, as I would not wish to take anything except what belongs to Enriquez.[12]

In Search of the Straits of Anian

His two vessels (the *Golden Hind* and Tello's bark) filled to the gunwales with booty, Drake boldly struck north into the uncharted Pacific Northwest, convinced he could reach England via the mythical Straits of Anian (the long-sought passage across North America then thought to lie in the latitude of today's Canada). Frustrated by storms, frigid temperatures, dense fog, and contrary winds, the company finally induced their resolute captain to abandon this hopeless pursuit after they stumbled upon the coast of Washington and came to anchor in a "bad bay" (Queenhythe Bay, Washington - see appendix E).

> . . .[because of the cold] a sudden and great discouragement seized upon the minds of our men, and they were possessed with a great mislike and doubting of any good to be done that way [in sailing further north] – yet would not our general be discouraged, but as well [as] by comfortable speeches of the divine providence and of God's loving care over his children out of the scriptures; as also by other good and profitable persuasions, adding thereto his own cheerful example, he so stirred them up to put on a good courage, and to acquit themselves like men, to endure some short extremity to have the speedier comfort, and a little trouble to obtain the greater glory; that every man was thoroughly armed with willingness, and resolved to the uttermost . . .[13]

Before departing, Drake put ashore as dead the fever-stricken Portuguese pilot, N. de Morena, who subsequently made his way on foot to Mexico.[14]

Nova Albion

Retreating south in search of warmer weather and a safe haven where they could repair and reprovision the *Golden Hind*, on June 17, 1579, they fell upon a good harbor in the vicinity of today's San Francisco, California. During their thirty-six-day stay, the English established friendly relations with the natives and explored the country inland. Drake claimed the land for Queen Elizabeth and

10. The information on María and the cargo the English looted from Zárate's ship is from the "Anonymous Narrative."
11. Zárate y Salmerón, in Lummis, ed., "Pioneers," pp. 184–85.
12. Wagner, *Drake's Voyage*, p. 172. The Drake quotation is from Wagner's translation of the declaration of Francisco Gómez de Rengifo, factor of the Port of Guatulco. Nuttall, *New Light on Drake*, p. 357, also translated this declaration.
13. Drake, *The World Encompassed*, p. 63.
14. Zárate y Salmerón, in Lummis, ed., "Pioneers," pp. 184–85.

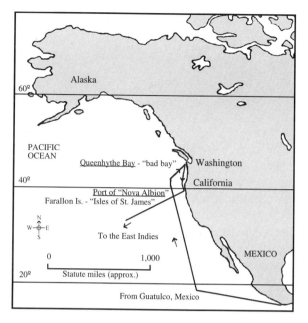

Map 5. Outline map of North America showing Drake's three landfalls. Two locations have been hotly disputed (underlined). For the first landfall in a "bad bay" on the Pacific Northwest Coast, *The World Encompassed* reports a latitude of "48. deg." This reading was undoubtedly made aboard ship. Assuming it is accurate to within thirty minutes, it places the anchorage at Queenhythe Bay, Washington, the only haven in the area that offers some protection from the northwest wind (*see appendix E*). *The World Encompassed* reports a latitude of "38. deg. 30. min. for the "Nova Albion" landfall, placing the anchorage in the vicinity of San Francisco, California. In all likelihood, the reading was made on land. Most authorities agree that the third landfall was at the Farallon Islands lying off the mouth of San Francisco Bay .

named the territory Nova Albion (New England). They scuttled Tello's bark, presumably using her for firewood.

Across the Pacific to the "Island of Thieves"

For sixty-eight days they were without sight of land as the Golden Hind lumbered westward across the Pacific. They finally came upon an island about eight degrees north of the equator, one of the Palaus in the Caroline Archipelago (Babelthuap Island). The Micronesian natives swarmed around the ship in large dugout canoes, initiating a lively trade with the crew. Soon, however, the situation turned ugly as the natives' demands became unreasonable and their demeanor threatening. Drake tried to warn them off with an artillery shot, but was eventually compelled to direct a shot into one of the larger dugouts, killing many.

Departing the next day with the islanders in pursuit, Drake made his way west to the southern Philippines, briefly putting into an unpopulated harbor on the east coast of Mindanao (the mouth of the Davao River in Davao Gulf).

A Spice Trade Alliance in the Moluccas

Passing between the Sarangani Islands and continuing south along the Sangihe Archipelago, the English confronted a large Portuguese merchant ship. Drake backed off, however, when the captain refused to surrender without a fight. Off one of the Sangihes, they were fortunate to pick up several native fisherman who agreed to show them the route to the Moluccas.[15]

The dream of many of England's imperial schemers was to break into Portugal's lucrative East Indies monopoly in the spice trade. Drake, ever the opportunist, pulled it off with an impressive display of diplomatic skill. Learning from some local fishermen that the powerful island of Ternate had recently expelled the Portuguese, he sent a messenger to its sultan, Babu, offering trade goods and explaining that the English and Portuguese were enemies. The sultan received them with enthusiasm, providing four large dugouts to tow the Golden Hind into port. With as much pomp and ceremony as he could muster under the circumstances, Drake won over the confidence of the dangerous Muslim ruler and entered into a series of tense negotiations on behalf of the Crown. They

---

15. The encounter with the Portuguese ship and Indian guides is from John Drake's second declaration in Nuttall, *New Light on Drake*, p. 52.

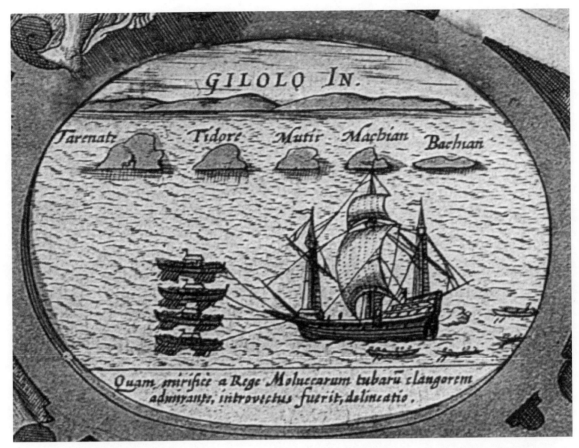

Fig. 9. The *Golden Hind*'s arrival at the Moluccas: an inset from the Hondius Broadside Map, attributed to Jodocus Hondius, drawn circa 1589. In the Moluccas, Drake negotiated a spice trade agreement with the sultan of Ternate that ultimately led to the foundation of England's enormously successful East India Company. *Courtesy The Bancroft Library, Berkeley, California.*

struck the following bargain: the sultan would allow the English to trade for cloves at Ternate; in return, the English would provide, within two years, a fleet to help expel the Portuguese.[16]

Four days after arrival, Drake departed with a fresh store of provisions and about six tons of cloves. Although the English were not able to follow up immediately on this agreement due to the war with Spain, it eventually bore fruit in 1600 with the founding of Great Britain's East India Company.

Crab Island and Farewell to María

Drake needed an uninhabited island with a good harbor to refit the *Golden Hind*, and five days out of Ternate, found one in the Sula Islands (Lifamatola Island) where they dined on large land crabs that climbed trees to escape their pursuit. With repairs and provisioning completed after a twenty-six-day stay, he bid farewell to the pregnant María, and left her there with the two other Africans he had captured from the Spanish, ostensibly to form a settlement.[17]

Almost Shipwrecked in the Banggai Archipelago

The English departed Crab Island, December 12, 1579, on a western heading, and four days later sighted the east coast of "Celebes" (Sulawesi). Without the benefit of accurate charts, Drake entered

---

16. Drake's treaty with the sultan of Ternate is summarized in the account of Francisco de Dueñas, a Spanish ambassador who journeyed from the Philippines to the Moluccas in the fall of 1581 (cited in Wagner, *Drake's Voyage*, p. 180).
17. From the first declaration of John Drake (Nuttall, *New Light on Drake*, p. 32). María's fate is unknown.

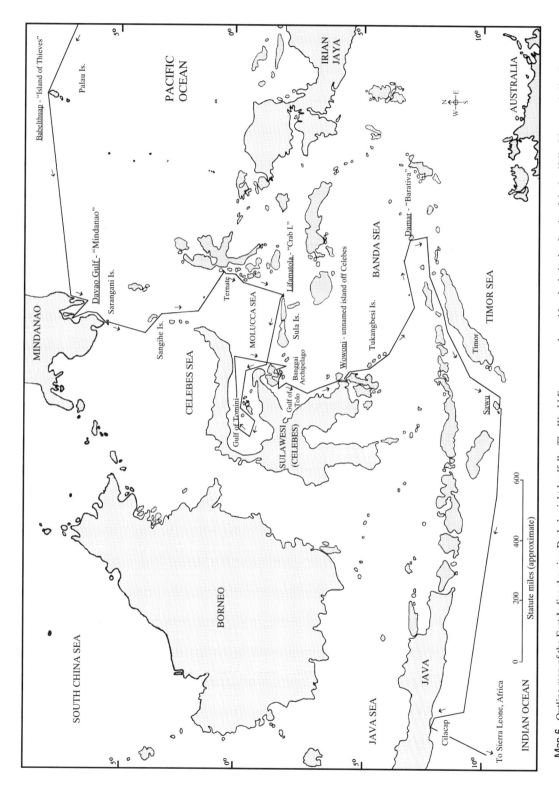

Map 6. Outline map of the East Indies showing Drake's eight landfalls. *The World Encompassed* provides the latitudes for six of the landfalls. I have underlined my solutions to six anchorage locations that have been disputed. The latitude readings reported for the Island of Thieves and Ternate were undoubtedly made aboard ship. The other four were probably made on land.

the Gulf of Tomini thinking he could round Sulawesi's north side. Finding himself at a dead end, he reversed course and veered south into the treacherous Banggai Archipelago. Three weeks later, having threaded his way into the Gulf of Tolo, Drake had the crew put up full sail and retired to his cabin for a peaceful night's sleep, convinced his worries were at least temporarily over. Instead, he awoke to his worst nightmare. Late that night running hard before a northeast gale, the ship struck fast upon a steep-sided reef lying off the extreme south end of the archipelago. Tossing out cloves, cannon and other valuable cargo in a fruitless attempt to get free, Drake and his crew beseeched their Lord for salvation and anxiously awaited their fate. They were delivered the next day when the wind changed at just the right point in the tidal cycle, allowing them to slip gently off the reef.

Visits to "Barativa" and Java

Over the next few months, the great mariner skillfully guided the *Golden Hind* down the east coast of South Sulawesi where they stopped for water and wood at an unnamed island (Wowoni Island). He then sailed through the Tukangbesi Islands, negotiating another maze of uncharted reefs and shoals. Finally breaking free, they entered the Banda Sea, and being forced southeastward by contrary winds, found themselves among today's Barat Daya Islands. They rested and provisioned for two days at "Barativa" (Damar Island), finding the land rich in mineral wealth and the Indonesians particularly attractive and friendly.

After navigating the Ombai Strait, they watered at a little island off the southwestern end of Timor (perhaps at Sawu Island). Entering the Indian Ocean, Drake's next port of call was on the south coast of Java (perhaps at Cilicap). There he socialized with several rajahs, reprovisioned, and refit the ship for what was to be the longest leg of the voyage. An approaching Portuguese fleet hastened their departure.

Across the Indian Ocean and Around the Tip of Africa

The distance from Java to their next landfall on the west coast of Africa is an incredible 9,700 miles. The accounts say little about this amazing feat of navigation that Drake may have accomplished without the benefit of any charts. By the time they had rounded the Cape of Good Hope, and found their way into the mouth of a river in Sierra Leone, they had only three jugs of water and a half a jug of wine for the fifty-nine persons aboard. They spent two days watering and provisioning during which time they were awed by their first encounter with elephants.[18]

Back to England

On September 26, 1580, Drake hailed some Plymouth fishermen at work on their nets as he brought the heavily laden, weather-beaten Golden Hind toward the mouth of the harbor:

"How goes the queen?"[19]

---

18. Ibid., p. 33.
19. From the second declaration of John Drake (Nuttall, *New Light on Drake*, p. 54).

CHAPTER 2
The California Sojourn

Let us go back over four hundred years to an isolated stretch of beach on the chilly, fog-drenched coast of northern California, where we find Francis Drake, the *Golden Hind*, its motley crew, and a beautiful young woman, surrounded by a host of bewildered Indians. Using firsthand accounts, a good deal of research, and a little imagination, I have recreated the scene, transposing it into the actual setting. You will find the contemporary account from which I drew the story in appendix B.

Should you have the chance, read this chapter at the fort site; there is a picnic area nearby. Try to ignore the modern changes. Then drive up the bluffs and walk the old Indian trails skirting the cove; the tourists are still using them. Imagine the *Golden Hind* below you propped up on the beach, the boom of her cannon reverberating off the cliff sides while the Indians peer down, awestruck. Continue your hike, scan the inland horizon (with a little luck the fog will allow this), and try to imagine what a dreary scene it was without the modern structures and landscaping additions. Go to the lookout point, feel the gusting wind, breathe the salt air, listen to the thunder of the surf, and admire the white cliffs as Drake did so many years ago. Now depart the cove and traverse the five miles to the scenic inland valley, taking the route directly over the hills. Looking down from the final hillside, imagine the town replaced with an Indian village, the cattle replaced with deer, the roads replaced with well-worn trails beaten by countless bare feet.

We began the story in the introduction with the *Golden Hind*'s approach to the harbor on June 17, 1579. To continue . . .

By God's Will

With night falling, exploration of the bay would have to wait for morning. Drake, with a large Bible and John Foxe's *Book of Martyrs* in hand, called the crew together for the customary evening prayer. The pages placed a small table on the poop deck, and at its head a kneeler covered with an embroidered cushion.

Laying the books down, Drake struck the table twice with the palm of his hand. Immediately, nine of the company joined him, seating themselves around the table. Each carried a small missal.

Once everyone had quietly assembled, Drake clasped his hands and, kneeling on the cushion, lifted his eyes to heaven, silently remaining in that attitude for a quarter of an hour. Finally looking down, he prayed aloud:

> By God's will hath we been sent into this fair and good bay. Let us all, with one consent, both high and low, magnify and praise our most gracious and merciful God for His infinite and unspeakable goodness towards us. By God's faith hath we endured such great storms and such hardships as we have seen in these uncharted seas. To be delivered here of His safekeeping, I protest we are not worthy of such mercy. Our most gracious sovereign, her poor subjects, and the church of God hath opened the heavens here as in divers other places, and pierced the ears of our most merciful Father, unto Whom, in Jesus Christ, be all honor and glory. So be it, Amen, Amen.[1]

Opening his book, he then gave the day's readings. Finishing, he nodded to his cousin and several other youths who strummed their lutes while Drake and his company filled the night air with a bullfrog chorus of their favorite psalms.

---

1. This prayer ceremony was observed aboard the *Golden Hind* and recorded in 1580 by Francisco Gómez de Rengifo, one of the Spanish Drake briefly held prisoner at the port of Guatulco, Mexico (Nuttall, *New Light on Drake*, p. 354). The prayer itself I adapted from a letter Drake wrote in 1588 to Queen Elizabeth's minister, Sir Francis Walsingham (cited in Sugden, *Sir Francis Drake*, p. 7).

Fig. 10. Sir Francis Drake at age forty-two: miniature portrait by Nicholas Hilliard, 1581. *Courtesy of the National Portrait Gallery, London.*

Drake and His Company

The man who led the twilight service aboard the *Golden Hind* that night was in his late thirties, short in stature, barrel-chested, and very robust. He had a fine countenance, a ruddy complexion, and piercing, hazel-blue eyes. His curly reddish-brown hair was close-cropped and receding. A short, well-groomed, pointed beard, his mustache, and his markedly raised eyebrows were a reddish-blonde, surprisingly fair for his hair color.

Aboard the *Golden Hind* that evening, a company of sixty assorted individuals had listened to Drake's short service. Francis Fletcher, the ship's chaplain and official chronicler, who normally gave the afternoon service, stood by Drake's side.

Along with Fletcher, there were nine officers, gentlemen or cavaliers including William Hawkins, the *Golden Hind*'s master; John Chester, the bark's captain; Thomas Drake, Drake's handsome, twenty-two-year-old brother; and John Drake, their fifteen-year-old cousin who served as the expedition's artist and head page. There were forty-one able-bodied seamen; five boys; two Cimarrones they had rescued from the Spanish at the ports of Guatulco and Paita; and Diego, Drake's longtime personal attendant. The most striking member of the party was María, Drake's African mistress, a stunningly beautiful young woman he had taken from a Spanish nobleman. John Doughty, one of the gentlemen, whose brother Drake had beheaded for inciting mutiny earlier on the voyage, was under close watch.[2]

A Singular Reception

The next morning they found themselves still engulfed in the dense fog that had descended upon them the evening before. When it finally began to lift as the winds picked up around ten o'clock in the morning, the crew prepared to launch the ship's boat. A sharp cry from the crow's nest interrupted them. Through the mists they could just make out a small group of naked Indians on the beach to their north.

One of the Indians, after much deliberation with the others, entered a reed canoe and started for the *Golden Hind*. He paddled all the way out to the ship, speaking continuously along the way. Arriving within a safe distance, he stayed himself, and delivered a long and solemn oration accented by many strange gestures of his hands and body. Finishing with a great show of reverence, he headed back to shore, continuing his oration as he paddled.

The Indian repeated this routine twice, the intensity of his address increasing with successive visits. On the third visit, he finally summoned the courage to come alongside the ship's boat that had been lowered by this time, and tossed in a small stick to which were attached several gifts. One was a feathered headdress made from a neatly tied bunch of cut black feathers. The other was a pretty little grass basket filled with an herb the English later learned was native tobacco.

Seeking to recompense him, Drake, now aboard the ship's boat, offered the native a variety of gifts including some colorful beads and silks. The Indian was emphatic, however, in his refusal to accept anything, even when they secured the gifts upon a board and floated it his way. Drake was beside himself with frustration and the crew was having great difficulty controlling their laughter at their commander's plight.

Finally, after one of the crew threw an old hat into the water, the Indian darted over to retrieve it for him. Now it was Drake's turn to frustrate the native. Each time he tossed the hat into the boat in an effort to return it, Drake promptly tossed it right back out. With a wry smile, the native reluctantly placed the sopping hat on his head, turned, and paddled for shore. Drake's booming laugh rippled across the water behind him.

---

2. Nuttall, *New Light on Drake*, pp. 425–28, referencing various contemporary sources, provides a partial list and description of the company that returned to England with Drake in 1580. I am speculating as to María's age and appearance.

Fig. 11. "Reed boat; double-bladed paddles": drawing of the Indians of San Francisco Bay by Louis Choris, 1816. *Courtesy The Bancroft Library, Berkeley, California.*

The Landing

After three full days exploring this bay and a larger one they discovered to the south, Drake found only one suitable location to bring the *Golden Hind* ashore for repairs. This was the small cove they could see lying just inside the bluff headland, about one-quarter mile due west of the anchorage. It was entered by a narrow channel scarcely half-a-cable's length (a hundred yards) across. Inside this cove, the channel wound into a shallow lagoon about two miles long and a mile across.

At dawn on the twenty-first, they weighed anchor and, in an eerie blinding fog, towed the *Golden Hind* closer to shore with the jolly boat, anchoring in about twenty feet of water, just a few hundred yards outside the channel's mouth. They then used the boat to bring ashore some of the crew with necessary supplies for constructing a fortified encampment, working until the tide began to ebb at about half-past eleven. At this point, Drake posted a guard of fifteen men and had the others start work on the camp.

The cove Drake selected for refitting the ship was a snug one. It had several hundred yards of sandy shore from which the beach extended back only about seventy-five feet from the high-water mark. This beach was walled in on all sides by steep bluffs that rose some fifty feet up to the hilly tableland that formed the end of the peninsula.

All the time they were unloading, armed Indians from a nearby village had been assembling along the edges of the cliffs that enclosed the cove. As he looked up at them, Drake's right hand went involuntarily to his face, his fingers tracing the jagged scar that ran across his cheek, a memento of his first encounter with the natives of America's west coast. He could never forget the white flash and agonizing pain as the arrow struck his head, or the odd taste and feel of that stone arrowhead embedded in his blood-drenched mouth. Nor could he forget the two good men he lost that day, and their screams of terror as the treacherous Indians carried them away.[3]

A Tense Moment

As soon as the crew started work on the camp, the Indians above recognized with certainty what the English were doing, and like men set to war in defense of their country, made a collective rush down the steep trail leading from bluff to beach.

---

3. I allude here to the November 26, 1578, incident at Mucho Island off the coast of Chile when native inhabitants ambushed Drake and his shore party after making peaceful overtures (Drake, *The World Encompassed*, pp. 47–48).

Fig. 12. Drake and his men landing on the shores of Nova Albion. The inset is a detail from an illustration in AMERICAE PARS VIII by Theodore de Bry, Frankfurt, 1599. *Adapted from a photograph taken by Mrs. Rose Gaffney, circa 1962, held in the California Public Utility Commission archives for Application Number 43808.*

Drake's men, their arquebuses and swords poised, were ready to meet them at the foot of the cliffs. "Fire into the air at my command!" Drake shouted, taking a position among the guard.

"Fire!" The collective roar of the explosions reverberated off the walls of the cove, leaving several dozen terrified Indians tumbling onto the sand, their hands cupped over their ringing ears.

Drake approached the nearest, picked up his bow and broke it over his knee, making signs to the rest to lay theirs down. Those on the beach quickly complied, and as the rest of their company of about fifty men and women descended to the beach, Drake and his guard went quickly among them compelling them to do likewise.

There was little danger now as the natives were disarmed and completely submissive, looking upon the English as if they were gods. To effect a friendly peace, Drake and his crew went among them presenting each a French bracelet made of small, red and white glass beads. Wishing to teach them to cover their nakedness, he had his men start distributing shirts, pants, and other types of clothing. To his dismay, however, the Indian women, some of them young and attractive, responded in kind by casting off their grass skirts and fur cloaks in front of his wide-eyed men.

Apparently pleased with the turn of events, the Indians filed back up the hill and returned to their village. Drake and his crew returned to their work with an added sense of urgency.

The Fort

Recognizing from past experience that this peace might be transient and that there were many Indians dwelling in the vicinity, Drake determined to construct a substantial defense works. He sent about twenty armed men up the bluffs, equipped with pickaxes. They pried loose an abundance of stones from the rock outcroppings scattered about the hilly tableland and then rolled them over the steep bluff at the cove's south side. Below, other men gathered and set them in place. In this fashion, over the next few days they completed a large three-walled stone fortress among the hillocks at the foot of the rear bluff, using the bluff itself for the rear wall. Within the stone walls the men pitched tents and constructed several storehouses.[4]

To isolate the camp from the Indian trail cutting down across the face of the bluff on the north side of the cove, they also constructed a bulwark or outer defense works stretching from the base of the bluff to the hull of the grounded ship. The bulwark was a five-foot-deep moat with soil, stones and sticks thrown up behind. Brass swivel guns were mounted on stone bastions at each end.

The Indian Village

Shortly after the Indians departed that first day ashore, the hazy afternoon air was pierced by a distant mournful wail. This cry, closely followed by another, emanated from a sandy point about three-quarters of a mile inside the harbor where the men could see smoke rising. Soon an eerie chorus of weeping and wailing filled the air.

Fletcher, who was among those who were gathering rocks below, gained Drake's permission to investigate. Escorted by an armed guard, he scaled the steep trail on the north side of the cove, trekked north across the headland, and then down into the sand dunes that formed the isthmus connecting the headland to the main. There the English came upon a group of natives gathered near the water's edge on the lee side of the dunes. The Indians had built a large fire into which they were tossing the bracelets, clothes and other trifles the English had given them.

As soon as they saw Fletcher, the Indians sat down, showing much fear and again began to humble and demean themselves. Horrified, the preacher did all he could to persuade them of their error, finally succeeding in getting them to interrupt their pagan rites.

This village had about a dozen underground lodges, each occupied by an extended family. Peering into a smoky opening, Fletcher observed a round pit about ten feet across, dug three to four feet into the ground, with a small fire in the middle. They had formed the roofs by driving strips of wood into the ground along the pit's perimeter, joining them close together at the top, steeplelike, and then covering the structure with grass and dirt. The cone-shaped roof was low-vaulted and the entrance resembled more a ship's scuttle than a doorway and also served as a chimney.

A Fabulous Horde

As planned, they drifted the lightened *Golden Hind* into the cove about eight that evening, anchored, and secured her to shore where the channel made a turn directly opposite the fort. Early next morning they began the arduous task of fully unloading the treasure-laden ship. Drake took a careful inventory of their more valuable booty as the work proceeded throughout the day: two large gold crosses—one studded with emeralds; a magnificent golden falcon with an emerald in its chest; several gold crucifixes; fifteen chests of silver coin; two chests of gold coin; fourteen hundred silver bars; eighty pounds of gold ore; a bag of emeralds—some as long as a man's finger; an abundance of jewelry; over a thousand jugs of good Chilean wine; six bales of beautiful linen and China silks; and four chests of artfully decorated porcelain dishes.

---

4. The fort's description is derived from the Portus Nova Albionis inset on the Hondius Broadside Map (circa 1589).

Fig. 13. The *Golden Hind* aground on the shores of Nova Albion. The inset is a detail from the Hondius Broadside Map, attributed to Jodocus Hondius, drawn circa 1589. *Adapted from a photograph taken by Mrs. Rose Gaffney, circa 1962, held in the California Public Utility Commission archives for Application Number 43808.*

One rather unusual item that came up from the bottom of the hold was a large tree trunk about fifteen feet long and several feet in diameter.[5]

More Indians

For the next two days, the Indians left Drake and his crew to their business, giving them ample time to complete the encampment and fully unload the ship. During this period, however, lookouts posted above reported that Indians arriving from the surrounding country were massing among the hills on the east side of the lagoon. Word of the English arrival was spreading.

About noon on June 23, a shout came down from above: the Indians were on the move. Drake ordered the crew into their assigned positions and scrambled up the bluff to watch their progress.

5. The contemporary reference to the tree trunk in the hold was by the Portuguese pilot, Nuño da Silva, whom Drake released in Mexico (Nuttall, *New Light on Drake*, p. 319, referencing the Simancas MS). According to da Silva, Drake intended to bring it back to his queen as a memento of his passing the strait. I am speculating that Drake opted to use it for his Nova Albion post.

Fig. 14. Indians welcoming Drake and his men to Nova Albion: detail from an illustration in AMERICAE PARS VIII by Theodore de Bry, Frankfurt, 1599. *Courtesy The Bancroft Library, Berkeley, California.*

Hundreds of Indian men, women, and children had descended the low cliffs at the northeast corner of the lagoon and were making their way up the dunes along the western shore. After passing the point where the small village was located, they momentarily disappeared as they started up the opposite side of the headland. Drake and the lookouts hurried back down the trail and joined the rest of the company.

From their defenses, they watched the Indians climb to the top of a hill to their north and then disappear as they came down towards them. When the Indians reappeared along the cliff tops on the opposite side of the cove, they made a stand. Their appointed leader came to the front and delivered a loud oration, gesturing wildly to emphasize his points. His words were falling so thick upon each other that it seemed he would suffocate before he could finish. To the amazement of the English, however, he went on like this for over an hour. When he finally concluded, the rest of the assembly collectively let forth a sort of protracted "amen," given in a dreamy manner and with a reverent waving of hands and bowing of bodies: "hooooiiii!"

The preliminaries finished, the Indian men laid down their bows and arrows and picked up the various gifts the women had carried along. Leaving the women and children behind, they descended the trail leading down the bluff at the north side of the cove. Drake, with an impressive guard of his own, passed through the bulwarks to meet them.

Any doubts that the Indians considered them gods were quickly dispelled. Presenting their offerings, the men groveled pitifully at Drake's feet while the women lining the cliffs above carried on in a monstrous manner. With dreadful wails and shrieks they began violently scratching their cheeks, tearing away at the flesh. Blood streaming from their faces, they tore off their crude fur cloaks and began throwing themselves chest-first onto the hard ground. Time and time again they hurled themselves down, their hands raised above their heads so they might not protect their breasts from the stones and prickling bushes. Fletcher, in horror, watched as directly above him a pregnant woman furiously repeated this violence upon herself until she finally collapsed in a swoon. The Englishmen found themselves the objects of a massive bloody sacrifice.

Over the din, Drake shouted for his company to gather behind him, raised his arms high and looked silently up to heaven. When the Indians had quieted somewhat, he began to pray aloud. When he finished his prayer, the Indians issued a collective "hooooiiii" as they had after their chief had spoken.

Fletcher, taking the lead, opened his Bible and began a reading. The Indians, now completely assuaged, sat down and attentively listened to every word, never neglecting to respond with a collective "hooooiiii" whenever he paused. They took particular pleasure when Fletcher's sweet voice carried the Englishmen through their favorite psalms, notwithstanding the "confusion of bawlings" that accompanied him.[6]

While Fletcher sang, Drake went about the natives, distributing sundry trifles in exchange for their baskets of herbs, feather plumes, and net caps. As soon as Fletcher finished, Drake ordered everyone back to the fort, making it clear that the ceremony was over. The Indians went on their way, leaving all of Drake's presents behind.

The Golden Hind

The *Golden Hind* was a hundred-ton, French-style merchantman that Drake had designed very aptly for his purposes. The square-rigger measured about a hundred feet stem to stern, twenty feet in beam, and drew about seven feet of water, thirteen feet fully laden. She had two main decks, and an afterdeck, or poop. The forecastle, and the round house at the stern were rather small affairs. Her three masts carried up to 4,000 square feet of sail including special fore and aft topgallant sails to lend extra speed in the event of a chase.

The ship was heavily armed, a virtual floating fortress. She had sixteen gun ports for the lower deck, seven on each side and two at the stern. To make the ship more seaworthy, Drake had these ports closed and caulked just before leaving the Spanish Main. The dozen or so nine-foot-long, one-ton cannon that could fire deadly nine-pound shot, were now stored securely about the deck. The upper deck had two gun ports on each side from which protruded the polished muzzles of 1,000-pound brass cannon. These had been specially cast for Drake and bore an impression of his coat of arms. Ready at hand, in the forecastle and other arsenals below, was a deadly array of small arms including swivel-mounted brass falconets for firing small shot at the enemy deck, arquebuses, calivers, pistols, pikes, fire-pikes, firebombs, bows, arrows and swords. Powder and shot were always kept under lock and key to discourage mutiny.[7]

The crew lived and slept willy-nilly among the armament, provisions, and personal sea chests secured on the two main decks. Some of the officers slept below the forecastle and others in small shelves suspended from the upper deck. There were two cabins situated under the poop at the ship's stern: a small private cabin for the captain, and the "great cabin." The great cabin had a large carved oak table surrounded by velvet-cushioned chairs. Here Drake would convene with his officers to dine and plot strategy. In the round house above, Drake spent hour after hour with his young cousin, drawing and painting the seascapes, landscapes, and land and sea creatures that they observed on their journey.

Drake's private cabin was an elegant affair lavishly decorated with a small carved oak table and bookcase, red velvet-cushioned chairs, a fancy sea chest, and a canopied bed. Expensive tapestries and rugs covered the sloped walls and floor. Here, not long ago, Drake had grandly entertained some very distinguished Spanish gentlemen while his crew was busy ransacking their ports and vessels.

---

6. Morison, *Southern Voyages*, p. 179. He cites Eugenio de Salazar, a Spanish officer who, after sailing from Spain to Santo Domingo in 1573, reported that "mariners divide the four winds into thirty-two . . . perverse, resonant, and very disresonant, as if we had a tempest of hurricanes of music. . ."
7. For the *Golden Hind's* description and refitting, I drew from Wilson, *The World Encompassed*, pp. 44–48.

Fig. 15. The refitting of the *Golden Hind*: an illustration from AMERICA SAILS THE SEAS by John O'Hara Cosgrave II. *Copyright © 1962 by John O'Hara Cosgrave II. Reprinted by permission of Houghton Mifflin Company. All rights reserved.*

Now in his snug little California port he had only María to entertain. This he did many nights in equally grand style. With the pages outside serenading them with their lutes, they sipped the highest quality Chilean wines from silver goblets and ate sumptuous dinners served on silver plates.

Graving and Breaming

After the Indians departed, Drake issued the order the crew was dreading. "We'll grave her tonight." The crew groaned at the thought of the dirty, back-breaking work that lay ahead.

About nine o'clock that night, with the tide at its highest point, they hauled the *Golden Hind* as far up on the beach as they could, propping her up on the port side. The next morning her double-sheathed bottom lay completely exposed, a slimy mess of barnacles and weed. The crew set immediately to work, beginning a familiar process that, with all the interruptions, would not be completed for almost three weeks.

The Indians watching from across the channel and the hills above were amazed at the antlike industriousness and peculiar behavior of these strange white spirits. They observed one snarling group violently attack the sides of the great vessel with fiery brooms and sticks, while another attacked the smaller ship with an even greater vengeance. Others on the beach built large fires over which they cooked a most odorous black gruel. The scraping, burning and tearing-apart went on hour after hour, day after day, with much shouting, grunting, and singing. The Indians agreed it was the most magnificent ceremony they had ever seen.

The carpenters took over once the exterior was clean. They carefully examined the timber for damage, drilling out and plugging worm holes, replacing cracked planks, and caulking and filling gaps. After the repairs were completed they applied a thick coating of tar and brimstone, and then a layer of grease.

The ship's boys, and those without a trade, had the most unpleasant task of all: cleaning the insides. "Slipping and stumbling over the slimy boards," by dim lantern light they bucketed out the putrid green water in the hold, scrubbed every square inch of planking, and peered closely at the timbers to identify areas in need of a carpenter's attention.[8]

Comes the King

When the fog thinned at mid-morning on the twenty-sixth, lookouts posted above once again sang out. Looking northeast across the lagoon, Drake watched what looked like an endless train of Indians descending from the hills. "There could be thousands of them," he roared as he hustled about making sure the defenses were ready. "By God's faith, men, we'll withstand them if we must!"

As they had three days before, they watched the lead group crest the hill at the center of the headland and then disappear as they headed down towards the fortress. This time, however, only two Indians reappeared along the cliff tops above the north side of the cove; they made their way directly down the trail. Drake and his guard went out of the bulwarks to meet them, recognizing one by his hat as the brave soul that had rowed out to greet them the morning after their arrival.

It became quickly apparent that the other Indian was an emissary representing the leader of the new group, and that the familiar Indian was serving as his translator. The emissary quietly issued a half-hour proclamation that the other translated line by line into the local language, still gibberish to the English. Drake, patiently waited for them to finish, then put his hands on his chest and said, "Friend!" and then in Spanish, "Amigo!"

At that last word, the emissary gave Drake a quizzical look, pointed up the hill and uttered the word "hioh." Then, by signs, he made it clear that he would like Drake to furnish him with a gift to deliver to his leader waiting above, apparently in this case a great Indian king. Drake gave him some colorful china silks and a painted porcelain dish. Satisfied, they both climbed back up the hill.

Soon thereafter, the king and a select guard of about a hundred warriors emerged into view above them. When they had formed in ranks along the cliff with the king at center front, they raised their hands above their heads, shouted out a common salute, and then stood quietly at attention. Meanwhile, a large warrior of "goodly aspect" who stood beside the king, stepped forward holding a scepter and was immediately joined by the interpreter. Between them, the scepter-bearer talking softly, the translator shouting, they delivered an oration lasting about half an hour. When they finished, the assembly gave their collective "amen" and, with the king leading the way, started down the trail.

On they came until many hundreds of Indian men and women had assembled on the beach on the north side of the cove. Only the children remained behind, peering down from the cliff tops with frightened eyes. As the throng gathered on the beach, they reverted back to their original order, with the king and his guard at front. Drake, during this period, had his men put on a very warlike show, marching behind the barricades and posing atop the bulwarks with their swords flashing.

The king was a portly yet attractive man. He wore a fancy headdress and a fur cape. The headdress consisted of a net cap decorated with fine white down and multi-colored feathers. The cape was made from the pelts of a small animal with a very long, ratlike tail. Around his neck he wore several necklaces made from strings of bone-colored disks made from clam shells. The rest of his guard were dressed similarly, but their head dressings were much less elaborate and their necklaces were fewer and shorter. The common sort of men that followed went mostly naked with their long black hair tied behind their heads. Their only dress was a headband with a single feather in the front and other feathers gathered in a bunch behind. All of the Indians had their faces painted, some white, some black, and some with reds, blues and yellows.

---

8. Ibid., p. 121.

Fig. 16. Drake crowned by the Indians of Nova Albion: detail from an illustration in GRANDE ET PETIT VOYAGES by Johann-Theodore de Bry, Frankfurt, circa 1617. *Courtesy The Bancroft Library, Berkeley, California.*

Finally, when all had assembled, the entire party started moving slowly towards the fort. To Drake's great relief, however, the warriors had put down their arms and now were bearing gifts.

As they came on, the scepter-bearer, with composed countenance and stately carriage, began to sing and perform a sort of skipping, shuffling dance: "Hodeli oh – heigh oh heigh oh – hadali oh." The entire assembly followed suit, except that the women, who took up the rear, danced but did not sing.

Drake ordered his men to hold their fire and allowed the king and his company to pass around the outer defense works.

Drake Crowned

Inside the bulwarks, the dancing and singing went on for some time. Finally, at a signal from the king, the entire party sat down and the king, scepter-bearer, and several other lesser chiefs addressed Drake with a series of eloquent speeches. When they had finished their talks, the scepter-bearer came forward carrying the royal mace: a short piece of polished black wood to which were attached two headdresses and three long disk necklaces similar to those worn by the king. With great ceremony, the king took one of the headdresses and all the chains off the mace and placed them over Drake's head. The entire party then rose up and, in a frenzy of dancing and singing, each presented him with a finely wrought and beautifully decorated rush basket filled with their herbs or foods: cooked fish, roots, seeds, and the like.

With as princely an attitude as he could muster, Drake received on behalf of his queen what he considered to be their apparent willful subjugation. To his every utterance, the Indians would respond with the cry of "hioh!"

A Strange Rite

When the coronation ceremonies ended, the common men and women divided into small family or tribal groups and dispersed themselves among Drake's men. Going from man to man, they carefully scrutinized each face as if they were looking for someone they knew. When they finally found a face they seemed to recognize, one usually belonging to a younger crew member, the little group would surround him to offer their all too familiar sacrifices. Crying out with their horrible shrieks and moans, the women and old men began tearing at the flesh on their faces. In this case, it seemed to Fletcher that the Indians were seeing in the pale faces of the Englishmen the ghostly apparitions of their dearly departed.

Drake's men did their best to stop them, forcefully holding back their bloodied hands, raising their own hands to heaven as Drake had done before, singing psalms, and finally retreating to their fort. Nothing could appease them this time, however, and the madness went on until the Indians had fully expended themselves.

To repair the damage, some of Drake's men now went among them, particularly the pretty young women, treating their wounds with lotions, plasters, unctions, and bandages. Soon, however, they had an army of older Indians begging for attention as well, revealing shrunken sinews, cankered ulcers, and a profusion of other maladies.

The Gracious Guest

After the coronation, there were few days that Drake and his men escaped the attention of the Indians and for a while, every third day they would come in large numbers to offer their sacrifices. In time, Drake and his crew were able to make them understand that such bloody rites were abhorrent to them and they ceased the practice. This did not keep them from coming, however, and in lieu of the hysterical bloodletting they simply brought presents.

Their regular visits became a problem for Drake, as they interfered with his reprovisioning efforts. He regularly found himself compelled to feed legions of hungry Indians from the mussels, sea lions and clams his foraging and hunting parties had gathered for the day. Nevertheless, he always treated his hosts with great courtesy and kindness.

Drake had become particularly friendly with Valenila, the local chieftain, or hoy-puh as a chief was called in the native tongue. This was the brave soul who had first greeted them and later served as the interpreter when the king arrived from the inland. Though they could only communicate through signs, the two leaders had developed a certain camaraderie that transcended the vast differences in their cultures.[9]

A Squalid and Unhandsome Shore

On July 1, fourteen days after their arrival, something unusual happened. The day dawned and they saw the sun rise. It was the first they had seen of sun or moon in fourteen days. The temperature remained very cool, however, in the low fifties, and the blustery northwest wind was blowing with even more fury than usual.

Drake, hand-in-hand with María, climbed the bluff and made his way south around the perimeter of the cove to a good vantage point looking out over the harbor. The top of this hilly headland was a bleak, windy place with just a few groves of gnarled pines. Its grass and brushwood covering was shriveled and black as if it had been recently burned.

Looking east, they could see below them a long sandspit extending in a gentle arc two miles out from the mainland. They had first anchored about a half mile off the sandy beach on the spit's south

---

9. This is fictional. From Drake and Nichols, *Sir Francis Drake Revived*, in Hampden, ed., *Francis Drake Privateer*, pp. 102–3, however, we know that Drake developed a similar bond with Pedro, the leader of the Cimarrones— escaped African slaves who lived in the rain forests of Central America—giving him a very valuable gold scimitar upon his departure from Panama in 1572. Drake was far ahead of his time in the empathy and respect he exhibited for other races and cultures.

side. The lagoon lay to the north forming a large inland body of water. They watched one of the hunting parties working side by side with Indians digging clams in the mud flats on the south shore of the lagoon. On the rocks directly below them, another group was harvesting mussels.

All the interior land they could see on either side of the spit was a scarred, wind-swept plain with low, rolling hills and shallow valleys. It was completely devoid of trees, presenting to their eyes "a most squalid and deformed aspect." It reminded Drake of the bleak moors of his homeland.

They followed a trail along the southern edge of the headland that took them to a jutting point of land forming its southwestern terminus. Along the way they stopped to watch the jolly boat chasing sea lions around the flat rocky island they had passed coming in. These creatures, after two weeks of murderous onslaught, were becoming increasingly difficult to catch. "See the white cliffs!" Drake shouted to María over the roar of the pounding surf as he pointed below them.

Looking south, some seven miles distant, they could see a line of breakers crossing the mouth of a river at the extreme south side of the harbor. The second day after their arrival, Drake had carefully examined a very promising inlet just inside the mouth. There were many Indians settled there—too many, Drake had concluded that day. They could see the smoke of their fires.

Separating the river from the sea, a large peninsula came to a point of land that marked the southern tip of the inner part of the bay. Following the line of the peninsula's steep outer shoreline as it fell away southwest from this point, they could even make out the huge headland on the extreme south side of the bay proper, about twenty miles distant.

At noon, Drake took advantage of the clear sky to make an observation of their latitude. "Thirty-eight degrees twenty-five minutes, as near as I can make it," Drake announced, handing the sextant to his cousin. Later, as he entered the reading in his journal, however, he recorded "38. deg. 30. min.," approximating on the high side to avoid confusion with the other harbors and bays located not far south. Drake repeated the observation that night and every clear day and night thereafter to verify the measurement. On the chance that he or his countrymen ever ventured this way again, he would not want any problems locating this most convenient and fit harbor.[10]

A Visit Inland

It took them almost three weeks, but the crew finally completed their repairs to the ship. A sparkling clean and well-trimmed Golden Hind, her treasures and provisions carefully stowed, rested comfortably at anchor within the cove at the channel's edge. By waiting another week, the tide would be high enough to carry them out almost fully laden, so Drake decided to take the opportunity to explore the country inland.

He, Fletcher, and ten others packed up some tents, a few days' worth of provisions, and an array of small arms, and with the king's guard serving as porters, set forth in the early morning fog. They retraced the route by which the Indians had arrived on the twenty-sixth: northward over the dunes on the west side of the lagoon, east over the muddy beach at the northern end, up the low cliffs in the northeast corner, and then directly into the hills.

Drake and his men marveled at the strength of the Indian men as they bore their heavy loads up the steep terrain. "It would take two Englishmen to lift such a load, much less carry it up this hill," Drake gasped, trying to ignore the pains in his straining legs and lungs. Up and down they went, cresting one rounded hillside after another. The land around them was arid and desolate and, except for a low heath in some of the rifts and valleys, it was completely barren of trees and bushes. "What a Godforsaken land," Drake wearily commented after they had traveled some three miles.

Over the next few miles, however, the climate and character of the land started to change for the better. The mists disappeared, the winds died down, and the air became much warmer. They could see pine forests covering the hills far to the north. As they started cresting another hillside about five

---

10. Drake, *The World Encompassed*, p. 64, reports the latitude as "38. deg. 30. min." I am speculating that Drake made the observation on land and repeated it to ensure accuracy. I am also speculating that he rounded the reading up.

Fig. 17. The hinterland: the view to which Drake and his men were treated after hiking five miles inland from the encampment. *Adapted from a photograph by the author, June 1995.*

miles inland, the Indian porters looked back smiling, gave a shout, and suddenly set off at a run, disappearing down the other side.

Gasping for breath, Drake hurried after them, his men close behind. When he reached the point from which he could see the land below, he stopped short in wonder. Before him lay a beautiful valley with a stream winding down the middle. On the north side, the hills were covered with a lush growth of pine. To the south and east, groves of oaks dotted the hillsides and swales. Herds of plump deer grazed the meadows.

They remained several days, their tents pitched by a large Indian village located near the stream. During their stay, they visited a second village located upstream, another five miles inland. The Indians at this village spoke the language of the great king, who lived among the distant mountains to the northeast. The further inland they explored, the more fertile and beautiful the country became.

California Primordial

In general, the Indians were of medium stature, well-proportioned, and very strong. There were some tall individuals among them. Their skin was brownish-red with no artificial coloration. Their long, straight, untamed hair was jet black. The younger men and women and the children had charming features.

The common men went naked for the most part, some wearing a deerskin cape with the hair still on. The women all wore loose skirtlike garments constructed either of marsh reeds or deerskin. Some women also wore deerskin cloaks about their shoulders and a variety of simple jewelry made from shells.

Their principal prey was deer and gophers; the latter, by evidence of their burrows, had completely infested the area. No animal or insect large or small, however, was safe within the Indians' sight.

The Indians captured the gophers with ease, either by smoking them out of their burrows or prodding them out with long, thin sticks. They ate them whole, after burning off the hair and roasting them in special holes bored under their simple stone hearths. They skinned the largest and left the pelts drying in the sun, using them to make their chief's fur coats.

Fig. 18. "Everyday Scene": painting by Mikhail Tikhanov, 1818. *Courtesy of the Scientific Research Museum, Academy of Arts, St. Petersburg, Russia.*

The day after they arrived, the Englishmen were able to watch a deer hunt in progress. The Indians were very clever, if not always successful, in the manner they hunted the skittish creatures. The hunter disguised his head and shoulders with a deer's head, horns and neck skin and set off toward a small herd, his bow and a few arrows in hand. Approaching, he crouched down and did all he could to act the part of the grazing animals, making his way among them. In this case, however, the deer spooked just as he was getting ready to launch an arrow, disappearing over the hills. Drake sent a hunting party after them with high expectations. Although sighting thousands, the men returned empty-handed and with all the noise of the guns, managed to clear the entire area for miles around. The highly anticipated feast never occurred.

The Indian men spent their time hunting, fishing, wood-gathering, constructing and maintaining the lodges, and repairing fishing nets. They loved to smoke. Their pipes were thin wooden tubes about a foot long, the ends carved into a crude bowl. With their heads tilted back in order to keep the herb from spilling out, they inhaled great puffs.

The women gathered berries, nuts and seeds, dug roots, prepared and cooked the food, and served meals. Those with newborns wrapped them in deerskins and carried them around in baskets. Whenever the infants cried, the mothers would promptly interrupt their labors to nurse or soothe them.

Fletcher picked up a pretty, bowl-shaped basket one day to observe its makings. Constructed of fine twigs and reeds, it was artfully decorated with tiny red feathers from a woodpecker's nape, pieces of seashells, and mother of pearl from abalone shells. Tiny black feathers stuck upright all along the rim. The women wove these baskets so tight that they held water.

Fig. 19. "A sweat and a cold plunge": drawing by A. W. Chase, circa 1872. *Reproduced from THE TRIBES OF CALIFORNIA by Stephen Powers, 1877. Courtesy The Bancroft Library, Berkeley, California.*

The creek sides and hillsides abounded in oak trees, from which the Indians had stored great quantities of acorns. From these and some of the roots they dug, the elder women made a nutritious bread they called *cheepe*. The acorns were first left out in the sun to dry, shelled, and then pounded into a meal with special stones. They placed the meal in a pit and repeatedly filled it with water, allowing the water to seep through the meal and into the ground. This flushing process removed the bitterness. The wet meal was taken from the pit, wrapped in broad leaves, and roasted on coals until it was black.

The Indians made much of a certain plant growing on the hillsides that resembled English lettuce. Standing about four to five feet tall, it had prickly oblong leaves along the base and long thin stems. The flower heads had already gone to seed and now appeared as fluffy white balls surrounded by sharp spines. The women collected these balls in their baskets, separating the seeds from the delicate white fibers to which they were attached. Such was their reverence for the sun-dried seeds, that only Drake and the king were allowed to partake of them.

The lodges were like the others they had already seen, each occupied by a single family. The inland villages, like the one near their encampment, had a special lodge, constructed similar to the others, but much larger. It served as a sweat house.

Late each afternoon, when the men returned from hunting, they all gathered together at the sweat house, each bearing an armload of wood. Soon after they crawled inside, smoke could be seen pouring out the top of the roof and the sounds of muffled chants and laughter filtered through the walls. After about half an hour, the Indians streamed from the entrance, their brown skins glistening with sweat, and broke into a dead run heading for the creek. Upon reaching the bank, they leaped, yelling, into the air and came down with a great splash in the frigid water.

The English could not help growing in their admiration for the people of Nova Albion. They found them an exceptionally gentle and peaceful people with a very tractable and loving nature.[11]

Fig. 20. Drake claims Nova Albion: detail from an illustration in GRANDE ET PETIT VOYAGES by Johann-Theodore de Bry, Frankfurt, circa 1617. *Courtesy The Bancroft Library, Berkeley, California.*

The Great Post

Back at the fort, Drake set about the task of taking formal possession of these new lands, putting to good use the large tree trunk they had brought from the Strait of Magellan. Originally intended as a souvenir for the queen, he felt certain that the trunk's newfound purpose would be even more to Her Majesty's liking.

With no shortage of cursing, the crew lugged the massive post up the steep bluff trail, and then up the low, sparsely wooded hill at the center of the headland to an area midway between the sea and the lagoon. There, at a level vantage point overlooking both the bay and lagoon, they planted it firmly in the ground.

The afternoon before their departure, Drake had most of his company dress in their finest military attire and follow him in solemn procession up to the post. Recognizing that a special ceremony was at hand, the Indians had already gathered about it.

Holding a hammer in one hand, and a small brass plate in the other, Drake stepped to the post and nailed the plate fast to it. Reaching into his pocket, he then pulled out an English sixpence bearing the queen's image, and nailed that also to the post, through a hole cut into the plate for that purpose. Stepping back and turning, he shouted out the following proclamation that he had inscribed on the plate:

> Be it known unto all men by these present:
>
> By the grace of God and in the name of Her Majesty Queen Elizabeth of England and her successors, I Francis Drake take possession of this kingdom whose king and people freely resign their right and title in the whole land unto Her Majesty's keeping, and now named by me and to be known unto all men as Nova Albion!

On his signal, the men fired their arquebuses over the lagoon. The reports of their guns triggered a series of great booms from below as the *Golden Hind* punctuated the salute with her brass.

The Treasure Trove

To lighten the burden of his ship, Drake decided to remove from his cabin a quantity of silver bars, most of which had come from the ship the *Grand Captain of the South* at Valparaíso, Chile. The English hauled the plunder up onto the bluff headland and buried it near the great post.[12]

---

11. The descriptions of native culture not taken directly from *The World Encompassed* are based on the accounts of some of the first European visitors to the area, including the Russians Fedor Lütke (1818), Vasili Golovnin (1818), Ferdinand Petrovich von Wrangell (1833) and Peter Kostromitinov (1830–38), and the Frenchman Cyrille LaPlace (1839). I have also used some descriptions provided by Stephen Powers (1871–72). These early accounts and Powers' descriptions can all be found in Lightfoot et al., *Archaeology and Ethnohistory*, pp. 125–36. The deer hunt is fictional.

12. There is no evidence in the contemporary accounts to suggest that Drake left a cache of silver in Nova Albion. The Drake Navigators Guild, however, has come into possession of evidence that he did. The Guild has a transcription of a letter that Drake reportedly left to his heirs. It states that Drake did indeed bury a trove of silver bullion at Nova Albion on "a low but sparsely wooded hill halfway between the sea and a large inland body of water." The Guild's Matthew Dillingham obtained a copy of the transcription from Allen Chickering, then president of the California Historical Society. Chickering, in turn, reportedly obtained the information from an

A Sorrowful Farewell

As the fog cleared on the morning of July 23, 1579, the *Golden Hind* lay at anchor just outside the mouth of the cove. They had drifted her out about eleven o'clock the night before when the tide was at the highest point of the lunar cycle. The crew was busy loading the remaining cargo that had been moved to a transfer point on the spit. Drake was still at the fort with his officers, completing their final loading.

The Indians were devastated at the prospect of Drake's leaving them. Since final preparations had commenced, they had become increasingly glum. The laughter, dancing, and singing that marked their previous visits to the camp were replaced with sighs, wringing of hands, and whimpers. They seemed to count themselves as forsaken castaways, and nothing Drake and his company could say or do could relieve them of their despair.

Now with the day of departure at hand, it became obvious that they were preparing a massive sacrifice. When they came down from the hill that morning, every man was carrying an armful of wood, while the women and children bore whatever valuable earthly possessions the family might have at hand.

The jolly boat was loaded with the final articles. As Drake's men started to climb aboard, flames shot into the air on the beach behind them as the Indians filled the air with their lamentations and shrieks. It was an especially heartbreaking sight to Fletcher, who had tried so hard to steer them from these dreadful rites. Running back among them, he beseeched them to stop, but they were already in a frenzy. Drake signaled his men to follow and they made their way to the sacrificial fire the Indians had constructed. He raised his eyes and hands to heaven and waited for them to become silent.

After Drake had finished a prayer, Fletcher, his sweet voice filling the misty air, soothed them with the psalms they loved so much to hear. The fire now smoldering, the disk chains and bunches of feathers half-consumed, Drake and his crew turned, walked slowly to the boat and cast off, leaving the tranquil Indians with their hands and eyes raised toward heaven.

"Weigh anchor!" Drake shouted as he climbed over the rail of the *Golden Hind*. Looking back toward the cove, he could see the sacrificial fires raging back on the beach and all around the hills on the headland.

---

unidentified British seaman in the form of a typewritten scrap of paper. Cited in an article by Jeff Greer, "$15 Million Treasure Buried in Marin?" *San Rafael Independent-Journal*, Aug. 11, 1984.

Fig. 21. The *Golden Hind* departing Nova Albion: painting by Charles J. Lundgren, circa 1980. *Copyright © 1983 Unicover Corporation, Cheyenne, Wyoming 82008. All rights reserved.*

PART II

Contemporary Evidence

THE PRINCIPALL
NAVIGATIONS, VOIA-
GES AND DISCOVERIES OF THE
English nation, made by Sea or ouer Land,

to the most remote and farthest distant Quarters of
the earth at any time within the compasse
of these 1500. yeeres: Deuided into three
seuerall parts, according to the positions of the Regions wherun-
to they were directed.
∵

The first, conteining the personall trauels of the English vnto *Iudæa, Syria, A-rabia,* the riuer *Euphrates, Babylon, Balsara,* the *Persian* Gulfe, *Ormuz, Chaul, Goa, India,* and many Islands adioyning to the South parts of *Asia:* toge-ther with the like vnto *Egypt,* the chiefest ports and places of *Africa* with-in and without the Streight of *Gibraltar,* and about the famous Promon-torie of *Buona Esperanza.*

The second, comprehending the worthy discoueries of the English towards the North and Northeast by Sea, as of *Lapland, Scriksinia, Corelia,* the Baie of *S. Nicholas,* the Isles of *Colgoicue, Vaigats,* and *Noua Zembla* toward the great riuer *Ob,* with the mightie Empire of *Russia,* the *Caspian* Sea, *Georgia, Armenia, Media, Persia, Boghar* in *Bactria,* & diuers kingdoms of *Tartaria.*

The third and last, including the English valiant attempts in searching al-most all the corners of the vaste and new world of *America,* from 73. de-grees of Northerly latitude Southward, to *Meta Incognita, Newfoundland,* the maine of *Virginia,* the point of *Florida,* the Baie of *Mexico,* all the In-land of *Noua Hispania,* the coast of *Terra firma, Brasill,* the riuer of *Plate,* to the Streight of *Magellan:* and through it, and from it in the South Sea to *Chili, Peru, Xalisco,* the Gulfe of *California, Noua Albion* vpon the backside of *Canada,* further then euer any Christian hitherto hath pierced.

Whereunto is added the last most renowmed English Nauigation,
round about the whole Globe of the Earth.

By Richard Hakluyt Master of Artes, and Student sometime
of Christ-church in Oxford.

Imprinted at London by George Bishop
and Ralph Newberie, **Deputies to**
Christopher Barker, **Printer to the**
Queenes most excellent Maiestie.

1589.

Fig. 22. The title page of *PRINCIPAL NAVIGATIONS, VOYAGES AND DISCOVERIES by Richard Hakluyt, published in 1589. Reproduced from LOST HARBOR by Warren Hanna, University of California Press, 1979, courtesy of Mrs. Warren Hanna, Kensington, California, and the University of California Press, Berkeley, California.*

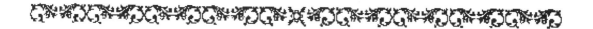

CHAPTER 3
Sixteenth-Century Documents

minent anthropologist, Robert Heizer, writing in 1974, offered this explanation for why the anchorage puzzle has proven so difficult to solve:

Every person who has addressed the question, despite the fact that he might be formally labeled as historian, anthropologist, climatologist, or geographer, has worked under the restriction of having the same limited documentary information. Thus far no specialist knowledge has succeeded in discovering the solution to the problem.[1]

In this chapter and the next, we will review the "limited documentary information" to which Professor Heizer referred—that is, all the important contemporary accounts and maps of the circumnavigation that have a significant bearing on the anchorage question. In chapter 5, I will show you how Drake researchers have consistently misinterpreted the most important of them.

As we go along, you will find that there was more than enough documentary information available for solving the Drake anchorage puzzle after all—and that no specialized knowledge was required to decipher it.

Fate of the Original Records

The California anchorage mystery began on September 26, 1580, not in California, but in England, when Drake nosed the heavily laden *Golden Hind* into Plymouth Harbor after completing the last leg of the circumnavigation. Among his personal belongings, Drake had various records containing ample information to disclose his whereabouts in June–July 1579. First and foremost was his journal: the ship's log in which he entered all his navigation, sketches and notes. Next were the numerous paintings made by Drake and his cousin John that, according to the captured Spanish nobleman Don Francisco de Zárate, were "so naturally depicted that no one who guides himself according to these paintings can possibly go astray." Finally there was the detailed journal of the expedition's chaplain, Francis Fletcher, that contained minute observations on all the important events that had taken place over the course of the voyage.[2]

What happened to these precious documents? According to Bernardo de Mendoza, Spain's ambassador to England at the time, Drake met with Queen Elizabeth shortly after his return and handed over his diary and a large map of the world. Presumably, he also gave her Fletcher's journal and all the paintings. No one knows with certainty what the queen did with these original records for it seems they disappeared before the end of the Elizabethan Age. Some Drake researchers have expressed hope that the California anchorage mystery might someday be solved by some new discovery in the archives of England.[3]

For reasons to which I alluded in chapter 1 and will fully present in chapter 5, I believe the queen gave the records back to Drake circa 1589.

The Queen's Censorship

From Drake's return in 1580 through the end of the decade, no one published any accounts of the circumnavigation. This circumstantial evidence, along with certain documentary evidence, suggests that the queen censored the release of information on the voyage. Indeed, according to Spain's ambassador, Mendoza, Drake's men were not to disclose the route they took on pain of death.[4] The

---

1. Heizer, *Elizabethan California*, p. 11.
2. Zárate in Nuttall, *New Light on Drake*, pp. 207–8.
3. Cited by Wallis, "The Cartography of Drake's Voyage," in Thrower, ed., pp. 121–22.
4. Cited by Wallis, " The Cartography of Drake's Voyage," in Thrower, ed., p. 133.

usual explanation is that Elizabeth was afraid of further provoking Spain in those precarious years just before the attack of the Spanish Armada (1588). It is also likely that she did not want to release any information that might prove that she had authorized Drake's raids, support Spain's claims for restitution, or provide England's enemies with any useful geographic information.[5]

Drake's Censorship

Even after England's defeat of the Spanish Armada rendered concerns over restitution or reprisal moot issues, published accounts of the great voyage were not immediately forthcoming. The first we know of with certainty was a very brief account published by John Stow in 1592.

Why the continued delay? The evidence suggests that while he lived, Drake was determined to publish his own accounts of his voyages and was vigorously resisting publishing attempts by Elizabethan chroniclers such as Stow and Richard Hakluyt.

The First Accounts

I list here and then briefly describe the most important early accounts of the circumnavigation that were prepared or published in Drake's lifetime (he died at sea in 1596):

"The Famous Voyage" – written in England circa 1589 and first published circa 1596 by Richard Hakluyt.

The John Cooke narrative – written in England circa 1589 and first published in 1854 by the Hakluyt Society.

The "Anonymous Narrative" – written in England circa 1589 and first published in 1854 by the Hakluyt Society.

The John Stow account – written in England circa 1591 and first published in 1592 by Stow.

Sir Thomas Blunderville's treatise – written in England circa 1593 and first published in 1594 by Blunderville.

The John Drake declarations (two) – taken in South America in 1584 and 1587 respectively. They were first published in 1911 by Lady Eliott-Drake.

The Nuño da Silva account – written in South America in 1578 and first published in 1600 by Richard Hakluyt. It covers the voyage from the Cape Verde Islands to Mexico.

The Lopez Vaz discourse – probably written in Spain circa 1583 and first published in 1600 by Richard Hakluyt. It covers the voyage from the Cape Verde Islands to Mexico and was reportedly based on information obtained in writing from Nuño da Silva.

The Edward Cliffe account – written in England circa 1578 and first published in 1600 by Richard Hakluyt. It covers just the first half of the voyage to the Pacific and then the return voyage of the *Elizabeth*.

The Richard Madox diary – written in England in 1582. Excerpts were first published in 1932 by E. G. R. Taylor.

The Francis Fletcher narrative – written in England circa 1590 and first published in 1854 by the Hakluyt Society (discussed in chapter 5).

"The World Encompassed" – written in England circa 1593 and first published in 1628 by Drake's nephew, Sir Francis Drake, the baronet (discussed in chapter 5).

---

5. By late 1580, Drake had succeeded in persuading most of his surviving crew to sign an affidavit that understated the financial losses that Spain was claiming (Nuttall, *New Light on Drake*, pp. 411–24). When the principal accounts of the voyage were finally published, they largely substantiated Spain's claims.

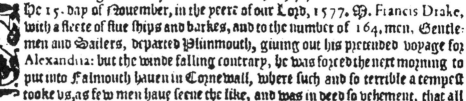

S. Fran. Drake. of the English nation.

The famous voyage of Sir Francis Drake into the South Sea, and there hence about the whole Globe of the Earth,

begun in the yeere of our Lord, 1577.

The 15. day of November, in the yeere of our Lord, 1577. M. Francis Drake, with a fleete of fiue ships and barkes, and to the number of 164. men, Gentlemen and Sailers, departed Plimmouth, giuing out his pretended voyage for Alexandria: but the winde falling contrary, he was forced the next morning to put into Falmouth hauen in Cornewall, where such and so terrible a tempest tooke vs, as few men haue seene the like, and was in deed so vehement, that all our ships were like to haue gone to wracke, but it pleased God to preserue vs from that extremitie and to afflict vs only for that present with these two particulars: The maste of our Admiral which was the Pellican, was cut ouer boord for the safegard of the ship, and the Marygold was driuen a

Fig. 23. The first lines of Richard Hakluyt's twelve-page account "The Famous Voyage," found inserted in most surviving copies of PRINCIPAL NAVIGATIONS, VOYAGES AND DISCOVERIES, 1589 edition. *Reproduced from LOST HARBOR by Warren Hanna, University of California Press, 1979, courtesy of Mrs. Warren Hanna, Kensington, California, and the University of California Press, Berkeley, California.*

"The Famous Voyage"

Richard Hakluyt, besides being one of the prime movers in England's colonial movement (see chapter 39), was the premier chronicler of the Elizabethan Age. In his lifetime, Hakluyt published a series of epic volumes covering the entire history of English maritime expedition and colonial expansion: *Divers Voyages* in 1582; *Principal Navigations, Voyages and Discoveries* in 1589; and a greatly enlarged *Principal Navigations, Voyages and Discoveries* in three volumes published respectively in 1598, 1599, and 1600. Historically, these publications have been collectively referred to as *Hakluyt's Voyages,* and here simply as *Principal Navigations*.

Hakluyt did not include an account of Drake's circumnavigation in his 1582 *Divers Voyages,* presumably due to the queen's censure. He intended to include an account in the 1589 edition of *Principal Navigations,* but changed his mind at the last minute. Here is what he says in its preface (*emphasis added*):

> For the conclusion of all, the memorable voyage of Master Thomas Cavendish into the South Sea, and from thence about the globe of the earth does satisfy me, and I doubt not but will fully content thee: which as in time it is later than that of Sir Francis Drake, *so in relation of the Philippines, Japan, China, and the Isle of Saint Helena it is more particular and exact*: and therefore the want of the first made by Sir Francis Drake will be the less: wherein *I must confess to have taken more than ordinary pains, meaning to have inserted it in this work*: but being of late (contrary to my expectation) *strongly urged not to anticipate or prevent another man's pains* and charge in drawing all the services of that worthy knight into *one volume,* I have yielded unto *those my friends which pressed me in the matter,* referring the further knowledge of his proceedings to those *intended discourses.*[6]

I see no reason to doubt that Drake was ultimately the one behind Hakluyt's sudden decision not to publish an account of the circumnavigation in 1589. From a letter Drake wrote to the queen, we know that by 1591, he was already immersed in a monumental effort to publish discourses covering all his enterprises. In 1589 he would have been just "setting sail" on this very "troublesome" voyage.

Though it could have been earlier, it was probably not until after Drake's death in January 1596 that Hakluyt finally published the first detailed account of Drake's circumnavigation. He printed on six unnumbered folio leaves, "The Famous Voyage of Sir Francis Drake into the South Sea, and

6. Hakluyt's 1589 preface is reproduced in Hanna, *Lost Harbor,* p. 89.

there hence about the whole globe of the earth, begun in the year of our Lord 1577" ("The Famous Voyage"), and inserted them in a second release of his 1589 *Principal Navigations*.[7]

While there has been some speculation about who actually compiled the account and its sources, it is likely that Hakluyt put it together himself using Francis Fletcher's journal, the John Cooke narrative, and the "Anonymous Narrative." He may have obtained Fletcher's journal from his employer at the time, Sir Francis Walsingham, Elizabeth's secretary of state, one of the heaviest investors in the voyage, and a likely guardian of the original records. He seems to have obtained the other two accounts from Stow, but it could have been the other way around.

In 1600, Hakluyt published a slightly revised edition of "The Famous Voyage." The revisions are of no significance to the anchorage controversy, other than to show that the chronicler was uncertain how far north Drake sailed before approaching the coast to search for the so-called Straits of Anian; his forty-two degrees north latitude in 1589 became forty-three degrees in 1600. He also published an account entitled "The Course . . ." which essentially repeats the text of the California section of "The Famous Voyage" and provides additional and corrected information concerning the events that took place off the Central American coast. In this volume he also published for the first time the Lopez Vaz discourse, the Nuño da Silva account, and the Edward Cliffe account.

"The Famous Voyage" contains concise and, for the most part, accurate descriptions of all the key events of the voyage, but little navigational detail, particularly with regard to the Far East. Because it was the first detailed account of the voyage and prepared and published by a writer of Hakluyt's stature, some anchorage sleuths have considered it the most trustworthy and important of all the contemporary accounts. Unlike some of the sources it was drawn from, it is complimentary toward Drake. Of its twelve pages, approximately two are devoted to the California sojourn (see appendix A).

The John Cooke Narrative

John Cooke served aboard John Winter's *Elizabeth*, which returned to England shortly after passing the Strait of Magellan. While accurate in most details, his narrative is overtly hostile to Drake, accusing him of the murder of his second-in-command, Master Thomas Doughty, whom Drake beheaded during the voyage for allegedly inciting mutiny. The manuscript, found in the collections of John Stow, is purportedly in Stow's handwriting, and may have been compiled by Stow circa 1589 during an interview with Cooke. Neither Stow nor Hakluyt ever published Cooke's narrative, presumably in deference to Drake.[8]

The "Anonymous Narrative"

The "Anonymous Narrative" picks up where the John Cooke narrative leaves off and covers the rest of the voyage. Containing few dates and very little navigational detail, it provides an interesting sidelight on the fate of Drake's mistress María:

---

7. Quinn (1984), "Early Accounts," in Thrower, ed., p. 34; and Kerr (1940), "The Treatment of Drake's Circumnavigation," argue that Hakluyt published "The Famous Voyage" in 1589 or early 1590 within months of the first release of *Principal Navigations*. They base this conclusion on the fact that most surviving copies have the six folio leaves and because one copy with the Drake leaves bears some internal evidence of being "strictly contemporary." In his *Report of Findings*, Raymond Aker speculated that Hakluyt published "The Famous Voyage" shortly after Drake's death in January 1596. I concur with Aker's thinking. This tribute to Drake would have made the 1589 *Principal Navigations'* second release an instant best seller explaining the preponderance of copies with the leaves. There are two pieces of contemporary evidence that strongly suggest that Hakluyt did not publish "The Famous Voyage" before 1594: (1) the track of Drake's voyage on the Molyneux Globe (1592) which shows Drake's route off the Pacific Northwest Coast significantly at odds with that set forth in "The Famous Voyage," and (2) *Blunderville's Exercises*, published in 1594, where he complains of the want of a detailed narrative. The fact that the first Dutch translation of "The Famous Voyage" was published in 1596 lends further support to the argument (Wallis, "The Cartography of Drake's Voyage," in Thrower, ed., p. 145). As mentioned above, the John Stow account published in 1592 appears to predate the Hakluyt account.
8. Wagner in *Drake's Voyage*, pp. 241–85, reprints the John Cooke narrative and "Anonymous Narrative" side-by-side with Hakluyt's "The Famous Voyage " and discusses their likely sources.

Drake left behind him on this Island [Lifamatola Island, Indonesia] the two Negroes which he took at Guatulco, and likewise the Negro wench María, she being gotten with child on the ship . . .[9]

It also provides the only information about a falling-out between Drake and Francis Fletcher that occurred in Indonesia:

. . . he said, Francis Fletcher I do here excommunicate thee out of the church of God and from all the benefits and graces thereof and I denounce thee to the devil and all his angels, and then he charged him upon pain of death not once to come before the mast for if he did he swore he should be hanged, and Drake caused a poesy to be written and bound about Fletcher's arm with charge that if he took it off he should be hanged. The poesy was: "Francis Fletcher, the falsest knave that livith."[10]

As with the John Cooke narrative, the "Anonymous Narrative" was found in the Stow collection, is antagonistic towards Drake, and seems to have been put together circa 1589. Neither Stow nor Hakluyt ever published it. Here is all it has to say with respect to Drake's visit to California:[11]

Drake watered his ship and departed sailing northwards till he came to 48 gr. of the septentrional [north] latitude still finding a very large sea trending toward the north – but being afraid to spend a long time in seeking for the strait he turned back again still keeping along the coast as near land as he might, until he came to forty-four degrees – and then he found a harbor for his ship where he grounded his ship to trim her, and here came down unto them many of the country people while they were graving of their ship and had conference with them by signs – in this place Drake set up a great post and nailed thereon a coin, which the country people worshipped as if it had been God – also he nailed upon this post a plate of lead and scratched therein the queen's name.[12]

The John Stow Account

John Stow was one of Hakluyt's principal colleagues and another renowned Elizabethan chronicler. In 1580, Stow published the *Chronicles of England, from Brute unto this present year of Christ*. In 1592, he published a third edition of his *Chronicles* that included a brief account of the circumnavigation, probably the first ever released. It is apparent that Stow compiled the account using the John Cooke narrative, the "Anonymous Narrative," and at least one other unknown source. There is no way of being certain whether he used the Hakluyt account which may have already been published by 1592, but based on internal evidence, it does not seem so. Stow's account has the following to say about the California visit including some details (*highlighted*) not found in any other account:[13]

. . . he passed forth northward till he came to the latitude of *forty-seven*, thinking to have come that way home: but being constrained by fogs and cold winds to forsake his purpose, came backward to the lineward the *tenth of June* 1579 and stayed in the latitude of thirty-eight to grave and trim his ship until the twenty-fifth of July . . .[14]

Sir Thomas Blunderville's Treatise

In 1594, Sir Thomas Blunderville published a book entitled *His Exercises, Containing Six Treatises*, by which he strove to help Elizabethan mariners improve the art of navigation. In one of these treatises he described the "great terrestrial globe lately put forth by M. Sanderson and M. Molyneux" and the track it showed of Drake's circumnavigation. Blunderville had this to say of Drake's movements along the Pacific Northwest Coast:

From thence he sailed northerly to the Cape Mendocino, which is the land called Quivira, and sailed still northward into a certain bay on the west part of Quivira, which he named Nova Albion (that is to

---

9. This information about María is corroborated in the second declaration of John Drake, 1587 (Nuttall, *New Light on Drake*, p. 53).
10. In Wagner, *Drake's Voyage*, p. 282. The passage on Fletcher's fallout with Drake is not actually part of the "Anonymous Narrative." It is among some collateral documents that are appended to the manuscript: in this case a "memorandum to appendix II."
11. The hostile tone of the "Anonymous Narrative" and John Cooke narrative may have come in the wake of the Portugal Enterprise of 1589 led by Drake and Sir John Norris—England's disastrous response to the Spanish Armada (see chapter 39 of this book).
12. In Wagner, *Drake's Voyage*, p. 277.
13. Unlike Hakluyt, Stow does not appear to have used the Fletcher journal, presumably because it was not available to him.
14. Stow, in Wagner, *Drake's Voyage*, p. 304.

say New England) having in north latitude 46 degrees, and from this bay Sir Francis himself (as I have heard) was of good will to have sailed still more northward, hoping to find passage through the narrow sea Anian, but his mariners finding the coast of Nova Albion to be very cold, had no good will to sail any further northward, wherefore Sir Francis was fain to come back again southward to Mendocino, which (as hath been said before) is distant from the foresaid bay of Nova Albion 140 leagues.[15]

The John Drake Declarations

Drake's cousin John, who had taken part in the circumnavigation at fifteen years of age, was captured by the Spanish during the disastrous Edward Fenton expedition of 1582 (see chapter 39). The Spanish, of course, were very anxious to tap this source of firsthand information on the voyage, and put John through the terrors of at least two interrogations. Two declarations have survived.

From the first declaration taken March 24, 1584:

He does not know the day they left Guatulco, only that it was in April. They sailed out at sea always to the northwest and north-northwest the whole of April and May until the middle of June, from Guatulco, which lies in 15 degrees north, until they reached 48 degrees north. On their voyage they met with great storms. All the sky was dark and filled with mist. On the voyage they saw five or six islands in 46 and 48 degrees. Captain Francis gave the land that is situated in 48 degrees the name of New England. They were there a month and a half, taking in water and wood and repairing their ship.[16]

From the second declaration taken January 8, 1587:

Then they left and sailed, always on a wind, in a northwest and north-northwesterly direction, for a thousand leagues until they reached forty-four degrees when the wind changed and he went to the Californias where he discovered land in forty-eight deg. There he landed and built huts and remained for a month and a half, caulking his vessel. The victuals they found were mussels and sea lions. During that time many Indians came there and when they saw the Englishmen they wept and scratched their faces with their nails until they drew blood as though this were an act of homage or adoration. By signs Captain Francis told them not to do that, for the Englishmen were not God. These people were peaceful and did no harm to the English, but gave them no food. They are the color of the Indians here [Peru] and are comely. They carry bows and arrows and go naked. The climate is temperate, more cold than hot. To all appearances it is a very good country. Here he caulked his large ship and left the ship he had taken in Nicaragua. He departed, leaving the Indians, to all appearances, sad.[17]

The Nuño da Silva Account

Drake took the Portuguese pilot Nuño da Silva off a merchant ship he captured in the Cape Verde Islands and put him ashore in Mexico, just before setting off on his search for the Straits of Anian. While in Spanish custody, da Silva prepared "a detailed account of all that occurred during his voyage until he reached Guatulco." Shortly after preparing this "first relation," da Silva was compelled to prepare a second, a copy of which fell into English hands. Hakluyt published a translation of the "second relation" in *Principal Navigations*, vol. 3 (1600). As indicated in appendix D, there is information in the da Silva account that helps identify some of Drake's other questionable landfalls, in particular, "Mucho Island" and the "southernmost cape" of South America. The Spanish accused da Silva of voluntarily assisting Drake and rigorously deposed him. His ultimate fate is unknown.[18]

The Edward Cliffe Account

Edward Cliffe, like John Cooke, was one of the mariners aboard the *Elizabeth*. Unlike the Cooke narrative, Cliffe's account is impartial toward Drake and filled with navigational detail. It appears to have been prepared with the *Elizabeth*'s logbook at hand soon after the vessel's return to England

---

15. Blunderville, in Wagner, *Drake's Voyage*, p. 312.
16. Drake in Nuttall, *New Light on Drake*, p. 31.
17. Ibid., pp. 50–51.
18. The quotation is from the letter of Don Martin Enriquez, viceroy of Mexico, to the viceroy of Peru, dated May, 17, 1579, translated in Nuttall, *New Light on Drake*, p. 244.

in 1580. It is evident that the compiler of *The World Encompassed* used either the Cliffe account or the logbook from which it was drawn as one of the principal source materials for the first leg of the circumnavigation. Hakluyt published the Edward Cliffe account in *Principal Navigations*, vol. 3 (1600).

The Madox Diary

In 1932, an English historian, E. G. R. Taylor, brought to light the diary of Richard Madox, the chaplain who accompanied Edward Fenton's 1582 expedition, just mentioned in the context of the John Drake declarations. This diary included some important information on the circumnavigation that Madox had gathered from conversations with those members of Fenton's crew who had served aboard the *Golden Hind*:[19]

> In Ships Land which is the back side of Labrador and as Mr. Hall supposed near thereunto, Sir Francis Drake graved and breamed his ship at forty-eight degrees to the north. The people are for feature, color, apparel, diet and "holo" speech like those of Labrador and is thought kingless for they crowned Sir Francis Drake. Their language is thus: "cheepe" – bread; "huchee kecharoh" – sit down; "nocharo mu" – touch me not; "hioghe" – a king.

> Their song when they worship God is thus – one dancing first with his hands up, and all the rest after like the priest and people – "Hodeli oh heigh oh heigh oh hadali oh . . ." [20]

Spanish Accounts

The Spanish officials Drake terrorized in sixteenth-century America were great record keepers. In 1908, while doing some unrelated research in the Mexican archives, scholar Zelia Nuttall stumbled upon a document dealing with the inquisition of Nuño da Silva, the captured Portuguese pilot Drake put ashore at Guatulco, Mexico. Over the next half decade she delved into the archives of Mexico, Spain, the United States and Europe, looking for more information on Drake. She came up with a wealth of eyewitness accounts shedding much "new light" on Drake's raids along the Spanish Main. Among the more important documents she turned up, translated, and published in 1914, were an account by Pedro Sarmiento de Gamboa, a celebrated sixteenth-century navigator; sworn depositions from over a dozen prisoners taken by Drake while in the South Seas; and a great deal of information furnished by the Portuguese pilot Nuño da Silva, including depositions, trial documents, a relation of his voyage, and a copy of his logbook entries.[21]

Aside from substantiating the accuracy of some of Drake's latitude readings for South American landfalls, these documents do not have any direct bearing on the California anchorage controversy.

Several other Spanish documents have emerged from the archives that do have some bearing. In 1900, historian Charles Lummis, in *Land of the Sunshine* magazine, published the translation of a "relation" written by Father Gerónimo de Zárate y Salmerón that covered events in California and Mexico up to 1626.[22] Zárate y Salmerón's history included a thirdhand account of a Portuguese pilot named N. de Morena who allegedly made his way on foot to Mexico after Drake put him ashore, presumably along the coast of today's Washington State. In 1924, historian Henry Wagner translated and published several accounts of the voyage of Sebastián Rodríguez Cermeño, whose Manila galleon was shipwrecked in Drake's Bay in 1595. In chapters 30 and 32 respectively, you will find excerpts of Wagner's translation of the Cermeño account and Lummis' full translation of the N. de Morena account.[23]

---

19. In 1976, The Hakluyt Society, Elizabeth Donno, ed., published the entire account: *An Elizabethan in 1582: The Diary of Richard Madox, Fellow of All Souls.*
20. Madox in Donno, ed., *An Elizabethan in 1582*, pp. 208–9.
21. Nuttall, *New Light on Drake*, 1914.
22. Lummis, ed., "Pioneers," pp. 184–85. It may have been Theodore Hittell, a contributor to the magazine, who provided the translation.
23. Wagner's translations of the Cermeño accounts appeared in the *California Historical Society Quarterly* in 1924 as part of a series of articles on the early navigators. The society collectively published the articles in 1929 under the title *Spanish Voyages to the Northwest Coast of California in the Sixteenth Century.*

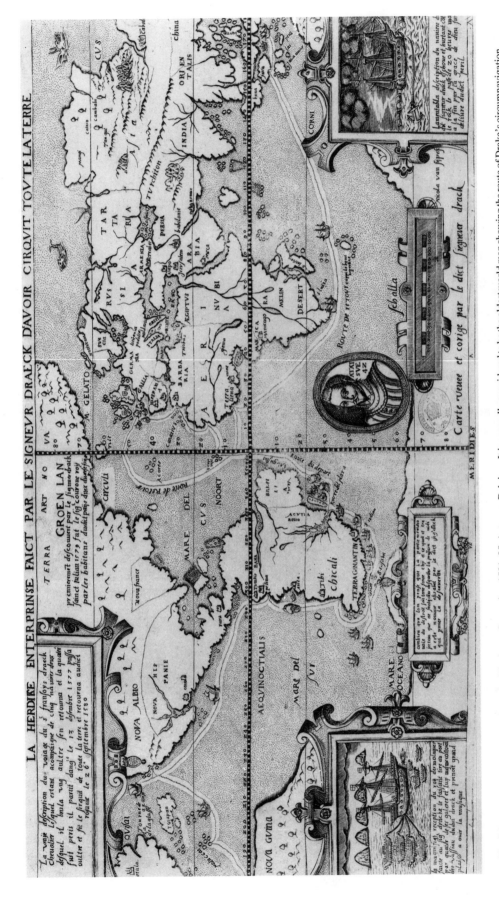

Fig. 24. The French Drake Map: drawn by Nicola van Sype, circa 1581–85, Antwerp, Belgium. It is generally considered to be the oldest world map showing the route of Drake's circumnavigation. It bears the notation "seen and corrected by the said Francis Drake." Most of the few surviving copies of the map are bound into copies of a French work entitled *Le Voyage Curieux faict autour du Monde par Francois Drach*, published in Paris in 1613, 1627 and 1641. The British Library copy here reproduced is from a free standing copy of the map. *By permission of the British Library, London.*

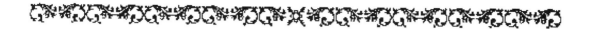

CHAPTER 4
Sixteenth-Century Maps

As mentioned in chapter 3, it is documented that shortly after disembarking the *Golden Hind* in September 1580, Drake handed over to the queen his diary of the voyage and a large map. It seems this map was not concealed as well as the written records of the voyage, for the chronicler Samuel Purchas reported seeing either the original or a copy in Whitehall Palace sometime around 1618. Furthermore, there is documentary evidence that Drake presented a copy of the map to his friend the Archbishop of Canterbury by 1583, and that in 1585 Sir Francis Walsingham, secretary of state, sent a copy to Henry of Navarre, later to become king of France.[1]

There exist today a number of late sixteenth-century maps showing the route of the circumnavigation, some apparently derived from the Whitehall Map Drake presented to the queen (the queen's map), including the following:

The French Drake Map – drawn circa 1581–85 in Antwerp by Nicola van Sype (figure 24).

The Dutch Drake Map – drawn circa 1581–85, author unknown. It may have been copied from the French Drake Map.

The Drake-Mellon Map – drawn circa 1581–87, author unknown. It bears the exact same inscription Purchas quoted in 1618 (reproduced inside the rear cover of this book).

The Silver Map – drawn 1589 in London by Michael Mercator. It is engraved on a silver medallion. The first Drake map published in England, it may have been commissioned by Drake himself.

The Hondius Broadside Map – drawn circa 1589 in London and published circa 1596 in Holland by Jodocus Hondius (reproduced inside the front cover of this book).

The Molyneux Globe – drawn 1591 in London by Emery Molyneux. It does not appear to have been derived from the queen's map.

These maps and others have been discussed at length by several anchorage researchers including George Davidson, Henry Wagner, John Robertson, Raymond Aker, and Warren Hanna. With the exception of the Portus Nova Albionis inset on the Hondius Broadside Map (discussed elsewhere), I found the purported evidence contained on these early maps too untrustworthy to be of much value in solving California's anchorage mystery; there are too many uncertainties about sources, authorship, and map details.[2]

Notwithstanding the uncertainties, I will introduce some interesting details from four of the above-referenced maps that at least appear to be pertinent to the anchorage question. In addition, I will cover three seventeenth-century maps drawn by the English cartographer, Robert Dudley, that include some information that others have considered important: unpublished manuscript chart number 85, and two maps from Dudley's *Arcano del Mare*: the Carta Particolara chart number 33 and the Carta Prima General del India . . . .

---

1. Wallis, "The Cartography of Drake's Voyage," in Thrower, ed., p. 123.
2. Davidson, *Drake's Anchorage*, pp. 35–54; Wagner, *Drake's Voyage*, pp. 405–37; Hanna, *Lost Harbor*, pp. 113–33; and Aker, *Report of Findings*, pp. 48–57.

The French Drake Map

The French Drake Map (figures 24 and 25) drawn by Nicola van Sype is generally believed to be the earliest of all the surviving sixteenth-century maps showing the route of Drake's circumnavigation. It may have been copied from the map that Sir Francis Walsingham sent to Henry of Navarre in 1585.

Fig. 25. Detail of North America from the French Drake Map, drawn circa 1581–85. The map's track shows Drake reaching as far north as approximately forty-seven to forty-eight degrees latitude before backtracking to about thirty-nine degrees. The map shows two bays on the coast of Nova Albion in the vicinity of thirty-eight degrees, with Drake departing from an island located just south of the more northerly inlet. *By permission of the British Library, London.*

The Drake-Mellon Map

The Drake-Mellon Map (reproduced inside the rear cover of this book), so-called because the only known copy is part of the Paul Mellon collection, is similar to the French Drake Map, but its captions are in Latin. While it is generally assumed to be of later issue than its French and Dutch counterparts in that it marks Sir Walter Raleigh's 1584–85 Virginia colony and the track of Drake's 1585–86 West Indies expedition, there are indications that this information was superimposed at a later date. For a number of reasons, I suspect the Drake-Mellon Map is both the earliest and most faithful rendition of the queen's map, and that Drake was personally connected with it.[3]

Fig. 26. Detail of North America from the Drake-Mellon Map, drawn circa 1581–87. In contrast to the French Drake Map, it does not show a diversion to forty-eight degrees, though it has a caption to the effect; and it shows only one small island off the coast of Nova Albion rather than four large ones. Like the French Drake Map, it depicts two bays on the coast of Nova Albion with Drake again departing from an island located just south of the more northerly inlet. In this case, the more northerly inlet is much smaller than the other. In addition, the more northerly inlet appears to be marked with a circular cartographic symbol, perhaps representing a landfall. *Courtesy of the Yale Center for British Art, Paul Mellon Collection, New Haven, Connecticut.*

---

3. Indications that the information on Raleigh's Virginia Colony and Drake's 1585–86 voyage were superimposed on the original include the following (see inside rear cover of this book): the letter "C" for the legend beginning "VIRGINEA Colonia . . ." is superimposed on the "O" for "FLO-RI-DA." The track of Drake's 1585–86 expedition is superimposed over the following map details for the circumnavigation: the depiction of Drake's fleet as it sets out from England; the first "o" in "Mogador;" and the "g" in "Santiago."

The Hondius Broadside Map

The Hondius Broadside Map (reproduced inside the front cover of this book) is the most famous of all the Drake maps. Jodocus Hondius apparently drew the map circa 1589 during a stay in London but did not engrave it until after he returned to Amsterdam circa 1595. It was published in Holland circa 1596.[4]

The broadside contains a highly controversial inset labeled the Portus Nova Albionis (the Portus or Portus plan) that ostensibly depicts Drake's California port-of-call (figure 40). The Hondius Broadside Map is the only contemporary Drake map that contains this particular inset, leading to a

Fig. 27. Detail of North America from the Hondius Broadside Map, drawn circa 1589, published as a broadside in 1596. The map shows Drake reaching forty-two degrees north latitude and then backtracking to about thirty-eight degrees where he enters a large bay similar in appearance to the more southerly bay of the two shown on the French and Drake-Mellon Maps. It is evident that the map originally had Drake reaching forty-eight degrees. The "correction" was presumably made in 1596 to make the track correlate with the text of the broadside's narrative, "The Famous Voyage." The Hondius map does not show the second, more northerly bay of the French Drake Map and Drake-Mellon Map, or the island(s). Comparing the California coastline as a whole for the three maps, it is apparent that the Drake-Mellon Map provides the greatest detail. *Courtesy The Bancroft Library, Berkeley, California.*

---

4. Wallis, "The Cartography of Drake's Voyage," in Thrower, ed., p. 145. I suspect the Hondius Broadside Map was part of a mass of memorabilia published in the wake of Drake's death (January 1596). When the map was published it was accompanied by the printed text of Hakluyt's "The Famous Voyage" (in Dutch) attached along the sides and bottom; hence the term broadside.

Fig. 28. The tip of South America from the Hondius Broadside Map. *Courtesy The Bancroft Library, Berkeley, California.*

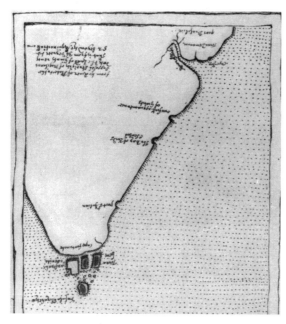

Fig. 29. The tip of South America from the Francis Fletcher narrative. *By permission of the British Library, London.*

great deal of speculation on how Hondius obtained it. Throughout the anchorage controversy there has always been doubt as to whether Hondius created the Portus from imagination or from an authentic painting or chart. As the other three insets Hondius provided on the broadside seem authentic, there is good reason to believe the Portus is also. The problem, however, has been that no one, up until now, has been able to match it convincingly to an existing harbor on the Pacific Coast.[5]

There is an interesting aspect of the Hondius map that was first noted by Raymond Aker of the Drake Navigators Guild. It appears that in drawing the map Hondius used what must have been Drake's sketch of the tip of South America, a copy of which turned up in the so-called Fletcher notes transcribed by John Conyers in 1677 (see chapter 5). This provides another strong indication that Hondius had at least some of Drake's original drawings in his hands when he prepared his Portus inset.[6]

The Molyneux Globe

In the preface to the 1589 edition of his *Principal Navigations*, Richard Hakluyt gave notice that the English cartographer Emery Molyneux was working on a pair of globes, celestial and terrestrial, England's first major contribution in the field of global cartography. These historic twenty-five-inch diameter globes, of which there are just two sets still extant, were engraved by Jodocus Hondius, forged in 1591, and completed in 1592. The terrestrial globe shows the track of Drake's voyage with a California landfall near Cape Mendocino at about forty-two degrees. This inaccurate depiction of the California landfall and the manner in which Molyneux depicts the tip of South America in the wake of Drake's discovery of Cape Horn and the Elizabeth Islands (figure 35) suggests that the famous mapmaker was not on friendly terms with Drake. As you will see in the next chapter, the apparent antagonism between Drake and Molyneux bears a small part in the anchorage controversy.

---

5. Aker, *Report of Findings*, pp. 186–209, discusses the inset drawings at great length.
6. We hear of this same sketch in the Madox diary (Donno, *An Elizabethan in 1582*, pp. 239–40). Madox questions how Drake could have delineated the islands so exactly with the little time he spent traversing the Strait of Magellan and speculates that Drake copied the drawing from one of the Portuguese charts he took from Nuño da Silva.

When Molyneux sold the printer's plates for the globes to Hondius circa 1596, Hondius, working with Richard Hakluyt and the English cartographer Edward Wright, promptly revised the plates not just to show Drake's geographic discoveries, but to duly credit him for them. The revised world map, published circa 1600, is the so-called Twelfth-Night Map (figure 36) immortalized by William Shakespeare in act 3, scene 2, of his play by that name with the lines, "He does smile his face into more lines than are in the new map, with the augmentation of the Indies."

Fig. 30. Detail of North America from the Molyneux Globe, drawn circa 1591 (facsimile). The track shows Drake reaching forty-eight degrees north latitude and backtracking to a landfall in the mouth of a river located at about forty-two degrees. According to Henry Wagner, as of 1926, the Molyneux Globe was preserved in the Middle Temple in London. *The map detail is from a sketch of the globe by Dr. J. G. Kohl, held by the Library of Congress, Washington, D.C.*

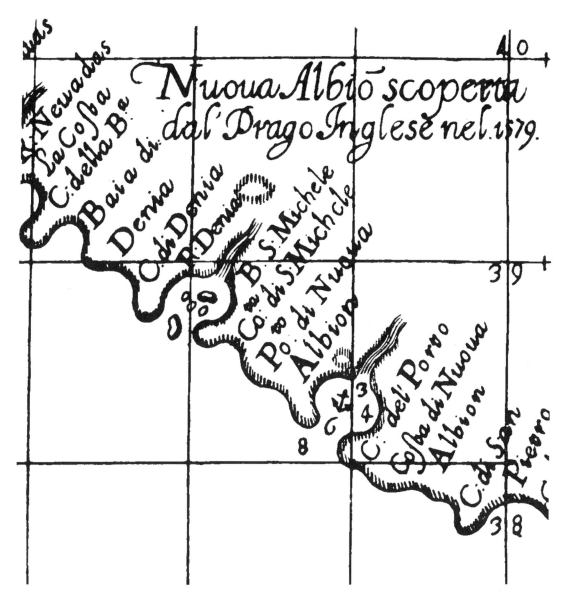

Fig. 31. California detail from Robert Dudley's Carta Particolara, chart number 33, drawn circa 1635. Dudley placed his "Bay of S. Michele" at 38º 45′ presumably based on the reports of the Sebastián Rodríguez Vizcaíno expedition of 1603. He placed Drake's "Porto di Nueva Albion" at about 38º 15′, presumably based on the recently released *The World Encompassed*. In my opinion, none of the inlets on this chart bear any significant resemblance to those found in the San Francisco area. *Reproduced from IDENTIFICATION OF SIR FRANCIS DRAKE'S ANCHORAGE ON THE COAST OF CALIFORNIA by George Davidson, California Historical Society, San Francisco, 1890. Davidson obtained his reproduction of the chart from the Harvard College Library, Cambridge, Massachusetts.*

The Dudley Maps

Robert Dudley, a prominent seventeenth-century English cartographer, was the illegitimate son of the carl of Leicester. Leicester was a member of Elizabeth's Privy Council, a staunch friend and supporter of Drake, and allegedly, the queen's longtime love. Because of this connection, some have speculated that Dudley, through his father, came to possess Drake's original drawings. Circa 1630, Dudley began work on an atlas that was published posthumously in 1646–47 under the title *Arcano del Mare*. Historically, anchorage investigators have used three of Dudley's drawings in promoting their cases: the Carta Particolara chart number 33; the Carta Prima General del India . . . ; and unpublished manuscript chart number 85, also prepared circa 1635.

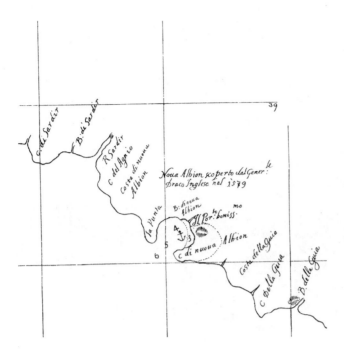

Fig. 32. California detail from Robert Dudley's unpublished manuscript chart number 85, drawn circa 1635. Dudley may have derived the principal cartographic details for this chart from the sketches produced during the 1603 Vizcaíno expedition. In my opinion, none of the inlets on this chart bear any significant resemblance to those found in the San Francisco area. *Reproduced from IDENTIFICATION OF SIR FRANCIS DRAKE'S ANCHORAGE ON THE COAST OF CALIFORNIA by George Davidson, California Historical Society, San Francisco, 1890. Davidson obtained his reproduction of the chart from the Royal Library at Munich, Germany.*

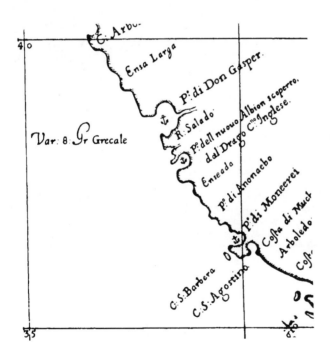

Fig. 33. California detail from Robert Dudley's Carta Prima General del India . . . , drawn circa 1635. The depiction of yet a third harbor in the immediate vicinity of Drake's port reveals more uncertainty by the mapmaker. In my opinion, none of the inlets on this chart bear any significant resemblance to those found in the San Francisco area. *Reproduced from IDENTIFICATION OF SIR FRANCIS DRAKE'S ANCHORAGE ON THE COAST OF CALIFORNIA by George Davidson, California Historical Society, San Francisco, 1890. Davidson obtained his reproduction of the chart from the Harvard College Library, Cambridge, Massachusetts.*

CHAPTER 5
The Secrets of "The World Encompassed"

The World Encompassed by Sir Francis Drake is a 108-page book published in London in 1628 by Drake's nephew, heir, and namesake, Sir Francis Drake, the Drake family's first baronet. It is by far the most detailed contemporary account of the circumnavigation and provides the majority of the information we have on Drake's California sojourn (see appendix B). I believe that there has been a universal misunderstanding about the authorship of *The World Encompassed*—a misunderstanding that has contributed significantly to the perpetuation of California's great maritime mystery.

Who Wrote "The World Encompassed"?

According to the first lines on its title page (figure 34), *The World Encompassed* was authored "by Sir Francis Drake."

Which Sir Francis Drake? Does "by Sir Francis Drake" refer to the deceased commander of the expedition, or his nephew, Sir Francis Drake, the baronet, who edited and published the book in 1628?

Most Drake researchers have credited the baronet as the author of *The World Encompassed*. Others have credited Francis Fletcher, the expedition's preacher. There is only one researcher I am aware of that has tentatively credited the great Elizabethan mariner.[1]

In this chapter I will demonstrate that Francis Drake, the great Elizabethan mariner, not the baronet, should be considered the true author of *The World Encompassed*.

How the Baronet Came to Publish Drake's Sagas

Sir Francis Drake died at sea in January 1596. His will directed much of his estate into the hands of his second wife, Lady Elizabeth, and his only surviving brother, Thomas. The unfortunate Thomas was to spend much of the ten remaining years of his life in court fighting to hold on to his inheritance. There were a series of ugly lawsuits involving Lady Elizabeth, Drake's close friend Jonas Bodenham, and Drake's cousin Richard Drake of Esher. From some of the court documents we learn that Thomas came into possession of Drake's personal records, journals and books and kept them at Buckland Abbey, near Plymouth. When Thomas died in 1606 most of his estate, including Buckland, eventually passed to his only son Francis.[2]

Educated in law, it did not take long for Francis to rid himself of the pending lawsuits that had plagued his father. After securing his inheritance, he finished his education, married, and settled his family into Buckland Abbey. King James I made the thirty-three-year-old Francis a baronet in 1622.

At the time of King James' death in 1625, the English were fully committed to another war with Spain. With soldiers and sailors pouring into Plymouth for an attack on Cadiz, the new king, Charles I, appointed the baronet among the commissioners charged with housing and provisioning the assembled forces. The English attack later that year was completely unsuccessful and the fleet returned "broken down with mutiny and disease." To make matters worse, with the town awaiting a counterassault by the Spanish, the plague broke out. As after the defeat of the Spanish Armada in 1588, the streets of Plymouth were paved with starving, sickly and despondent men who had not been paid for their services.[3]

---

1. This is readily confirmed by checking their bibliographies to note the designation "Bart." (for baronet) in *The World Encompassed* citations. Quinn ("Early Accounts," in Thrower, ed., p. 38) is the one researcher that tended toward the belief that Drake himself took an active role in preparing the account.
2. Eliott-Drake, *Family and Heirs*, pp. 184 and 187.
3. Ibid., p. 215 (unreferenced quotation).

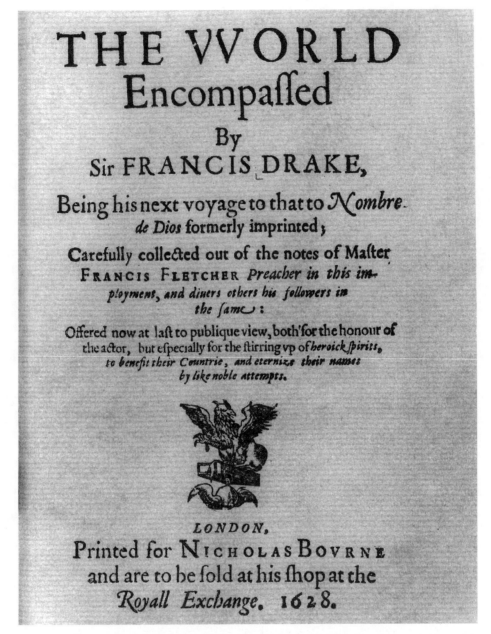

Fig. 34. The title page of THE WORLD ENCOMPASSED by Sir Francis Drake, 1628. *Courtesy The Bancroft Library, Berkeley, California.*

It was in these most trying times that the resolute baronet found time to assemble a neglected manuscript that had long been gathering dust in the Drake library. In 1626 he published *Sir Francis Drake Revived*, "calling upon this dull effeminate age to follow his noble steps for gold and silver." He dedicated the book to Charles I, presumably to honor and commemorate the king's visit to Plymouth that year. It recounted Drake's 1572 raid on Panama by which he helped Queen Elizabeth I finance her war with Spain.

The baronet was elected to a seat in Parliament in 1628 and spent the first six months of the year in London. It was during a six-month recess in the second half of the year that he found time to dust off another neglected manuscript and publish his second book, *The World Encompassed*. Once again it was with the expressed intent of stirring up the spirits of his beleaguered countrymen "to benefit their country" by robbing the king of Spain to help finance the war effort. He dedicated this work to the earl of Warwick, then lord high admiral of England.[4]

When Did Drake Prepare His Manuscripts?

Let us closely examine a letter Drake prepared in late 1591 that I believe contains the keys to unlocking the secrets of *The World Encompassed*. The baronet used this letter as the dedicatory epistle to his first book, *Sir Francis Drake Revived* (*emphasis added*):[5]

> To the Queen's most excellent Majesty, my most Dread Sovereign:
>
> Madam,
>
> Seeing divers others have diversely reported and written of *these voyages* and actions which I have attempted and made, every one endeavoring to bring to light whatsoever inklings or conjectures they have had – *whereby many untruths have been published, and the certain truth concealed* – as I have thought it necessary myself, as in a card [navigational plan] to prick the principal points of the counsel taken, attempts made and success had, during *the whole course of my employment in these services against the Spaniard*, not as setting sail for maintaining my reputation in men's judgment, but only as sitting at helm, if occasion shall be, for conducting like actions hereafter . . . So I have accounted it my duty to present *this discourse* to Your Majesty, as of right, either for itself, *being the first fruits of your servant's pen*, or for the matter, being [not only] service done to Your Majesty by your poor vassal against your great enemy at such times, in such places, and after such sort as it may seem strange to those that are not acquainted with the whole carriage thereof, but [also because it is a relation that] will be a pleasing remembrance to Your Highness, who take the apparent height of the Almighty's favor towards you by these events, as truest instruments – humbly [I am] submitting myself to your gracious censure, both in writing and presenting, [so] that posterity be not deprived of such help as may happily be gained hereby, and our present age, at least may be satisfied in the rightfulness of these actions, *which hitherto have been silenced*; and [also that] your servant's labor not seem altogether lost, not only in travel by sea and land, but also in writing the report thereof, *a work to him no less troublesome* yet made pleasant and sweet in that it hath been is, and shall be for Your Majesty's content; to whom I have devoted myself, live or die.
>
> Francis Drake, January 1, 1592.[6]

Here we have Drake's own testimony that in the early 1590s he was hard at work compiling discourses covering "the whole course of his employment against the Spaniard." Although the baronet used the letter to dedicate *Sir Francis Drake Revived* and not *The World Encompassed*, from such terms as "these voyages" and "these services" it is evident that when he signed the letter in January 1592, Drake was preparing a monumental discourse covering all of his expeditions: an autobiography of sorts. Furthermore, in requesting the queen's input and stating that he was submitting just "the first fruits" of his pen, Drake implies that the discourse he had submitted was just in draft form and that there was much more to come. Note also how these observations correlate with Richard Hakluyt's statement in the preface to the 1589 edition of his *Principal Navigations,* to the effect that someone was then endeavoring to compile the accounts of all Drake's exploits in a single volume (see chapter 3).[7]

Most significantly, from his letter to the queen we have very strong indications that in 1591 Sir Francis Drake was putting together an account of his most controversial voyage, the circumnavigation. When he states that he has been compelled to prepare his own accounts because

---

4. The information on the baronet and his father is from *The Family and Heirs of Sir Francis Drake*, by Lady Elizabeth Eliott-Drake, published 1911, pp. 197–229.

5. Drake dated the letter January 1, 1592. Because the Elizabethan legal year (new style) starting about this time was changed to begin on March 25, the date January 1592 could be equivalent to January 1593 by current reckoning. It is impossible to know with certainty whether Drake recognized the old or new calendar system in his letter.

6. I have attempted to clarify the letter's meaning by adding a few [phrases] and separating the text with dashes at a few of the punctuation marks. The letter is otherwise rather difficult to decipher due to Drake's rambling writing style.

7. The queen appears to have lifted the ban on publication in 1589. This also appears to be the time at which other firsthand accounts began to appear (for example, the John Cooke narrative). It was also in 1589 that Drake was working with the Florentine historian, Petruccio Ulbalino, on an account of the Armada campaign (Corbett, *Drake and the Tudor Navy*, vol. 2, pp. 446–47).

"divers others" have written "conjectures" and "many untruths" about his voyages and actions, he is very likely referring to the circumnavigation. As discussed in chapter 3, circa 1589 "divers others" had indeed prepared some inaccurate or unflattering accounts of this voyage: specifically the Lopez Vaz discourse, the John Cooke narrative and the "Anonymous Narrative."[8]

The Mysterious Sloane Manuscript No. 61

It so happens that there is extant today in the British Library a half-completed narrative of the circumnavigation transcribed in the seventeenth century that, by all indications, was taken from an authentic contemporary document written, at least in part, in 1591. This antique manuscript (Sloane MS No. 61) is the only known original copy of the narrative and by all appearances it is unique. The account covers the voyage from England to Mucho Island off the coast of Chile, is written in a very rambling style, contains numerous illustrations, and is obviously a rough draft. You will find excerpts in appendix C.[9]

According to its flyleaf, the account was transcribed from an original manuscript in 1677 by one Joh. (John) Conyers, "pharmacopolist, citizen, and apothecary of London." The flyleaf bears the following title identifying Francis Fletcher as the narrative's author:[10]

> The first part of the second voyage about the world attempted, continued and happily accomplished within the time of three years by Mr. Francis Drake, at Her Highness' command and his company, written and faithfully laid down by Francis Fletcher, minister of Christ and preacher of the gospel, adventurer and traveler in the same voyage.

London's Hakluyt Society first brought the Fletcher narrative to light in *The World Encompassed by Sir Francis Drake* (Vaux, 1844), in which it reprinted the baronet's 1628 publication and other contemporary documents. W. S. Vaux referred to the account as the "Fletcher Notes" derived from the following statement on the title page of *The World Encompassed* (figure 34): "carefully collected out of the notes of Master Francis Fletcher, preacher in this employment, and divers others, his followers in the same."[11]

As Vaux demonstrated by his footnotes, in comparing the first part of *The World Encompassed* to the Fletcher notes (hereafter referred to in this chapter as the Fletcher narrative), it is evident that the author of *The World Encompassed* did indeed draw heavily from the narrative transcribed by Conyers. Furthermore, from similarities in style, it is also apparent that in compiling the second half of *The World Encompassed*, its author drew largely from the missing second half of the Fletcher narrative. One can also readily discern where the author added information from divers other accounts, specifically, the Edward Cliffe account, the Nuño da Silva account and "The Famous Voyage."[12]

There is no question, however, that the author of *The World Encompassed* drew primarily from the complete text of the Fletcher narrative in compiling the account.

A Closer Look at the Fletcher Narrative

The circumstantial evidence immediately suggests that Sir Francis Drake was intimately involved in the preparation of the Fletcher narrative that was eventually transcribed by Conyers. First and foremost, look at the timing of its preparation. In his January 1, 1592, letter to the queen, Drake

---

8. There is only one account of Drake's 1572 expedition that is known to have been published in the sixteenth century. Hakluyt released Lopez Vaz's short and inaccurate discourse of Drake's 1572 voyage in his 1589 *Principal Navigations*.

9. Sloane MS No. 61, British Library. The manuscript is in the collection of Sir Hans Sloane.

10. Conyers provided no information on how or where he obtained the original document.

11. Barrow (1843), *The Life, Voyages and Exploits*, appears to have been the first to cite the Conyer's manuscript. Vaux (1854) was the first to publish it in extenso. In attempting to correlate the Fletcher narrative with a reprint of *The World Encompassed*, he presented it in lengthy footnotes. Penzer (1926) was the first to publish the Fletcher narrative text in full but did not include all the illustrations. Wagner, in *Drake's Voyage* (1926), was the first to publish all the illustrations and also correlated excerpts of the text with *The World Encompassed*. The best reproductions I have found of the illustrations are in Roche (1973), *The Golden Hind*.

12. Wagner, *Drake's Voyage*, pp. 294–302, demonstrates the close correlation with the Edward Cliffe account.

revealed that circa 1591 he was in the process of compiling the accounts of his controversial voyages, presumably including the circumnavigation. From certain statements in the Fletcher narrative (discussed elsewhere), we know the account was written in part at this very time, i.e., circa 1591. Secondly, look at when and where the Fletcher narrative eventually turned up. It was in 1628 at Buckland Abbey, that Drake's heir and namesake, the baronet, "offered now at last to public view," a circumnavigation discourse "carefully collected" from the "notes of Master Francis Fletcher." It is evident that the baronet found what appears to have been the one and only copy of the Fletcher narrative's manuscript among the personal records Drake left behind at Buckland—it seems to have come from the great mariner's private library.

Despite this circumstantial evidence, the Drake researchers who have discussed the Fletcher narrative have generally concluded that Francis Fletcher produced the account on his own and intended to publish it on his own.[13] They base this contention on the fact that the Conyers' flyleaf includes the statement "written and faithfully laid down by Francis Fletcher," and because the account, at first glance, seems to be somewhat hostile to Drake. While some thus credit Francis Fletcher as the author of *The World Encompassed*, most seem to feel that the baronet modified his source materials to the extent that he should be considered its rightful author.[14]

I disagree. What I see in the Fletcher narrative is Fletcher, circa 1589, transcribing his on-board journal and adding commentary while working under Drake's employ. I then see Drake, circa 1591, taking Fletcher's manuscript and editing and enlarging the text with his own commentary. Specifically, I think Drake added some navigational detail; the discussion on his fellow English mariner, Thomas Cavendish; the laudatory language about the queen; the discussion of plans to erect a monument after successfully passing the Strait of Magellan; all discussion on Magellan; the discussion of renaming Terra Incognita (the unknown land) to Terra Nunc Bene Cognita (the land well-known); some caustic remarks about his critics; and a good deal more (see appendix C). It could have been either Fletcher or Drake who edited in the illustrations, but more likely the latter.[15]

The circumstantial evidence (timing of preparation) also suggests that the Fletcher narrative was part of the draft discourse Drake submitted to the queen under his January 1, 1592 cover letter. I suspect he submitted a full draft of the 1572 voyage to Panama along with it—in other words, the first chapter-and-a-half of his autobiography of voyages.

I do not think the queen refused to let Drake publish his discourse when he submitted it to her in January 1592. It is more likely that the task of compiling his entire life's story was simply too much for the busy statesman. Nevertheless, by the time he set forth on his last voyage in August of 1595, Drake would have had plenty of time to thoroughly redraft the first part of his circumnavigation discourse and add the second. He may have enlisted the help of Philip Nichols in this effort, who we know was then helping Drake prepare the account of his earlier expedition to Panama.

In the following ten points, I defend my contention that circa 1591, Sir Francis Drake coauthored the Fletcher narrative that exists today in the British Library as Sloane MS No. 61.

1. *The account is not hostile to Drake* – From the time the Fletcher narrative was first published in 1854, Drake scholars have been describing it as overtly hostile to Drake, in effect dismissing out of hand the notion that Drake could have played any part in its preparation. Upon close review of the rambling text, however, I find nothing derogatory about Drake, but a considerable amount that is complimentary. What emerges is that Fletcher thought very highly of Thomas Doughty, believed he was innocent of the charges that led to his execution, and insinuated that he had been

13. Those who have discussed the Fletcher narrative's authorship in some detail include Vaux, ed., *The World Encompassed*, pp. xi–xii; Penzer, ed., *The World Encompassed*, p. 87; Wagner, *Drake's Voyage*, pp. 290–91; Aker, *Report of Findings*, pp. 44 and 152; and Wallis, "The Cartography of Drake's Voyage," in Thrower, ed., p. 127.
14. Those who have discussed *The World Encompassed*'s authorship at some length include Wagner, *Drake's Voyage*, pp. 286–93; Aker, *Report of Findings*, pp. 149–52; Hanna, *Lost Harbor*, pp. 96–102; and Quinn, "Early Accounts," in Thrower, ed., pp. 37–39. Quinn expresses uncertainty on the issue and advances some of the same theories that I expound in my arguments.
15. From Nuño da Silva's sworn deposition we know that Drake carried a copy of Magellan's journal aboard the *Golden Hind* (Nuttall, *New Light on Drake*, p. 303).

unjustly accused by several individuals of questionable character, namely "John Brewer, Edward Bright, and some others of their friends." While he places no blame for the affair on Drake, he does on several occasions express his opinion that God was punishing the entire company for the injustice committed at Port St. Julian where Doughty was executed. [16]

2. *The account is consistent with Drake's feelings for Thomas Doughty* – Here is an example of the lavish praise Fletcher heaped on the unfortunate gentleman Drake executed at Port St. Julian:

> . . . he feared God, he loved his word, and was always desirous to edify others and confirm himself in the faith of Christ – for his qualities in a man of his time they were rare and his gifts very excellent for his age as sweet orator, a pregnant philosopher, a good gift for the Greek tongue and a reasonable taste of Hebrew, a sufficient secretary to a noble personage of great place, and in Ireland an approved soldier and not behind many in the study of law for his time.[17]

When considering if Drake would have allowed Fletcher to state such things about a man he had beheaded, bear in mind that Doughty was Drake's very close friend before the voyage, that they shared the sacrament and dined together the night before the execution, that they embraced on the gallows, and that Doughty with his last words called Drake a good captain, begged the company to forgive him, and recalled their once close friendship.[18]

Now weigh the following statement from a letter written in April 1579 by Francisco de Zárate, the Spanish nobleman whose vessel (and mistress) Drake had captured off the coast of Nicaragua (*emphasis added*):

> . . . a gentleman who he had brought with him said to him, "for a long time we are in this strait and you have placed near death all who follow and serve you. It would be well to order us to return to the north sea where we are certain of booty and not hunt out new discoveries which you see to be so difficult." This gentleman must have sustained this opinion with more spirit than seemed good to the general, who answered him by ordering him to be taken below deck and to have some shackles put on him. The next day at the same hour he ordered him to be taken out and in the presence of all to have his head cut off . . . He himself told me this, *saying many good things about the dead man* but that he could not do anything less because this was necessary in the service of the queen . . . [19]

3. *There is dialogue that very strongly suggests Drake's hand* – Since Drake almost certainly gave Fletcher's on-board journal to the queen in 1580, he was not at liberty in his discourse to tamper with its content, only to offer insight and commentary. Let us take what seems to be an obvious example. In the following passage Fletcher tells us how, just before leaving Port St. Julian, he gave a sermon warning the company of God's pending wrath for the injustice committed there:

> Now the time of our departure drawing near it was desired of my hand that we might have a general communion and some necessary doctrine tending to love and Christian duty . . . which as the Lord enabled me I performed with exhortation to repentance every man as he felt the guilt of his own conscience lest our hope of joy we assured ourselves [of finding] in . . . Mare Pacificum be turned into sorrow . . . for even there also could God mett us jump [forsake us] as well as in other places to punish our sins . . . [20]

Fletcher then sees his prophecy fulfilled in a violent storm that befell them after exiting the Strait of Magellan and entering the Pacific:

> . . . fifty-six days of so unspeakable setting of the Heavens, earth, seas, and winds against us, forced us to confess that God had mett jump with us [forsaken us] in Mari Pacifico [the Pacific] . . . I doubt not by the Spirit of God in at Port Julian [because of what happened at Port St. Julian].[21]

Beginning the last paragraph of the account, note the distinctively different style and tone as the author, now Drake it seems, puns the following witty retort to Fletcher's cryptic statements, while

16. Penzer, ed., *The World Encompassed*, p. 97. Those who have cited the hostility of Fletcher's narrative toward Drake include Penzer, ed., p. 87; Wagner, *Drake's Voyage*, p. 291; Aker, *Report of Findings*, p. 44 and 152; and Quinn, "Early Accounts," p. 39.
17. Penzer, ed., *The World Encompassed*, p. 125.
18. From the John Cooke narrative, in Penzer, ed., *The World Encompassed*, pp. 159–61.
19. Zárate in Wagner, *Drake's Voyage*, pp. 376–77.
20. Penzer, ed., *The World Encompassed*, p. 127.
21. Ibid., p. 134.

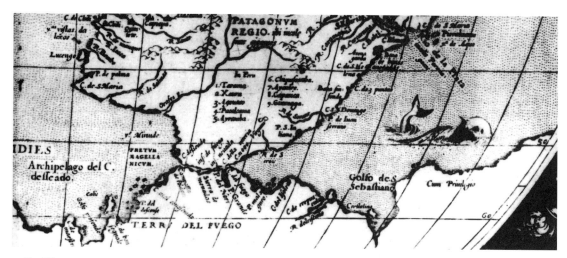

Fig. 35. The tip of South America from the Abraham Ortelius Map of America, drawn 1587. The Molyneux Globe of 1592 depicted the area in a similar fashion. The great cartographers of Drake's day rejected his claim that the Atlantic and Pacific were joined. *By permission of the British Library, London.*

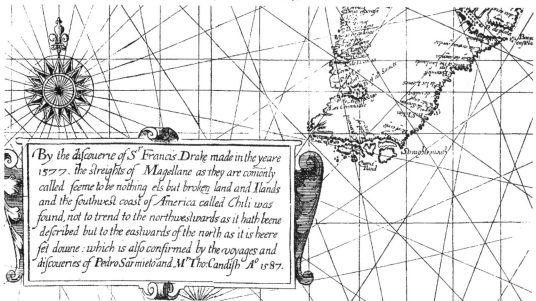

By the diſcouerie of St Francis Drake made in the yeare 1577. the ſtreights of Magellane as they are comonly called ſeeme to be nothing els but broken land and Ilands and the ſouthweſt coast of America called Chili was found, not to trend to the northweſtwards as it hath beene deſcribed but to the eaſtwards of the north as it is heere ſet downe: which is alſo confirmed by the voyages and diſcoueries of Pedro Sarmieto and Mr Tho: Candiſh Aᵒ 1587.

Fig. 36. The tip of South America from the Wright-Molyneux-Mercator-Hakluyt Map, drawn 1599 – "The Twelfth-Night Map." Note how Richard Hakluyt and Edward Wright appear to be making amends for the slights of the earlier cartographers. *By permission of the British Library, London.*

providing his queen with strategic information on what they found soon after entering the Pacific (*emphasis added*):

> *Wherein whether God mett us jump in the South Sea or no we ourselves might easily judge* [when we consider that] this Island [Mucho Island] is most rich in gold and silver – it abounds in many good things necessary for the maintenance of God's good people, flourishing with trees and fruit, continually wanting nothing but a God-fearing people to enjoy it to glorify his name. For such excellent blessings bestowed upon it, I may compare it aptly to Her Majesty's island named Wight upon the coast of England which in respect to its strategic position is called a door bar [doorjamb] to the land on that side of the country – this island lays in like sort against a golden province of the world named Valdavia. The island being possessed of one prince would make the one [the island] and the other [Valdavia] invincible.[22]

---

22. Ibid., p. 141.

Fig. 37. Birds, trees and a sea lion: details of three illustrations from the Francis Fletcher narrative. I believe these to be second-hand copies of the drawings illustrating Drake's personal logbook. *By permission of the British Library, London.*

4. *Evidence revealing the timing of its preparation suggests Drake's hand* – I have already pointed out that his January 1592 letter strongly suggests that in 1591 Drake was working on a circumnavigation discourse. Internal evidence demonstrates that the Fletcher narrative was in preparation at this very time. In the following colorful passage the author refers to the Molyneux Globe (forged circa 1591) as "new-forged" while caustically ridiculing the mapmaker:

> . . . which many of our featherbed milksops boasting of their deep judgment in cosmography at a smith's forge hammering out a globe to make a childish brag amongst simple people, do laugh and mock-at and say it is a lie to report such things of God's great and marvelous works and further to confirm their blind error and ignorance, their own new-forged globe must be showed . . . [23]

What are the chances that Fletcher would have been granted an interview with the great cosmographers of the time much less have had the audacity to denigrate them like this in his writings? From the globe itself we see that Molyneux had flatly rejected Drake's claim that the Atlantic and Pacific Oceans were joined south of Cape Horn, a monumental slight (figures 35 and 36). There can be little doubt that Drake was absolutely seething at the mapmaker, and the passage reflects just this type of indignant rage. Conversely, there is no reason for Fletcher to be so enraged about the matter; why would he care?

5. *The illustrations strongly suggest Drake's hand* – There is no evidence that Fletcher, while aboard the *Golden Hind*, made any pictures or charts such as those illustrating the Fletcher narrative. On the other hand, there is an eyewitness account that "Francis Drake kept a book in which he entered his navigation and in which he delineated birds, trees, and sea lions," just as they appear in the account (figure 37). Assuming that they were Drake's pictures that turned up in the document, how could Fletcher have possibly gotten his hands on them without Drake's approval? [24]

6. *It was Drake, not Fletcher, who inscribed a monument in the strait* – According to page 38 of *The World Encompassed*, in September 1578, Francis Drake engraved a monument that he

---

23. Ibid., p. 96.

24. Nuño da Silva's sworn deposition, in Nuttall, *New Light on Drake*, p. 303. We know from the Madox diary that Drake provided copies of some of his sketches to the Fenton expedition; in particular, the one showing Tierra del Fuego at the tip of South America (in Donno, ed., *An Elizabethan in 1582*, p. 239).

intended to erect immediately after exiting the Strait of Magellan. For this to be so, Drake must have written at least some of this commentary in the Fletcher narrative:

> Now God in mercy at the last brought us through this labyrinth which so long had entangled us with so many extremities and imminent dangers . . . wherein as we exceedingly rejoiced: so were we resolved there to stay a while to do as we had done to our good God and in duty to Her Majesty in other places and to have set up a monument for Her Highness upon the cape for a witness of our passing that way and arrival at that place—which monument I had engraved in metal for the same purpose the like of that which you may see in the end of this book . . . [25]

7. *It was Drake, not Fletcher, who trekked to the southernmost point of Tierra del Fuego –* According to the Fletcher narrative, after separating from the *Elizabeth* in October 1579, northwest gales drove the *Golden Hind* southward to the "uttermost part" of Tierra del Fuego. The author provides this interesting anecdote of what transpired after they came to anchor on the lee side of the southernmost island (*emphasis added*).

> Myself being landed, did, with my bag, travel to the *southernmost point of the island, to the sea on that side where I found that island to be more southerly three parts of a degree* [three minutes] than any of the rest of the islands. Where having set up on end a stone of some bigness, and with such tools as I had of purpose ever about me when I went on shore, had engraved Her Majesty's name, her kingdom, the year of Christ, and the day of the month, I returned again in some reasonable time to our company. We departing hence and taking our farewell from the southernmost part of the world known, or as we think to be known here, we altered the name of these southerly islands from Terra Incognita (for so it was before our coming thither, and so should have remained still with our good wills) to Terra Nunc Bene Cognita, that is broken islands; which in coasting it again on that side in returning to the northward, we proved to be true, and were thoroughly confirmed in the same.

Now consider this story obviously based on the same event. In 1593, Richard Hawkins, son of John Hawkins, took an expedition through the Strait of Magellan attempting to repeat Drake's exploits. The Spanish captured him off the coast of Peru in July 1594 and brought him back to Spain. Having finally gained his freedom, Hawkins wrote an account of the expedition that was first published in 1622. In the account of his travails, after exiting the Strait of Magellan, he relates the following story that he reportedly obtained firsthand from Sir Francis Drake (*emphasis added*):[26]

> And moreover, he [Drake] said, that standing about, when the wind changed, he was not able to double the southernmost island, and so anchored under the lee of it; and going ashore, carried a compass with him, and seeking the *southernmost part of the island*, cast himself down on the uttermost point groveling, and so reached out his body over it. Presently he embarked, and then recounted to his people, that he had been upon the southernmost known land in the world, and more further southwards upon it, than any of them, yea, *or any man as yet known* [including Fletcher].[27]

8. *The account was obviously prepared with great pains –* In his January 1592 letter to the queen, Drake complained of how difficult he found it to write his discourse. The Fletcher narrative bears testimony to how tortuous the effort must have been.

9. *The last edition of the Fletcher narrative appears to have come from Drake's personal library at Buckland Abbey –* The Fletcher narrative seems to have disappeared at the same time as did the discourse Drake submitted to the queen in January 1592, and it did not resurface until after *The World Encompassed* was published in the seventeenth century. The first signs we see of it are in *The World Encompassed*.

10. *The unlikelihood of an independent Fletcher publication –* Fletcher was a minister of God and was well aware of the potential fury of the great admiral. It is extremely unlikely that he would have attempted to bootleg such an account, particularly with Drake alive.

---

25. Penzer, ed., *The World Encompassed*, p. 131.
26. Williamson, ed., *The Observations of Sir Richard Hawkins*, includes the 1622 account, a Spanish deposition, and letter from Richard Hawkins to his father John.
27. Wagner, *Drake's Voyage*, pp. 90–91 and 473. Wagner discusses the quote at length and notes the close correlation with the Fletcher narrative.

A Closer Look at "The World Encompassed"

If you accept the contention that Drake did indeed sponsor and co-write the Fletcher narrative, he then emerges as the effective author of *The World Encompassed* (hereafter cited as *W.E.* with page numbers from the Readex Microprint facsimile). As pointed out earlier, the baronet indicates on the title page that he used this account as his primary source, and internal evidence confirms this.

I think the baronet found what I will now call the Drake-Fletcher narrative somewhat too rambling to publish as he found it, even with all the improvements Drake (or Nichols) likely provided in subsequent revisions. What I see in *The World Encompassed* is the baronet's paraphrasing and editing of the account to make it more readable and deleting the illustrations, presumably to facilitate the publication effort. It is clearly evident that the baronet added some excerpts from Hakluyt's "The Famous Voyage": the out-of-sequence Tarapacá episode (*W.E.*, pp. 54–55) for instance, that follows Hakluyt's text almost verbatim. In the first half of the discourse, the baronet may have also added information he paraphrased from various other accounts, in particular the Edward Cliffe account. I think it just as likely, however, that it was Drake himself or someone under Drake's employ (Nichols perhaps), not the baronet, who added the verbiage from the Cliffe account. Unlike Henry Wagner and others, I do not think the baronet added anything original to the documents he had on hand.[28]

Now let us say that one rejects the contention that Drake coauthored the Fletcher narrative that Conyers transcribed in 1677. For the sake of discussion, let us also say one is right in so doing, i.e., that Fletcher wrote the account by himself. My contention then follows that in preparing *The World Encompassed*, the baronet must have used a later version of the Fletcher narrative that Drake coauthored sometime between 1592 and 1595.

I offer the following seven points to support the contention that Sir Francis Drake coauthored the "notes of Master Francis Fletcher" referenced on the title page of *The World Encompassed*, irrespective of whether or not he coauthored the source document for the Conyers' manuscript.

1. *The amount of navigational detail strongly suggests Drake's hand* – The best indication we have that Drake himself contributed heavily to *The World Encompassed* is in the amount of navigational detail it provides, especially for the Far East. Consider the following passages—did they and many others like them come from Fletcher's rambling journal, or were they from Drake's logbook? The passages most certainly did not come from any of the other contemporary accounts extant today:[29]

> The 18, we cast anchor under a little island, whence we departed again the day following; we wooded here, but other relief except turtles, we received none.

> The 22, we lost sight of three islands on our starboard side, which lay in 10 deg. and some odd minutes.

> After this, we passed on to the westward without stay of anything to be taken notice of till the 9 of March, when in the morning we espied land, some part thereof very high, in 8 deg. 20 min. south latitude; here we anchored that night, and the next day weighed again, and bearing farther north, and nearer shore, we came to anchor a second time. (*W.E.* pp. 105–6.)

> In 38 deg. 30 min. we fell with a convenient and fit harbor, and June 17 came to anchor therein: where we continued to the 23 day of July following. (*W.E.* p. 64.)

2. *There is information that could not have come from Fletcher's journal* – We know from the "Anonymous Narrative" that Drake was furious with Fletcher after an incident on January 9, 1580, when the *Golden Hind* grounded on a reef off Celebes Island in the Banda Sea. In their despair the next day, Fletcher apparently gave a sermon that contained language to the effect that

---

28. For the Cliffe material, Drake would have been using the *Marigold's* original records rather than the account Hakluyt published in 1600. This provides a possible explanation for the uncertain latitude entry for "Cape DeGuerre." Wagner, *Drake's Voyage*, p. 287, was convinced that the baronet contributed liberally to the text. He observed that "numerous passages are of suspicious origin which were probably not found in any of the accounts of the voyage used." I contend that those passages of "suspicious origin"—presumably the extensive navigational detail—were added by Sir Francis Drake.

29. In 1589, Hakluyt mentioned that his accounts of Drake's voyage had little navigational detail for the Far East. Based on the amount of navigational detail *The World Encompassed* provides for the Far East, it seems that Drake went out of his way to prove Hakluyt wrong.

God must have "mett jump" from them (forsaken them), presumably because of the sins committed at Port St. Julian. According to the account, the livid Drake not only took it upon himself to excommunicate the preacher, but he also locked him in the forecastle and told him that if he dared show his face on deck he would hang him.[30] There just happens to be a month's gap in Hakluyt's "The Famous Voyage" that lasts from January 10 to February 8, 1580, at which time they disembarked at the island of "Barativa." While there is no such gap in *The World Encompassed*, there is a distinctive change in the style of the narrative. Fletcher's commentaries are gone and I again get the clear impression that I am reading from the entries in the *Golden Hind's* log:[31]

> Jan. 12, being not able to bear our sails, by reason of the tempest, and fearing of the dangers, we let fall our anchors upon a shoal in 3 deg. 30 min.

> Jan. 14, we were gotten a little farther south, where, at an island in 4 deg. 6 min., we again cast anchor, and spent a day in watering and wooding. After this we met with foul weather . . .

> After we had gone hence, on February 8 we descried two canoes [of Indians], who having descried us as it seems before, came willingly unto us and talked with us, alluring and conducting us to their town not far off, named Barativa; it stands in 7 deg. 13 min. south the line. (*W.E.* p. 104.)

On February 8, Fletcher's rambling commentary picks up again:

> The people are gentiles, of handsome body and comely stature, of civil demeanor, very just in dealing, and courteous to strangers, of all which we had evident proof, they showing themselves most glad of our coming, and cheerfully ready to receive our wants with whatever their country could afford. (*W.E.* p. 104).

3. *The dialogue regarding the plundering of the Spanish suggests Drake's hand* – First witness the witty sarcasm Drake used in a letter he wrote to Sir John Wolley on May 8, 1589, after sacking the port of Corunna, Spain: "We have done the king of Spain many pretty services here at this place, and yet I believe he will not thank us."[32] Now note the same tone in one of several similar passages in *The World Encompassed*: "In two barks here we found some forty-odd bars of silver . . . of which we took the burden on ourselves to ease them." (*W.E.* pp. 56–57.)

4. *There is contemporary evidence that Drake prepared a circumnavigation discourse circa 1593* – From Sir Thomas Blunderville, writing circa 1594: "it might please Sir Francis to write a perfect diary of his whole voyage . . . I doubt not but that he hath already written, and will publish the same when he shall think most meet."[33]

5. *It is inconceivable that Drake did not contribute to this discourse* – Why would Drake have taken the trouble to work on the account of the Nombre de Dios expedition and then neglect the circumnavigation, still considered one of the greatest voyages in history? As demonstrated in his January 1592 letter to the queen, Drake was very concerned about securing his just place in history and was most anxious to set the record straight for posterity.

6. *There is an obvious correlation with the way "Sir Francis Drake Revived" was prepared* – On the title page to *Sir Francis Drake Revived*, the baronet reveals that Drake employed the seafaring preacher Philip Nichols to "faithfully" transcribe the journals of some of the ship's officers (Christopher Ceely, Ellis Hixon, and others). The baronet further notes that Drake then took the Nichols' manuscript and edited and enlarged the text with his own commentary. This makes Drake both the owner and coauthor of the discourse. This correlates very closely to the method that I contend Drake used in preparing the draft manuscript for *The World Encompassed*.[34]

7. *The baronet unambiguously credited Sir Francis Drake with authoring "The World Encompassed"* – First let us examine how the baronet credited his famous uncle with coauthoring

---

30. The story is actually found in a contemporary document found with the "Anonymous Narrative" and presumed to be associated.
31. Andrews in *The Last Voyage*, pp. 263–65 (plates v-x), provides some examples of log entries for Drake's last voyage.
32. Tenison, 8: 114. Cited in Sugden, *Sir Francis Drake*, p. 274.
33. Blunderville, in Wagner, *Drake's Voyage*, p. 304.
34. Quinn provides biographical information on Philip Nichols in "Early Accounts," Thrower, ed., pp. 44–45. He notes that Nichols lived in the West Country in the early 1590s, within easy reach by water of Drake's residence in Plymouth.

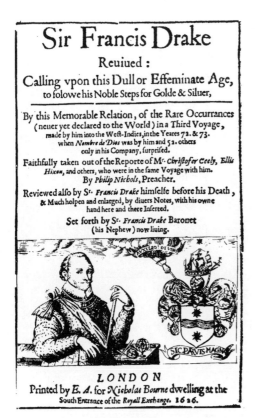

Fig. 38. The title page of SIR FRANCIS DRAKE REVIVED, 1626 (facsimile). *Reproduced from FRANCIS DRAKE PRIVATEER by John Hampden, University of Alabama Press, 1972.*

Sir Francis Drake Revived. On the title page he states that the source manuscript was "much helped and enlarged by divers notes with his [Drake's] own hand here and there inserted." Then, to emphasize his uncle's authorship role, the baronet provides the following statement in the book's dedicatory epistle (*emphasis added*):

Most Gracious Sovereign, that this treatise is yours, both by right and succession, will appear by the *author's* and actor's ensuing dedication. [i.e., Drake's January 1, 1592, letter to the queen] . . . Caesar wrote his own commentaries, and this doer was partly the *inditer* [writer].[35]

Now let us examine the similar manner in which the baronet credits authorship for *The World Encompassed.* On the third line of the title page, the baronet credits authorship to "Sir Francis Drake," which, as already mentioned, could refer to either his famous uncle or to himself. From the fourth line, however, which states "Being *his* next voyage . . ." the baronet unambiguously indicates that "by Sir Francis Drake" on the line above refers to the deceased commander. He reinforces this notion in the fourth paragraph of the title page by implying that he is offering a long-lost Drake manuscript "now at last to public view." In the preface the baronet again unequivocally credits his famous uncle as the author, just as he did in his first book (*emphasis added*):[36]

Fame and envy are both needless to the dead because unknown, sometimes dangerous to the living when too well known; reason enough that I rather choose to say nothing, than too little, in the praise of the *deceased author* . . .

Where he states on the title page "carefully collected from the notes of Master Francis Fletcher . . . and divers others . . . ," the baronet is telling us, not how he prepared the account, but rather, how his famous uncle did.

Where he states in the introduction "it shall for the present be deemed a sufficient discharge of duty to register the true and whole history of that his voyage, with as great indifferency of affection as a history doth require, and with the plain evidence of truth, as it was left recorded by some of the chief [i.e., Drake and Fletcher], and divers other actors in that action," the baronet is simply being careful to explain that he compiled the story from firsthand accounts and did not doctor them.

Authorship of Hakluyt's "The Famous Voyage"

If Drake did indeed coauthor the principal source material for *The World Encompassed*, does it then follow that he also coauthored the principal source document for Hakluyt's "The Famous Voyage"?

Others have commented on the similarities between *The World Encompassed* and "The Famous Voyage." Indeed, from Guatulco on, the two accounts appear to have been drawn largely from the same source, ostensibly the so-called complete Fletcher notes. This has led some Drake researchers

---

35. Drake, baronet, in Hampden, ed., *Francis Drake Privateer*, p. 50.
36. Lady Eliott-Drake, in *Family and Heirs*, speculated that the baronet was referring to Drake here as the author of the voyage, not the book; i.e., that "by Sir Francis Drake" is actually part of the title.

Fig. 39. Buckland Abbey, Drake's former country estate in Devonshire, England (view from the southwest, 1986). Constructed as a monastery in 1278, the abbey was Drake's home from 1581 to his death in 1596. There is a bit of irony in the fact that Drake purchased the estate from Sir Richard Grenville using proceeds from the circumnavigation; Grenville, Drake's lifelong rival, was originally slated to command the expedition. Buckland remained in the Drake family until 1946 and was soon after donated to the National Trust. It is now administered as a maritime museum by the city of Plymouth. *Courtesy of the National Trust, London, © National Trust.*

to conclude that the differences are due to Hakluyt's need to condense "The Famous Voyage," that he simply edited out a lot of the extraneous detail included in *The World Encompassed*.[37]

The evidence suggests to me, however, that Hakluyt used either Fletcher's original on-board journal prepared in 1577–80 or, less likely, Fletcher's unedited manuscript prepared circa 1589 while under Drake's employ. In closely reviewing the first half of "The Famous Voyage," I see no internal evidence that Hakluyt used what appears to be Drake's input on the document transcribed by Conyers.[38]

Respecting the second half of "The Famous Voyage," remember that Hakluyt had complained of the lack of navigational detail for the East Indies. The second half of *The World Encompassed* is loaded with navigational detail.

"The Famous Voyage" and *The World Encompassed* were prepared from several of the same firsthand accounts, but only the latter used a source document that included Drake's own commentary.

The Lost Records Mystery

Why is it that Richard Hakluyt, even after Drake's death, never could get his hands on anything more than Fletcher's narrative or journal of the circumnavigation, and even that for what was apparently just a brief period in 1589 "by more than ordinary pains"? And why is it that no one has been able to find so much as a page of the original records of the voyage, notwithstanding what must have been some very diligent searches of Buckingham Palace and the archives of England?

---

37. Hanna, *Lost Harbor*, pp. 91–93, and Quinn, "Early Accounts," in Thrower, ed., pp. 41–42.
38. Wagner, *Drake's Voyage*, p. 290, observed that Hakluyt made little, if any, use of the Fletcher narrative for the first half of the voyage.

It seems to me that in the aftermath of the Spanish Armada, there was but one person who had both the means and motive to conceal the records of the voyage: Sir Francis Drake. And if Drake did indeed conceal them, there is but one location that comes immediately to mind: Buckland Abbey.

Through whose auspices do we see the only tangible signs that at least some of these records did indeed survive the sixteenth century? We know that thirty years after Drake's death, the baronet, while living at Buckland, was able to lay his hands on some version of the Fletcher narrative. How curious that all of these records immediately disappeared again after the baronet completed the two books.

It seems well within the realm of possibility that at least some of the original records of the voyage are still gathering dust in a secret recess at Buckland Abbey where Sir Francis secreted them in August 1595 just before embarking on his last voyage. Before he departed, Drake surely would have passed such a secret on to his only surviving family member and heir, Thomas. Thomas, in turn, would have passed it along to his only son and heir, the future baronet. If so, the baronet, it seems, took this secret with him to his grave.

Summary

Circa 1589, Sir Francis Drake set forth on the ambitious project of compiling accounts of all his former expeditions into a single volume. Toward this end, he hired Francis Fletcher to help him prepare his own account of the circumnavigation from the original records, by then in Drake's possession. Fletcher transcribed his personal journal, adding a running commentary and perhaps some of the illustrations. Starting around 1590, Drake began editing Fletcher's draft manuscript using his own journal to add navigational details, illustrations and commentary. During this period he also had Philip Nichols similarly helping him prepare an account of the 1572 voyage to Panama. In January 1592, Drake submitted to the queen a partial draft of the circumnavigation and perhaps a full draft of the 1572 expedition. Drake continued work on the two manuscripts between 1592 and 1595, largely completed them, but met his end before he could carry out his dream of compiling the accounts of all his voyages in one great volume. Before departing on his last voyage in the fall of 1595, Drake secreted the two manuscripts and original records at Buckland, and passed the location on to his brother Thomas in the event he did not return. Some thirty years later, Thomas' son, with the intent of "stirring up the heroic spirits" of his beleaguered countrymen in their war with Spain, briefly retrieved the old documents from their hiding place, polished up the manuscripts and published them, offering them "at last to public view."

What Conyers transcribed in 1677 was just part of a complete discourse Fletcher had prepared circa 1589. It was the portion that Drake had edited for the queen's review circa 1591.[39]

Richard Hakluyt did not have access to the Drake-edited manuscript when he wrote "The Famous Voyage" in 1589, or when he published it circa 1596 and again in 1600. "Using more than ordinary pains," he did manage to briefly get his hands either on Fletcher's personal journal or, possibly, the original discourse Fletcher had prepared for Drake sometime around 1589.

Because past anchorage researchers assumed that the baronet authored *The World Encompassed*, they underestimated its value as a contemporary source while overestimating the value of Hakluyt's "The Famous Voyage." It was due largely to this misunderstanding that prior anchorage sleuths went astray in interpreting the all-important evidence on latitudes. Although Drake's own journal of the great voyage may have been lost, it is more than likely that *The World Encompassed* contains all the pertinent navigational detail it provided on the California landfall.

---

39. Quinn, "Early Accounts," in Thrower, ed., pp. 37–38, offers some of the same logic.

PART III

The Great Debate

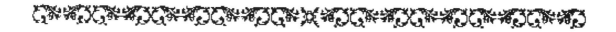

Fig. 40. The Portus Nova Albionis inset from the Hondius Broadside Map. It is the most controversial piece of all the contemporary evidence. *Courtesy The Bancroft Library, Berkeley, California.*

CHAPTER 6
The Center of Controversy

he Drake anchorage conundrum has been aptly labeled "the grandfather of all questions" in California history.[1] Indeed, with Drake's California sojourn providing such a spectacular first chapter in its modern age, speculation and debate over where, exactly, the scene was played out, has become an indelible part of California lore, especially for the San Francisco area.

The anchorage controversy started in October 1775 when Spain first attempted to establish a mission outpost at the site of today's city of San Francisco. During the two-plus centuries since, the controversy has alternately raged and smoldered. By my count, more than a hundred individuals have published express opinions on the location of Drake's anchorage based on careful analysis of the evidence. In addition, hundreds of others have simply cast opinions with little or no explanation for their reasoning. Though I have tried to be thorough in my research, it is possible that I have missed some of the more serious published attempts to solve the mystery. I also suspect that there have been others who have silently pursued the hunt.

Before I introduce you to the anchorage sleuths and opinion-makers, first examine this list of the principal clues around which the debate has centered; you will need to recognize them to fully understand the discussions in the ensuing chapters (*emphasis added*):

The fair and good bay at thirty-eight degrees – "We came *within 38 degrees* towards the line. In which height it pleased God to send us into a *fair and good bay*, with a good wind to enter the same" (Hakluyt's "The Famous Voyage," 1589 edition [hereafter cited as *F.V.*].

The fit harbor at thirty-eight degrees thirty minutes – "In *38. deg. 30. min.* we fell with a *convenient and fit harbor*" (Drake's *The World Encompassed*, p. 64 [hereafter cited as *W.E.* with page numbers from the Readex Microprint facsimile]).

The cold and foggy weather – ". . . yet were we continually visited with like *nipping colds* as we had felt before . . . neither could we at any time, in whole fourteen days together, find the air so clear [due to the most vile and *stinking fogs*] as to be able to take the height of sun or star" (*W.E.*, p. 64).

The white cliffs that lie toward the sea – "Our general called this country, Nova Albionis, and that for two causes: the one in respect of the *white cliffs and banks which lie toward the sea*; and the other, because it might have some affinity with our country in name, which was sometime so called" (*F.V.* and *W.E.*, p. 80).

The long-tailed conies – "We found the whole country to be a warren [breeding ground] of *a strange kind of conies*, their bodies in bigness as be the Barbary conies, their heads as the heads of ours, the feet of a want [mole?], and the tail of a rat being of great length" (*F.V.* and *W.E.*, p. 80).

The plate of brass – "At our departure hence our general set up a monument of our being there, as also of Her Majesty's right and title to the same, namely a plate, [a *plate of brass* or lead] nailed upon a fair great post" (*F.V.*, *W.E.*, p. 80 and "Anonymous Narrative," in Wagner, *Drake's Voyage*, p. 277).

The Isles of Saint James – "Not far without this harbor did lie certain islands (we called them *the Islands of Saint James),* having on them plentiful and great store of seals and birds, with one of which we fell July 24" [after departing July 23] (*W.E.*, pp. 81–82*).

---

1. Todd, *San Francisco Sunday Examiner and Chronicle*, Jan. 22, 1989, p. B1.

The Indian language – "The Indians' language is thus: . . . 'cheepe'– bread, 'hioghe' – a king" (Madox, in Donno,ed., p. 209).

The Portus plan – The inset on the Hondius Broadside Map titled the *Portus Nova Albionis* that ostensibly depicts Drake's California port of call (figure 40).

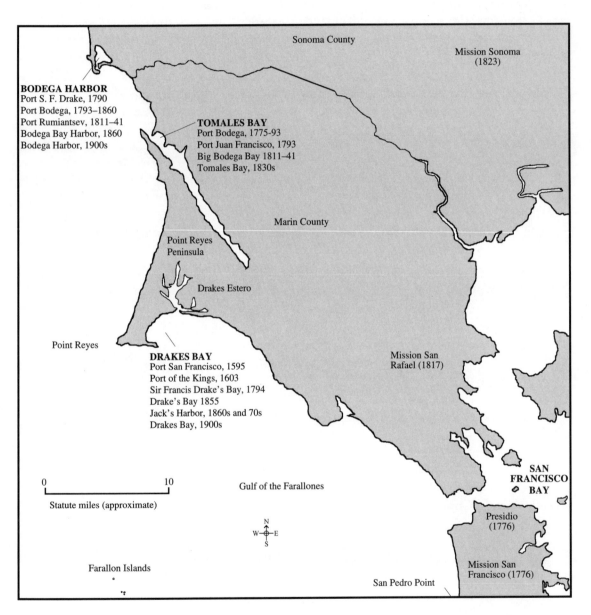

Map 7. Outline map of the area just north of San Francisco Bay showing how the coastal bays were named in the Colonial Era. The area's first explorers and settlers tangled the names applied to its three principal coastal bays, a circumstance that muddled the anchorage debate throughout its history. Today's Drakes Bay was originally named Port San Francisco. Today's Tomales Bay was originally named Port Bodega. Today's Bodega Harbor was originally named Port Sir Francis Drake.

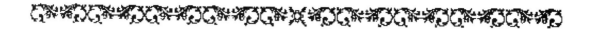

CHAPTER 7
Colonial Period Opinions: 1775–1845

Conflicting opinions concerning the whereabouts of Drake's California anchorage arose as soon as Europeans began exploring and settling the San Francisco Bay area. This was not until the mid-eighteenth century, almost two centuries after Drake's California sojourn. By then, it would seem, time had erased all overt signs of his visit.

Bodega y Quadra, 1775, for Puerto de la Bodega (Tomales Bay)

It is not surprising that the Spanish, ever fearful of foreign encroachment, had little to say about Drake's California visit. In 1718, the mission friar Juan Amando Niel, for reasons unknown, speculated that Drake entered the Carmel River just south of the Monterey Peninsula.[1]

The only other documented Spanish opinions and the true beginning of the anchorage debate came in 1775 when Spain first attempted to establish a mission outpost on the San Francisco Peninsula. Looking for the entrance to San Francisco Bay, Captain Juan Francisco de la Bodega y Quadra mistakenly sailed into the mouth of today's Tomales Bay. From his journal (*translated and quoted courtesy of The Bancroft Library, as holder*), it is clear that Bodega y Quadra thought Tomales Bay, which he named "Puerto de la Bodega," was Drake's lost harbor while his pilot, Don Francisco Antonio Mourelle, thought Drake's harbor was San Francisco:

> In the diary of the pilot [Mourelle] you find a dissertation about what is said about that place [San Francisco Bay] by the chart of Bellin [a French map of 1766 depicting an amorphous "Sir Francis Drake's Harbor" in the approximate position of San Francisco Bay]. He says: "the port of Francis Drake ought to be called Saint Francis." The pilot said this having sensibly considered the placement of that port [the Drake port on Bellin's chart] and the one in this day called San Francisco, still having been persuaded by the error of Mr. Bellin and even of the Spanish: neither the port of Francis Drake nor that which we call San Francisco is that which the first navigators so named. That of Francis Drake is doubtless Bodega and that of San Francisco [as named by Cermeño in 1595] is to the southeast of Point Reyes a very short distance from it from whence run the Farallones to the south-southwest. Point Reyes is distant six leagues to the south from the port of Bodega or Drake. The port that we now call San Francisco is eight leagues to the east of Point Reyes and it is not easily discerned unless entered some distance. The merchant ship *San Agustin* of the Manila trade would not have been driven ashore with the southeast wind if it [the first port of San Francisco] had been discovered in the San Francisco of today or in that of Drake; and from this chance circumstance [Cermeño's coming to anchor under Point Reyes and naming it "San Francisco"?] so happens the three situations of the three distinct ports that we and the French have confused as one.[2]

Bodega died without ever seeing the hidden lagoon on the north side of the bay that came to bear his name, today's Bodega Harbor.

---

1. Cited by H. H. Bancroft in *History of the Northwest Coast*, p. 62. Padre Niel's account of events in the early 1700s, the "Apuntamientos" (1718), is appended to Gerónimo de Zárate y Salmerón's *Relacions del Nuevo Mexico* in *Documentos para la historia de Mexico*, (1626).
2. Bodega y Quadra, "Viaje de 1775," p. 21. I believe this to be the first published translation of the entire passage. Copies of both the Bodega y Quadra and Mourelle diaries are held by the Bancroft Library in a manuscript that George Davidson copied from the archives of Spain in 1874: *Viajes en la Costa al Norte de las Californias, 1774–1790* (BANC MSS P-B 26). The Mourelle diary, however, is a summary of events and quite different from the detailed account translated and published by Daines Barrington in 1781 in *Miscellanies*, and reprinted by Thomas Russell in 1920 as *Voyage of the Sonora*. Davidson, *Identification*, p. 24, by providing a misleading translation of just a part of the Bodega y Quadra passage, implied that the pilot Mourelle was the source of the opinion that Drake harbored at Puerto de la Bodega. This mistake was passed along by several anchorage researchers including John Robertson.

Fig. 41. The Tomales Bay candidate anchorage site (*above*) showing the author's Portus correlation for the cove just inside the mouth (called Smith's Landing in the late 1800s). The first Spanish navigators to examine Tomales Bay (members of the Sebastián Vizcaíno expedition in 1603) reported seeing an island at the mouth, making a good safe port inside. The second Spanish navigator to examine Tomales Bay, Juan Francisco de la Bodega y Quadra, anchored off Sand Point in October 1775, naming the cove inside "Puerto de la Bodega" (Port of the wine cellar). Bodega y Quadra's sketch of the port shows what is labeled "sand banks" correlating with the position of the Portus Island (*inset*). Bodega y Quadra was convinced that Drake had harbored in this cove, presumably based on the correlations for latitude and approach. The cove detail (*above*) is from a chart prepared by the Spanish navigator Juan Martinez y Zayas during his visit in 1793. Surprisingly, with the exception of John Dwinelle's blundering comparison of today's Bodega Harbor to Bodega y Quadra's earlier sketch of the cove (discussed in chapter 7), not one anchorage detective has ever cited the strong resemblance Spain's Puerto de la Bodega once bore to the Portus plan, both in appearance and orientation. Starting in the 1850s, landslides from the huge sand dune (Sugarloaf Hill) at the base of Sand Point began filling the cove in, ostensibly burying the site of Drake's encampment under a mountain of sand (see figure 66). Today this cove bears little resemblance to the Portus. *The chart detail is reproduced from VOYAGE OF THE SONORA by Thomas C. Russell, ed., San Francisco, 1920.*

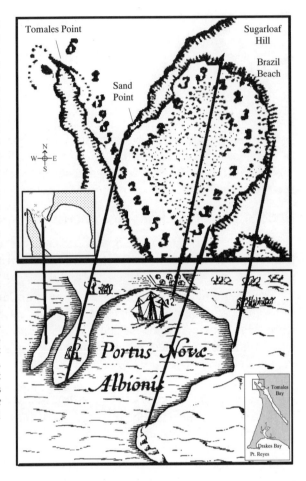

James Colnett, 1790, for Bodega Harbor

It was an English sea captain, James Colnett, who in 1790 first stumbled on today's Bodega Harbor. He must have thought he sensed Drake's presence, for in his journal and accompanying sketch he named it Port Sir Francis Drake.[3]

George Vancouver, 1792, for Drakes Bay

In 1792–94, British Captain George Vancouver explored and charted the Pacific Northwest Coast during his famous voyages of discovery. He reported that the Spanish believed Drake harbored under Point Reyes Peninsula and on his chart of the area named it the "Bay of Sir Francis Drake" (figure 42). This chart influenced the opinions of the early nineteenth-century English, Russian, American and French mariners who subsequently used it.[4]

James Burney, 1803, for San Francisco Bay

Vancouver's chief pilot, James Burney, took issue with the notion that Drake harbored in today's Drakes Bay, doubting that he would have hazarded the grounding of his ship in such a relatively open roadstead. He found "much more reason to conclude that the 'Port of Drake' was that which is now known by the name of Port San Francisco."[5] British Captain Frederick Beechey, after visiting

---

3. Colnett, *Journal*, pp. 174–80.
4. Vancouver, *Voyage of Discovery*, p. 701.
5. Burney, *History of the Discoveries*, pp. 354–55.

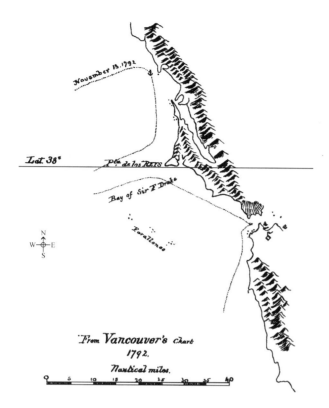

Fig. 42. Detail from George Vancouver's map of Bodega, San Francisco and Monterey, 1798 edition, showing a "Bay of Sir Francis Drake" under Point Reyes. *Reproduced from IDENTIFICATION OF SIR FRANCIS DRAKE'S ANCHORAGE IN THE YEAR 1579 by George Davidson, California Historical Society, San Francisco, 1890.*

San Francisco in 1826 in His Majesty's sloop *Blossom*, also arrived at the conclusion that today's Drakes Bay was "too exposed to authorize the conjecture of Vancouver."[6]

Alexander von Humboldt, 1811, for Puerto de la Bodega (Tomales Bay)

This brilliant Prussian naturalist, who visited San Francisco in 1802, tried to clear up the confusion created by Vancouver and Burney. He dismissed Burney's notion that Drake landed in San Francisco Bay, and explained that Spanish Californians thought Drake landed in Puerto de la Bodega (Tomales Bay), not under Point Reyes.[7] Burney allegedly retorted that Humboldt did not know what he was talking about.[8]

Vasili Golovnin, 1817, for Drakes Bay

In 1811 the Russians established the first colonial settlement north of San Francisco Bay. The settlement included port facilities at today's Bodega Harbor (their Port Rumiantsev), and a fortified outpost, Fort Ross, about twenty miles to its north. During their twenty-year stay, the Russians apparently found no evidence to indicate that Drake had visited Bodega Harbor; all their charts and maps for the period show Drake's name under the protective arm of the Point Reyes Peninsula. According to the Russian sea captain Vasili Golovnin, who stopped in at Port Rumiantsev in 1817 during his circumnavigation of the globe (Russia's second):

> The name Nova Albion is applied to that section of the northwest coast of America between latitudes thirty-eight degrees and forty-eight degrees because the English seafarer Drake, the second circumnavigator of the world, gave it this name when he observed this stretch of the coast from forty-eight degrees to thirty-eight degrees latitude. At the latter point he found the open bay, subsequently named Drake's Bay, in which he anchored.[9]

---

6. Beechey, *Voyage to the Pacific*, vol. 1, p. 343.
7. Humboldt, *Political Essay*, pp. 355 and 366.
8. Cited without reference by Hanna, *Lost Harbor*, pp. 67–68.
9. Golovnin, *Around the World*, p. 162.

Alexander Forbes, 1839, for San Francisco Bay

In one of the first English-language histories of California, renowned British historian, Alexander Forbes, reported that Drake "remained some time in the harbor of San Francisco, and explored the country inland for some distance."[10]

General Mariano Vallejo, circa 1840, for Drakes Bay

While California's colonial Russians were content to accept Vancouver's appellation for the bay under Point Reyes, the Spanish, to whom the name Drake was a scourge, were not. They duly credited the bay's discovery to Sebastián Rodríquez Cermeño, who spent over a month there in 1595 with the crew of his shipwrecked Manila galleon, the *San Agustin*. Cermeño named it the "Puerto de San Francisco" (Port of San Francisco) for the day of its discovery. After a 1769 Spanish land expedition stumbled upon the "Estero de San Francisco" (estuary of San Francisco – today's San Francisco Bay), the bay under Point Reyes lost its original Spanish name. From that point it seems the Spanish referred to it as the "Puerto de los Reyes" (Port of the Kings), and considered it part of their "Baia de los Farallones" (Bay of the Farallones—today's Gulf of the Farallones which extends from Point Reyes to San Pedro Point, Pacifica). Although the Spanish established a mission in southern Marin County in 1817, they never settled the Point Reyes Peninsula and rarely used its harbors. Spain lost its California colonies when Mexico declared independence in 1822.

General Mariano Vallejo was the military commander of the Mexican settlements immediately north of San Francisco when the area was first settled in the 1830s and 1840s. During the centennial celebration of the founding of San Francisco's Mission Dolores, the old general opined that Drake might have landed in the "Port of Tomales," by which he seems to have been referring to either the cove or estuary under Point Reyes (Davidson's Cove or Drakes Estero):

> In 1578, Sir Francis Drake, an English buccaneer, anchored and remained a month, perhaps, in the small bay on the northern extremity of the ocean or open bay of the Farallones, [or?] at the same place which was called by us the Port of Tomales. Drake gave this latter bay his name, and the surrounding country he called New Albion. There is a bare possibility of Drake's entering the present bay of San Francisco, but the weight of evidence is against him.[11]

Eugène Duflot de Mofras, 1841, for Drakes Bay

Taking note of Mexico's political unrest and inept military, in the late 1830s the French government cast imperial designs on California and the Pacific Northwest. In 1839 they recalled Eugène Duflot de Mofras from their embassy in Madrid, and sent this highly educated and articulate young envoy on a diplomatic tour of Mexico to assess opportunities for trade and colonization. In 1841 Duflot de Mofras spent five months at Monterey during which time he visited the ports of San Francisco and Bodega. In recounting the history of Pacific Coast exploration in the fascinating journal of his travels, he had this to say about Drake in Nova Albion:

> . . . In 1579, after devastating the coast of Guatemala, Sir Francis Drake appeared off the shores of New Spain, sailing up as far as forty-five degrees or forty-six degrees north latitude. Approaching the coast, he dropped anchor in a small bay, not clearly defined, where he failed to find supplies. Forced as a result to turn back as far as latitude thirty-eight degrees, he anchored in Los Reyes Harbor, between the ports of Bodega and San Francisco, of whose existence he was in ignorance.[12]

---

10. Forbes, *California: A History*, p. 5.
11. Vallejo, cited in W. H. Davis, *Seventy-Five Years in California*, pp. 354–55. It appears that in Vallejo's day, today's Drakes Bay was called Tomales Bay.
12. Duflot de Mofras, *Exploration of Oregon and California*, p. 52.

CHAPTER 8

The Original Anchorage Sleuths:
1846–1935

The following discussions cover most, but not necessarily all, of the historians and scholars who expressed educated opinions during the period 1846–1935, on the whereabouts of Drake's California landfall.

Fig. 43. George Davidson (1825–1911). This eminent nineteenth-century scientist participated in the U.S. Coast Survey expeditions that first charted the Pacific Coast in the 1850s. Davidson, the best qualified of all the anchorage sleuths, was convinced that Drake harbored under the protective arm of Point Reyes Head. In my opinion, Davidson's work on the subject significantly contributed to the perpetuation of California's great maritime mystery. *Photograph by Taylor & Brown, Philadelphia. Courtesy of the California Historical Society, San Francisco. FN-27385.*

Robert Greenhow, 1845, for San Francisco or Bodega Bay

Having decided that nothing could be learned from the early accounts of the expedition except that the harbor was situated near thirty-eight degrees and had a cluster of small islands just outside the mouth, historian Robert Greenhow concluded that the descriptions "apply equally to the great bay of San Francisco, and to the small bay of Bodega, a few leagues farther north."[1]

Frank Soule, 1855, for Drakes Bay

One of the San Francisco area's earliest historians, Frank Soule recognized the value of the white cliffs in identifying Drake's lost harbor: "The cliffs about this part of the coast [Drakes Bay], for a space of nearly forty miles, resemble in height and color, those of Great Britain in the English Channel, at Brighton and Dover."[2]

Fig. 44. The white cliffs and banks lining the north shore of today's Drakes Bay. They bear strong resemblance to the famed Seven Sisters of Dover in Sussex County, England. Advocates of Marin County anchorage sites consider these white cliffs benchmark evidence of Drake's visit. *Courtesy of the Drake Navigators Guild, Palo Alto, California.*

Franklin Tuthill, 1866, for San Francisco Bay

From the unabashedly provincial historian, Franklin Tuthill, we learn that "from time immemorial, until lately," it was presumed that Drake had landed in San Francisco Bay. Discounting the negative evidence in the early accounts respecting weather, he concluded that San Francisco Bay not only sufficiently matched the descriptions in the contemporary narratives, but was the only bay in the area worthy to bear Drake's name; he considered the name of the closely adjoining "dent in the coastline" [Drakes Bay] "a feeble memorial" to the great Elizabethan navigator.[3]

---

1. Greenhow, *History of Oregon and California*, p. 74.
2. Soule et al., *The Annals of San Francisco*, p. 3.
3. Tuthill, *History of California*, p. 24.

J. D. B. Stillman, 1868–77, for San Francisco Bay

Historian J. B. D. Stillman made much of the "strange kind of conies" that are described in *The World Encompassed*. He concluded that they could only refer to the California ground squirrels that abounded in the San Francisco Bay environs, but were absent at the cool coastal bays. He also pointed out that the white cliffs lining Drakes Bay indeed lie toward the sea from San Francisco Bay and echoed Burney's opinion that Drake could not have safely beached the *Golden Hind* at any of the open roadsteads of Drakes Bay.[4]

Edward Hale, 1873–78, for San Francisco Bay

Eminent nineteenth-century historian, Edward Hale, made two huge contributions to the anchorage debate. In a paper he read to the American Antiquarian Society in 1873, and in an essay in William Bryant's *Popular History of the United States* in 1878, Hale introduced America to Robert Dudley's 1630 *Arcano del Mare* atlas and the 1596 Hondius Broadside Map (see chapter 4). Little did he know how profoundly these antique maps would add to the controversy.

Hale thought the Dudley maps, Carta Particolara chart number 33 and manuscript chart number 85 (figures 31 and 32) offered some proof of Drake's entering San Francisco Bay. Since he could not find a good correlation between the Portus Nova Albionis inset on the Broadside (the Portus plan) and any harbor on the Pacific Northwest Coast, he concluded it was purely imaginary.[5]

John Dwinelle, 1878, for Bodega Bay

John Dwinelle, a prominent attorney, political figure (mayor of Oakland), and author, was the first in a long line of anchorage sleuths to convince himself that he had identified Drake's lost harbor from the Hondius Portus plan and the Dudley Carta Particolara. In his arguments for Bodega Harbor, however, Dwinelle was actually comparing the Portus plan with Bodega y Quadra's 1775 sketch of Puerto de la Bodega at the mouth of Tomales Bay (figure 132). He further contended:

> All the indications called for by the Drake narrative exist here . . . the Indian villages; the shellfish; the seals; the deciduous trees; and the conies which honeycombed the soil; the elevation of the coast which commenced at about that latitude; the white sand hills that suggest the name Albion.[6]

Writing in 1884, Edward Hale, who had introduced the Portus, scoffed at Dwinelle's comparison and reiterated his opinion that the little sketch was imaginary.[7]

Jules Verne, 1879, for San Francisco Bay

In his *Exploration of the World, Famous Travels and Travelers*, the famous French science fiction novelist, Jules Verne, included a brief summary of Drake's circumnavigation that includes this statement, remarkable for its inaccurate information about Bodega y Quadra's alleged sixteenth-century discovery of San Francisco Bay:

> . . . he [Drake] therefore put out to sea, reached thirty-eight degrees of north latitude, and landed on the shore of San Francisco Bay, which had been discovered three years previously by Bodega.[8]

Hubert H. Bancroft, 1884–1915, for Drakes Bay

One of Western America's most illustrious historians, Hubert Howe Bancroft, provided a lengthy discussion on the identity of Drake's anchorage in his monumental *History of California*. Based on an initial review of the early accounts and maps, Bancroft could not decide between Drakes Bay and

---

4. Stillman, *Seeking the Golden Fleece*, pp. 332–37.
5. Hale, in Bryant, *Popular History*, pp. 575–77.
6. Dwinelle, *San Francisco Evening Bulletin*, Oct. 5, 1878, p. 1.
7. Hale, in Winsor, ed., *Critical History*, pp. 75 and 80.
8. Verne, *Exploration of the World*, p. 189. The Spanish commander Bodega y Quadra discovered and named Bodega Bay in 1775.

Bodega Bay because the early accounts "abound in discrepancies and inaccuracies, as is shown still more clearly in parts not relating to California."[9]

Bancroft also denigrated the value of the early maps:

I have not found a single map of the California coast of earlier date than 1769 bearing the slightest indication of having been founded on anything but the narratives still extant and the imagination of the mapmaker.[10]

Bancroft found it inconceivable that Drake could have entered San Francisco Bay, providing the most pointed and colorful arguments of all the anchorage investigators on this issue (see chapter 33). He was also unimpressed with Stillman's arguments regarding ground squirrels, pointing out that Fletcher reported seeing them on the visit inland, not along the coast.[11]

Despite his uneasiness about the evidence, as of 1914 Bancroft had relented to George Davidson's arguments for a Drakes Bay anchorage.[12]

George Davidson, 1890–1911, for Drakes Bay

Professor George Davidson, "pioneer West Coast scientist,"[13] began as an assistant and ended up in charge of the ambitious U.S. Coast Survey program that first charted the Pacific Coast. As they began the project in the early 1850s, Davidson and his colleagues became interested in learning what the old navigators had seen and written of the coast. Having first accepted the then locally prevalent notion that Drake had entered San Francisco Bay, Davidson changed his mind after more carefully researching the matter:

I have carefully studied the narratives of Drake's voyage, and the manuscript charts copied from his sketches, or drawn from his personal descriptions; have located his first anchorage; know every foot of the shore he coasted; have tried to see it with his eyes; have sailed the *Fauntleroy* over the very track he pursued; have conned the shoreline, and the crest line, and the landfall from seaward, under varying circumstances of weather; have surveyed Bodega Head, and anchored in Bodega Bay; have been over every rod of Point Reyes Head several times; and have frequently anchored in Drake's Bay in pleasant weather, and under stress of weather, even as late as last year (1886). I have visited the South and North Farallones, measured their heights, and studied their relation and visibility to the harbor in which Drake anchored. I have also collated some of the narratives of the discoveries of the Spaniards with that of Drake. Long before I had gathered all this information, my early judgment was corrected, and I saw the great circumnavigator anchored in Drake's Bay; could almost point out the spot where he careened his ship; and today there remains not the shadow of a doubt in my mind as to the exact locality.[14]

Davidson, without question, was the most distinguished and best-qualified individual to have ever taken on the anchorage question. His career with the U.S. Coast Survey spanned a full fifty years, during which time he authored numerous related works, including the monumental *Pacific Coast Pilot* published first in 1869 and for the final time in 1889. He held a sixteen-year tenure as president of the California Academy of Science and was instrumental in establishing the society's museum of natural history and Lick observatory. He helped found the Geographic Society of the Pacific in 1881 and presided over the organization for three decades. He was a charter member of the Sierra Club and vice president of the California Historical Society. He was professor of geodesy and astronomy at the University of California, Berkeley, from 1870–98, and professor of geography to 1905. He was a prolific writer in various scientific disciplines and a list of his published works contains well over one hundred items. His true passion, however, was maritime history: he amassed a huge collection of manuscripts and books by and about the pioneer navigators and wrote extensively on the subject. In his later years, he became increasingly obsessed with the Drake anchorage question.

9. H. H. Bancroft, *History of California*, vol. 1, p. 85. He alludes here to the latitude evidence.
10. Ibid., p. 88.
11. Ibid., pp. 91–93.
12. H. H. Bancroft, *The New Pacific*, 3d ed., p. 452.
13. After the title of the Oscar Lewis biography, *George Davidson, Pioneer West Coast Scientist* (1954).
14. Davidson, *Identification*, p. 4.

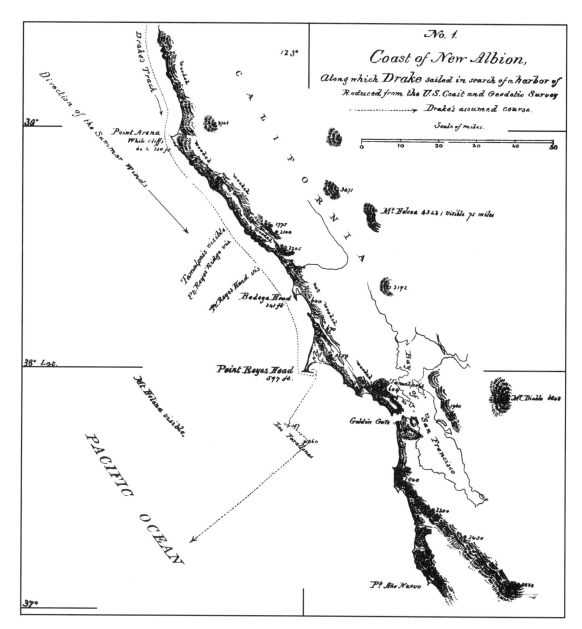

Map 8. Relief map of Northern California showing Drake's alleged course along the coast of Nova Albion as envisioned by George Davidson. *Reproduced from IDENTIFICATION OF SIR FRANCIS DRAKE'S ANCHORAGE ON THE COAST OF CALIFORNIA IN THE YEAR 1579 by George Davidson, California Historical Society, San Francisco, 1890 (photo-edited for clarity).*

Davidson published his first thoughts on Drake's anchorage in the appendix to the 1869 *Pacific Coast Pilot*. In 1889, he authored a monograph entitled, *Identification of Sir Francis Drake's Anchorage on the Coast of California in the Year 1579,* that was published by the California Historical Society in 1890. Such was Davidson's reputation as a scientist and navigational expert that for almost thirty-five years thereafter, there was almost no important research on the matter. Most historians simply accepted at face value his conclusion that Drake harbored under Point Reyes Head.

Nevertheless, by 1908, Davidson felt compelled to reaffirm his opinion with a second monograph, this time published by his own Geographic Society of the Pacific: *Francis Drake on the Northwest Coast of America in the Year 1579—The "Golden Hind" Did Not Anchor in the Bay of San Francisco.* In this paper he launched piercing attacks on any writers or historians who had lauded Drake's career

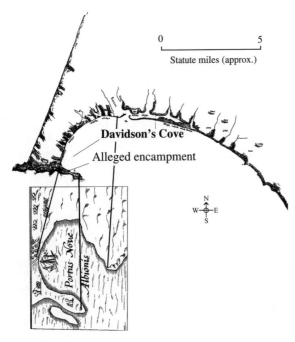

0   5
Statute miles (approx.)

Davidson's Cove

Alleged encampment

N
W—⊕—E
S

Fig. 45. The Davidson's Cove candidate anchorage site showing George Davidson's Portus correlation. *The chart detail is derived from illustration No. 4 in IDENTIFICATION OF SIR FRANCIS DRAKE'S ANCHORAGE ON THE COAST OF CALIFORNIA by George Davidson, California Historical Society, San Francisco, 1890.*

or pronounced in favor of a San Francisco landfall. The subjects of his attacks included Clements Markham, James Froude, Miller Cristy, Julian Corbett, Gardiner Hubbard, Justin Winsor, Edward Hale, and Jules Verne.

Given his "long life of activity on this coast in the line of geodesy, astronomy, geography, and navigation, with conditions and opportunities that will not again fall to the lot of one man,"[15] Davidson had little difficulty reconstructing Drake's movements as he sailed down the California coast:

> From Trinidad Head to Bodega Head or to Point Reyes, Drake was absolutely disappointed in his search, and in his increasing anxiety he would certainly seek for shelter under the lee of Point Reyes before he went farther. Here he found a "fair and good bay with a good wind to enter the same" and he was satisfied . . . [16]

For all his qualifications and nautical expertise, however, it seems that Davidson relied heavily on the Portus correlation: "I assumed the orientation and the scale of the long point of the Portus to be the same as the eastern promontory of Point Reyes, and reduced and compared them. The result was so reasonable that it seemed as if the problem was solved."[17]

Davidson also put great stock in a legend among Marin County's Nicasio Indians to the effect that Drake had indeed landed at Drakes Bay where some of his crew deserted and lived among the natives. The legend also held that Drake left them a dog, pigs, ship's biscuits, and seeds.[18]

Ironically, it was in his treatment of Drake's navigational measurements that Davidson led almost all subsequent researchers astray (discussed in chapter 14). As his contemporary Edward Berthoud put it:

> We leave it to the eminent hydrographer, Professor George Davidson, who has most clearly and sagaciously worked out the devious and puzzling questions involved, from the expeditions of Cabrillo and Ferrelo, and he alone is competent to sit in judgment over the positive value of Drake's nautical astronomy.[19]

In very carelessly comparing some of the latitudes *The World Encompassed* provides for South American landfalls with correlating positions on modern charts, Davidson observed large discrepancies, noting that similar discrepancies were to be found in the journals left behind by sixteenth-century Spanish navigators like Juan Rodríguez Cabrillo, Sebastián Rodríguez Cermeño, and Sebastián Vizcaíno. "At best," he concluded "the instrument for obtaining the latitude was liable to error reaching one or two degrees."[20]

---

15. Davidson, *Francis Drake on the Northwest Coast,* p. 2.
16. Davidson, *Identification*, p. 56.
17. Ibid., p. 38.
18. By my estimation (and Robert Heizer's), this legend was born of Juan Bautista Matute's 1793 visit to Bodega Harbor.
19. Berthoud, "Sir Francis Drake's Anchorage," p. 213.
20. Davidson, *Francis Drake on the Northwest Coast,* pp. 2 and 11.

After his death in 1911, Davidson's postulations on the Drakes Bay anchorage were enthusiastically upheld by a group of Marin County residents, The Sir Francis Drake Association, established circa 1915. Its honorary president, Bishop William Lord Nichols, erected a large wooden cross marking the Point Reyes landfall, and, each June, the association sponsored a Drakes Bay pilgrimage concluded by a chilly outdoor commemorative service sometimes well attended by state and foreign dignitaries. Climaxing a thirty-year campaign of fund raising and appeals, in August 1949, the construction of the Sir Francis Drake Memorial on the shores of Drakes Bay County Park was witnessed by the four surviving members of the association: Mrs. Robert Menzies of San Rafael; and Marion Leale, Paige Monteagle, and Millen Griffith of San Francisco. The memorial, still standing, consists of a 3,600-pound granite cross, a podium and an amphitheater.[21]

Edward Berthoud, 1894, for Bodega Harbor

National Geographic magazine's Edward Berthoud seems to have been the only contemporary researcher who challenged

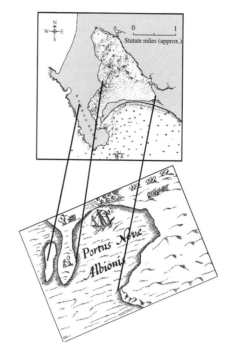

Fig. 46. The Bodega Harbor candidate anchorage site showing Edward Berthoud's Portus correlation. The few advocates of this site argued that in Drake's day the harbor was historically navigable and wide open at its mouth. The chart detail reveals that such was not the case. *The chart detail is derived from an 1862 U.S. Coast Survey Chart.*

Davidson's opinion on Drakes Bay without inciting his wrath. In an article entitled, "Sir Francis Drake's Anchorage," he made two points: (1) [the Portus] "has a strong resemblance to Bodega Harbor and Romanzoff Point, now known as Bodega Head"; and (2) "the preponderance of locality and distance [to the Farallon Islands] seems to point to Bodega Bay."[22]

Alexander McAdie, 1918, for Drakes Bay

Dr. Alexander McAdie (1863–1943), longtime professor of meteorology at Harvard University, has been accredited as "the father of the modern science of meteorology."[23]

In the late 1800s and early 1900s, while employed as professor of meteorology with the National Weather Bureau at San Francisco, McAdie became friendly with George Davidson. Having accompanied Davidson on some of his surveying expeditions and on "many a lonely trip" of his own along the foggy and wind-swept shores of Northern California, McAdie decided that Davidson's theory on Drake's landfall was confirmed by Francis Fletcher's descriptions of Nova Albion's weather. In 1902, he made an abstract of the weather records for Point Reyes corresponding to the period of Drake's visit. McAdie presented his views to the American Antiquarian Society in a paper entitled "Nova Albion—1579." Approaching the problem from the standpoint of an "aerographer," rather than a historian, he deduced that due to low-lying fog, Drake could not have seen the entrance to San Francisco Bay even if he had ventured just a few miles outside its mouth. He further explained that this dense summer fog was responsible for Fletcher's observations of snow-covered hills: "it is

21. The story of the Drake memorial is featured in the *San Rafael Independent-Journal*, including well-illustrated articles on Aug. 13 and Aug. 22, 1949.
22. Berthoud, "Sir Francis Drake's Anchorage," pp. 210 and 212. Berthoud did not provide an illustration for his Portus correlation.
23. Biographical information on McAdie is from his obituary in *Proceedings of the American Antiquarian Society*, Apr. 1944.

Fig. 47. Henry Wagner (1862–1957). Wagner's *Sir Francis Drake's Voyage Around the World* is still considered the definitive work on the circumnavigation. *Photograph by Edgar M. Kahn. Courtesy of the California Historical Society, San Francisco.* FN-23779.

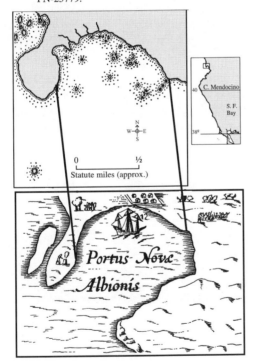

Fig. 48. The Trinidad Bay candidate anchorage site at 41° 03′ showing Henry Wagner's Portus correlation. *The chart detail is derived from a sketch made during the Bruno de Hezeta expedition of 1775.*

a common occurrence today for tourists on coasting vessels to call attention to what they think is snow on the mountains."[24]

Henry Wagner, 1926–57, for Trinidad Bay

Henry Wagner, another of Western America's eminent historians, gave up a very successful career in mining and metallurgy to follow his passion for collecting old books, conducting scholarly research on the exploration of the Pacific Coast, and translating the accounts of the old Spanish navigators. An extremely prolific writer, he is credited with 180 published works and with helping to revive and reorganize the California Historical Society. He also founded the Society's quarterly magazine in 1922.

Wagner wrote his epic *Sir Francis Drake's Voyage Around the World* in 1926, and with its publication inherited Davidson's position as America's foremost authority on the great Elizabethan mariner. He devoted an entire chapter of the book to his California landfall theories, reluctantly taking issue with Davidson's findings. After discovery of the Shinn plate in 1936 (discussed in chapter 10), he wrote a monograph, *Drake on the Pacific Coast*, that was published posthumously in 1970.

Writing in 1926, Wagner took the following logical point of view: "In the necessity in which he found himself, [Drake] would have stopped at the first suitable location he encountered." He deduced that Drake probably explored most of the coastal bays in Oregon and Northern California, but spent most of his time first at Trinidad Bay (about 100 miles north of San Francisco) and then at Bodega Harbor.[25]

Wagner held the Portus as "the most important piece of evidence known to us," and compelling proof that Drake anchored at Trinidad Bay. In 1926, he argued that Fletcher's ethnological observations and descriptions of coastal terrain indicated that Drake also spent some time in Bodega Harbor. In his later paper, however, he changed his mind on this, concluding that the ethnological descriptions applied to Trinidad's Yurok Indians, and that in his

24. McAdie, "Nova Albion–1579," pp. 190 and 194.
25. Wagner, *Drake's Voyage*, pp. 143 and 169.

coastal descriptions, Fletcher was simply reporting what the English had observed of the coast in the Bodega area as they sailed by.[26]

Wagner repeated Davidson's careless analysis of the latitudes reported for Drake's stops in Chile and Peru, reinforcing Davidson's conclusions with his own: "the observations on the ships were extremely inaccurate." A persistent critic of *The World Encompassed* the most untrustworthy of all accounts written about the voyage," Wagner concluded that Drake's nephew adopted the "38 deg. 30 min." position from a contemporary English map, the 1625 Henry Briggs Map (figure 124), rather than from firsthand information.[27]

Wagner believed the 1595 Cermeño account provided convincing proof that Drake did not spend any time in Drakes Bay:

> Cermeño evidently found no trace of this [Drake's camp] or anything else to indicate that any Europeans or civilized people had been there [on the Point Reyes Peninsula] before, for if he had he certainly would have mentioned it in his report, stock instructions to Spanish explorers at that period being to look for signs of civilized people.[28]

John Robertson, 1927

The second prominent Drake scholar to emerge in the 1920s, John Robertson contented himself with criticizing previous anchorage theories rather than formulating his own opinion. In his *Harbor of St. Francis* he sums up the arguments of Hale, Dwinelle, Bancroft, Stillman, and Davidson and aptly points out: "no selection has resulted in a choice that, after disinterested study, can be considered so fully proved as to be beyond controversy."[29]

He then vigorously attacked each of Davidson's arguments—except the latitudes. His observations on the Portus plan's value as a clue are worth repeating:

> This sketch map plays queer tricks with those who too fixedly gaze upon its bewitching contour. . . . The dreams that this "Portus" may induce are past psychoanalysis. For my part I refuse to gaze into its crystalline depths and, with wax-filled ears, I give no heed to the siren-song. Not recognizing the tangible resemblance it bears to any known Pacific harbor, I regard the Hondius sketch, not as a serious attempt to carry out a description given either by Drake or by a member of the crew; merely an illustrator's design to accompany the verbiage of *The World Encompassed*, published in 1628, the only account of the voyage narrating these incidents.[30]

Robertson scoffed at Davidson's argument regarding the Nicasio Indian legend, arguing that the Indians could not possibly have passed down the name of Francis Drake through eight generations, or distinguished the English from the Spaniards who occasionally stopped in the area. He similarly found it inconceivable that Drake would have carried pigs, much less left them behind, at this stage of the voyage.[31]

Like Henry Wagner before him, Robertson pointed out that the white cliff evidence provided no proof that Drake harbored under Point Reyes:

> White cliffs are a marked feature of the California coast . . . Nothing in either of the narratives . . . can be so construed as to make it a necessary conclusion that the Harbor of St. Francis was so marked; it is merely an identification of that coastline "which lies towards the sea."[32]

26. Ibid., pp. 154–56; and *Drake on the Pacific Coast*, pp. 14 and 15.
27. Wagner, *Drake's Voyage*, pp. 161, 162, and 474.
28. Ibid., pp. 168–69.
29. Robertson, *The Harbor of St. Francis*, p. 1.
30. Ibid., p. 45.
31. Ibid., p. 58.
32. Ibid., pp. 51–52.

Other Opinions 1846–1935

Other notables of this era who offered opinions on the anchorage locale with little or no explanation for their reasons include the following:

Edmond Randolph for Drakes Bay in his oration before the Society of California Pioneers, September 1860, in *Hutchington's California*, vol. 5, December, 1860, p. 265.

J. H. S. for Drakes Bay, in *Hutchington's California*, vol. 5, December, 1860, p. 277.

Alexander Taylor for Drakes Bay in "The Indianology of California," *California Farmer*, August 29, 1862.

Titus Fey Cronise for Drakes Bay in *The Natural Wealth of California*, 1868, pp. 5–6.

Samuel G. Drake for San Francisco Bay in the *American Historical Record*, August 1874.

B. E. Lloyd: "a question of dispute until the 'pillar' erected by the admiral has been unearthed and identified by its inscriptions," in *Lights and Shades in San Francisco*, 1876, p. 499.

Cleveland Rockwell, Assistant U.S. Coast Survey, for Drakes Bay, in J. D. B. Stillman's, *Seeking the Golden Fleece*, 1877, p. 294.

D. L. Purves for San Francisco Bay in *A Voyage Round the World by Sir Francis Drake and William Dampier*, 1880, p. 49.

James Froude for San Francisco Bay in *History of England*, 1881, p. 418.

George Towle for San Francisco Bay in *Drake the Sea-King of Devon*, 1883, p. 154.

Theodore Hittell for Drakes Bay in *History of California*, 1885, p. 89.

Gardiner Hubbard for San Francisco Bay in *National Geographic Magazine*, vol. 5, 1893, p. 16.

Julian Corbett for Drakes Bay in *Drake and the Tudor Navy*, vol. 1, 1898, pp. 307 and 311.

Miller Christy for San Francisco Bay in *The Silver Map of the World*, 1900, p. 20.

J. M. Guinn for Drakes Bay in *History of the State of California*, 1904, p. 39.

Irving Richman for Drakes Bay in *California Under Spain and Mexico*, 1911, p. 18.

John McGroarty for Drakes Bay in *California, Its History and Romance*, 1911, p. 36.

James Dixon for Drakes Bay in "Drake on the Pacific Coast," *Overland Monthly*, vol. 63, 1914, p. 540.

Zoeth Skinner Eldredge for Drakes Bay in *History of California*, 1915, p. 92.

Percy Valentine for Drakes Bay in *California: The Story of Our State*, 1916, p. 6.

C. Hart Merriam, anthropologist, for Drakes Bay in "Indian Names in the Tamalpais Region," *California Out-of-Doors*, April 1916, p. 118.

William Wood for Drakes Bay in *Elizabethan Sea Dogs*, 1918, p. 138.

Thomas Russell for Drakes Bay in *Voyage of the Sonora*, 1920, pp. 100–1.

Charles E. Chapman for Drakes Bay in *History of California: The Spanish Period*, 1921, p. 104.

Herbert E. Bolton for Drakes Bay in *California's Story*, 1922, p. 22.

Harr Wagner for Drakes Bay in *California History*, 1922, p. 31.

Margaret Synge for San Francisco Bay in *A Book of Discovery*, 1925, p. 255.

Honoria Tuomey for Drakes Bay in *History of Sonoma County*, vol. 1, 1926, p. 46.

Sir Richard Temple for Drakes Bay in Penzer, ed., *The World Encompassed*, 1926, p. lii.

E. F. Benson for Drakes Bay in *Sir Francis Drake*, 1927, p. 160.

Holland Rose for Drakes Bay in *The Cambridge History of the British Empire*, 1929, p. 101.

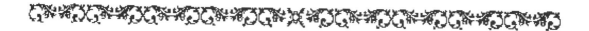

CHAPTER 9
The Anthropological Researchers

The English interaction with the natives dwelling near their encampment is the main topic of *The World Encompassed*'s chapter on Nova Albion. Approximately fourteen of the eighteen pages concerning the California sojourn are devoted to this colorful story. Fletcher provides a wealth of accurate ethnological details including: physiognomy, food, dress, dwellings, basketry, weapons, gender roles, religion, ceremony, economics, chieftainship, and language. Tantalizing clues to the anchorage location emerge from these descriptions.

It was not until the early twentieth century that serious attempts were made to use Fletcher's ethnological descriptions to identify the anchorage site. At this point, however, three noted anthropologists associated with the University of California Department of Anthropology have rigorously analyzed the data and offered opinions and conclusions.

Fig. 49. Archaeologist Aubrey Neasham looking for evidence of Drake's visit to California at an unidentified site in Marin County. Photographer and date unavailable. *Courtesy The Bancroft Library, Berkeley, California.*

Fig. 50. Indian of the Guimen tribe (presumably Coast Miwok): detail of a drawing titled "Long-time neophytes of Mission San Francisco" by Louis Choris, 1816. *Courtesy The Bancroft Library, Berkeley, California.*

Samuel Barrett, 1908, for Drakes Bay

Samuel Barrett, a renowned expert on the Pomo and Coast Miwok Indians of Marin and Sonoma Counties, was the first to use ethnological evidence in attempting to identify Drake's landfall. Working under Alfred Kroeber, in 1903 Barrett conducted a detailed study of the remnants of both tribes. In his 1908 "Ethno-Geography of the Pomo and Neighboring Indians," Barrett found ample support for George Davidson's then prevalent theory that Drake had anchored under the arm of Point Reyes Head.[1]

Alfred Kroeber, 1925, for Drakes Bay

Professor Alfred Kroeber's monumental *Handbook of the Indians of California* (1925) was long considered the standard reference on the native tribes of California. Starting around 1900, Kroeber and his research assistants systematically studied and recorded California tribal ethnogeography through detailed interviews with native informants. His chapter on the Coast Miwok summarized *The World Encompassed*'s detailed ethnological observations, ending with the following statement:

---

1. Barrett, "Ethno-Geography," p. 37.

The ethnologist can only conclude that Drake summered on some piece of the coast not many miles north of San Francisco, and probably in the lagoon to which his name now attaches . . . and he has tolerable reason to believe that the Indians with whom the great explorer mingled were the direct ancestors of the Coast Miwok.[2]

Robert Heizer, 1940–74, for Drakes Bay

Professor Robert Heizer, California's preeminent anthropologist of the modern era, was long considered the state's foremost expert on the ethnological evidence associated with Drake's California visit. He wrote his first paper on the subject in 1940 while still a graduate student at the University of California, Berkeley: Francis Drake and the California Indians, eventually published in 1947. He coauthored a second paper in 1942 with William Elmendorf, entitled "Francis Drake's California Anchorage in Light of the Indian Language Spoken There." In 1974, Heizer authored one of the most important books on the subject, Elizabethan California: A Brief, and Sometimes Critical, Review of Opinions on the Location of Francis Drake's Five Weeks' Visit with the Indians of Ships Land in 1579. In it, he provides a concise review and analysis of the then raging debate, and reevaluates all the evidence in the wake of new findings and theories.

In concluding his first paper in 1940, Heizer went beyond the ethnological evidence to come up with his solution to the problem: "the white cliffs which face the sea are at Drakes Bay and not at Bodega . . . Drake probably landed in what is now known as Drakes Bay."[3]

By 1974, Heizer had temporarily changed his mind based on the discovery of what was then considered a key Drake artifact, the alleged brass plate upon which Drake had inscribed his Nova Albion proclamation: "if we accept the plate of brass as the genuine article, the San Quentin Cove site seems indicated."[4]

Archaeological Explorations on the Point Reyes Peninsula

Since the turn of the century, California anthropologists have conducted a great deal of archaeological research north of San Francisco Bay, none of which has produced any definitive proof of Drake's California sojourn. The following is a brief summary for the area encompassing today's Point Reyes National Seashore.[5]

In 1907–8, Nels C. Nelson located and recorded 425 aboriginal shellmounds along the coasts of San Francisco Bay and Marin County. In 1911–13, Jesse Peter of Santa Rosa surveyed additional sites on the coasts of Marin and southern Sonoma County. In 1927–34, Commander S. F. Bryant, U.S. Navy (retired), surveyed the west shore of Tomales Bay and Drakes Estero by boat and excavated a burial site.

The Shinn plate discovery in Greenbrae, believed to be Drake's plate of brass, and resulting publicity and controversy that followed in the late 1930s, focused attention on the search for Drake artifacts. In 1938–40, Robert Heizer conducted the first scientific excavations of several Drakes Bay shellmounds, turning up porcelain shards and bent iron spikes attributed to the San Agustin, the Spanish galleon driven ashore at Drakes Bay in 1595. In 1941–42, University of California field crews directed by Richard Beardsley resumed the search, exploring thirteen additional sites. In 1945, Adan Treganza headed investigations and salvage operations at thirteen other Drakes Bay sites. Between 1948–65, Aubrey Neasham directed excavations at four Bolinas Lagoon sites and one Drakes Estero site. In 1949–50, Clement Meighan supervised excavations on Heizer's former estero sites, turning up many additional historic artifacts. In 1951–58, members of the Drake Navigators Guild conducted limited explorations of four estero midden sites under the direction of Aubrey Neasham, and "considerable probing" in the area of their proposed Drake's Cove anchorage site. In

2. Kroeber, Handbook of the Indians, p. 278.
3. Heizer, Elizabethan California, p. 81.
4. Ibid., p. 22.
5. Derived from Schenk, ed., Archaeology of Point Reyes, pp. 98–104.

1959, Treganza explored several new sites on the shores of Tomales Bay, and after testing several estero sites, supervised thorough investigations on one of the more productive estero mounds as well as a new site on Limantour Spit. In 1961–65, Edward Von der Porten, using Santa Rosa Junior College field crews, conducted extensive new digs on the Limantour Spit site. In 1964, Treganza and his assistant Frank Rackerby explored a Marshall Beach site on the west side of Tomales Bay. In 1966–69, Ward Upson directed excavations into additional areas of Limantour Spit and several other estero sites. During this period, Thomas King explored seven sites in the area of Bolinas Bay. In 1967–68, Robert Edwards conducted a comprehensive survey of the Point Reyes National Seashore, summarizing information for 121 archaeological sites.

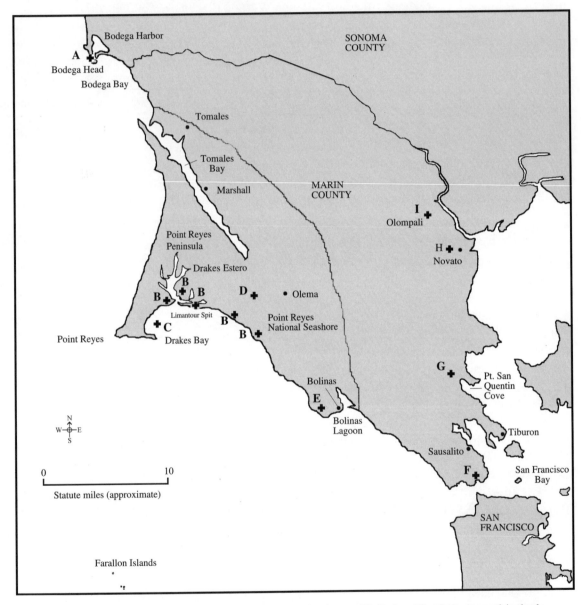

Map 9. Outline map of the area just north of San Francisco Bay showing possible Drake artifact finds: *A*, porcelain shards (unconfirmed) 1950–60; *B*, stoneware fragments and/or porcelain shards, 1949–70; *C, anchor, 1887; D*, Caldeira plate, 1933; *E*, Bolinas plate, ship's timbers, armament, ballast and other relics, 1936–80; *F*, Sausalito plate, 1892; *G*, Shinn plate, 1938; *H*, brass mortar, 1974; *I*, Elizabethan sixpence, 1974.

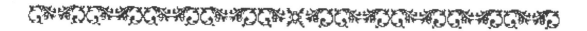

CHAPTER 10
The Plate of Brass and Other Finds

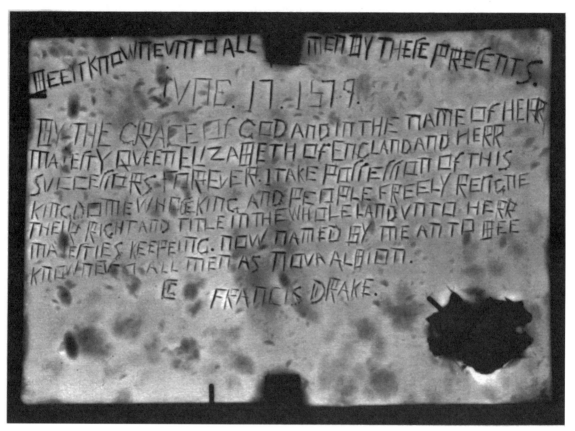

Fig. 51. The Shinn plate discovered 1936 in Greenbrae, California. The plate is currently on exhibit at the Bancroft Library in Berkeley accompanied by a discussion on the doubtfulness of its authenticity. Reproduced from a positive made from a radiograph taken at a distance of five feet in THE PLATE OF BRASS REEXAMINED 1977, Bancroft Library, 1977. *Courtesy The Bancroft Library, Berkeley, California.*

The Shinn Plate Discovery, 1936

The following statement signed by Beryle Shinn, an Oakland, California, department-store clerk, describes how he accidentally discovered an inscribed brass plate about five inches wide, eight inches long, and an eighth of an inch thick. Shinn signed the statement on October 31, 1956, twenty years after the fact.

> In the summer of 1936 I was traveling south on highway 101 from San Rafael, and when coming down the ridge approaching Greenbrae one of my tires was punctured. Veering to the side of the road, I stopped my car. On the ridge above was a likely picnic spot. I climbed under a barbed wire fence and climbed to the top of the ridge. There an extensive view presented itself. To the east was Point San Quentin and upper San Francisco Bay, bounded on the southwest by the Tiburon Peninsula. Below was the tidal estuary of Corte Madera Creek. Approaching an outcrop of rock near the top of the ridge, I picked up rocks and rolled them down the hill. As I pulled a rock from the soil I saw the edge of a metal plate which was partly covered by the rock. When I pulled the plate free from the ground I noticed that it was about the size to repair the frame of my automobile. So when I returned to my car I took it along and tossed it in. Several months later I thought of repairing the frame. While handling the plate I noticed that it seemed to have some inscription on it. I scrubbed the plate with a brush and soap and noted a date, 1579, near the top of the plate. This interested me, so I showed it to a few of my friends, but none

Fig. 52. The Shinn plate discovery site. The view is from the top of Greenbrae Hill looking south over San Quentin Cove. Walter Starr is the gentleman pictured near the rock outcropping where Shinn found what was originally believed to be the brass plate Drake posted in July 1579. Today, most authorities consider the Shinn plate a clever hoax. *Reproduced from "Evidence of Drake's Visit to California, 1579," by Walter Starr, CALIFORNIA HISTORICAL SOCIETY QUARTERLY, March 1957. Courtesy of the California Historical Society, San Francisco, California.*

could make out what it was until one, a college student, deciphered the word Drake and suggested that the metal plate be shown to Dr. Herbert E. Bolton of the University of California. This was done and Dr. Bolton discovered that it was Sir Francis Drake's plate of brass.[1]

On February 28, 1937, Dr. Bolton and Allen Chickering, president of the California Historical Society, accompanied Shinn to the spot on the ridge where he reportedly found the inscribed plate (figure 52). In April, Bolton read a paper before the California Historical Society at the Sir Francis Drake Hotel in San Francisco announcing the finding and the story behind it. The Society published Bolton's paper and another short paper by Douglas Watson in a special publication entitled *Drake's Plate of Brass.* Watson pointed to evidence in the contemporary accounts matching Drake's anchorage site to the locale where the plate was found. He also noted a positive correlation with the Portus Nova Albionis inset:[2]

> The jutting peninsula of the Hondius map has more than a faint resemblance to the Tiburon Peninsula, while the island lying at its side may be taken to be Angel Island [he obviously meant Belvedere Island]. The anchorage of the *Golden Hind* would be at the mouth of Corte Madera Creek, and the other enclosing arm protecting the Portus would be Point San Quentin.[3]

The William Caldeira Story, 1937

Soon after Bolton made public the discovery of the Shinn plate, William Caldeira, a chauffeur for Bank of America's then vice-chairman Leon Bocqueraz, came forward claiming that in 1933 he had discovered the same plate on the Laguna Ranch, a property bordering the east side of Drakes Estero. He further contended that after carrying the plate in his car for several weeks, he threw it away near what is now the intersection of Highway 101 and Sir Francis Drake Boulevard, about a mile from where Shinn found his plate. Allen Chickering, then president of the California Historical Society, who interviewed and cross-examined Caldeira shortly thereafter, published Caldeira's story

---

1. From Starr, "Evidence of Drake's Visit," p. 31.
2. Watson, *Drake and California*, p. 20.
3. Ibid., p. 23.

in the 1937 *California Historical Society Quarterly*.[4] It came out later that Chickering was actually very skeptical of Caldeira's story.[5]

The Plate of Brass Authenticated, 1938

The California Historical Society paid Shinn $3,500 for the plate and donated it to the University of California, Berkeley, where it can be seen to this day at the Bancroft Library. The University promptly submitted it to Colin G. Fink, Ph.D., head of the Division of Electrochemistry at Columbia University, then considered the best qualified individual in the United States to determine if it was genuine. After enlisting the services of George R. Harrison of the Massachusetts Institute of Technology (MIT), a recognized expert in spectroscopy, and E. P. Polushkin, consulting metallurgical engineer, the investigators conducted a series of tests in an attempt to date the artifact. Their report summarizes their findings and conclusions as follows:

- There is no doubt whatsoever that the dark coating on the surface of the plate is a natural patina formed slowly over a period of many years.

- Numerous surface defects and imperfections usually associated with old brass were found on the plate.

- Particles of mineralized plant tissue are firmly embedded in the surface of the plate. (This accounts for the carbon found in the patina).

- Cross sections of the brass plate show an excessive amount of impurities . . . these characteristics indicate a brass of old origin.

- Among the impurities found in the brass of the plate is magnesium, far in excess of the amount occurring in modern brass.

- There are numerous indications that the plate was not made by rolling but was made by hammering, as was the common practice in Drake's time.

  On the basis of the above six distinct findings, as well as other data herewith recorded, it is our opinion that the brass plate examined by us is the genuine Drake plate referred to in the book, *The World Encompassed* by Sir Francis Drake, published 1628.[6]

Questions on Authenticity, 1937–47

When the California Historical Society released the Fink-Polushkin report in another special publication entitled *The Plate of Brass Authenticated*, its conclusions were immediately challenged by other experts in the related disciplines. Professor Earle Caley of the Frick Chemical Laboratory at Princeton University thought the zinc content of the plate strongly suggested modern brass. Pointing out that fraudulent metal tablets frequently turn up in Europe, Captain R. B. Haseldon, curator of manuscripts at the Huntington Library in San Marino, California, was highly skeptical of the plate's tallying to such a great extent with the language in the contemporary accounts. Vincent Harlow of Oxford University thought the lettering too crude and found the structure of the text too much like twentieth-century writing. Dr. Robin Flower, the deputy keeper of manuscripts at the British Museum stated that "on the whole the spelling, like the letter forms, leaves me rather uneasy."[7]

From his 1970 paper, we learn that Henry Wagner was extremely skeptical of the authenticity of the brass plate. In fact, the reason his paper was published posthumously was because: "I do not wish to publish an article now as the people [from the California Historical Society] who put up money to buy the plate are all intimate friends of mine and I do not wish to hurt their feelings."[8]

---

4. Chickering, "Some Notes," pp. 275–81.
5. Hanna, *Lost Harbor*, p. 422.
6. Fink and Polushkin, "The Report on the Plate," p. 25.
7. Bancroft Library, *The Plate of Brass Reexamined*, pp. 8–11.
8. Wagner, *Drake on the Pacific Coast*, p. 5.

Investigation at the Shinn Plate Site and Additional Support, 1947–53

In April 1947, the University of California Department of Anthropology (under Robert Heizer) in cooperation with the National Park Service (under Aubrey Neasham) made a series of excavations at the Shinn plate discovery site in Greenbrae, looking for other evidence of Drake's visit. They found no other artifacts and no evidence of a subsurface disturbance associated with a former post hole within a twenty-foot radius of the spot where Shinn had picked up the plate.[9]

In 1953 the California Historical Society issued a third special publication on the plate entitled *The Plate of Brass, Evidence of the Visit of Francis Drake to California in the Year 1579*. In it, the Society provided reprints of their first two publications (*The Plate of Brass*, 1937 and *The Plate of Brass Authenticated*, 1938), commented defensively on some of the questions that had been raised on the plate's authenticity, and reported on the University of California's recent archaeological work at Drakes Estero and the Shinn plate site.

Ethel Sagen, circa 1965

Historical researcher/writer Ethel Sagen of Bakersfield, California, after closely inspecting the plate, noted that its three "j's" appeared to have been gouged out with hammer blows to look like "i's." From an inscription on a family heirloom, she recognized that the letter "j" was not used in the Elizabethan Age and reasoned that the forger had learned of his mistake and tried to correct it. She filed her findings in the Bancroft Library.[10]

The Plate of Brass Reexamined and Branded a Forgery, 1977

Because so many doubts had been raised, Dr. James Hart, who assumed the Bancroft Library's directorship in 1975, decided to have the plate thoroughly reexamined for authenticity. He submitted the plate to leading experts in various disciplines including metallurgy, wording and lettering.

All the historical and lettering experts expressed skepticism. Their various concerns and misgivings included the crudeness of the engraving, the spelling, the wording, the style and the terminology. It was the metallurgical experts, however, who most severely challenged the legitimacy of the plate.[11]

At Oxford, British scientists compared the metal content of the plate to that of eighteen brass instruments and four brass memorials dating from 1540–1720. They found the plate had much higher zinc concentrations and much lower levels of the impurities lead and tin. They concluded that "the analytical evidence cannot be used to support the contention that the brass is of the Elizabethan period."[12]

At MIT, Dr. Cyril Stanley Smith, professor emeritus and noted authority on the history of metallurgy, carefully examined the plate with a microscope. He concluded it had been made by a nineteenth- or twentieth-century rolling process, not by hammering. "All of the features that I have noted make me incline to the opinion that the plate is a modern forgery."[13]

The Lawrence Berkeley Laboratory, in addition to running comparative chemical analyses similar to those at Oxford, also ran gamma-ray absorption tests to measure variations in plate thickness, and X-ray diffraction studies to determine grain orientation. In their forty-six page report entitled, "Chemical Study of the Plate of Brass," Helen Michel and Frank Asaro arrived at the following conclusions:

> The low level of lead and iron in the plate of brass indicates it was made from zinc metal which had probably been produced by the retort process and possibly redistilled. This indicates a date of the

9. Chickering and Heizer, eds., *The Plate of Brass*, pp. iv–v.
10. Heizer, *Elizabethan California*, p. 20; and Wiegand, "Pirates Cove," p. 8.
11. Bancroft Library, *The Plate of Brass Reexamined*, pp. 12–16.
12. Ibid., pp. 18, 37 and 43.
13. Ibid., pp. 21–22 and 75.

nineteenth or twentieth century although a late eighteenth-century date might be possible. The low level of antimony, arsenic, nickel, cobalt, silver, and gold in the plate of brass indicates it was made from high purity copper which would not have been generally available until about the middle of the nineteenth century, although an eighteenth-century date is possible. The high degree of chemical uniformity of the plate of brass, better than 0.45% from the zinc to copper ratio, and the small variations in thickness in the area away from the edges, 0.001 inch, are consistent with twentieth-century rolled brass. The average measured thickness of the plate of brass away from the edges, 0.129 +/- 0.002 inches, is consistent with the specifications for No. 8 gauge brass of the American Wire Gauge standard used in the 1930's. The plate of brass therefore was made in the eighteenth to twentieth centuries. The most probable period is the late nineteenth and early twentieth centuries.[14]

In the conclusion of Hart's 1977 report, *The Plate of Brass Reexamined*, the Bancroft Library's director pointed out that all the evidence brought forth in the studies had been negative. The plate continues to be displayed at the library, but posted with it is Hart's two-page summary of investigations on authenticity ending with the following statement: "Scientific studies have shown that what was thought to be an ancient artifact is evidently but a modern creation, yet they have left unanswered the intriguing questions about who made it and why."[15]

Additional Testing, 1991

In July 1991, the Bancroft Library submitted the plate to the University of California's Davis campus, for some additional testing. A research group headed by physicist Bruce Kusco placed the plate in an atom smasher and bombarded it at seventy-five different spots with a high-energy beam of protons. By measuring the X-ray emissions generated from the brass, the researchers were able to precisely determine the uniformity of the copper-to-zinc ratio. The testing revealed a fairly uniform composition, leading the researchers to conclude that the brass was probably made in the twentieth century.[16]

Other Plate Sightings

Casting further doubt on the authenticity of the Shinn plate is the fact that other such plates allegedly have been discovered.

The Sausalito plate – The November 7, 1956, *Oakland Tribune* includes the report of an interview with a Mrs. Grace Limback who claims she saw a metal plate bearing the date 1579, "embedded in a tree" overlooking Sausalito's "Tide Gate Cove." She told the press that she had made this observation in 1892 as a teenage girl, some sixty years before the article was printed.[17]

The Bolinas plate – The September 25, 1978, *San Rafael Independent-Journal* carried a report reminiscent of the Caldeira story. A Mr. Jefferson Graves, then forty-nine, of San Anselmo told them that at the age of seven or eight he had found a metal plate buried in the dirt while playing in a field near Bolinas. This was in 1936, just about the time Shinn discovered his plate. Graves noticed some writing on the surface to the effect that "someone was claiming the land for Her Majesty somebody." Thinking it was junk, another boy threw it out the window of his grandfather's car as they were driving home, somewhere in the vicinity of today's Larkspur Landing shopping center.[18]

Local anchorage sleuth, George Epperson, believes the Drake plate, Bolinas plate, Shinn plate, and Caldeira plate are one and the same, adding a few new names and interesting twists to the plate of brass saga:

In 1934 the Drake plate was pulled out of the soft earth on a remote hill at Agate Beach, Bolinas, by Jefferson Graves of Bel Marin Keyes. A large post found laying along side it was carbon-dated to

14. Michel and Asaro, "Chemical Study," p. 23.
15. Morison (*Southern Voyages*, pp. 677–80) speculates that the hoax was perpetrated by students of Professor Herbert Bolton, University of California, Berkeley, who purportedly "begged his students to keep their eyes open for it; and if it were found, bring it to him."
16. Horowitz, *San Rafael Independent-Journal (SRIJ)*, July 18, 1991, p. A1; and July 21, 1991, p. 1.
17. From Heizer, *Elizabethan California*, pp. 36–37. I have not been able to ascertain the location of Tide Gate Cove.
18. Smith, *SRIJ*, Sept. 25, 1978, p. 1.

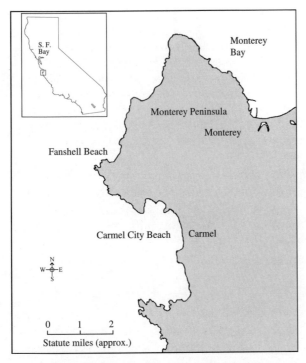

Fig. 53. The Fanshell Beach, Monterey, California, candidate anchorage site. This is the site where Monterey plate 1 (lead scroll) was reportedly discovered in 1934. *The map detail is redrawn from USGS 7.5-minute Monterey Quadrangle, 1983.*

Drake's time by the Department of Interior. Jeff's buddy threw the plate out the car on Tiburon Boulevard in San Rafael. Tom Parsons and Florence Paganetti picked it up while watching construction of Highway 101. Toms' mother, a historian, told Tom the plate should be given to authorities for posterity. This was done by his father, editor of the *Marin Recorder*. The plate was planted at Drakes Estero on property owned by the president of the Drake Association for Leon Bocqueraz, a banker and historian, to find. No one would doubt his word. I was told about this by a coast navigator, Matthew Ryan, who was disgusted with the affair. It was known to him that William Caldeira, chauffeur for Bocqueraz, beach-combed while waiting for him to hunt. Caldeira found the plate and put it in the car door pocket. Later he was driving from Richmond to San Anselmo. On the ferryboat he sorted out the junk he had collected and trashed the plate on the side of the road by the first valley north of Highway 101 on Sir Francis Drake Boulevard. Joe Cattaneo of San Rafael found the plate as he crawled over the fence to go up to his favorite rock on the Greenbrae Hill. He wiped off the mud and saw it was brass; he decided to make a monument on his rock and hid it in a crevice. He planned to come back later with cement. Meanwhile Beryle Shinn of Oakland discovered the plate while rolling rocks down the hill.[19]

Monterey plate 1 – One day in 1934, while walking the picturesque stretch of beach bordering today's Cypress Point Country Club in Monterey, retired U.S. Army Colonel Warren Clear found an old, barnacle-encrusted bottle protruding from the sand. As the bottle appeared old and unusual, he kept it and used it as a paperweight. Fifteen years later, in 1949, he noticed something inside and brought it to his friend Myron Oliver, Monterey antiquarian, artist and craftsman. Within the bottle, they discovered an English sixpence, doubled over to fit inside the bottle neck, and a thin, tightly rolled lead plate bearing Drake's name and the following inscription:

> In Nomine Elizabeth Hib et Britanna Riarum Regina: I do claim this great land and the seas thereof, there being no inhabitants in possession to witness thereto this bottle at great tree by small river at lat. 36 D. 30 M. beyond Hisp. four our most fair and puissant Queen and Her heirs and successors forever unto their keeping.
>
> By God's Grace this first day of May 1579.
>
> Francis Drake
>
> Fletcher, Scribe.[20]

Clear delivered the lead scroll to Robert Heizer at the University of California. Heizer only had the chance to examine it for one day, however, before the colonel abruptly demanded its return. Based on his cursory examination, Heizer considered the plate a crude fake.[21]

English authorities reportedly came to similar conclusions when they examined the object.[22]

19. Epperson, "Drake's Port of Nova Albion," p. 21.
20. Staff writer, *SRIJ*, July 6, 1956.
21. Heizer, *Elizabethan California*, pp. 38–40.
22. The information on the scroll is from an article by Randall Reinstedt entitled "Monterey's Mysterious Drake Plate," that I found in the Monterey Public Library California Room clipping file under "Historic Objects." The clipping, pulled directly from the magazine,

Fig. 54. A 1567 English sixpence found in 1974 at the former Olompali village site near Novato, California. Note the small nail-hole. *Courtesy The Bancroft Library, Berkeley, California.*

Some locals were unimpressed by the opinions of the experts. Myron Oliver, who had opened the bottle and flattened out the foil, was convinced the scroll was genuine: "There is every evidence that it is genuine and no expert has been able to disprove its authenticity."[23]

Monterey Peninsula Herald columnist, Ed Kennedy, writing in 1956, found the scroll to be conclusive proof that Drake and his company, on their way north, had put ashore at Monterey for the purpose of lightening their burden: "our guess is that there is an enormous treasure buried here; we ask only ten percent from the finder for calling it to attention."[24]

Monterey plate 2 – A second Monterey plate sighting is of even more questionable integrity. Thomas French of Auburn, California, wrote to the University of California in September 1962, reporting that he had observed a bronze plaque printed half in English – half in Spanish, lying in the sand. Heizer tried without success to reach French for further details.[25]

Other Elizabethan Artifacts

In modern times, two Elizabethan artifacts have been discovered in Marin County that may very well be authentic Drake relics: a 1567 English sixpence and a brass mortar. The story of four other possible finds, a Drake anchor allegedly discovered in Drakes Bay in the late 1800s, a 1573 sixpence, a sixteenth-century breast plate discovered near Bolinas in 1914, and the "Guns of Goleta" are of more questionable origin or authenticity. The porcelains and terra cottas of Drakes Bay; and the remarkable discoveries on Agate Beach, Bolinas, by beachcomber extraordinaire, George Epperson, are discussed in chapters 11 and 25.

The 1567 English sixpence – In June 1974, Charles Slaymaker, then a thirty-three-year-old archaeologist and lecturer at the University of California, uncovered an Elizabethan sixpence dated 1567 at Olompali, the ancient Coast Miwok Indian village near today's Novato. Slaymaker, who had been digging at the site for two years, reportedly found it two to three feet into the debris while hand-troweling a ceremonial dance floor. He also found nearby: two glass beads, a remnant from a silver rosary, and a Ming porcelain shard. The coin bears the image of Queen Elizabeth I, and has a

---

did not include the magazine title or date. According to the reference librarian, the magazine in question is probably *Monterey Savings Locale*, which has not been published for many years. The article includes a color picture of "a rare photo" of the scroll.

23. Staff writer, *SRIJ*, July 6, 1956.
24. Ibid.
25. Heizer, *Elizabethan California*, p. 40.

Fig. 55. A brass mortar inscribed with the date 1570 found in 1974 at a church in Novato, California. *Courtesy of Edward Von der Porten, Santa Rosa, California.*

small nail-hole at the top, giving one the impression that it was once nailed fast to a post. It is currently held by the Bancroft Library in Berkeley, California.[26]

The brass mortar – Just a few months before Slaymaker discovered the sixpence, the Drake Navigator Guild's Edward Von der Porten announced that he had found a sixteenth-century brass mortar at a Novato church. While he was able to determine that the crudely inscribed date 1570 was probably authentic, he was completely frustrated in his attempts to establish the object's historical significance. Katherine Ebright, who had donated it to the church in 1950, was already deceased and he could find no one with any reliable knowledge of how she had come to possess it.[27]

The DRAKE anchor – In February 1887, Captain Charles Raynor, of Seattle, Washington, put his small seal-hunting schooner into Drakes Bay to ride out a winter storm. When it finally abated and he attempted to raise his anchors, he snagged and brought up a very old and unusual anchor. Having lashed it on deck, some members of his crew chiseled off the rust, exposing what appeared to be the name DRAKE inscribed in the iron. According to Hart North, a member of the California Historical Society in the 1920s, Raynor left the anchor at the Katz Wharf in Port Townsend, Washington, before departing for the Bering Sea. Although Captain Raynor had died by the time North began his research, he was successful in tracking down several of the crew that had been aboard the vessel and obtained signed affidavits attesting to the anchor's discovery. He was unsuccessful, however, in finding a trace of the anchor itself. It was purportedly confiscated by a local halibut fisherman.[28]

The 1571 English Sixpence

Circa 1937, The *San Rafael Independent-Journal* reported that one of the exhibits at San Francisco's Panama-Pacific Exposition of 1915 included an English sixpence of 1571 alleged to have been discovered near Drakes Bay by some Marin school children.[29]

The Bolinas Breastplate

In the late 1910s, the citizenry of Marin County were buzzing with the news that a sixteenth-century breastplate had turned up near the town of Bolinas. In 1919, however, *The San Rafael Independent-Journal* quashed all speculation that the breastplate had been left behind by Drake when it reported that the armor, by then in the possession of Jack Wisby, "the artist of Bolinas," had come from the ashes of a little bungalow that once belonged to "Wildcat" Pearson. The armor, it seems, was one of the curios decorating the wall of his home when it burned to the ground.[30]

According to George Epperson, however, there is more to the story. He believes that the breastplate was just one piece of the "pile of armor and swords" that Wildcat had collected on Agate

---

26. Leydecker, *SRIJ*, Nov. 8, 1974, p. 1; and Champion, *San Francisco Chronicle*, Nov. 9, 1974, p. 3.
27. Staff writer, *SRIJ*, Jan. 10, 1974, p. 21.
28. Dana, *Pony Express Courier*, Placerville, California, June 1937; also Hanna, *Lost Harbor*, pp. 236–37.
29. I found a copy of the article, "Drake's Sixpence" in the Marin County Civic Center Library scrapbook but it was missing the date.
30. Staff writer, *SRIJ*, Oct. 30, 1919.

Beach, Bolinas circa 1904. He further disclosed that it was Donald Perry who in 1914 first discovered the Bolinas breastplate in the ashes of Wildcat's bungalow, actually located in Copper Canyon, east of Dogtown, about three and a half miles north of Bolinas. Epperson reports that the plate was authenticated by New York University as sixteenth-century Spanish. The breastplate is currently in the possession of the Phoebe Hearst Museum in Berkeley, California.[31]

The Guns of Goleta

In January 1981, a series of intense storms battered the coast of Southern California, leaving five heavily encrusted, iron cannon lying in open view on the exposed bedrock of Goleta County Beach Park, just north of Santa Barbara. Nolan Harter, the beachcombing jogger that discovered them, alerted archaeologists at the nearby campus of the University of California who promptly claimed the relics for the county. Initial examinations conducted by Robert Madden, professor of metallurgy at the University of Pennsylvania, indicated that the anchors may have been cast in England sometime between 1540 and 1750. Out of these findings emerged a group of "Santa Barbara advocates" led by Jim Gilmore, a local writer, and John Foster, a state archaeologist, who began to raise speculation that the cannon had been left behind by Drake.

X-rays of four cannon were first taken at the U.S. Naval Weapons Station at Concord in January 1982, with the ghostly images revealing some of the hidden details. Through the joint efforts of the University of California Santa Barbara, the county, the Goleta Valley Historical Society, and others, the cannon were painstakingly cleaned, preserved and by 1986, fully restored. With some diligent research into ordnance identification and regional maritime history, history buff Justin Ruhge was able to trace the likely source of the cannon to the Yankee schooner *Eagle* that went aground at the mouth of Goleta Slough circa 1830. As discussed at the end of the next chapter, the Santa Barbara advocates were not convinced by Ruhge's findings.[32]

Fig. 56. A sixteenth-century breastplate found circa 1914 near Bolinas, California. *Courtesy of the Phoebe Hearst Museum of Anthropology, University of California at Berkeley.*

Fig. 57. "The guns of Goleta": one of the five iron cannon discovered near the mouth of the Goleta Slough in 1981, now on display at the Goleta Valley Maritime Museum. *Photograph by Mary L. Williamson, curator.*

31. Epperson, "Drake's Port of Nova Albion," p. 27.
32. The guns of Goleta story is covered in Ruhge, *Gunpowder and Canvas*, chapter 14.

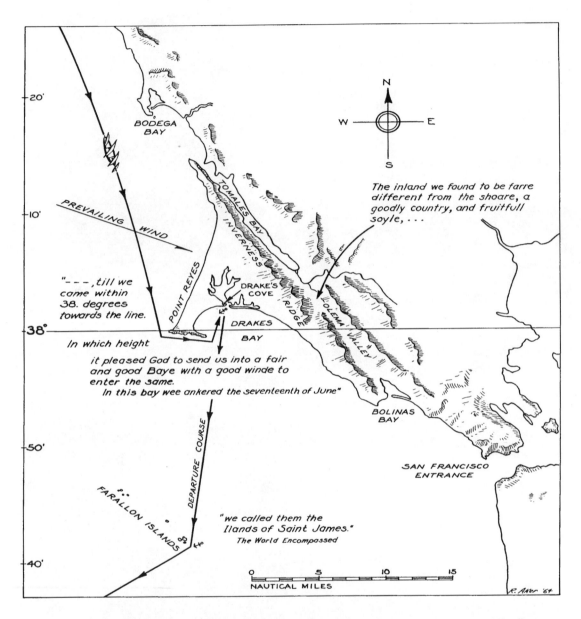

The inland we found to be farre
different from the shoare, a
goodly country, and fruitfull
soyle, . . .

"- - -, till we
came within
38. degrees
towards the line.

In which height

it pleased God to send us into a fair
and good Baye with a good winde to
enter the same.
In this bay wee ankered the seventeenth of June"

"we called them the
Ilands of Saint James."
The World Encompassed

PREVAILING WIND

DEPARTURE COURSE

FARALLON ISLANDS

BODEGA BAY

TOMALES BAY

INVERNESS

POINT REYES

DRAKE'S COVE

DRAKES BAY

INVERNESS RIDGE

OLEMA VALLEY

BOLINAS BAY

SAN FRANCISCO ENTRANCE

N
W   E
S

20'

10'

38°

50'

40'

0   5   10   15
NAUTICAL MILES

R. Aker '64

Map 10. Relief map of the greater Point Reyes area showing the *Golden Hind*'s alleged course en route to a Drakes Estero
anchorage. *Courtesy of the Drake Navigators Guild, Palo Alto, California.*

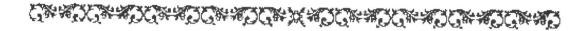

CHAPTER 11
Modern-Day Sleuths: 1936–96

The publicity surrounding the Shinn plate stimulated a great deal of interest in the anchorage question. For some, the matter became more than just historical curiosity. As it had for George Davidson — for Robert Power, Walter Starr, Aubrey Neasham, George Epperson, Warren Hanna, and the members of the Drake Navigators Guild — it became a passionate lifelong quest.

The Drake Navigators Guild, 1950–96, for Drakes Estero

The Drake Navigators Guild, formed in 1950, is a nonprofit organization that conducts historical research and disseminates information relating to Sir Francis Drake's 1579 visit to California. Its members have been fully committed to the task of locating the exact site of Drake's California encampment based on the premise that a group of experts in the field of navigation and seamanship might successfully join forces to resolve the problem.

Fig. 58. Raymond Aker, longtime president of the Drake Navigators Guild, dedicating a commemorative plaque at Drake's Cove in July 1979. *Courtesy of the Drake Navigators Guild, Palo Alto, California.*

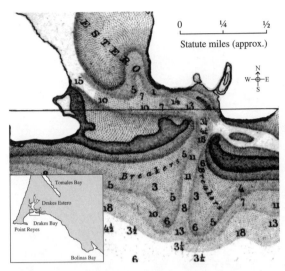

Fig. 59. The mouth of Drakes Estero in 1860. *Detail from an 1860 U.S. Coast Survey chart.*

Circa 1960, Guild members included:Chester Nimitz, famous World War II Fleet Admiral (honorary chairman); Robert Marshall, Laguna Ranch (onetime president); Captain Adolph Oko, master mariner (onetime president); William Hall, Hall Ranch, site of Drake's Cove; Raymond Aker, master mariner (current president); Edward Von der Porten, archaeologist; Captain David Edwards, U.S. Navy; Robert Parkinson; Robert Anderson; Robert Allen; Matthew Dillingham; Daniel Dillon; and Earnest Michelson. In 1979, the Guild's membership totaled thirty-six.[1]

After reviewing all available maps and documents bearing on the visit, considering the practical requirements of the case, and weighing the conclusions reached by George Davidson, the Guild concluded that Davidson's identification for Drakes Bay was correct, but that the *Golden Hind*'s careening place must be sought for in the sheltered waters of Drakes Estero. By late 1952 the Guild had tentatively identified the location of Drake's careenage and encampment in a cove near the mouth of the estero that they named "Drake's Cove." Following four more years of investigations to confirm the theory, on June 14, 1956, Admiral Nimitz announced the findings to the academic community in an address to the California Historical Society, concluding with these remarks:[2]

> It is unlikely that any recharting of the course the Guild has followed could lead to a different site, unless Drake's calculation of latitude be proved wrong; Fletcher's observation of weather conditions be proved fraudulent; the landmark of "white cliffs" be termed falsification; the naming of the land "Nova Albion" be considered a figment of nostalgia; the avowed purpose of the landing be declared a fiction; or the charting of Portus Nova Albionis be declared an imaginative invention only coincidentally true to periodic physical fact.[3]

During the near half-century of its existence, the Guild has published an enormous amount of literature on the Drakes Estero theory. In 1954, Dillingham, Parkinson and Aker prepared a preliminary identification report that evolved into the final report, "Nova Albion Rediscovered," that Nimitz presented to the California Historical Society in 1956. Between 1964 and 1970, Aker compiled a 461-page work entitled, *Report of Findings Relating to Identification of Sir Francis Drake's Encampment at Point Reyes National Seashore*, summarizing the results of all the Guild's investigations. He followed this up in 1978 with a short book entitled *Sir Francis Drake at Drakes Bay: A Summation of Evidence . . .* making an eighteen-point correlation, matching the Drake's Cove anchorage site with the contemporary evidence. In 1979, Aker coauthored a 33-page pamphlet with Von der Porten entitled *Discovering Portus Novae Albionis*, providing a concise summary of the Guild's research, findings and conclusions.

Other Drake Navigators Guild publications include the articles "Drake's Cove," by Chester Nimitz (1958); "Our First New England," by Edward Von der Porten (1960); "A Safe Harbor at Point Reyes," by Raymond Aker (1980); *The Drake and Cermeño Expeditions' Chinese Porcelains at Drakes Bay, California, 1579 and 1595,* by Clarence Shangraw and Edward Von der Porten (1981); "The Drake Puzzle Solved," by Edward Von der Porten (1984); and five research reports: (1) *Drake-Cermeño: An Analysis of Artifacts* (1965); (2) *The Porcelains and Terra Cottas of Drakes Bay* (1968); (3) *Identification of the Nova Albion Conie* (1971); (4) *An Examination of the Botanical*

---

1. Aker and Von der Porten, *Discovering Portus Nova Albionis*, introduction and p. 33.
2. Aker, *Report of Findings*, p. iv.
3. Aker, *Summation*, pp. 1–2.

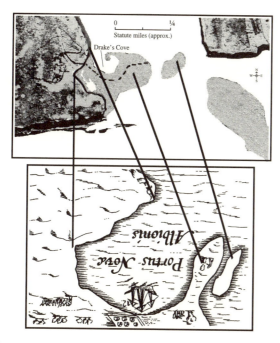

Fig. 60. The Drake's Cove candidate anchorage site showing the Drake Navigators Guild's Portus correlation. The gray areas correspond to sand flats and the dashed line corresponds to the high-water mark. Pointing out that the configuration of the sand flats is altered by storms, the Guild contends that the Portus correlation was much better in June 1579. *The illustration is adapted from an aerial photo taken September 11, 1952, and is based on a similar illustration in Raymond Aker's "Summation," 1978.*

References in the Accounts Relating to Drake's Encampment at Nova Albion in 1579 (1971); and (5) *Identification of an "Herb Much Like Our Lettuce"* (1971).

Of all the evidence they presented to support their case, the Guild rates the Portus correlation as the most important. In 1952, the Guild started its search for Portus-matching shorelines, making extensive field trips along the estero and taking an abundance of photographs. On November 24 that year, Dillingham examined a photograph he had taken of the west side of the estero entrance, and noticed a cove within the sandspit in the entrance that resembled the seal-head shape of the cove in the Portus plan. A return visit to the site revealed that the shores of the inner terrain not visible in the photograph bore a close comparison with other Portus shorelines (figure 60). The point and island of the Portus plan were seen to correlate with the types of sandspits and sandbar islands that typically form at the estero entrance (figure 61). In 1956, Aker came upon some aerial photographs taken in September 1952 by the U.S. Department of Agriculture that showed Drake's Cove features in a way that could not be accurately observed from the ground, particularly the true relationship and configuration of the sandspit and sandbar, as well as the navigable channel leading into the estero and to the cove. These observations led the Guild to conclude that the Portus provided "the key clues that pinpoint Drake's encampment site."[4]

Wagner truly stated: "The most important piece of evidence known to us is the plan of the Portus." Its influence has been subtle, but once seen, its agreement with the small cove on the west side of Drakes Estero cannot be denied, nor can it be keyed to any other site within the entire suspect area of Drake's landing except by gross distortion.[5]

Like George Davidson, the Guild's longtime president Raymond Aker, degreed in navigation and seamanship, was another individual very well qualified to speak on the latitude issue. Long after the Guild had become firmly entrenched in its Drakes Estero theory, he made a rigorous study of the accuracy of Drake's nautical astronomy. After carefully checking some forty-odd latitudes provided in *The World Encompassed* against a modern world atlas, he came to a startling conclusion: Drake's latitude readings, when likely measured on land, appeared to err from the actual latitudes on the modern atlas by an average of only nine minutes (nine nautical miles/eleven statute miles); where

---

4. Aker, in Holliday, "The Francis Drake Controversy," p. 251.
5. Aker, *Report of Findings*, p. 284.

Fig. 61. The Drake's Cove sandspit in 1955. The photograph, looking east across the cove at low tide, shows a sandspit and outlying sandbar islands. Although the spit does indeed bear a strong resemblance to the point of land depicted on the Portus plan, it is not oriented in the corresponding direction (see figure 60). *Courtesy of the Drake Navigators Guild, Palo Alto, California.*

the readings were likely taken at sea, the average discrepancy was only twenty-five minutes (twenty-five nautical miles/thirty statute miles).[6]

Not only did Aker's findings contradict the findings of the venerable Davidson, but according to *The World Encompassed*, on which the Guild relied heavily for most of its corroborating evidence, Francis Drake did not appear to land in Drakes Bay! *The World Encompassed* reports Drake's fit harbor was at "38. deg. 30. min." With Drake's Cove located at thirty-eight degrees two minutes, according to Aker's findings, the twenty-eight-minute discrepancy is outside the expected average error for a measurement made either on land or sea.

To rationalize this negative evidence, the Guild argued that "within 38 degrees" from "The Famous Voyage" reflected an exceptionally precise navigational reading taken on land, while "38. deg. 30. min." from *The World Encompassed* was an imprecise sea-reading taken by dead reckoning during a period of fog.[7] Here is more of their reasoning on the matter:

> The question as to whether to consider 38º 30′ or 38º is almost conclusively resolved in favor of the lower figure by the evident fact that Hakluyt, who may have consulted with Drake and had access to the same source as the compiler of *The World Encompassed*, chose to use a totally different description of the discovery of Drake's haven and apparently rejected "38. deg. 30. min." in favor of "38 degrees towards the line." [8]

Starting in 1953, the Guild made a monumental effort to confirm their Drake's Cove theory through archaeological investigations. Deducing that Drake's fort site was to be found buried within a large filled area on the south side of the cove, they hand-excavated a series of trenches in search of the fort and Elizabethan artifacts. It did not take them long to come upon what they thought were the slumped stone walls of the fort. As they continued their trenching, however, their excitement abated as they did not find any parallel or cross-connecting walls or Elizabethan artifacts. By 1958 it was concluded that the stones had been washed in by storms, possibly from Drake's fort, and that the fort had stood on an outer beach that had eroded away since Drake's time. Although additional trenching did confirm that a substantial portion of the cove had indeed been filled in by storms, in

---

6. Aker, *Summation*, pp. 11 and 12.
7. Ibid., p. 12.
8. Ibid., p. 11.

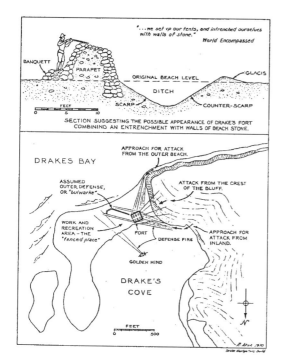

Fig. 62. The Drake's Cove fort site as envisioned by the Drake Navigators Guild. *Courtesy of the Drake Navigators Guild, Palo Alto, California.*

almost ten years of work in completing over ninety pits and trenches the Guild had produced no Elizabethan artifacts.[9]

As discussed in the previous chapter, University of California anthropologists have conducted numerous archaeological investigations within the boundaries of the Point Reyes National Seashore looking for evidence of Drake's visit. Although they have turned up many sixteenth-century artifacts, most have been unequivocally attributed to the wreck in 1595 of Sebastián Rodríguez Cermeño's Manila galleon, the *San Agustin*. Nevertheless, there have been several attempts to link some of these artifacts to Drake's visit.

In 1949, Dr. Clement Meighan reopened one of the estero middens Robert Heizer had originally worked in 1940. His field crews unearthed eleven fragments of a glazed stoneware vessel that experts identified as pieces of a large water or wine jar produced in China in the late sixteenth century. Dr. Heizer created a stir in 1952 when he published an opinion that the stoneware shards "clearly reached the site before the Cermeño visit" of 1595.[10] The Guild's own archaeologist, Edward Von der Porten, however, rejected Heizer's conclusions.[11]

In 1980, Doctor Clarence Shangraw, senior curator of the Asian Art Museum at Golden Gate Park, helped Von der Porten make a careful examination of some seven hundred plus fragments of Ming Dynasty porcelain that researchers had recovered from some of the previous archaeological investigations at Drakes Bay. Although it had been generally presumed that all these shards were associated with the Cermeño visit, Shangraw concluded that the porcelain pieces actually came from two shipments, one of which is "fairly attributable to Drake," and "in all likelihood" arrived on the *Golden Hind*. Dr. Michael Moratto, an anthropologist/archaeologist at San Francisco State University, voiced his support for this archaeological confirmation. Based on Shangraw's findings, it is the Guild's position that no question remains: "seventy-seven objects that had once formed part of the treasure of Francis Drake's *Golden Hind* were final confirmation of the location of his landing—the truly named Drakes Bay."[12]

In recent years the Guild has taken to the political front in attempting to procure official acceptance for their landfall theory. In 1992, the state Historical Resources Commission rejected their second formal petition for registration of Drakes Bay as a historical landmark. In 1994, the Guild successfully petitioned the Marin Board of Supervisors to approve a two-part resolution: the first part declaring that Drake indeed encamped at Drake's Cove; the second requesting the National Park Service to approve the placement of a plaque at Drake's Cove, declaring it the landing site. According to Supervisor Gary Giacomini, "We've been looking at this controversy for twenty years. This is the most likely place. . . . Only God could prove us wrong."[13]

9. Ibid., pp. 343–76.
10. Meighan and Heizer, "Archaeological Exploration," p. 104.
11. Von der Porten, "Drake–Cermeño: An Analysis of Artifacts," pp. 52–55.
12. Von der Porten, "The Drake Puzzle Solved," p. 26; Creib, *San Francisco Chronicle (SFC)*, May 5, 1980, p. 3; and Irving, *San Francisco Examiner (SFE)*, July 6, 1984, p. A1.
13. Horowitz, *San Rafael Independent-Journal (SRIJ)*, June 16, 1994.

Fig. 63. Robert Power (1926–91), staunch proponent of the theory that Drake entered San Francisco Bay. *Courtesy The Bancroft Library, Berkeley, California.*

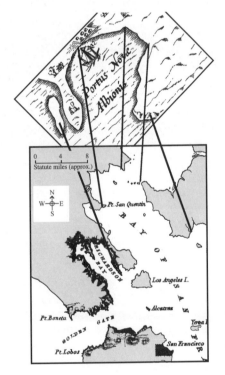

Fig. 64. The San Quentin Cove candidate anchorage site showing Robert Power's Portus correlation. *The chart detail is derived from an 1851 U.S. Coast Survey chart (preliminary).*

Robert Power, 1954–91, for San Quentin Cove

Robert Power, a partner in the popular Nut Tree Restaurant of Vacaville, California, became interested in the anchorage controversy in the early 1950s after reading Henry Wagner's book on the circumnavigation. Power's evaluation of the evidence left him utterly convinced that Drake had harbored at San Quentin Cove in San Francisco Bay. He published his first article on the subject, "Portus Nova Albionis Rediscovered?" in the June 1954 issue of *Pacific Discovery*, the magazine of the California Academy of Sciences. In a 1968 paper entitled "Early Discoveries of San Francisco Bay," Power challenged the U.S. Senate parks and recreation subcommittee on proposed language for a plaque intended to mark the location on Sweeney Ridge in Pacifica from which the Spanish ostensibly "first discovered" San Francisco Bay. In April 1973, Power presented a paper to the State Historical Landmarks Advisory Committee, unsuccessfully petitioning them to erect a monument commemorating Drake's entry through the Golden Gate. The California Historical Society published this paper in 1974: *Drake's Landing in California: A Case for San Francisco Bay*. Power followed this up with a short booklet entitled *Francis Drake & San Francisco Bay*: *A Beginning of British Empire* in which he repeated his arguments and discussed the significance of the landfall in British-American history.

Power went on to become president of the Solano County Historical Society, president of the California Historical Society, chairman of the California Heritage Preservation Commission, and a member of the Sir Francis Drake Commission. His interest in the anchorage question spurred Power to amass a remarkable collection of antique printed books, maps and Drakiana—centuries of literature on the Drake theme. Power's collection included the extremely rare Jodocus Hondius "Vera Totius Expeditionis Nauticae . . . " (the Hondius Broadside Map with the Portus inset) (1589); Hakluyt's three-volume *Principal Navigations, Voyages and Discoveries* (1598, 1599, 1600); and *The World Encompassed* by Sir Francis Drake (1628). In 1977, he exhibited the collection at the Drake 400 celebration in Plymouth, England, where he was introduced to Queen Elizabeth II.[14]

---

14. The Power collection described in the "Drake 400 Exhibition" catalogue, 1977, is now in the possession of the Bancroft Library.

Arguing that the Portus was the principal key to locating the site, Power made an eleven-point comparison with a section of the north shore of San Francisco Bay (the same comparison Douglas Watson made in 1937). Power established eleven points of correlation, noting that in fingerprint identification, ten points are considered proof beyond doubt. He contended that the "good wind" mentioned in "The Famous Voyage" had to have been the "unique bernoulli wind" commonly experienced when entering San Francisco Bay. He also pointed out that a strong northwest wind would have created the visibility Drake needed to see the narrow entrance to the Golden Gate, but hazardous conditions for a vessel trying to enter the coastal bays.[15]

Power introduced a new piece of evidence to the anchorage controversy. He maintained that a 1671 illustration by Arnold Montanus, a Flemish craftsman, entitled "The Crowning of Drake," had geographic features that bore a "startling resemblance to those of the Point San Quentin area." He noted that Montanus was related to Hondius and could have inherited whatever written records on Drake that Hondius had obtained.[16]

In the August 1978 issue of *California History* (formerly *California Historical Society Quarterly*), Power rebutted the findings of Dr. Hart's report on the plate of brass. He argued that the metallurgical tests were inconclusive and that the researchers, notwithstanding their conclusions, had actually detected additional positive clues in the inscription. He remained convinced the plate was authentic, and proof positive of Drake's stay at San Quentin Cove.[17]

Power was unimpressed with the Guild's claims to have archaeologically confirmed Drake's landing at Drakes Estero by analysis of Chinese pottery shards. He found it logical to expect that in loading Cermeño's galleon, Manila merchants would have shipped crates of goods manufactured in different parts of China at different times.[18]

A. Bray Dickinson, 1954, for San Francisco Bay

In a five-part series that appeared in the *Marin County Independent-Journal* in October 1954, Tomales' postmaster, A. Bray Dickinson, aired his views on the Drake landing, "pooh-poohing" the Drake Navigators Guild's theories that the *Journal* had been featuring in recent articles. After summarizing all the documentary evidence, Dickinson, by process of elimination, concluded that Drake dropped anchor in upper San Francisco Bay. He felt that the Guild's alleged Drake's Cove site and all the other coastal inlets and backwaters in the area were too shallow for the *Golden Hind* to have safely entered. He accepted Robert Power's Portus correlation but labeled the Shinn plate a fake. He decided that upper San Francisco Bay was the only site that correlated with the reported latitudes in *The World Encompassed* in that the account had Drake backtracking north to thirty-eight degrees thirty minutes, after exploring the coast south to thirty-eight degrees. At the time he wrote the articles, Dickinson, a descendant of one of Marin's prominent pioneer families, was director of the Marin County Historical Society and a well-respected local historian.[19]

Walter Starr and Francis Farquhar, 1956–62, for San Quentin Cove

Walter Starr earned a doctorate degree in business at the University of California, Berkeley, in 1897 and capped a successful career as director of the Scott Paper Company. In 1942, he served a term as president of the California Historical Society. Like Robert Power, Starr was a tireless supporter of the San Quentin Cove anchorage theory. Endeavoring to refute the presentation the Drake Navigators Guild made to the California Historical Society in June 1956, he and historian

---

15. Power, in Holliday, "The Francis Drake Controversy," pp. 209 and 214; and Power, "Drake's Landing," pp. 123–24.
16. Power, "Drake's Landing," p. 119. Robert Heizer was actually the first anchorage researcher to publish the Montanus illustration in his *Francis Drake and the California Indians*, 1947.
17. Power, "A Plate of Brass," pp. 172–85.
18. Greer, *SRIJ*, May 5, 1980, p. 1.
19. From the title of an article in the *SRIJ*, Sept. 25, 1954. The five-part series appeared on Oct. 2; Oct. 9; Oct. 16; Oct. 23; and Oct. 30, 1954. Dickinson wrote another piece lambasting the Guild's "propaganda" and "beating of drums" that appeared July 28, 1956.

Francis Farquhar presented papers in November of that year. Farquhar simply summarized the various types of evidence and the current status of research. Starr rehashed the plate of brass evidence in light of the current controversies, explaining why he thought the plate was authentic: "only by being where it was found for a very long period of time could it have become so embedded." In 1962, Starr made his last pitch with another article in the *California Historical Society Quarterly* entitled "Drake Landed in San Francisco Bay in 1579, the Testimony of the Plate of Brass." He compiled not only the testimony of the brass plate, but all the evidence he could muster to support his contention, including Stillman's arguments for the ground squirrels and white cliffs, and Power's arguments for the Isles of Saint James, the Portus, the "fit harbor," the terrain, and the visit inland. In rebutting the Guild's arguments for Drakes Bay, he carefully analyzed Caldeira's story, noting that he reported throwing his metal plate out of the car two miles from the spot where Shinn found his.[20]

Alan Brown, 1960–96, for Half Moon Bay

Alan Brown, then a graduate student at Stanford University, contributed a short essay on Drake's landing to the February 1960 issue of *La Peninsula*, journal of the San Mateo County Historical Association. This was a special issue on Half Moon Bay in which Brown espoused an opinion on the question, "Did Drake Land Here?" First pointing out that "the sum of historical evidence so far found fits this bay as well as any of them, and in some respects better," he pointed to a strong correlation between the Portus, and the way Half Moon Bay may have looked in Drake's day:[21]

> The stumbling block in identifying the harbor is the island on the seaward side of the point, for none of the harbors mentioned [Bodega, Bolinas, Drakes] has such an island. But Half Moon Bay once had an island that could be it. Only recently, and within the possible time span of living memory, "the Sail" or Pillar Rock outside Pillar Point was reduced from an islet 500 feet long, with two peaks 150 feet high, to an insignificant stump. In the same period the sea washed away about a million cubic yards of rock out of the reef off the point, exposing the inner bay to ocean surf. It would not take a great deal more material than this to build the old Pillar Point reef and Sail Rock into a point and an island . . .[22]

Unable to locate the February 1960 issue of *La Peninsula* in the local libraries or from interlibrary loan, I ventured into the association's museum in August 1994, hoping to find it at their

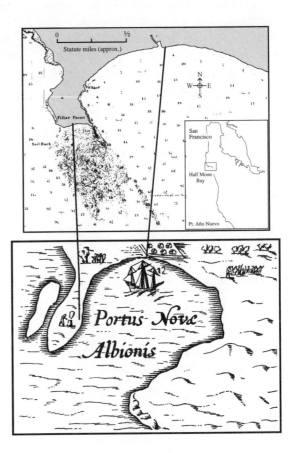

Fig. 65. The Half Moon Bay candidate anchorage site showing Alan Brown's Portus correlation. By comparison with a current chart of the area, there have been no overt geologic changes in the past century. *The chart detail is derived from an 1863 U.S. Coast Survey chart.*

---

20. Starr, "Evidence of Drake's Visit," p. 34; and "Drake Landed in San Francisco," p. 19.
21. Brown, "Did Drake Land Here," p. 3.
22. Ibid.

reference desk. Arriving just before it closed for the evening, I found a scholarly looking gentleman absorbed in some files. Upon my request, the librarian quickly produced the journal from a file drawer, and then looked at me with a very surprised expression as I confirmed that I was seeking Brown's essay on Drake. "Why, you are standing behind its author!" she exclaimed.

The puzzled gentleman who turned around was indeed Dr. Alan Brown, now a professor of English at Ohio State University. Twenty-odd years after leaving California, he had arrived just that day for the express purpose of visiting the museum in order to procure the reference materials used in preparing his article—in particular, the early U.S. Coast Survey drawing that led him to his theory about the island. Brown explained that he was writing a section on the settlement of the Pacific Coast for an upcoming book, and planned to include his theory on Drake's landfall. He told me, "I will not be so bold as to conclude that Drake landed at Half Moon Bay, mind you, but just to present the information so people can form their own opinions."

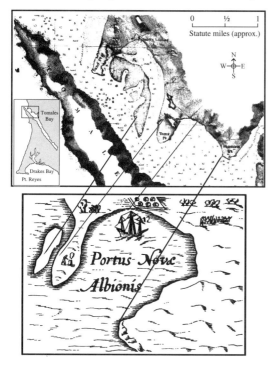

Fig. 66. The Tomales Bay/Tom's Point candidate anchorage site showing Robert Becker's Portus correlation. *The chart detail is reproduced from an 1863 U.S. Coast Survey chart (photo-edited for clarity).*

Robert Becker, 1960 for Tomales Bay

A renowned local historian, Robert Becker contributed to a U.S. National Park Service report entitled *Land Use Survey: Proposed Point Reyes National Seashore* (1960). In a chapter dealing with the early explorers, he provided an objective discussion about the controversy over Drake's landing, summing up the arguments for all the candidate sites in the area. His conclusion:

> . . . all the indications suggest that Drake and his crew saw the white cliffs of Drakes Bay and they landed somewhere in present Marin County. But the actual determination of the site must await proof from the shovel of the archaeologist, and, perhaps, corroboration from some documentary sources as yet undiscovered."[23]

He labeled the Portus plan of minor importance and noted that five different Drake scholars had, in their minds, successfully compared it to five different local harbors. He then made it six, however, after observing "a reasonable similarity between the Hondius drawing and the embayment on the northeast side of Tomales Bay, between Tom's Point and Preston Point."[24]

Robert Pate, Richard Dobson, and Ethel Sagen, 1967–78, for Pirates Cove

In the late 1950s, Margie Mallagh, of San Luis Obispo Bay in Southern California, told a newcomer to the area about a local legend concerning the small cove where her great-grandfather had built Pirates Cove Wharf: today's Mallagh Landing located at approximately thirty-five degrees north latitude. According to this legend, Sir Francis Drake had landed there in 1579, leaving behind a fabulous treasure. Based upon this tip, Robert Pate, an engineer and retired U.S. Air Force bomber pilot from Carmichael, and his friend Richard Dobson, a salesman from Rancho Cordova, would spend the next ten years pursuing the dream of discovering Drake's horde.

---

23. Becker, "Historical Survey of Point Reyes," p. 39.
24. Ibid., p. 39.

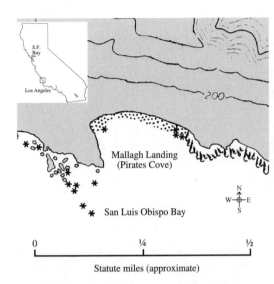

Fig. 67. The Pirates Cove/Mallagh Landing candidate anchorage site at 35º 11´ north latitude in San Luis Obispo Bay. *The map detail is redrawn from USGS 7.5-minute Pismo Beach Quadrangle, 1994.*

Pate began to suspect the legend was based in fact when he noticed some mysterious holes and a large square impression bored into solid rock at the point that commanded the entrance to the cove. Suspecting the holes may have been from Elizabethan gun emplacements, he began researching the anchorage mystery and found plenty of evidence to support his theory. He observed an exact match with the Portus plan, concluded the square impression near the gun emplacement had once held the plate of brass, and identified Drake's "Isles of Saint James" as the present-day Channel Islands, citing some of the contemporary maps, in particular the French Drake Map. In explaining away the conflicting latitude evidence that had Drake landing far to the north in thirty-eight degrees, he reasoned that the Elizabethan chroniclers deliberately misreported his position to confuse the Spanish. The two treasure hunters were fully convinced that Drake had left a cache behind, having calculated that his plunder would have dangerously overburdened the *Golden Hind*.[25]

Historical researcher/writer, Ethel Sagen of Bakersfield who played an instrumental role in debunking the plate of brass, seconded Pate and Dobson's views. She pointed out that Pirates Cove lies approximately 140 leagues south of Cape Mendocino, which roughly corresponds to the distance mentioned in Thomas Blunderville's 1594 account of the circumnavigation (chapter 3). [26]

The Pirates Cove anchorage story was featured in the December 1967 issue of *True West* magazine. It also appeared on national television in 1967 in a NBC production called "The Drake Story," directed by veteran newsman Elmer Peterson.[27]

Don Viles and Wayne Jensen, 1971, for Nehalem Bay, Oregon

Hanna reported that he found the Don Viles-Wayne Jensen paper in *Vanguard*, a publication of Portland State University, November 16, 1971, and in a pamphlet entitled *The Northern Mystery*. I could not locate either.

Harold Gilliam, 1973

A well-known figure in the San Francisco area from his work as an environmental activist, local author, and columnist for the *San Francisco Chronicle*, Harold Gilliam included a chapter in his 1973 *Island in Time: The Point Reyes Peninsula* that provides an excellent summary of the then raging debate. After reviewing the various theories that had lately been espoused, he arrived at the following conclusion:

> Doubtless the controversy will continue to vex historians, both amateur and professional, and complicate the plans of dinner hostesses for generations to come . . . Unless, perhaps, someone strolling along a lonely beach should stumble across the ruins of an Elizabethan rock fort, exposed by shifting sands. [28]

---

25. Wiegand, "Pirate's Cove," pp. 6–9 and 48.
26. Sagen, *Francis Drake's Secret Landfall*, pp. 3 and 5.
27. Ibid., p. 14.
28. Gilliam, *Island in Time*, p. 25.

Samuel Eliot Morison, 1974, for Drakes Bay

America's eminent twentieth-century historian, Samuel Eliot Morison, devoted a short chapter to the controversy in his epic, *The European Discovery of America*. After examining all the evidence and sailing the Marin coast north from San Francisco to Bodega Bay, he arrived at the following conclusion: "Drakes Bay is correctly named . . . The white cliffs like those of Albion 'on the side of the sea' (from the estero) were for us the determining factor." He dismissed Bodega Harbor for lack of white cliffs, but noted that cliffs on the outer shores of Tomales Peninsula are of a pale gray color that might "by a stretch of imagination be called white." Morison branded the Shinn plate a complete and crude hoax.[29]

Fig. 68. Aubrey Neasham, proponent of Bolinas Lagoon anchorage site (1975). *Reproduced from LOST HARBOR by Warren Hanna, University of California Press, 1979. Courtesy of Mrs. Warren Hanna, Kensington, California, and the University of California Press, Berkeley, California.*

Aubrey Neasham and William Pritchard, 1974, for Bolinas Lagoon

Professors Aubrey Neasham and William Pritchard, though apparently the only sleuths to have supported Bolinas Lagoon for anchorage honors, were not lacking in qualifications. A professional historian and professor of environmental resources at Sacramento State University, Neasham served with the National Park Service, was vice-president of the California Historical Society, president of Western Heritage, and a member of the Sir Francis Drake Commission. Pritchard was one of California's foremost archaeologists. Together, they coauthored a short book published in 1974 entitled *Drake's California Landing: The Evidence for Bolinas Lagoon*.[30]

In 1948, during his tenure with the U.S. Park Service, Dr. Neasham supervised archaeological investigations looking for evidence of Drake's visit at the Davidson's Cove anchorage site under Point Reyes Head, and in the vicinity of the Shinn plate find in Greenbrae. His interest in the subject aroused, he began a systematic search for the encampment at other possible sites in the San Francisco area. In the 1960s he participated in archaeological digs at two potential sites on the west shore of Bolinas Lagoon and several other Indian sites in the Bolinas area. All these investigations were unproductive.[31]

Through the aid of aerial photographs, Neasham noted another possible fort site on the lagoon's western shore: an active reservoir purportedly constructed in 1872 that, according to local knowledge, had been used as a fish pond until the end of the nineteenth century. Postulating that the original fish pond had been built upon the remains of Drake's fort, Neasham personally financed and supervised archaeological tests at the site in 1973. In constructing two trenches they found indications of two distinct strata comprising the earthen banks of the old fish pond which, as far as Neasham was concerned, established the discovery of Drake's fort. He admitted, however, that to provide 100 percent proof, it would take "every resource of the federal, state, and county

29. Morison, *Southern Voyages*, pp. 672, 677, and 680.
30. According to historian John S. Hittell, *History of the City of San Francisco*, 1878, p. 27, Dwinelle originally thought Drake landed in Bolinas Bay.
31. Neasham and Pritchard, *Drake's California Landing*, pp. 6 and 12.

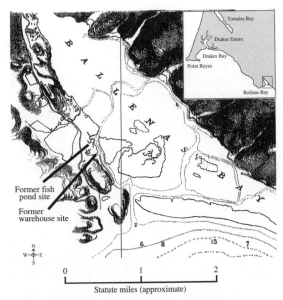

Fig. 69. The Bolinas Lagoon candidate anchorage site showing the locations of a former fish pond and a former warehouse where Aubrey Neasham successively claimed to have unearthed the remains of Drake's fort. *The chart detail is from an 1855 U.S. Coast Survey chart (preliminary).*

governments, as well as private donations," to finance the archaeological investigations needed to confirm the discovery.[32]

Neasham further strained his credibility by arguing that the shape of the jagged coin hole in the bottom of the Shinn plate was the clue that first led him to discover the Bolinas Lagoon fort site. He observed a "surprising similarity" that matched the upper portion of the coin hole not only to the Bolinas Lagoon shoreline, but also to the "Port of Nova Albion" depicted on Dudley's Carta Prima General.[33]

In June 1979, in a paper he presented at the international conference for the quadricentennial of Drake's landing, Neasham startled the academic community by announcing that he had finally unearthed some Drake artifacts in new digs on the shores of Bolinas Lagoon—but not at the fish pond. The new alleged fort site was at the foot of a hill where the old Leonard Warehouse once stood in the 1850s. "We have found nothing with Drake's name on it, but we have found what appears to be a portion of a massive structure having all the characteristics of a fort." Neasham's principal finding was a layer of hard pan earth located three feet below the ground surface, ostensibly representing the floor of Drake's old fort. Within the buried hard pan layer, he found a half-dozen tent pegs and a couple dozen iron spikes. In the strata above the hard pan, he found nineteenth-century American and Indian artifacts.[34]

Alan Villiers, 1975, for Drakes Bay

Captain Alan Villiers, sea dog, circumnavigator, and "celebrated chronicler of the sea," recreated the life and achievements of Sir Francis Drake in the February 1975 issue of *National Geographic* magazine. After providing a short summary of the California anchorage debate, he came to the following conclusion:

> Taking in the extraordinary "Englishness" of those white cliffs and thinking back into the mind of an old-time shipmaster groping along the shoreline, desperately in search of a safe careening berth, I know I would have been very interested as my ship rounded Point Reyes and the lonely inlet—easily visible from the open sea—came into view. I'd round to, bring her up to the wind, ghost in, sound, and examine. And thank God.[35]

Kenneth Holmes, 1979, for Drakes Bay

Dr. Kenneth Holmes, professor of history at Oregon College of Education, spent over twenty years trying to "iron out the details of Drake's voyage in our part of the globe." Focusing most of his efforts on determining how far north Drake sailed in his search for the mythical Straits of Anian, he made a detailed analysis of the ocean currents and prevailing winds (see appendix E). After concluding that Drake made it as far north as the coast of Vancouver Island in forty-eight degrees, he then turned his attention to the California landfall. He found Drakes Estero "the most plausible

32. Ibid., p. 17.
33. Ibid., p. 8.
34. Greer, *SRIJ*, June 14, 1979, p. 1; and Champion, *SFC*, June 15, 1979, p. 4.
35. Villiers, *Sir Francis Drake*," p. 223.

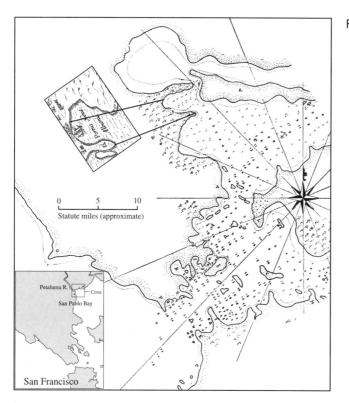

Fig. 70. The San Pablo Bay candidate anchorage site showing Robert Thomas' Portus correlation. This crude depiction of the north shore of San Francisco Bay is reproduced from a chart drawn in 1775 by Juan Manuel de Ayala during the first Spanish entry into San Francisco Bay. Thomas correlated the Portus plan cove with the large cove Canizares depicted at a location that appears to correspond with the mouth of today's Petaluma River. Judging from the earliest detailed charts of the area, however, it is probable that the cove, as depicted by Canizares, never existed. The chart detail is reproduced from CALIFORNIA UNDER SPAIN AND MEXICO, 1535–1847 by Irving Richman, Houghton Mifflin, Boston and New York, 1911.

locale for New Albion on our Pacific Coast," based on the practical refuge value of the setting, the white cliffs, the cool climate, and the presence of the Coast Miwok Indians.[36]

Roger Verran, 1979, for Drakes Bay

In his "Francis Drake's Silent Witness," published in the March 18, 1979, *San Francisco Sunday Examiner and Chronicle*, Roger Verran forcefully argued that the weather conditions described in *The World Encompassed* point strongly to Drakes Bay and preclude a landfall at San Quentin Cove. He pointed out that there are ninety-one fog signals around San Francisco Bay, but none at Point San Quentin where, as the sailors say, "the sun always shines." Backing this point up with some diligent research, Verran reported that operating records for the foghorn closest to San Quentin Cove showed almost no foghorn activity for June and July over a forty-one-year period. He concluded that "if Drake had landed there, all Francis Fletcher would have complained about was sunburn." He also accurately pointed out that, even in the unlikely event that Drake discerned the Golden Gate and was foolhardy enough to attempt an entrance without careful reconnaissance, the combination of current, tidal flow, and wind would have forced him to seek shelter on the south side of San Francisco Bay. To enter San Quentin Cove he would have had to turn into the teeth of the wind.[37]

Robert Thomas, 1979, for San Pablo Bay

A descendant of Chief Marin and Chief Ynita of the Coast Miwok nation, Robert Thomas became interested in the Drake's Bay controversy as he started researching a book he was writing on the history of his people. In *Drake at Olompali* (1979) he adopted Power's arguments that Drake entered the Golden Gate, and then used the Portus plan and ethnological evidence to make the case that Drake anchored in San Pablo Bay at the mouth of the Petaluma River—about three-quarters of a mile from his ancestors' village. Thomas thought Fletcher's description of low-vaulted, earth-covered Indian

---

36. Holmes, "Francis Drake's Course," p. 34.
37. Verran, *San Francisco Sunday Examiner and Chronicle*, Mar. 18, 1979, *This World*, pp. 28–30.

lodges proved that Drake had not encamped at Drakes Bay. He pointed out that when Cermeño visited Point Reyes in 1595, the Spanish reports had the Indians living in grass-roofed lodges.[38]

Curiously, one clue that Thomas neglected to mention was the sixteenth-century sixpence that Charles Slaymaker had recently unearthed at an Olompali ceremonial dance floor. Imagine the ecstasy in the Guild camp if their shovels had unearthed this gem at Drakes Estero!

In the presentation of his arguments, Thomas gives a very enlightening interpretation of the events that transpired at Nova Albion from the Native American standpoint: "The interesting task has been to extract from Drake's story and to isolate and use what appears to be basic information about the Olomko and other California Indians."[39]

Warren Hanna, 1979

Warren Hanna of Kensington, California, had a lifelong interest in Drake's lost harbor. He became fascinated with the mystery after reading John Robertson's *The Harbor of St. Francis* (1927). Like Power, he was a great collector of literature on the subject, even procuring a copy of Hakluyt's rare 1589 *Principal Navigations*. Hanna was a member of the Sir Francis Drake Commission and his book was published as part of the quadricentennial celebrations. He had a very successful legal career, managing a major California law firm and writing extensively in the field of worker's compensation.[40]

Hanna published *Lost Harbor* (1979) as a follow-up to the 1974 anchorage debate sponsored by the California Historical Society (see chapter 12). Like Robertson before him, Hanna played the neutral observer, critiquing and rating the arguments offered by the three debaters: Raymond Aker of the Drake Navigators Guild, Robert Power, and Aubrey Neasham.

Of the twenty tenets of the Society's debate, Hanna could only rate the participants' answers on six. On the others, he decided the debaters'

Fig. 71. Warren Hanna (1898–1987). Just as John Robertson critiqued the original anchorage sleuths, Hanna played the neutral observer and rated the arguments of the latter-day sleuths. *Reproduced from LOST HARBOR by Warren Hanna, University of California Press, 1979. Courtesy of Mrs. Warren Hanna, Kensington, California, and the University of California Press, Berkeley, California.*

arguments did not justify a score. For the six he rated, Hanna scored the Guild (Drakes Bay) eleven, Power (San Francisco Bay) nine, and Neasham (Bolinas Lagoon) four. In his conclusions, he lamented: "After four hundred years of mystery, including nearly two hundred years of controversy, we still have no solution to the Drake anchorage riddle."[41]

Hanna provided one poignant comment in his book. Troubled by the way the debaters had curtly dismissed *The World Encompassed* evidence on latitude, he warned us: "*The World Encompassed* clue, on careful analysis, may have more to offer than meets the casual eye!"[42]

38. Thomas, *Drake at Olompali*, pp. 67–72.
39. Ibid., p. 8.
40. Hanna, from the rear cover leaf of his book, *Lost Harbor*.
41. Ibid., p. 344.
42. Ibid., p. 161.

The Landmark Debacle

The intensity of the anchorage debate during the 1970s is epitomized by the bitter struggle that ensued when the state attempted to commemorate the location of Drake's California landfall with an official plaque. Robert Power was at the center of the controversy, and without his interaction, the state probably would have officially proclaimed Drakes Bay the anchorage site.

The struggle began in 1973 when the state legislature first proposed an act for establishing the Sir Francis Drake Commission. Power, recognizing that the majority of the commission members would be enthusiastic supporters of the Drakes Bay landfall theory, lobbied a restriction into the act that prevented the commission from designating an official landing site without prior action by the state Historical Landmarks Advisory Committee. When Power began earnestly petitioning the landmarks committee to designate San Quentin Cove as Drake's landing site, the Guild responded with "full broadsides."[43]

The matter came to a head in October 1978, when the California Historical Resources Commission conducted hearings regarding the landmark question. "Deluged by mountains of arguments and besieged by dozens of witnesses," the commission members found it impossible to decide the question and voted to do nothing.[44]

In the spring of 1979, at the continued urging of the Sir Francis Drake Commission, the historical commission again tried to resolve the question. Once again the commission deadlocked on the issue of whether or not to erect a plaque, "not because we're afraid to make a leap onto one of these landing beaches, but because it would be a leap of faith rather than fact." They did, however, reach a consensus on the most likely site of the anchorage: "it is the consensus of the commission that Drakes Estero is at this time, the most generally acceptable and probable landing site."[45]

Notwithstanding the lack of official state approval, two North Bay groups succeeded in getting the necessary funds and approvals to erect two unofficial monuments for the quadricentennial celebration: one sponsored by the Marin Coast Chamber of Commerce at Drakes Bay; and the other sponsored by the Redwood Empire Association at Vista Point overlooking the entrance to San Francisco Bay. While the wording on the two brass plaques was intended to leave the exact location of the "Marin" landfall open, it is suggestive of Drakes Bay.[46]

Arthur Davies, 1980–81, for San Francisco Bay

England's *Geographical Magazine* is the counterpart to our own *National Geographic* magazine. In the May 1980 issue, the editors included an article entitled "The Drake Conundrum," in which they published two papers advancing different theories on the location of the California anchorage: one by Robert Power, the other by Raymond Aker. In the July 1980 issue, Arthur Davies, professor emeritus of geography at the University of Exeter, after analyzing the views of the two protagonists, declared firmly for "Drake at San Francisco."

In 1981, Davies further espoused his San Francisco Bay theory in an article entitled "Drake in California," that appeared in *Pacific Discovery*, the quarterly magazine published by the California Academy of Sciences in San Francisco. Based on cartographic evidence, Davies also put forth the concept that in sailing north from Acapulco to the Pacific Northwest Coast, Drake did not follow the course set forth in *The World Encompassed*. He has Drake exploring and charting the entire coast of California.[47]

43. Nolan, *SFE*, Feb. 16, 1979, p. 37.
44. Irving, *SFE*, Oct. 24, 1978, p. 6; and Mar. 1, 1979, p. 9.
45. Irving, *SFE*, May 8, 1979, p. 3; and staff writer, *SFC*, Mar. 3, 1979, p. 2.
46. Smith, *SRIJ*, May 12, 1979, p. 3.
47. Davies, "Drake and California," p. 24.

Fig. 72. George Epperson (*right*) and the author (*left*) near Drake's alleged fort site on the shores of Agate Beach, Bolinas. *Photograph by R. B. Lucas, February 1997.*

George Epperson, 1980–96, for Agate Beach, Bolinas

Based on decades of investigation and research, retired San Rafael tile contractor George Epperson, has developed a passionate conviction that Drake careened the *Golden Hind* off today's Agate Beach, a stretch of alternately sandy and rocky shoreline located between Duxbury Point and Bolinas Point on the north side of Bolinas Bay. Since the early 1980s, Epperson has hosted annual Drake celebrations at his alleged Drake fort site, conducted numerous presentations, and written scores of letters attempting to convince government officials to recognize and protect the area. Epperson is now proposing to erect a monument on Duxbury Point using the *Golden Hind's* "tar baby ballast" that Drake ostensibly left on Agate Beach in 1579 to make room for treasure and supplies.

A dedicated and skilled beachcomber, Epperson's interest was sparked in 1958, when he found a brass boarding hook embedded in a piece of wood with the letter "D" inscribed on its end allegedly in a similar appearance to the "D" in Drake's own signature. Since then, he has turned up thousands of other artifacts he believes were associated with Drake's visit including the following: "two fancy shoulder pieces" belonging to the Bolinas breastplate (see chapter 10); armament from the *Golden Hind* (cannon balls inscribed with the letter "D," pieces of a sixteenth-century swivel gun, and thousands of lead, copper, and silver bullets); a silver mariner's dial; "Magellan's scaffold" (spruce mast); "parts of the *Golden Hind*" (oak timber and five deadeyes); two pieces of broken pewter; scraps of brass allegedly generated when Drake cut the sixpence hole out of the plate of brass; "the skeleton of Rodrigo Tello's frigate" (fifteen pieces of black ebony beam, a copper gun door, and a broken boat oar); a thousand pounds of flintstone ballast similar to that used on early British sailing ships; and "the great firm post" to which the brass plate had allegedly been nailed.

Notwithstanding all the evidence to the contrary, Epperson is convinced that not only is the Shinn plate authentic, but that it was made at Agate Beach. He has provided a detailed scenario for its

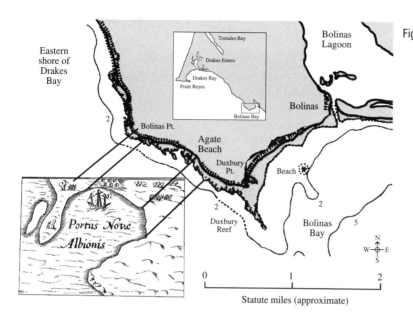

Fig. 73. The Agate Beach, Bolinas, candidate anchorage site showing George Epperson's Portus correlation. Epperson contends the correlation is a close match allowing for erosion. He also contends that the Portus peninsula and island are both visible at very low tides. *The chart detail is redrawn from National Oceanic Service chart 18645, Gulf of the Farallones.*

Statute miles (approximate)

movements from Bolinas, first to San Rafael, then to Drakes Estero, and finally to the Shinn discovery site in Greenbrae (see chapter 10).

Bolstering his archaeological evidence with the cartographic clues, Epperson argues that all the features depicted on the Portus inlet can be discerned from a Duxbury Point vantage point, "no more, no less." He also contends that the Drake-Mellon Map, Dudley Carta Prima General del India . . ., and Dudley's unpublished manuscript chart number 85, all depict a Drake encampment in an area correlating to Agate Beach.

Regarding the contemporary documents, Epperson contends that "all the known clues about Drake's stay in Marin fit Agate Beach."[48]

While Epperson concedes that the channel leading to his proposed careening site is now too shallow to afford entry to the *Golden Hind* and too exposed to ensure total safety, he attributes this to geologic changes that have taken place over the past four centuries: "the combination of quakes since Drake's time would easily cause a total rise [in the ocean floor] of two fathoms [twelve feet]." Similarly, he believes that Agate Beach was once protected by a Portus-like peninsula that has since disappeared into the sea, the remnants of which are still visible at very low tides.

Epperson has Drake's initial anchorage at the Davidson site under Point Reyes Head, but feels Drake passed up both Davidson's Cove and Drakes Estero due to concerns about Indians. He also believes that Drake made an excursion into San Francisco Bay using Tello's frigate.

As one might expect, other anchorage detectives have taken incredulous views of Epperson's theories. According to Raymond Aker, "he [Epperson] is absolutely convinced he is right, but he's totally wrong. The stuff he's finding comes from various shipwrecks." Robert Power had this to say: "Agate Beach isn't a harbor or a bay or an anchorage. It's a severely preposterous idea."[49]

Notwithstanding the views of the competing sleuths, Epperson has his supporters. They include Joe Hood, state historian for the California Department of Beaches and Parks; and Dr. Thomas Sneed, president of the Marin Historical Society who headed a committee charged with investigating Epperson's findings. Warren Hanna, though not convinced, was impressed by some of the relics Epperson had gathered. Other locals who inked support for Epperson's Agate Beach site as of 1992 include: Besby Holmes of San Rafael, Lt. Col. USAF (retired); Dick Murdock of San Rafael,

---

48. Epperson, *SRIJ*, Sept. 8, 1987.
49. Todd, *SFC*, Jan. 22, 1989, p. B1; and Merchant, *Point Reyes Light*, Dec. 1, 1988, p. 1.

author/columnist/historian; Jess France of Bolinas; Lester Ferris of Tiburon; and Nancy Heizer Cushing of Mill Valley—sister of Professor Robert Heizer (deceased).[50]

The Marin Cultural Center & Museum published Epperson's arguments and findings in a pamphlet that contains a paper entitled "Drake's Port of Nova Albion." The pamphlet is a summary of an exhibit Epperson displayed January–June 1994 at the Cultural Center's Exhibition Hall, located at the Kentfield campus of the College of Marin.

Jim Gilmore and Justin Ruhge, 1982–96, for the Goleta Slough

Undeterred by some rather convincing proof to the contrary, Santa Barbara advocates Jim Gilmore and archaeologist Jack Hunter, remain steadfastly convinced that the controversial guns of Goleta, discovered in 1981 at Goleta County Beach Park, are five iron cannon that were shipped aboard the *Golden Hind* when it left port in 1577, but missing upon its return three years later. They also point to an ancient iron anchor found at Goleta Slough in 1893 as further proof of Drake's visit: it measures in English inches.[51]

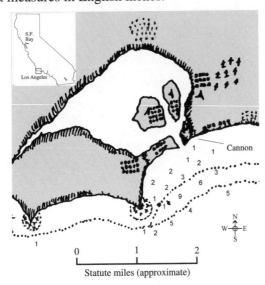

Fig. 74. The Goleta Slough candidate anchorage site at 34° 25′ north latitude on the north shore of Santa Barbara Bay. *The map detail is redrawn from Juan Pantoja Arriaga's map of the Central California coast, 1782.*

Justin Ruhge, while maintaining his belief that the Goleta guns are of eighteenth-century origin, is a convert to the Gilmore theory that Drake made port some 270 miles south of the San Francisco area. In *Drake In Central California*, 1990, Ruhge raises the following arguments in support of a Goleta Slough anchorage: (1) close correlation between the Channel Islands lying off Santa Barbara Bay and the four islands and east-west shoreline depicted on the French Drake Map; (2) a Portus correlation so well fitted that "there is no other place on the coast of California where such a close match can be made"; (3) a good fit with the Montanus illustration; (4) white cliffs on both sides of the entrance to the slough; and (5) a local Chumash Indian population that dwelled on the surrounding bluffs whose manners and customs are documented to have been strikingly similar to those described by Fletcher.[52]

Although the early maps suggest that the barred entrance to the slough was too shallow for the *Golden Hind*'s draft, the Goleta champions argue that it was navigable in Drake's day. They also conclude that the latitude evidence is unimportant since the readings reported in the various accounts are clouded and contradictory, with errors ranging from one to four degrees.[53]

Louise Welshons Buell, Harry Morrison, Richard Scott, and Gerald Weber, 1982–1996, for Point Año Nuevo Cove

Harry Morrison, a Catholic priest from Watsonville, and Richard Scott, a Berkeley video producer, were drawn into the Drake controversy in December 1981 during a dinner hosted by Scott's mother-in-law, Louise Welshons Buell. Having conducted fifteen years of diligent research on the anchorage question, Ms. Buell was convinced that Drake had harbored at Point Año Nuevo Cove

50. Keown, *SRIJ*, June 10, 1980, p. 15.
51. Sullivan, *Santa Barbara News-Press*, Mar. 21, 1982.
52. Ruhge, *Drake in Central California*, pp. 23, 41, and 44.
53. Ibid., pp. 50–56.

located at the extreme southern tip of the San Francisco Peninsula, some sixty miles south of the city of San Francisco. Her own interest in the subject had been sparked in 1965 when, at a seminar put on by Robert Power, she learned that the white cliffs of Drakes Bay were considered benchmark evidence due to their resemblance to the white cliffs of Dover. During a visit to Dover, England the year before, she had noted a striking similarity between the Dover cliffs and the white cliffs she had seen lining the shore just south of Point Año Nuevo (Waddell Bluffs) while sailing with her father. After laying out all the evidence she had compiled on the subject for her dinner guests, she found herself with two enthusiastic supporters of her anchorage theory.

Morrison immediately set to work putting the theory to ink. By October 1982, he had produced a scholarly report that to my knowledge was never published: "An Investigation of the Louise Welshons Buell Theory of the California Anchorage of Francis Drake, 1579." Scott contributed an essay on the Portus correlation that was appended to the report.

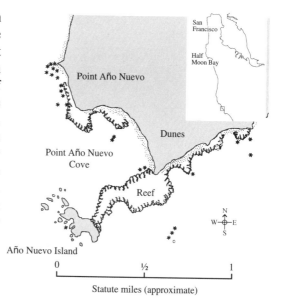

Fig. 75. The Point Año Nuevo Cove candidate anchorage site at 37º 6′ north latitude, forming the southern tip of the San Francisco Peninsula. *The map detail is redrawn from USGS 7.5-minute Año Nuevo Quadrangle, 1968.*

Using the twenty tenets of the California Historical society as a framework for analysis, Morrison/Buell/Scott argued that all the evidence in the contemporary maps and accounts support the Año Nuevo theory. They argued Drake would have been too far off the coast to discern Bodega Bay, and would have bypassed today's Drakes Bay for fear of offshore rocks extending from Point Reyes Head. Based on Davidson's evaluation of the latitudes reported by early Spanish navigators, they argued that Drake's latitude readings were probably running about a degree too high, which correlates well with Point Año Nuevo Cove at thirty-seven degrees, six minutes. They argued that the linguistic clues better support contact with the Costanoan Indians than Coast Miwok. For example, they contended the Nova Albion Indian word for king, "Hioh" correlates better with the Costanoan word for chief, "huck," than the Coast Miwok word for chief, "hoipu." They also pointed out that the Nova Albion Indian word for bread "cheepe," correlates very closely with the Costanoan word for knife, "chippi."

Realizing that the Portus correlation was a weak point in the theory, circa 1991, Morrison and Scott approached Santa Cruz geologist Gerald Weber, with the question as to whether, in Drake's day, Point Año Nuevo and its small offshore island could have more closely resembled the Portus. After carefully examining the area's geology and erosion patterns, Weber concluded that in the recent geologic past, the cove lying under the point may indeed have closely resembled the harbor shown in the Portus plan. In the offshore deposits he found evidence that a large sand island, correlating to the Portus island, once lay just north of what was once a much more pronounced peninsula, correlating to the Portus peninsula. Weber presented his findings at a meeting of the Geological Society of America and National Association of Geology Teachers, for the purpose of demonstrating how geology can help settle historical issues. When the Point Año Nuevo theory finally came to public attention in the San Francisco Bay area, however, the Guild's Raymond Aker quickly branded it "a foolish, very farfetched idea."[54]

---

54. Petit, *SFC*, May 28, 1993, p. B3.

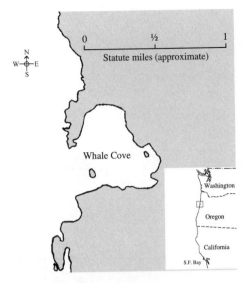

Fig. 76. The Whale Cove candidate anchorage site at 44° 48´ just south of Depot Bay, Oregon. *The map is redrawn from USGS 7.5-minute Depot Bay Quadrangle, 1984.*

Robert Ward, 1987–96, for Whale Cove, Oregon

A British engineer/writer/amateur historian from Liverpool, Robert Ward, created a stir when he proposed that California prepare to relinquish to Oregon its claim for the site of Drake's Nova Albion proclamation and the birth of the British Empire. In conducting research on the Pacific Northwest Coast, Ward noted a close correlation between the Portus plan and Whale Cove, a small inlet located at forty-four degrees forty-eight minutes north latitude just south of Depot Bay. He also contends that Fletcher's description of Indians and animals favored an Oregon landfall over a California landfall, particularly the Indian dwellings. Like Robert Pate, Ward argued that the Elizabethan chroniclers deliberately falsified the latitudes provided in the early accounts in order to mislead the Spanish. When Edward Von der Porten of the Drake Navigators Guild was asked to comment on Ward's theory, he replied "that's garbage."[55]

As of 1996, Ward has taken to the Internet in his effort to drum up interest and support for his Whale Cove anchorage theory. He is offering a visual tour and inviting anyone interested in helping solve the mystery to join the "Drake in Oregon Society." In his February 1996 update, Ward reports that the foundations of what may have been Drake's wooden stockade have been unearthed and that carbon-dating is in progress.

Other Opinions Since 1936

Continuing the trend from the era of the original sleuths, most of the casual opinions voiced since the late 1930s have been strongly in favor of a Drakes Bay landfall. Here is a smattering of additional opinions for the last sixty-odd years:

James Williamson: "not unanimously identified," in *The Age of Drake*, 1938, p. 191.

Felix Riesenberg for Drakes Bay in *Cape Horn*, 1939, p. 68.

James Hedges for Drakes Bay, *An Encyclopedia of World History*, 1940, p. 507.

Joseph Knowland, publisher of the *Oakland Tribune*, for Drakes Bay in *California a Landmark History*, 1941, p. xii.

Donald Peattie for Drakes Bay in "Pirate-Knight: Sir Francis Drake," *The Readers Digest*, July 1953, p. 122.

Rockwell Hunt for Drakes Bay in *California Firsts*, 1957, pp. 7–8.

Duncan Gleason for Drakes Bay in *The Islands and Ports of California*, 1958, p. 194.

David Pesonen for Drakes Bay in *A Visit to Atomic Park*, 1962, p. 11.

Neill Wilson: "the riddle of Drake's landing place remains," in *Here is the Golden Gate*, 1962, p. 18.

R. Coke Wood for Drakes Bay in *California History and Government*, 1962, p. 12.

Maurice Holmes, for Drakes Bay in *From New Spain by Sea to the Californias*, 1963, p. 22.

Ernle Bradford for Drakes Bay in *The Wind Commands Me*, 1965, p. 136.

Harper & Row Publishers, Richard Morris, ed., for San Francisco Bay in *Encyclopedia of American History*, 1965, p. 23.

---

55. Akre, *SRIJ*, Feb. 2, 1987, p. 1.

Professor John Hawgood: "the still-vexed question" in "The True and Fabulous History of Nova Albion," an inaugural lecture delivered at the University of Birmingham, May 6, 1965, p. 7.

Lawrence Kinnard: "the greatest mystery in California History" in *History of the Greater San Francisco Bay Region*, 1966, p. 24.

Kenneth Andrews: "somewhere near San Francisco," in *Drake's Voyages*, 1967, p. 78.

Walton Bean, for Drakes Bay in *California and Interpretive History*, 1968, p. 19.

George Sanderlin for San Francisco Bay in *The Sea Dragon*, 1969, p. 168.

Erwin Gudde: "never established" in *California Place Names*, 1969, p. 94.

Jack Mason for Drakes Bay in *Point Reyes*, 1970, pp. 4–6.

George Malcolm Thompson for Drakes Bay in *Sir Francis Drake*, 1972, p. 141.

David Lavender for Drakes Bay in *California: Land of New Beginnings*, 1972, p. 28.

T. W. E. Roche for Drakes Bay in *The Golden Hind*, 1973, pp. 148 and 152.

T. H. Watkins for San Francisco Bay in *California, an Illustrated History*, 1973, p. 23.

William Hamilton, former Mill Valley mayor, for Drakes Bay in "The Drake Question, and One Answer," *San Rafael Independent-Journal*, April 5, 1975.

The Reader's Digest Association for San Francisco Bay or Drakes Bay in *Family Encyclopedia of American History*, 1975, p. 357.

Elizabeth Goodman: "one of the most fascinating and controversial questions in western American history . . . the search back through history continues" in "He Founded the First New England," *The American West*, January/February, 1976, pp. 4 and 13.

Charles Scribner's Sons for Drakes Bay in *Dictionary of American History*, 1976, p. 369.

Derek Wilson for Drakes or Bodega Bay in *The World Encompassed*, 1977, p. 156.

O. H. K. Spate for San Francisco Bay in *The Spanish Lake*, 1979, p. 258.

Brian McGinty: ". . . as much mystery as a half-a-dozen Agatha Christie novels . . . the search for Nova Albion goes on," in "Search for Nova Albion," *Oceans*, July–August 1979, p. 27.

W. H. Hutchinson: "only foolhardiness will permit making a positive selection" in *California, the Golden Shore by the Sundown Sea*, 1980, p. 37.

David Beers Quinn for Davidson's Cove/Drakes Bay in "Drake's Circumnavigation of the Globe: A Review," monograph published by the University of Exeter in 1981, p. 7.

Arthur Quinn for Drakes Bay in *Broken Shore*, 1981, p. 7.

Worldmark Press, Ltd. for Drakes Bay in *Worldmark Encyclopedia of the States*, 1981, p. 48.

Grolier, Inc. for "the San Francisco area" in *Academic American Encyclopedia*, 1983, p. 256.

James Hart: "never determined" in *A Companion to California*, 1987, p. 118.

Merriam–Webster, Inc., for Drakes Bay in *Webster's New Geographical Dictionary*, 1988, p. 343.

Los Angeles Times for Drakes Bay in "Stalking Sir Francis," Sunday, March 5, 1989, p. 4.

John Sugden for Drakes Bay in *Sir Francis Drake*, 1990, p. 135.

John Faragher, ed., for Drakes Bay in *The Encyclopedia of Colonial and Revolutionary America*, 1990, p. 116.

Sylvia Barker Thalman for Drakes Bay in *The People of the Good and Fair Bay*, 1991.

Macmillan Publishing Company, Richard Bohlander, ed., for Drakes Bay in *World Explorers and Discoverers*, 1992, p. 157.

Jacques Legrand International for San Francisco Bay in *Chronicle of America*, 1993, p. 34.

The H. W. Wilson Company for Drakes Bay in *Facts About the States*, 1993, p. 47.

Columbia University Press for San Francisco Bay in *The Columbia Encyclopedia*, 1993, p. 792.

Encyclopedia Britannica, Inc.: "just north of San Francisco" in *The New Encyclopedia Britannica*, 1993, p. 212.

Rand Richards for Drakes Bay in *Historic San Francisco*, 1993, p. 16.

Sumerset Publishers for Drakes Bay in *Encyclopedia of California*, 1994, p. 67.

P. F. Collier (publisher) for Drakes Bay in *Collier's Encyclopedia*, 1994, p. 377.

Gladys Hansen for Drakes Bay in *San Francisco Almanac*, 1995, p. 41.

Mick Sinclair for Drakes Bay in *Fodor's Exploring San Francisco*, 1995, p. 30.
John Cummins for Drakes Bay in *The Lives of a Hero*, p. 117.
Simon Winchester for Drakes Bay in "Sir Francis Drake is Still Capable of Kicking Up a Fuss," *Smithsonian*, January 1997, p. 87.

Summary

It is evident that today's Drakes Bay has received the lion's share of support over the course of the anchorage controversy. San Francisco Bay is a very distant second choice. Here is the tally for the one hundred plus anchorage detectives I have covered in detail.

Table 1 - Summary of Opinions		
Bay	**Supporter**	**Tally**
Carmel River	Padre Niel	1
Half Moon Bay	Brown	1
San Pablo Bay	Thomas	1
Trinidad Bay	Wagner	1
Whale Cove, Oregon	Ward	1
Bolinas Lagoon	Neasham, Pritchard	2
Fan Shell Beach, Monterey	Oliver, Kennedy	2
Nehalem Bay, Oregon	Viles, Jensen	2
Bodega Harbor	Colnett, Dwinelle, Berthoud	3
Goleta Slough, Santa Barbara Bay, Southern Calif.	Gilmore, Hunter, Ruhge	3
Pirates Cove, San Luis Obispo Bay, Southern Calif.	Pate, Dobson, Sagen	3
Tomales Bay	Bodega y Quadra, Humboldt, Becker	3
Point Año Nuevo Cove	Welshons-Buell, Morrison, Scott, Weber	4
Agate Beach, Bolinas	Epperson, Hood, Sneed, Holmes, Murdock, Heizer-Cushing, France, Ferris	8
San Francisco Bay	Burney, Forbes, Tuthill, Stillman, Hale, Verne, Watson, Power, Starr, Farquhar, Dickinson, Davies	12
Drakes Bay/Drakes Estero	Vancouver, Duflot de Mofras, Soule, Vallejo, Bancroft, Davidson, McAdie, Barrett, Kroeber, Heizer, Sir Francis Drake Association (say 30), Drake Navigators Guild (say 50), Shangraw, Moratto, Morison, Villiers, Holmes, Verran	96
Total		**143**

With respect to the other opinions I cited in the lists found at the end of this chapter and chapter 8, there were fifty-nine votes for Drakes Bay and twelve for San Francisco Bay.

Now let us see who was right!

PART IV

Assembling the Pieces

Fig. 77. Sir Francis Drake: portrait attributed to Jodocus Hondius, circa 1589. *Courtesy of the National Portrait Gallery, London.*

CHAPTER 12
Tenets of the California
Historical Society Debate

n the summer of 1974, the California Historical Society brought two centuries of speculation and argument to a fitting climax. The editors of its periodical, the *California Historical Society Quarterly*, faced off the three most-dedicated anchorage sleuths of the twentieth century and refereed a written debate on the whereabouts of Drake's lost harbor. The contestants were Raymond Aker of the Drake Navigators Guild, proponent of the Drakes Estero theory; Professor Aubrey Neasham, sponsor of the Bolinas Lagoon theory; and Robert Power, staunch defender of the San Francisco Bay theory.

The Society sponsored the debate with the goal of clarifying, if not resolving, the controversy. Its staff formulated twenty tenet statements, mutually agreed to by the debaters, that covered all the evidence and arguments they had used to promote their respective theories. They allowed each debater to present 4,000-word opening arguments on each tenet, rebut their opponents' arguments in a total of 4,000 words, and then provide counter-rebuttals in 4,000 words. Here were some of the more colorful statements that came from this scholarly encounter:

> *Power, counter-rebutting Neasham on the identification of the "conies"*: "Dr. Neasham is so unspecific that it isn't worth the space in this limited-word debate to respond to his weasel-worded statement."

> *Power, rebutting Aker on the Portus Nova Albionis correlation*: "the Guild tries to make the Portus plan fit their area by creating a nonexistent anchorage out of a changeable sand shoreline . . . It is a geographical myth and a geologic improbability."

> *Aker, rebutting Power and Neasham on the plate of brass*: "Mr. Power's contention that William Caldeira was lying when he claimed original discovery of the plate at Drakes Bay is irresponsible. . . . As for Dr. Neasham's argument, we fail to see that the hole in the plate is anything but a provision for retaining the sixpence."

> *Neasham responding to Power's rebuttal on the conies*: "Ground squirrels!"

The Society published the debate transcripts in a special Fall 1974 issue of its magazine entitled "The Francis Drake Controversy: His California Anchorage." In *Lost Harbor* (1979), Warren Hanna provided an in-depth evaluation of the debaters' arguments, giving Aker (Drakes Bay) an indecisive edge over Power (San Francisco Bay), ranking Dr. Neasham (Bolinas Lagoon) a distant third.

The Society's twenty tenets provided a convenient framework for presenting the evidence supporting my own anchorage theory. In chapters 13 through 32, I address each of the tenets and counter-rebut past arguments others have raised against the location hereafter proposed for Drake's landfall. In chapter 33, I rebut the arguments for the other candidate sites.

So point S.F. DRAKES B.

as taken at ↔

A VIEW of the Entrance to Port S.F DRAKE

from ↔ in the Bay

Fig. 78. "The south point of Sir Francis Drake's Bay as taken at the anchorage" (*above*) and "A view of the entrance to Port Sir Francis Drake from the anchorage in the bay" (*below*): two sketches from the journal of Captain James Colnett, September 1790 (photo-edited and partly redrawn). Note the Indian approaching in his canoe to greet the English ships. *Courtesy of the Public Record Office, Richmond, England. ADM 55/142 f 156 LH.*

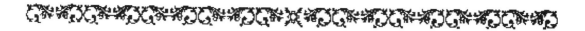

CHAPTER 13
Tenet 1 – The Approach

Reaching the coast of what is now called Southern Oregon early in June, 1579, Drake searched for the Northwest Passage. Not finding it, he sailed south some 300 miles to about thirty-eight degrees north latitude where he stopped to repair the "Golden Hind." What relevance does Drake's search for the Northwest Passage have on the landing-site controversy?

Map 11. Outline map of the coasts of Washington, Oregon and Northern California showing Drake's course from Queenhythe Bay, Washington, at about forty-eight degrees north latitude, south to Bodega Bay, California, at about thirty-eight degrees. According to John Stow, Drake covered the approximately 700-mile course in seven days (see appendix E).

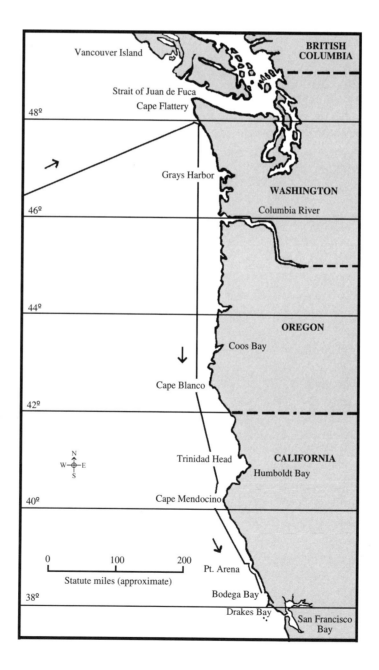

Clues

- They were at sea sixty-three days between landfalls: from April 16, 1579, to June 17, 1579 (Drake's *The World Encompassed* [hereafter cited in part IV of this book as *W.E.* with page numbers from the Readex Microprint facsimile], pp. 62 and 64).

- They had barely enough water on board to last fifty days when they left Guatulco, Mexico, on April 16 (Nuño da Silva's sworn deposition, in Nuttall, p. 303).

- They had a Spanish chart aboard showing the sailing route that Spain used to conduct trade in the Philippines. In showing the return route from Manila to Acapulco, Mexico, this chart would have depicted the Pacific Coast from about forty-three degrees (Spain's Cape Mendocino in 1579 – today's Cape Blanco, Oregon) southward to Mexico (Sarmiento de Gamboa, in Nuttall, p. 81).

- They were between forty-three and forty-eight degrees north latitude when they stopped looking for the strait and started south (*W.E.*, Hakluyt's, "The Famous Voyage," 1589 or 1600 editions [hereafter cited as *F.V.* in part IV of this book], and Stow, in Wagner, *Drake's Voyage*, p. 304).

- By the time they abandoned the search the crew was "utterly discouraged" by the bitter cold, violent winds, and fog (*W.E.*, p. 64).

- They headed south on June 10, 1579 (Stow).

- In sailing down the coast, they had great difficulty with the fog (*W.E.*, p. 66).

- After turning south they coasted along the shore without landing until they came "within 38 degrees toward the line" where "it pleased God" to send them into a "fair and good bay" (*F.V.*).

- As they sailed south they observed no disturbance in coastal currents that would mark the presence of a large strait (*W.E.*, p. 67).

- The ship had sprung a leak (*W.E.*, p. 68).

Answers

Approaching from the north – First and foremost, the search meant that when Drake fell upon his "fair and good bay," he was approaching from the north. On this basis alone, it is likely that Drake's lost harbor is the most northerly good bay lying "within 38 degrees toward the line": Bodega Bay.

In desperate need of a port – Sixty-three days out from the previous landfall, water and provisions must have been getting dangerously low and the ship was leaking. Under such conditions, Drake probably would have seized the first good opportunity that it "pleased God" to afford him. This again points to his entering and remaining in the most northerly bay within the thirty-eighth line: Bodega Bay.

In possession of a Spanish chart – A Spanish chart of the Manila route presumably showed Juan Rodríguez Cabrillo's Baia de los Pinos (Bay of Pines) at about thirty-nine degrees north latitude. With such a chart in hand, Drake would have been scouring the coast very carefully as he passed thirty-nine degrees, again increasing the probability that he fell upon the most northerly bay within the thirty-eighth line: Bodega Bay.[1]

In search of better weather – Because Drake's crew was "utterly discouraged" by the cold weather encountered during the search, it seems that one paramount objective in turning south was to find warmer environs as quickly as possible. Accordingly, with the wind at his stern, Drake probably sailed directly down the coast when he first set out, keeping a safe distance offshore. This explains why they came to anchor "within" thirty-eight degrees rather than farther north, and why the accounts

---

1. It is uncertain whether Drake's captured Spanish charts would have shown the results of Cabrillo's expedition though it seems likely; Cabrillo's was the only documented voyage of exploration that far north. The 1592 Molyneux Globe, presumably based on information brought back by Drake or Cavendish, has a "B. de Pinos" at thirty-nine degrees (figure 30).

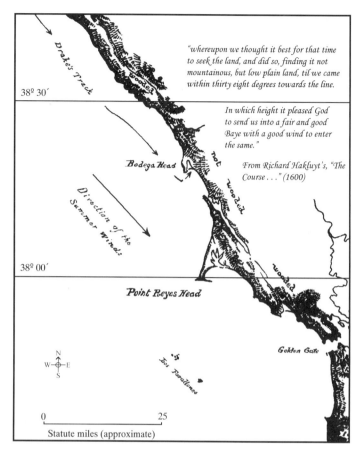

Map 12. Relief map of the San Francisco area showing the true course of Drake's approach from the north as he fell upon his fair and good bay. Based on the maritime record, the probability is high that Drake put in to Bodega Bay on June 17, 1579. Three out of the next six navigators to sail within sight of this section of the California coast, after Drake, ended up anchoring in Bodega Bay. Like Drake, they all approached from the north without the benefit of accurate charts, and under similar conditions of fog and wind. In all, eleven out of the next sixteen ships to coast these shores after the *Golden Hind* ended up at anchor in Bodega Bay. *Redrawn and adapted from Chart no. 1 in IDENTIFICATION OF SIR FRANCIS DRAKE'S ANCHORAGE ON THE COAST OF CALIFORNIA IN THE YEAR 1579 by George Davidson, California Historical Society, San Francisco, 1890.*

"whereupon we thought it best for that time to seek the land, and did so, finding it not mountainous, but low plain land, til we came within thirty eight degrees towards the line.

In which height it pleased God to send us into a fair and good Baye with a good wind to enter the same."

From Richard Hakluyt's, "The Course . . ." (1600)

do not mention the suitable bays they passed up along the coasts of Washington, Oregon, and Northern California. It also explains why Fletcher's descriptions of coastal terrain seem inaccurate, and why they relied on observations of coastal currents to provide negative evidence that the Straits of Anian existed: they obviously saw very little of the shorelines they passed.

The fourteen-day interval – In appendix E, I demonstrate that Drake's reported movements for the two weeks preceding a California anchorage at Bodega Bay correlate with the records of other early navigators.

Counter-Rebuttals

Too difficult to see? – Some have argued that sailing down the coast approaching the thirty-eighth line, Drake either would not have seen Bodega Bay, or would have passed it by. As Hanna put it:

> From a coasting position, Bodega Bay would have appeared as only a minor break in a long, low plateau, while the inner waterways offered by its lagoon, or by Tomales Bay, would have been screened from view by their headlands. The difficulty in perceiving either Bodega Lagoon or Tomales Bay from a coasting position is borne out by the fact that neither Cermeño nor Vizcaíno left any record of them in exploring that area.[2]

George Davidson argued that in sailing close by Bodega Head, Drake would have barely taken notice of the harbor at hand because "it is not a conspicuous place from the offing." He further reasoned that with "Point Reyes looming up and stretching well to seaward," Drake could have only arrived at one conclusion: "to steer for the western and highest part of Point Reyes and to seek shelter under its southern and leeward side."[3]

---

2. Hanna, *Lost Harbor*, p. 147. Hanna was reiterating the arguments that Raymond Aker made in *Report of Findings*, p. 264.
3. Davidson, *Identification*, p. 28.

Such arguments are not supported by the maritime record. Following Drake's 1579 California sojourn, three out of the next six ships that ventured into the area while it was still uncharted ended up at anchor in Bodega Bay. James Colnett, the next English commander to find himself in need of a port in this area, entered Bodega Bay in September 1790 under circumstances nearly identical to Drake's; so did Martin de Aguilar of the Vizcaíno expedition in January 1603; and so did Juan Francisco de la Bodega y Quadra of the Bruno de Hezeta expedition in November 1775. There is no indication that any of these commanders were distracted by Point Reyes Peninsula looming up to the southwest.[4]

For the three vessels that did pass by, the records suggest that the reasons for not entering Bodega Bay were other than the alleged difficulty in discerning the bay from a coasting position. In 1595, Sebastián Rodríguez Cermeño in the *San Agustin* narrowly missed entering Bodega Bay due to an extreme turn in the weather (see figure 129). When Sebastián Vizcaíno sailed the *Santo Tomas* by Bodega Bay in January 1603 in his desperate retreat to Mexico, the crew was so decimated by scurvy that "there were not more than six men aboard who were well and up."[5] In approaching San Francisco Bay in August 1775, Hezeta in the *Santiago* seems to have passed by Bodega Bay in the wee hours of the morning while out of sight of land; his accounts make no mention of either Point Reyes or Bodega Head.[6]

Too dangerous to enter? – Professor Davidson contended that even if Drake had seen Bodega Head, he would have declined to enter because it "was guarded by dangers," and had drawbacks "patent to all seamen." [7]

Davidson again neglected to check the maritime record. From Colnett's 1790 expedition to the turn of the nineteenth century, I have found records of only nine other expeditions that sailed within sight of the coast in the greater Point Reyes-Bodega Bay area. All but three of the ships involved in these expeditions entered and anchored in Bodega Bay. The ships that did enter included: George Vancouver in the *Discovery* in November 1792; William Brown in the *Jackal* in early 1793; Captain Sharp in the *Prince Lee Boo* in early 1793; Juan Matute in the *Sutil* in April 1793; Salvador Valdés in the *Aránzazu* in July 1793; Don Francisco Eliza in the *Activa* in August 1793; Juan Martinez y Zayas in the *Mexicana* in September 1793; and Peter Puget of Vancouver's second expedition in the *Chatham* in October 1793. The pertinent records of these voyages, where available, are devoid of complaints about the difficulty or danger of the approach to Bodega Bay. So are the descriptions in Davidson's *Pacific Coast Pilot* that reads as follows:

> The anchorage of Bodega Bay [off Doran Beach on the north side of the bay] is protected on the north, west, and east, and is open to the south-southeast. The southwest and west swell is partly broken by the low, rocky islet, called Bodega Rock . . . Between Bodega Rock and the Head there is a narrow four-and-a-half fathom passage opening directly upon the anchorage. In coming from the northwest in summer this channel is available . . . The best position for anchorage is in four fathoms of water with the southeast face of Bodega Head bearing southwest [about a half-mile off the bar at the mouth of the lagoon] . . . Small vessels may anchor in three fathoms halfway between the above position and the bar at the mouth of the lagoon . . . The bottom is hard and composed of coarse sand and patches of clay. . . . .During the heavy northwest weather in summer we have counted sixteen vessels anchored in this bay at one time, and as many as thirty to forty schooners and several steamers have been reported windblown there in one blow.[8]

---

4. I exclude here the first Don María Bucareli y Ursúa expedition in 1774 under Juan Perez, since he did not approach the coast in the Bodega area (H. H. Bancroft, *History of the Northwest Coast*, vol. 1, pp. 152–56). I also exclude the third Bucareli expedition in 1779 under Ygnacio Arteaga and Bodega y Quadra since Bodega y Quadra had already entered and charted Bodega Bay in 1775.
5. Fray Antonio de la Ascensión, from Wagner's translation of his account in *Spanish Voyages*, p. 252.
6. Fray Miguel de Campa, in Galvin, ed., *A Journal of Exploration*, p. 58.
7. Davidson, *Identification*, p. 28.
8. Davidson, *Pacific Coast Pilot* (1889), pp. 252–53.

CHAPTER 14
Tenet 2 – The Latitudes

Several early accounts of the Drake voyage reported conflicting latitude designations for Drake's landing site. How are these to be reconciled with the proposed landing site?

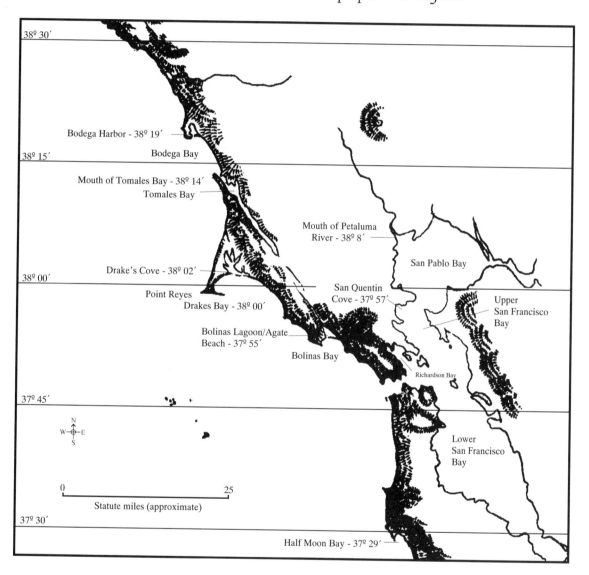

Map 13. Outline map of the San Francisco area showing the actual latitudes of the candidate anchorage sites from a modern atlas. Assuming "The Famous Voyage's" "within 38 degrees toward the line" is accurate to within thirty minutes, there are nine possible bays: Bodega, Tomales, San Pablo, Drakes, Bolinas, Richardson, San Quentin, San Francisco, and Half Moon. Assuming *The World Encompassed*'s, "38 deg. 30 min." is accurate to within fifteen minutes, however, there is only one possibility: Bodega Harbor. *Redrawn and adapted from Chart no. 1 in IDENTIFICATION OF SIR FRANCIS DRAKE'S ANCHORAGE ON THE COAST OF CALIFORNIA IN THE YEAR 1579 by George Davidson, California Historical Society, San Francisco, 1890.*

Clues

- They entered a "fair and good bay" when they "came within 38 degrees towards the line" (*F.V.*).
- At "38 deg. 30 min." Drake came to anchor in a "convenient and fit harbor" (*W.E.*, p. 64).
- They "stayed in the latitude of thirty-eight" (Stow).
- They "discovered land in forty-eight deg." (John Drake's second declaration).
- They "came to forty-four degrees . . . and then he found a harbor" ("Anonymous Narrative").
- The bay was "in north latitude 46 degrees" (Blunderville).

Answers

There is no conflict in latitude designations for the two principal accounts of the voyage or the Stow account. On the contrary, they confirm one another.

The latitudes reported in the other early contemporary accounts (John Drake, Blunderville and "Anonymous Narrative") are not credible. There is no good reason to believe that their authors had access to any of the official records of the voyage at the time the accounts were prepared. By all indications, they recounted the latitude of the California landfall from hazy memories, years after the fact.

Considering the accuracy of the latitude positions reported by Richard Hakluyt in "The Famous Voyage" 1589 edition, we should find Drake's "fair and good bay" between 37 degrees 30 minutes and 38 degrees 30 minutes. As shown in map 13, there are nine bays within this range.

Given the above, and then considering the accuracy of the latitude readings reported in *The World Encompassed*, we should find Drake's lost harbor lying somewhere between 38 degrees 15 minutes and 38 degrees 30 minutes. Bodega Bay on its north side contains the only suitable harbor located precisely within this range: Bodega Harbor at 38 degrees 19 minutes. The mouth of Tomales Bay, at 38 degrees 14 minutes, is close.

The latitude readings in the two principal accounts (*The World Encompassed* and "The Famous Voyage") are, by far, the most powerful pieces of evidence we have on the location of Drake's California landfall. Viewed objectively, they provide very compelling proof that Drake entered Bodega Bay and repaired his ship within Bodega Harbor.

Key Points

Every researcher who addressed the anchorage question prior to 1970 assumed that Drake's sixteenth-century navigational observations were patently too untrustworthy to pinpoint the anchorage location. As Raymond Aker discovered in the late 1960s, this assumption was incorrect:[1]

> In summary, the average error in latitude for places that can be reasonably pinpointed (thirty-six), excluding the doubtful latitudes of Valparaíso and Arica, is about sixteen minutes. Of these, the average error for latitudes that were probably obtained on shore is about nine minutes with nearly half of them being of less than ten minutes error. The average error of latitudes probably taken at sea is about twenty-one minutes.[2]

When Aker released these surprising findings in 1970, the Drake Navigators Guild was firmly entrenched in its Drakes Bay theory, Aubrey Neasham was fully committed to his Bolinas Lagoon theory, and there was no way Robert Power could be shaken from his staunch belief that Drake landed at San Quentin Cove. They were not looking at the matter objectively.[3]

---

1. Corbett (*Drake and the Tudor Navy*, vol. 1, pp. 271–72 [footnote]) was the first to comment positively on the accuracy of Drake's measurements. Riesenberg (*Cape Horn*, 1939, p. 80) also noted "remarkable evidence of the excellence of his [Drake's] observations, his instruments, and tables."
2. Aker, *Report of Findings*, pp. 436 and 449–50. In his recent (1997) article, "Francis Drake at Cape Horn," p. 12, Aker rebuts Henry Wagner's theory that Drake landed at Henderson Island (55º 39′ S) as opposed to Horn Island (56º S) with the following statements relative to the accuracy of *The World Encompassed* latitudes: "the very accurate latitude of 56º South for Drake's cape was recorded in *The World Encompassed*"; and "21′ is a large error for a latitude obtained on shore, as Drake's undoubtedly was."

My own analysis of *The World Encompassed*'s latitude data confirmed Aker's findings (see appendix D). Of the fifty latitudes provided in the account, I found twenty-seven that identify locations that can be reasonably pinpointed without inciting the wrath of most current Drake scholars. Of these, eleven of the associated measurements were likely made on land and the other sixteen aboard ship. By reference to a modern atlas, all eleven of the likely land-based readings appear to be accurate to within plus or minus fifteen minutes with a standard deviation of eleven and a half minutes. For the sixteen readings likely made aboard ship, fourteen appear to be accurate to within plus or minus thirty minutes with a standard deviation of nineteen minutes. The two other shipboard readings (Arica and Valparaíso) were off by several degrees, but the discrepancies appear to have resulted from transcription errors, not faulty navigation. Based on this simple statistical analysis, assuming "38 deg. 30 min." on page 64 of *The World Encompassed* came from a latitude measurement that was made on land, as seems highly likely, there is better than a 95 percent probability that Drake made port either at Bodega Harbor or Tomales Bay. Conversely, the probability is less than 5 percent he made port at Drakes Estero or points south.[4]

Counter-Rebuttals

Reliability of Sources – Proponents of other anchorage locales have argued that the latitudes provided in *The World Encompassed* are not credible or reliable, specifically pointing to the positions the account reports for Mucho Island and the landfalls Drake made farthest south (fifty-six degrees) and north (forty-eight degrees) on the Pacific Coast. Because Drake's nephew published the book almost fifty years after the voyage, they contend there is significant doubt as to the legitimacy of his source materials.[5]

I refer the reader to part II of this book where I discuss the contemporary accounts and their likely sources, and appendices D and E where I address the matter of Mucho Island and Drake's farthest points north and south. Here I will simply point out that there is every indication that Drake himself provided the navigational information in *The World Encompassed*. Aker came to the same conclusion in 1970: "it is probable that most [latitudes] given in *The World Encompassed* were his [Drake's]."[6]

Accuracy of "The Famous Voyage" latitudes – Proponents of Drakes Bay have argued that we should accept Hakluyt's California latitudes because he was "a man of education, a man in holy order, a collector and publisher of old voyages and travels, an enthusiast and withal he lived in London in the days of Drake's exploits."[7]

Such arguments are contrived. When Hakluyt originally published "The Famous Voyage" sometime between 1589 and 1596, he provided only seven latitudes. It appears that he used them to help identify either major turning points in the voyage or important geographic landmarks, not to pinpoint the location of landfalls. They are typically accurate only to the nearest degree. When Hakluyt published a slightly revised version of the account in 1600, he provided six additional latitudes in the margins that do serve to pinpoint landfalls along the coast of South America. These additional latitude readings, however, do not correlate with latitudes given in *The World Encompassed* or any of the other early accounts. It is likely that Hakluyt obtained them from contemporary maps such as the so-called Twelfth-Night Map that he provided with his publication, rather than a journal or informant.[8]

---

3. As noted by Morison (*Northern Voyages*, p. 512), Drake's fellow sixteenth-century mariner, Martin Frobisher, was making equally accurate latitude readings in the late 1570s.
4. The latitudes of Bodega Harbor and Tomales Bay are well within two standard deviations of the datum, while the latitude of Drakes Estero is barely within three.
5. For example, Wagner, *Drake's Voyage*, pp. 161, 286–87; and Aker, *Summation*, p. 11.
6. Aker, *Report of Findings*, p. 436.
7. Davidson, *Francis Drake on the Northwest Coast*, p. 2; also Aker, *Summation*, p. 11.
8. Hakluyt's "The Famous Voyage," 1589 edition, provides latitudes for the following turning points: (1) the first landfall on the coast of Brazil; (2) the mouth of the then legendary River of Plate; (3) the farthest point south; (4) the farthest point north; (5) the farthest point south on the coast of Nova Albion; and (6) the first landfall after crossing the Pacific.

How can anyone reasonably argue that "within 38 degrees toward the line" denotes anything more than "approximately 38 degrees"? In the Richard Madox diary of 1582, we find Fletcher's counterpart on the Edward Fenton expedition using almost the very same language in a situation where the mariners had approximated their latitude position by dead reckoning: ". . . we were *within four degrees of the line* by account of the ships, for the sun nor star had not been taken in ten days before" (*emphasis added*).[9]

Fig. 79. A squalid, barren and deformed shore with white cliffs and banks that lie toward the sea: aerial view of the coastal terrain looking south over Bodega Harbor, Tomales Bay, and the Point Reyes Peninsula. From a Bodega Head vantage point, Point Reyes (*upper right*) forms Bodega Bay's southern headland and the white cliffs lining the north shore of Drakes Bay and the south tip of Bodega Head do indeed lie towards the sea. *Photograph by William Garnett, 1959. Courtesy of the California Academy of Sciences, San Francisco, California.*

9. Donno, ed., *An Elizabethan in 1582*, p. 292.

CHAPTER 15
Tenet 3 – White Cliffs and Coastal Terrain

Drake discovered "white banks and cliffs" which lie toward the sea rimming what is today called Drakes Bay on the south side of Point Reyes. They reminded him of Old England (Albion) and inspired the name Nova Albion. What course did Drake follow after sighting this remarkable coastal formation?

Let us rephrase this tenet to read as follows: Explain how Fletcher's description of the coastal landscape helps identify Drake's California landfall.

Fig. 80. White cliffs that lie towards the sea opposite Drake's former encampment. They can be found on the southwest side of Bodega Head directly opposite Campbell Cove. *Photograph by the author, 1994.*

Clues

- Along with the reference to "white cliffs and banks which lie towards the sea" (*F.V.* and *W.E.*, p. 80), the two principal accounts give a few other important details on what Drake and his companions saw of the shores of Nova Albion. Here is all the information they provide for a shoreline that, for the most part, is dominated by lofty, tree-covered mountains:

- They found the land "not mountainous," but "low plain land" (*F.V.*).

- "We found the land by coasting along it to be but low and reasonable plain: every hill (of which we saw many but none very high) though it were in June . . . being covered with snow" (*W.E.*, p. 64).

- "How unhandsome and deformed seemed the face of the earth" (*W.E.*, p. 65).

- "Hence comes the general squalidness and barrenness of the country" (*W.E.*, p. 66).

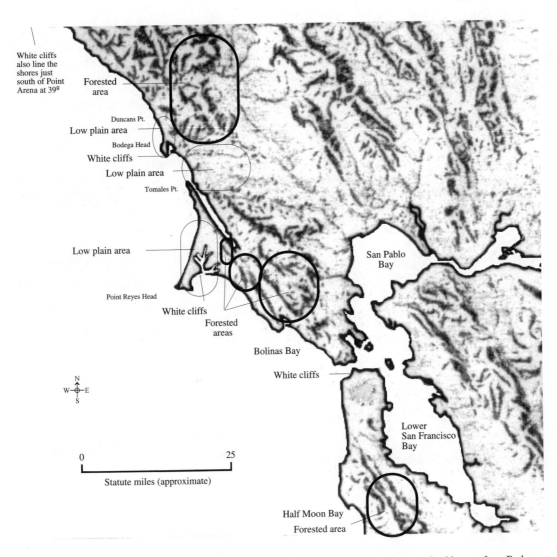

Map 14. Relief map of the greater San Francisco Bay area showing coastal terrain features. Looking out from Bodega Head, all one can see of the coastal terrain in any direction are barren, heavily scarred, wind-swept plains with rolling hills. Such can not be said for any of the other candidate anchorage sites in the San Francisco area. There is no shortage of white cliffs that lie toward the sea from a Bodega Head vantage point.

Answer

The descriptions of coastal terrain provided in "The Famous Voyage" and *The World Encompassed*, taken in their entirety, exactly fit the geographic features of the greater Bodega Bay area and no other bay or harbor in the vicinity of thirty-eight degrees.

Key Points

Because the descriptions "not mountainous," "low plain land," "barren," "squalid," and "deformed" do not apply to the Pacific Northwest's coastal terrain in general, they must apply to the California anchorage locale. There, Fletcher would have observed the same vista day after day for over a month. In addition, as discussed for Tenet 1, there are good reasons to believe that Drake and his company saw very little of the shoreline as they retreated south to escape the inclement weather. With these considerations in mind, examine map 14 which shows the coastal region just north of San Francisco including the topographic relief, forested zones, and white cliff zones.

White Cliffs – There are three sets of conspicuous white cliffs that "lie toward the sea" in the greater Bodega Bay area: (1) on the south side of Point Arena (figure 81); (2) on the ocean side of Bodega Head (figure 80); and (3) lining the shores of Drakes Bay on the south side of Point Reyes Peninsula (figure 44).

This is how George Vancouver, in September 1792, described Point Arena's white cliffs that are located within a day's sail of Bodega Bay: "low sandy or clayey cliffs, remarkably white, though interspersed with streaks of a dull green color."[1]

Fig. 81. The white cliffs of Point Arena. These white cliffs and banks made a vivid impression on George Vancouver when he sighted them in September 1792. It is very likely that Drake also saw them the same day he entered the "fair and good bay." *From the U.S. Coast Survey's Reconnaissance of the Western Coast of the United States, Middle Sheet, 1854.*

The white cliffs on Bodega Head "lie toward the sea" directly opposite Campbell Cove (figure 80). To find them, from the town of Bodega Bay take Bay Flat Road to Westshore Road and drive south to Campbell Cove. Drive up the bluff, park, and then make your way on foot to the most likely spot that Drake would have posted lookout to watch for enemy sails. Though none of the other anchorage sleuths ever appear to have noticed them, it may well have been these white cliffs that came to remind Drake, in his reveries, of Old England. He probably spent many hours at this vantage point scouring the horizon and reflecting on adventures past and to come.

As illustrated in figure 78, the white cliffs of Drakes Bay are very aptly described as "lying toward the sea" from a Bodega Head vantage point. As pointed out for Tenet 4, Point Reyes from the sixteenth-century seaman's viewpoint actually forms the southern headland for Bodega Bay. Given a California landfall at Bodega Bay, it is probable that Drake would have seen Drakes Bay and its

---

1. Vancouver, *Voyage of Discovery*, p. 699.

conspicuous white cliffs on the day of his departure. He also could have seen them by making an excursion in the little bark he had captured off Panama (Tello's bark), perhaps in the course of the explorations made during the first three days of arrival. A careful reconnaissance of the surrounding coastal area would be most prudent for a mariner in Drake's precarious situation, never mind the inquisitive and adventurous nature that commanded him to trek inland (Tenet 9) and to pursue so many grand and reckless schemes throughout his life. Indeed, *The World Encompassed* describes or alludes to reconnaissance expeditions of one sort or another for his stops at Cape Horn (*W.E.* p. 45), Philips Bay, (*W.E.* p. 49), and La Salada Bay (*W.E.* p. 54).

Low and reasonable plain with many hills – This description precisely fits the shoreline between Duncans Point just north of Bodega Head and Tomales Point to its south. From map 14, one can see how California's coastal mountain range is interrupted in this region. Several Drakes Bay advocates, among others, have provided some pertinent descriptions of how this stretch of the coast looks from Bodega Head or when sailing along the shoreline:

Fedor Lütke, 1818: From the inner part, as much as from the sea, Cape Rumiantsev [Bodega Head] differs markedly from the surrounding coastline. . . . The opposite shore rises in a moderate slope to low hills, beyond which lies a vast tundra.[2]

George Davidson, 1889: On account of the general depression of the hills behind Bodega Bay and to the northward some distance, as well as over Tomales Point and through Tomales Bay, the summer winds from the west-northwest draw in toward Petaluma Valley with great force.[3]

Raymond Aker, 1971: Bodega Head . . . appeared as only a minor break in a long, low plateau, . . . Bodega Bay is situated in the midst of moorlike country.[4]

Samuel Morison, 1974: From Bodega Head . . . The neighboring country resembles the downs along the English Channel.[5]

Squalid, barren, deformed – Look at the aerial views in figures 79 and 82. What an apt description for a partially submerged coastal valley rent from the earth by California's great San Andreas fault!

Snow-clad hills nearby – Sebastián Vizcaíno in January 1603 and George Vancouver in November 1792 both reported seeing snow-covered hills in the area between Cape Mendocino and Point Arena, about a day's sail north of Bodega Head.[6]

According to Vancouver:

As we proceeded, a distant view was obtained of the inland country, which was composed of very lofty rugged mountains extending in a ridge nearly parallel to the direction of the coast. These were in general destitute of wood, and the more elevated parts were covered with perpetual snow.[7]

As noted by Alexander McAdie, these early navigators were probably fooled by very dense low-lying fogs.[8]

2. Lütke, "Observations on California," in Dmytryshyn et al., eds., *Russian American Colonies*, p. 280.
3. Davidson, *Pacific Coast Pilot* (1889), p. 254.
4. Aker, *Report of Findings*, p. 264. Moors, according to Webster, are defined as tracts of "open rolling wasteland [hilly plains], usually covered with heather and often marshy."
5. Morison, *Southern Voyages*, p. 677.
6. Vizcaíno's observation of snow-covered hills can be found in Wagner's translation of the Father Antonio de la Ascensión account, *Spanish Voyages*, p. 252.
7. Vancouver, *Voyage of Discovery*, p. 698.
8. McAdie, "Nova Albion–1579," p. 194. Other early navigators reported seeing snow along the coasts of Washington and Oregon. Juan Perez in the first Bucareli expedition (1774), and Bruno de Hezeta and Bodega y Quadra in the second (1775), both made note of distant snow-clad hills (the Olympus range) when they approached shore at forty-eight degrees. James Cook was just north of Umpqua River on March 11, 1778, when he observed what appeared to be snow on the Oregon coast: "the bare grounds toward the coast were all covered with snow, which seemed to be of considerable depth between the little hills and rising grounds." Vancouver, in coasting the same area a few years later, discovered that Cook had been fooled by sand dunes comprised of extremely white sand (cited by Aker, *Report of Findings*, pp. 253–56).

CHAPTER 16
Tenet 4 – The Good Bay

Identify the "fair and good bay" mentioned in Richard Hakluyt's accounts and the "convenient and fit harbor" described in "The World Encompassed."

Fig. 82. The fair and good bay. *Photographed by the U.S. Army Map Service, August 1954. Courtesy of the University of California, Berkeley, Map Room.*

Clues and Considerations

- Where they first came to anchor on June 17, 1579, was "a great way from shore" but within earshot of the Indians standing upon it (*W.E.*, p. 67). For the voices to carry so far, they had to have been anchored in a very calm area downwind (southeast) of the shore.

- Where they first came to anchor, the sea was calm enough so that an Indian in a grass canoe could steady himself "moving his hands, turning his head and body in many ways" (*W.E.*, p. 67). This again indicates that they were behind a headland that well protected them from the northwest winds.

- Drake spent three full days at the outer anchorage (June 18, 19 and 20) before bringing the *Golden Hind* closer to shore on the twenty-first. During this time he and his men were exploring the area in the ship's boat (*W.E.*, p. 68). This again suggests that the initial anchorage was well sheltered and also suggests that there were many potential careening sites to be evaluated.

- To approach the inner harbor on June 21, the ship was simply "brought to anchor nearer the shore" (*W.E.*, p. 68). This implies that the inner anchorage was on a direct line of sight from the outer anchorage.

- As they departed on July 23, they observed the Indians running to the tops of the hills "making fires before and behind, and on each side of them" (*W.E.*, p. 81). This strongly suggests that they had doubled a hilly headland. This evidence is also set forth in the Portus Nova Albionis inset; the caption talks of Indians offering sacrifices in the hills, while the drawing depicts the scene.

- The very paucity of description in *The World Encompassed* suggests the locale was rather nondescript.

Answers

Fletcher's "fair and good bay" is Bodega Bay as it was observed and charted by the second English sea captain to enter it after Drake: George Vancouver in November 1792. Broadly considered, it is the shallow bay formed by Mussel Point at thirty-eight degrees twenty minutes on the north, and by the southwestern extremity of Point Reyes Head at thirty-eight degrees zero minutes on the south (map 13 and figure 79). More narrowly defined, it is the bay formed by Bodega Head at thirty-eight degrees nineteen minutes on the north and Tomales Point at thirty-eight degrees fourteen minutes on the south (map 13 and figure 82).[1]

Drake's "convenient and fit harbor" is "Port Sir Francis Drake" as it was observed, charted, and aptly named by the first English sea captain to enter it after Drake: James Colnett in September 1790. It includes a well-protected anchorage off Doran Beach at the northern end of Bodega Bay and the small cove on the lee side of Bodega Head, just outside the mouth of Bodega Harbor. When the Spanish attempted to establish a presidio in this cove in 1793 they called it Port Bodega. When the Russians occupied the area and constructed harbor facilities there in 1812 they named it Port Rumiantsev. On today's charts, it is called Campbell Cove (figure 83).

Key Points

A good bay and fit harbor – According to sixteenth-century usage, the term "bay" indicates an open inlet located between two headlands. The term "harbor" suggests a well-sheltered inlet. The impression from the accounts is that the initial anchorage was at a location that satisfies the meaning of both terms. The terms "bay" and "harbor" apply equally well for the anchorage off Doran Beach on the north side of Bodega Bay.[2]

---

1. The following excerpt from George Vancouver's account demonstrates that from the standpoint of early English navigators, Point Reyes at thirty-eight degrees forms greater Bodega Bay's southern terminus: "From the south point of Port Bodega (Tomales Point), which is formed by steep rocky cliffs with some detached rocks lying near it, the coast makes a shallow open bay, which is bounded by a low sandy beach [Point Reyes Beach] toward the southeast part of which the elevated land of Point de los Reyes again commences, and stretches like a peninsula to the southward into the ocean." Vancouver considered the bay lying between Bodega Head and Tomales Point (today's Bodega Bay) an "inner bay that seemed to be divided into two or three arms" (Vancouver, *Voyage of Discovery*, p. 699 –700).

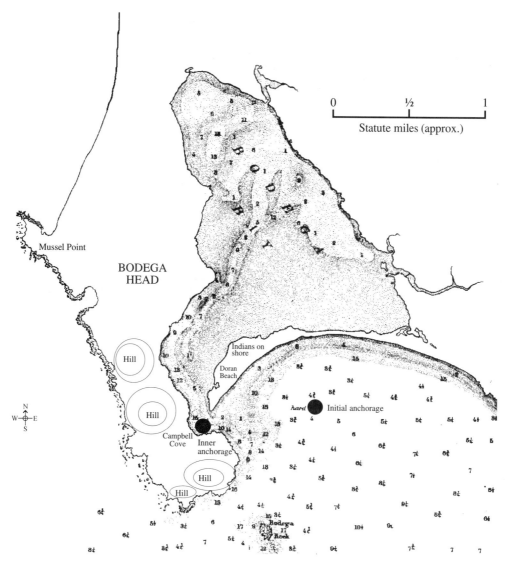

Fig. 83. The convenient and fit harbor (chart detail). I am showing Drake's initial anchorage at the site George Davidson recommended in the *Pacific Coast Pilot* (1889): approximately a half mile off Doran Beach in about four fathoms of water (twenty-five feet). *Adapted from an 1862 U.S. Coast Survey chart of Bodega Bay.*

A well-protected initial anchorage within earshot from shore – The best anchorage in Bodega Bay is off Doran Beach on the lee side of Bodega Head in about twenty-five feet of water.[3] There is a hard sandy bottom and excellent protection from the northwest wind. Any Indians on Doran Beach would have been not only a great way from the ship, but with a northwest wind, well within earshot. This anchorage is among the best havens on the California coast during the summer season.

The inner anchorage on a direct line of sight from the outer – Based on the earliest descriptions and charts, Campbell Cove is the only conceivable inner port Drake could have used in Bodega Harbor, and this cove is actually located just outside the mouth of the harbor, not in it. John Dwinelle and Henry Wagner raised arguments that Bodega Harbor in Drake's time was wide open at the entrance and much deeper throughout. They arrived at this conclusion, however, by mistakenly

---

2. Definitions from Sir Henry Mainwaring's *Seaman's Dictionary* (circa 1620) in Aker, *Report of Findings*, p. 387.
3. Davidson, *Pacific Coast Pilot* (1889), p. 253.

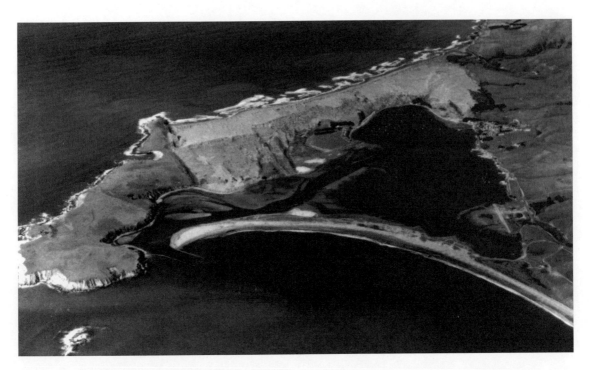

Fig. 84. The convenient and fit harbor (aerial view). For the summer season, the anchorage off Doran Beach on the northern end of Bodega Bay has historically been considered one of the best havens on the California coast. *Photograph © by Ed Brady, February 1959. Courtesy of Grethe Brady, Aero Photographers, Sausalito, California.*

comparing the harbor to Bodega y Quadra's 1775 plan of the mouth of Tomales Bay, the original Port Bodega of Spanish California. Campbell Cove would have been in direct line of sight from the initial anchorage, completely satisfying the textual association that Drake simply brought the *Golden Hind* closer to shore (figure 83).[4]

The three-day delay – Drake could easily have spent three days exploring the ins and outs of Bodega Harbor and the many options offered by Tomales Bay. If he noticed Point Reyes Peninsula looming to the south at his arrival, Drake may have been curious enough to explore Drakes Bay in Tello's bark. Had he done so, he might have been impressed by its white cliffs, but certainly not its harbors which offered no advantages over those at Bodega Bay. Assuming Drake did enter Bodega Harbor, the eight-inch rise in the high-water mark over this three-day period would have been highly significant. As discussed in the next chapter, with the *Golden Hind* fully laden, it was going to be difficult getting her over the barred entrance.

The hilly headland and Indian fires – Leaving Campbell Cove, Drake and his company could have seen Indians lighting fires on all four sides of the hills on Bodega Head, just as depicted on the Portus plan (see chapter 28).

Nondescript – Here is how George Davidson described Bodega Bay: "There is nothing remarkable about it";[5] and Raymond Aker: "Bodega Bay . . . is not impressive."[6]

---

4. Munro-Fraser, *History of Sonoma County,* p. 189, also reported that the harbor was deeper when the Americans first arrived, and that it had silted up because of the settlers' agricultural use of the head. His observation is refuted by the Russian accounts.
5. Davidson, *Identification,* p. 27.
6. Aker, *Report of Findings,* p. 264.

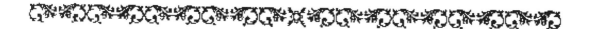

CHAPTER 17
Tenet 5 – The Inner Harbor

"The World Encompassed" reported that Drake's ship, *"having received a leak at sea,"* needed to be brought to shore and unloaded for repair. Discuss the suitability of the proposed landing site for careening and graving.

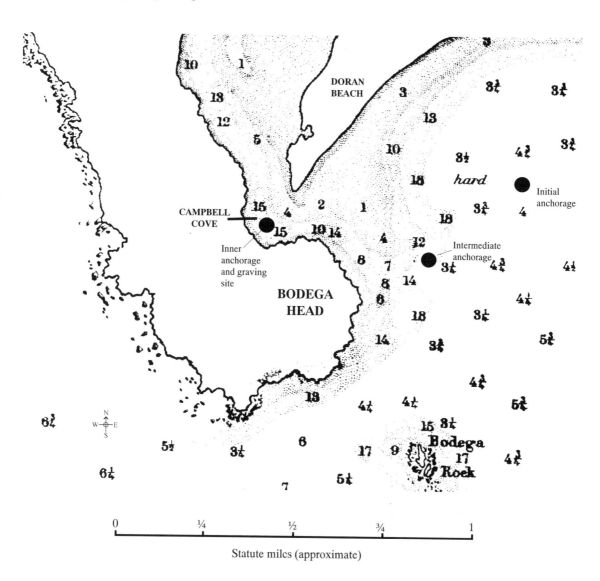

Fig. 85. The inner harbor. There was eight feet of water over the bar at mean low-water when George Davidson's group sounded Campbell Cove's entrance circa 1860. They observed a mean rise and fall of about four feet. Under those conditions, at high water, Drake could have brought the *Golden Hind* into Campbell Cove fully laden. *Redrawn from an 1862 U.S. Coast Survey chart of Bodega Bay.*

Clues

- They brought the ship in to "grave and trim" (Stow).
- Drake "grounded his ship to trim her" ("Anonymous Narrative").
- They "graved and breamed" the ship (Madox).
- Here "he caulked his large ship" (John Drake's second declaration).
- Drake entered the inner port on June 21, 1579, and departed on July 23 (*W.E.*, p. 64).
- The *Golden Hind* was about 100 tons and drew 13 feet fully laden (*W.E.*, pp. 2 and 102).

Answers

Campbell Cove is not only a fit location for Drake to have "graved" his ship, it is the only candidate anchorage location in the area where he likely could have graved her except, hypothetically, for the mouth of Tomales Bay and the Guild's estero site, though they are now completely shoal.[1]

There is no evidence in the accounts that Drake "careened" the *Golden Hind* at his California port.

Counter-Rebuttals and Key Points

Graving versus careening – According to sixteenth-century usage, "graved" means the ship was literally hauled out of the water and propped upright or leaning, leaving her high and dry at low tide and still grounded at high tide.[2] To do this requires near ideal conditions that are somewhat unusual for a coastal harbor. There must be little, and preferably no, surf. There must be a quick transition from navigable channel to hard sand or gravel beach. The shoreline cannot be comprised of sticky muck. In "careening," the vessel is still afloat, at least at high tide. The ship is simply keeled over and propped up in calm, shallow water exposing her side and bottom. It is done when the vessel cannot be dragged close enough to shore to grave her; for example, when the shoreline slopes gradually or when the bottom is muddy rather than hard.

At Campbell Cove, the conditions are perfect for graving a ship: it is dead calm and there is a very quick transition from navigable channel to hard sandy beach. Aker correctly argued that Drake would have found Campbell Cove unsuitable as a careening basin because "the cove itself forms a bend in the tidal channel and is exposed to the tidal current."[3] It is because of these very conditions that Drake was able to grave his ship there. At all the other candidate sites except Tomales Bay, the conditions are unsuitable for graving a ship; Drake would have had to careen the *Golden Hind* at all of them.

Could Drake have entered Campbell Cove? – The Guild members did not believe so. According to Aker, "the depths of water over the bar as reported by later explorers, and in recent times, show it unlikely that he would have found sufficient water for the *Golden Hind* to enter."[4]

The Guild members were mistaken. The Russians had a port facility at Campbell Cove for twenty-nine years between 1812 and 1841. Here is their objective assessment of its merits (*emphasis added*):

> Little Bodega Bay, or as our settlers call it, Rumiantsev Bay [Bodega Harbor]. . . has an entry no more than forty sazhens [280 feet] wide and is a fairly large bay about ten miles in circumference. On the

---

1. Based on the Juan Martinez y Zayas chart of 1793 (figure 134), I suspect that Drake would have found conditions suitable for graving the *Golden Hind* on today's Brazil Beach in the cove inside Sand Point at the mouth of Tomales Bay. According to Aker, Drake could ostensibly have grounded the *Golden Hind* on the sand banks of Drake's Cove as an alternative to careening. He feels he would not have elected to, however, in that it would not have totally exposed the keel or bottom (Aker, in Holliday, "The Francis Drake Controversy," p. 218).
2. Definitions for graving and careening from Sir Henry Mainwaring's *Seaman's Dictionary* (circa 1620) in Aker, *Report of Findings*, pp. 388–89.
3. Aker, *Report of Findings*, p. 370.
4. Ibid., p. 370.

shore there is an Indian settlement. . . . The bay [Bodega Harbor] would be the most wonderful harbor in the world if ships could enter it, but the depth permits only small rowing vessels to enter. A large part of it beginning from the spit and extending to the middle is dry during low tide. Thus it is of no use for seafarers. From the end of the spit a sandbar extends southeast almost parallel to the shore. *The shore itself at this point has a rather deep area [Campbell Cove] and thus makes it possible to have a small but rather good anchorage where two or three vessels can anchor safely.* But there are many disadvantages. The entry into it between the end of the sandbar and the shore of Cape Rumiantsev is narrow and winding, and furthermore the depth in this channel during low tide is no more than four feet, and at high tide between ten and eleven feet. *At full and change [of the moon, i.e., the highest of high tides; i.e. spring tides] it rises to fourteen or fifteen feet.* Thus this harbor can be used only by small vessels [under 200 tons], but in order to enter the harbor, *even small vessels must choose a time when the water is high and ideal circumstances prevail.*[5]

These navigational conditions that Fedor Lütke described in September 1818 were applicable when Davidson's group first charted Bodega Harbor in the 1860s (figure 85) and were presumably applicable in Drake's day. As such, Drake would have found Campbell Cove a fit location for his purposes if not altogether convenient. He would have found it advantageous to enter and leave on the ebb and flow of the highest high tides (spring tides). Thanks to the work of the Drake Navigators Guild, we are able to confirm that this is exactly what he did.

On behalf of the Guild, the United States Coast & Geodetic Survey (USGS) prepared a tide table for the months of June, July and August 1579 (figure 86) that, for our purposes, is sufficiently applicable to both Drakes Bay and Bodega Bay. According to their report, when Drake arrived on June 17, the tide in Drakes Bay was at its maximum high at 6:00 P.M. and had a rise of 5.4 feet. Four days later when he took the *Golden Hind* in to shore, the rise was 6.2 feet at 8:00 P.M.

The data for the July 23 departure is even more revealing. The July spring tide (6.0 feet) occurred on the twenty-second, the day before Drake set off for the Farallones. This clearly shows that Drake was "be-neaped"; that he had to wait for the maximum tidal rise to cross a barred entrance.[6]

The 100-ton *Golden Hind* drew about thirteen feet of water fully laden and about seven to eight feet when unladen.[7] To dispel any doubts as to whether or not Drake could have entered and repaired

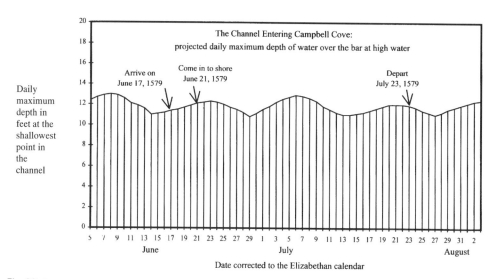

Fig. 86. Bar graph showing approximate daily maximum depths of water over the barred entrance to Bodega Harbor in June–July 1579 (based on a mean low-water depth of six feet). It is evident that Drake timed the entrance and exit from his California port to take advantage of the lunar cycle. *Derived from tidal data prepared in 1953 by the U.S.*

5. Lütke, "Observations on California," in Dmytryshyn et al., eds., *Russian American Colonies*, p. 279.
6. Definition from Sir Henry Mainwaring's *Seaman's Dictionary* (circa 1620) in Aker, *Report of Findings*, p. 393.
7. Aker, in Holliday, "The Francis Drake Controversy," p. 218.

his ship in Bodega Harbor in June 1579, here is what the record shows for vessels that were brought into Campbell Cove for refitting:

Sutil, 46 tons, Spanish, under Commander Juan Matute: April–August 1793 (Wagner, "Last Spanish Expedition," pp. 339–41).

Peacock, 108 tons, American, under Captain Oliver Kimball: March–May 1807 (Ogden, p. 161; and Baranov, 1808, pp. 169–71).

Kodiak, probably 100–200 tons, Russian, under Commander Ivan Kuskov: December 1808 – August 1809 (Ogden, p. 161).

Isabella, 209 tons, American, under Captain William Davis: September–October 1810 (Ogden, p. 163; and Moraga, p. 277).

Chirikov, probably 100–200 tons, Russian, under Commander Ivan Kuskov: February–June 1811 (Ogden, p. 163).

Rumiantsev, 80–160 tons, Russian, under Commander Ivan Kuskov: 1818 (Lütke, p. 281; and Khlebnikov, *Reports*, p. 116).

Buldakov, 200 tons, Russian, under Supercargo Kirill Khlebnikov: June–July 1820 (Khlebnikov, *Travel Notes*, p. 67; and *Reports*, p. 116).

Volga, 140 tons, Russian, under Supercargo Kirill Khlebnikov: December 11–13, 1822 (Khlebnikov, *Travel Notes*, pp. 124–25; and *Reports*, p. 117).

Baikal, 230 tons, Russian, under Supercargo Kirill Khlebnikov: June 3–27, 1824 (Khlebnikov, *Travel Notes*, pp. 130 and 152; and *Reports*, p. 78).

Kiakhta, 180 tons, Russian, under Supercargo Kirill Khlebnikov: August 10–20, 1824 (Khlebnikov, *Travel Notes*, p. 175; and *Reports*, p. 117).

Rover, 83 tons, Mexican, under Captain John Cooper: September–November 1824 (Khlebnikov, *Travel Notes*, pp. 177 and 199; and Ogden, p. 171).[8]

In the entry for the day of his arrival at Stephen Smith's Port Bodega in February 1851, Stephen Fowler reports "a large schooner" lying on shore in front of the two Russian buildings.[9]

Would Drake have graved the "Golden Hind" at Campbell Cove? According to Raymond Aker, Drake would not have elected to ground the *Golden Hind* at his California port because the tide's average rise and fall along this part of the Pacific Coast is only about six feet, and he needed to expose the ship's bottom for repairs.[10] Considering that when completely unloaded the *Golden Hind* probably drew only about seven or eight feet of water, I fail to see the problem. Assuming the English grounded the ship within a few days of bringing her closer to shore, there would have been a spring tide with a maximum rise and fall of seven feet. The English would have "graved and breamed" the *Golden Hind* by laying the lightened ship parallel to shore, pulling her aground at high tide, and then heeling her either to landward or seaward depending on the strength of her hull.[11]

---

8. Cooper was a Boston man that married into the Vallejo family. I have seen a reference to the *Rover*'s journal (in Ogden, *Sea Otter Trade*, p. 225) but have not yet found it.
9. Fowler, journal of, p. 160.
10. Raymond Aker, conversation with author, Palo Alto, California, January 1997.
11. Definition from Sir Henry Mainwaring's *Seaman's Dictionary* (circa 1620) in Aker, *Report of Findings*, p. 390

CHAPTER 18
Tenet 6 – The Victuals

Conscious that his crew would need to supplement their meager supplies with food and water from the land while the "Golden Hind" was undergoing repair, Drake chose to camp at a site at which such supplies were readily accessible. Discuss the suitability of the proposed site for watering and victualing [and explain how the victualing clues help identify the anchorage location].

Clues

- They observed "each woman bearing against her breast a round basket or two, having within them divers things, as . . . broiled fishes, like a pilchard [sardine]" (*W.E.*, p. 74).

- There was "one thing we observed with admiration, that if at any time they chanced to see a fish so near the shore that they might reach the place without swimming, they would never, or very seldom, miss to take it" (*W.E.*, p. 79).

- The Englishmen provided themselves with "mussels, seals and such-like" (*W.E.*, p. 79).

- The victuals they found were "mussels and sea lions" (John Drake's second declaration).

Fig. 87. "Sketches of sea lions": drawing by Louis Choris at San Francisco Bay, 1816. *Courtesy of the Oakland Museum, Oakland, California.*

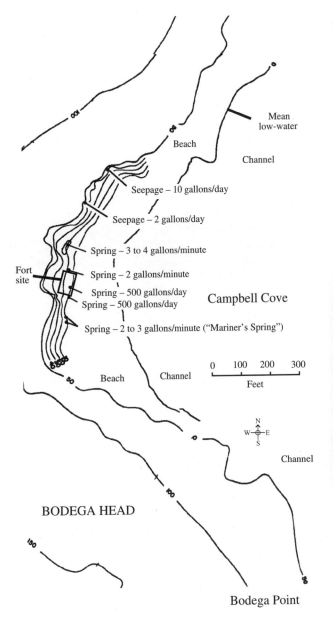

Fig. 88. The gurgling springs at Drake's former campsite (*left*). Drake would have selected the placement of his fort and outer defense works to take advantage of these springs. When the Russians arrived in 1811, they built a bathhouse at one of the springs. They also built a warehouse that fronted the buried walls of Drake's fort. *Adapted from a drawing prepared by Dames & Moore, San Francisco, for Pacific Gas & Electric Company, circa 1962, contained in the Public Utility Commission files for Application Number 43808.*

Fig. 89. "Mariner's Spring." *Photograph © by Karl Kortum, October 1962. Courtesy of Jean Kortum, San Francisco, California.*

Answers

Watering – Campbell Cove has excellent fresh water springs that drew praise from many of its earliest European and Russian visitors. Here is how the Russian sea captain Fedor Lütke, who observed the cove in September 1818, described this supply:

> At the present time a seafarer can find nothing in the Rumiantsev port but water, which is pure, delicious, and empties into the interior harbor from a small creek which flows from the hills. It is very easy to obtain water. All one needs is to build a trough from the water to a cask, and wait until it fills up.[1]

---

1. Lütke, "Observations on California," in Dmytryshyn et al., eds., *Russian American Colonies*, p. 280.

Victualing – Before the arrival of its restaurateurs, Bodega Harbor, unlike the other candidate sites, was a poor place for victualing. As Captain Lütke aptly pointed out in his 1818 journal, the Port Rumiantsev (Bodega Miwok) Indians (like Drake) had to rely on "whatever the sea provides."[2] With the exception of birds, there was virtually no worthwhile game in the vicinity, large or small. It was these demerits as a victualing site that provide surprisingly strong evidence that Drake and his men stayed at Bodega Harbor rather than at some other harbor in the area (see Tenet 6 rebuttals in chapter 33).

Key Points

Pilchards – We know from Colnett's and other early accounts that the Indians living within the Bodega Harbor environs were adept at snatching sardines from the surf with small nets (see Tenet 12). Thanks to the talented Russian painter Mikhail Tikhanov (1817), we can see exactly what Fletcher was referring to when he described the Indian women with their baskets of fish (figure 6).

Mussels and such-like – Bodega Harbor, even in Indian days, was known for its shellfish. Mussels are found in abundance all along Bodega Head's seaward shores, and in particular at Mussel Point. The mud flats within the harbor provide for some of the best clamming in the state. In the pre-colonial period, Coast Miwok, Pomo and Wappo Indians living within perhaps an eighty-mile radius would make annual pilgrimages to harvest the Bodega Harbor clam flats. They used the shells to make disk-beads, a form of Indian money.[3]

Fig. 90. A man digging clams at the approximate location where Drake beached the *Golden Hind* in June 1579. *Photograph by the author, Memorial Day, 1995.*

---

2. Ibid., p. 280.
3. Writing in 1897, Hudson ("Pomo Wampum Makers," pp. 10–13) described a Pomo tribe living near Ukiah that were the only Indians in Mendocino County still making this annual pilgrimage. Like Fletcher, he was very much impressed with the burdens the Pomo could carry on their backs.

Fig. 91. View looking south from Bodega Head at the rocky island where the Elizabethans hunted seals and sea lions. Drake's men must have wreaked a terrible carnage on this rocky islet—as did the first Russian-American fur hunting expedition to visit Bodega Bay in 1807. This island was first named the Farallon of Padre Sierra by the Bodega y Quadra expedition of 1775 and then Gibson's Island by the Colnett expedition of 1790. Today it is called Bodega Rock. *Photograph by Teri Fung, November 1996.*

Seals and sea lions – It was these furry mammals that brought Russian colonists to Bodega Harbor in the early nineteenth century. In the Russian American Company's first solo hunting expedition to Bodega Bay in December 1808, they took 1,453 large, 406 medium, and 491 small sea otters.[4]

Other game – The Bodega Harbor environs were never a good place for hunting. Perhaps it was its isolated location, the presence of Indians or a combination of factors. For whatever reason, the early accounts consistently confirm that the area lacked decent game. As I emphasize in chapter 33, the same cannot be said for the other candidate sites in the area.

---

4. Khlebnikov, *Reports*, p. 107.

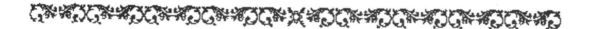

CHAPTER 19
Tenet 7 – The Fort Site

According to "The World Encompassed," Drake's men constructed a fort "for the defense of our selves and goods" while they repaired his ship. Discuss the suitability of the anchorage site for location of a fort and how the fort was constructed [and then explain how the clues pertaining to the site of the encampment help identify the anchorage location].

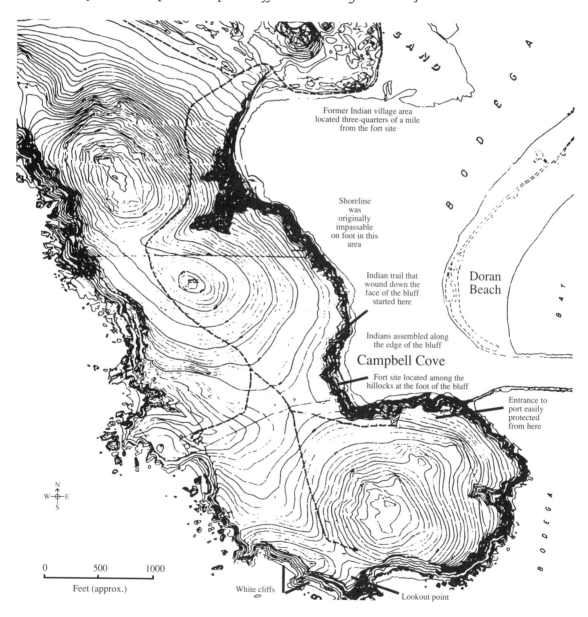

Fig. 92. The site of Drake's fort (topo map). Campbell Cove features a former Coast Miwok village close by the water's edge, three-quarters of a mile away, upwind from the fort, and separated by a hill. On foot, the cove could only be reached by descending a steep trail leading down the bluff. *Adapted from a drawing prepared by Dames & Moore, San Francisco, for Pacific Gas & Electric Company, circa 1962, contained in the Public Utility Commission files for Application Number 43808.*

Clues

- Fearful of the Indians, Fletcher and his companions "entrenched ourselves with walls of stone: that so being fortified within ourselves, we might be able to keep off the enemy (if they should so prove)" (*W.E.*, p. 70).
- They built the fort at the foot of a hill (*W.E.*, p. 71).
- The fort was close enough to the top of the hill that looking up from inside it, they could clearly see and hear an Indian chief addressing them (*W.E.*, p. 71).
- The fort was located close to where the Indian trail descended to the foot of the hill (*W.E.*, pp. 74–75; and the Portus plan). There was an Indian village located about three-quarters of an English mile away from the fort, close by the water's edge and apparently separated by the hill (*F.V.*; and *W.E.*, pp. 68 and 70–71).
- Although it was separated by a hill, the Indian village was within earshot of the fort (*W.E.*, p. 70).
- The Indians could reach the fort by descending the hill but apparently not by simply walking along the shoreline (*W.E.*, pp. 68, 71 and 75–76).
- There were knobby hillocks (small hills or earth mounds) nearby (*W.E.*, p. 72).
- The English observed "no part of the earth here to be taken up, wherein there is not some special likelihood of gold and silver" (*F.V.*).

Answers and Key Points

Fitness of the location – Campbell Cove was an excellent location for an encampment. Here Drake could easily command the narrow entrance to Bodega Harbor, secure in the knowledge that he was completely hidden from the most diligent lookout on a passing enemy ship. In addition, Bodega Head provided an ideal spot to post his own lookouts, particularly the one vantage point already discussed under Tenet 3.

In the colonial period, the English, Spanish, Americans, Russians, and French all recognized the port's advantages. Early in 1793, the English mounted guns at the entrance. Later that year the alarmed Spanish attempted to construct a presidio at Campbell Cove. Russian and American fur trappers were pitching their tents on the beach at Campbell Cove as early as 1807.[1] The Russians constructed port facilities and a warehouse there in 1812. Writing in 1841, Count Eugène Duflot de Mofras gave his government the following advice in urging them to take advantage of the Russian withdrawal to establish a French colony at Bodega: "I am of the belief that to possess the port of Bodega is of the utmost importance either to a European power, or the United States. . . ."[2]

Method of construction – In constructing their crude fort for defense against Indian attack, Drake and his men took advantage of the natural features of the cove. They built the front wall on the beach by piling stones parallel to and near the foot of the rear bluff for a length of perhaps a hundred feet. They ran the side walls back into the landslide debris at the foot of the bluff, burrowing through the debris and a considerable distance back into the bluff. Rising some fifty feet above the beach, the surrounding bluffs provided a natural rear wall, too steep for the Indians to descend directly, but not so steep that they could effectively shower the fort with stones and arrows from the precipice. In addition, the rugged cliff sides were so overgrown with prickling bushes that any attempt to descend directly upon the fort would be a tortuous affair.[3]

The English also constructed outer defense works (bulwarks) that ran across the beach on the north side of the fort to isolate the camp, the stretch of beach where they had graved the *Golden*

---

1. Ogden, *Sea Otter Trade*, p. 161.
2. Duflot de Mofras, *Exploration of Oregon and California*, p. 258.
3. For additional protection on the rear side, the fortifications presumably included some sort of defense works along the top edge of the bluff and/or perhaps some sort of overhang.

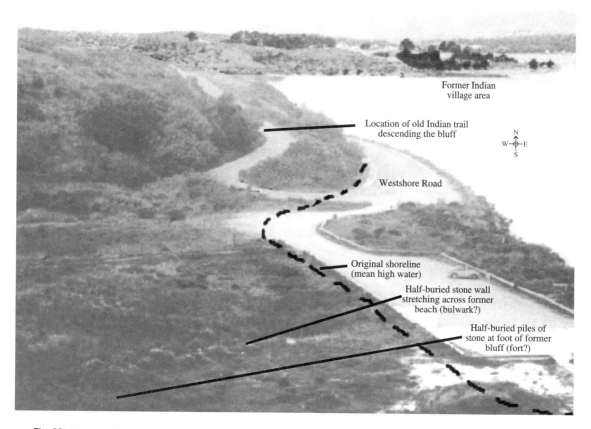

Fig. 93. The site of Drake's fort (photo). Note how the former Indian village area lay on a direct line of site and upwind of the encampment. The road winding up the face of the bluff follows the route of the old Coast Miwok trail. *Adapted from a photograph by the author, May 1995.*

Hind, and the cove's bountiful springs. The bulwarks probably consisted of a trench or moat with stones and excavated soils thrown up high behind. This bulwark would have faced a well-trodden Indian trail cutting down across the face of the bluff.

There was no shortage of stones to construct the walls of the fort. They pried them out of the rock outcroppings on the hilly headland to their south, rolled them over the edge of the cliffs, and collected them on the beach below.

Remnants of the stone walls – Incredible as it may sound, it appears that many of the stones of Drake's fort are still piled or scattered about in open view near where the Elizabethans assembled them more than four centuries ago (see Tenet 13). Looking northward from these rocks, you can see the hilltop where the Coast Miwok and Pomo chieftains addressed their European visitors. You can also see the modern rendition of the old Indian trail along which the king and his guard descended the bluffs; it has been converted to a paved road providing access to the headlands.

The Indian village about three-quarters of a mile away, close by the water's edge, and separated by a hill – Unlike the Indians in Drake's day, in departing Campbell Cove you can now follow Westshore Road northward along the west shoreline of the harbor. If you continue for three quarters of a mile, you will be in the general vicinity of a location where Coast Miwok Indians constructed quasi-permanent villages (see map 25). With the wind blowing from the northwest, you could shout out from this area to someone back at the cove and be heard. Before the road was built, you could not reach the cove from this village area unless you went up over the headland and followed the old Indian trail down across the face of the bluffs. Before Pacific Gas & Electric Company constructed

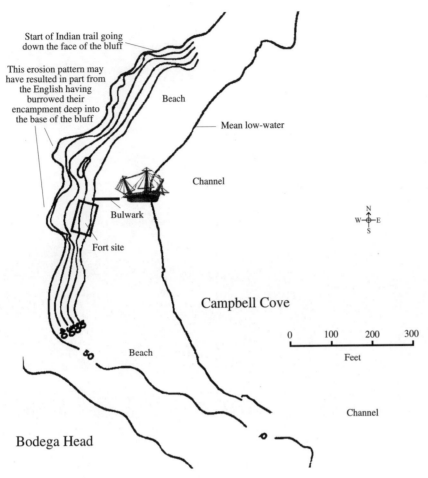

Start of Indian trail going
down the face of the bluff

This erosion pattern may
have resulted in part from
the English having
burrowed their
encampment deep into
the base of the bluff

Beach

Mean low-water

Channel

Bulwark

Fort site

Campbell Cove

N
W—◆—E
S

0   100   200   300

Feet

Beach

Channel

Bodega Head

Fig. 94. The layout of Drake's defense works. Drake would have positioned the bulwarks to isolate the ship, the camp and the springs. *Adapted from a drawing prepared by Dames & Moore, San Francisco, for Pacific Gas & Electric Company, circa 1962, contained in the Public Utility Commission files for Application Number 43808.*

Westshore Road, the channel cut right into the cliffs at the shoreline in between Campbell Cove and the village site (see figure 92).

Gold and silver – Bancroft ridiculed Fletcher for his glowing accounts of Nova Albion's mineral wealth (see chapter 33, Tenet 4 rebuttal). Considering this statement by nineteenth-century historian J. P. Munro-Fraser, however, perhaps there was something to Fletcher's remarkable claims:

> From time to time there has been more or less excitement in the vicinity of Bodega Port in regard to gold. Men have been known to wash out one dollar per day, although they would not average that. Considerable prospecting has also been done along the quartz ledges above, and traces of gold have been found. . . . That found was known as miner's shot, each grain being about the size of a pinhead, and round. In prospecting, the black sand is found in abundance, but the gold is wanting.[4]

4. Munro-Fraser, *History of Sonoma County*, p. 190.

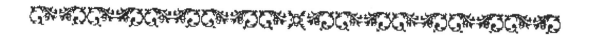

CHAPTER 20
Tenet 8 – The Weather

According to "The World Encompassed," Drake's men experienced severe and uncomfortable weather on the west coast of North America. What do the account's descriptions of weather and climate contribute to the identification of Drake's anchorage?

Clues

• During their entire stay in California, "notwithstanding it was in the height of summer and so near the sun, we were continually visited with like nipping colds as we had felt before" [in the search for the northwest passage] (*W.E.*, p. 64).

• Even the Indians who lived there were so bothered by the cold that they would "come shivering to us in their warm furs, crowding close together, body to body, to receive heat from one another, and sheltering themselves under a lee bank, if it were possible, and as often as they could, laboring to shroud themselves under our garments also to keep them warm" (*W.E.*, p. 65).

• During their stay in California it was so foggy that "neither could we at any time in whole fourteen days together, find the air so clear as to be able to take the height of sun or star" (*W.E.*, p. 64).

• The fogs were continuous "unless sometimes when the sudden violence of the winds doth help to scatter and break through, which thing happens very seldom, and when it happens is of no continuance" (*W.E.*, pp. 66–67).

• The fogs were "most stinking" (*W.E.*, p. 66).

• In California, "the north and northwest winds are here constant in June and July" (*W.E.*, p. 66).

Answers

Fletcher's complaints were justified; they very accurately describe the harsh weather conditions typically encountered at Bodega Harbor in the months of June and July. This is also true for Drakes Bay, outer Tomales Bay, Half Moon Bay, Point Año Nuevo, and perhaps Bolinas Lagoon, but is not true for San Quentin Cove or the mouth of the Petaluma River in San Pablo Bay.[1]

Key Points

Since there is a marine research laboratory located at Bodega Harbor, its weather conditions are well documented. Typical summer weather can be described in three words: chilly, windy, and foggy. Mean air temperature for June and July is in the low 50s. The northwest winds that average eight to ten miles per hour with gusts to twenty-five miles per hour can make it seem much colder. June and July are the foggiest months. In June about 70 percent of the days are foggy, in July about 60 percent. Often the fog lies along the grasslands and dunes until around ten o'clock in the morning when the winds pick up and scatter it. It then typically returns in the late afternoon and stays throughout the night. There are many days and nights when the head is completely engulfed, especially in mid-June. The fog is brackish to the taste and smell and "it can corrode cars, buildings and plants alike."[2]

The Russians did not like the weather at Bodega Harbor any more than the English. According to Fort Ross administrator Kirill Khlebnikov writing in the early nineteenth century: "In summer the

---

1. According to Aker, in Holliday, "The Francis Drake Controversy," p. 229, the weather at Bolinas Lagoon is not as severe as at Drakes Estero.
2. Barbour et al., *Coastal Ecology of Bodega Head*, pp. 9–16.

south winds stop and the wind comes from the northwest . . . It often happens that one can not see the sun for two or three months, and the weather is quite cold."[3]

According to George Davidson, "fogs are considered more frequent in Bodega Bay than under Point Reyes, as the low land allows the wind to draw very strongly over the coastline."[4]

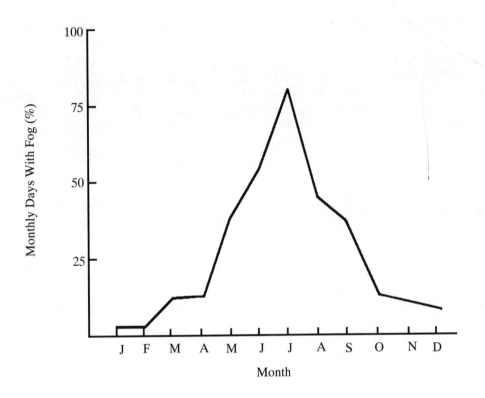

Fig. 95. Line graph showing the average incidence of fog on Bodega Head for the years 1968 and 1969. *Redrawn from COASTAL ECOLOGY, BODEGA HEAD by Michael Barbour et al., University of California Press, Berkeley, California, 1973.*

3. Khlebnikov, *Reports*, p. 124.
4. Davidson, *Pacific Coast Pilot* (1889), p. 253.

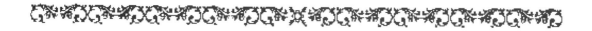

CHAPTER 21
Tenet 9 – The Visit Inland

Discuss the comparative observations from "The World Encompassed" that when Drake and his men "made a journey up into the land . . . to be better acquainted with the nature and commodities of the country . . . the inland we found to be far different than the shore."

Fig. 96. View looking inland from the site of Drake's former encampment. At Bodega Harbor the inland is indeed far different than the shore—a goodly country, and fruitful soil. *Photograph © by Ed Brady, February 1959. Courtesy Grethe Brady, Aero Photographers, Sausalito, California.*

Clues

- In making a journey "up into the land" their Indian porters carried their heavy burdens "up hill and down hill an English mile together" (*W.E.*, p. 79).

- The inland they found to be "far different from the shore, a goodly country, and fruitful soil, stored with many blessings fit for the use of man"(*W.E.*, p. 79).

- They found that the "riches and treasures" of Nova Albion abounded in the "upland country" (*W.E.*, p. 77).

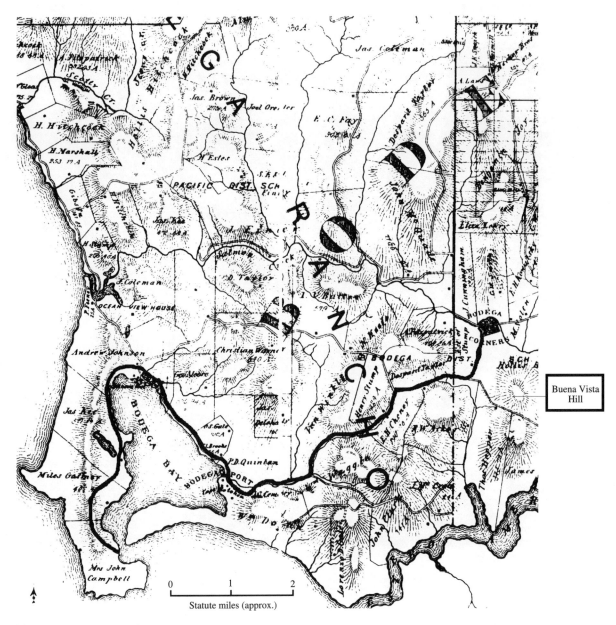

Buena Vista Hill

Map 15. Topographic map detail showing Drake's route inland (***highlighted in bold***). In journeying inland, the Elizabethans followed one of the most important Indian trade routes in Northern California. To recreate their trip to the hinterland, take Route 1 south from the town of Bodega Bay. After winding your way through Cheney Gulch, turn left on the Bodega Highway. As you approach the top of a hill, look for a cemetery on your left and park. Climb the hill and you can look down upon Salmon Creek Valley as Drake did in July 1579. *Detail from the HISTORICAL ATLAS MAP OF SONOMA COUNTY, CALIFORNIA, compiled, drawn and published by Thomas H. Thompson & Co., Oakland, California, 1877.*

Fig. 97. View of the hinterland looking north from what was once called "Buena Vista Hill." Today's picturesque town of Bodega —originally named Bodega Corners—lies in the heart of Salmon Creek Valley near the former Coast Miwok village of Tsüwütena (from pocket gopher). In aboriginal times, "the corners" was the junction of important Indian trails leading into Pomo, Wappo and Coast Miwok territories. *Photograph by the author, November 1996.*

Answer

These descriptions correspond exactly with the fertile upland valley lying about five miles due east of Bodega Harbor, just on the other side of the barren coastal hills. Today this area is known as Salmon Creek Valley and includes the townships of Bodega and Freestone.

Key Points

To journey inland, the Elizabethans hiked a well-worn Indian trail linking at least two former Indian villages located on either side of the coastal hills: Tókau and Tsüwütena as they were called as of the late 1800s. As illustrated in map 15, the existing thoroughfares linking the towns of today's Bodega Bay and Bodega (originally Bodega Corners) follow the same route. Starting at the northern end of the harbor, the trail took Drake and his company south along the top of the bluffs lining the east side of Bodega Harbor. They then followed a winding gulch (today's Cheney Gulch) that took them eastward along the sides of a series of rolling, barren hills. After cresting the side of a hill located about five miles inland, the Elizabethans observed an abrupt change in scenery. They looked down upon what an early visitor described as a "very pretty valley" with a salmon-laden creek running down the middle. The further inland they traveled from the village(s) near today's town of Bodega, the more beautiful the scenery became. Assuming they made their way to the former Indian village(s) near today's Freestone, about ten miles inland from the port, they were in some of the most pleasant and scenic environs in the entire state.[1]

---

1. The quotation is from the journal of Stephen Fowler, vol. 1, p. 161: the entry for the day he arrived at Bodega on Friday 21 February, 1851. He further records that Stephen Smith had appropriately named the hill on which he built his sawmill "Mount Pleasant."

Fig. 98. The former Chernykh/Gorky Ranch located in today's Coleman Valley: drawing by the Russian scientist Il'ia Voznesensky, circa 1840. It very aptly depicts the area's "many blessings fit for the use of man." *Reproduced from FORT ROSS: INDIANS-RUSSIANS-AMERICANS, Fort Ross Interpretive Association, 1981.*

The colonial Russians were just as impressed with this area as was Fletcher. In the late 1820s, Fort Ross administrator Kirill Khlebnikov advised his superiors that "we should turn our attention to the meadow near little Bodega Bay. There is a meadow land there with rich arable soil and it is not too far from the port."[2]

Khlebnikov had much to say about the "blessings fit for the use of man" that could be found at Salmon Creek Valley. In addition to its potential for agriculture, he specifically mentions grazing and livestock production, lumbering from the adjacent forests, oak bark for the production of shoes, clay of good quality for the production of pots, and elk and wild goats for the production of suede and chamois. Following up on his recommendations, the colony's next manager, Peter Kostromitinov, established two ranches there in the 1830s: the Vasili Khlebnikov Ranch in 1833 near today's town of Bodega and the Chernykh/Gorky Ranch in 1838 in today's Coleman Valley, located about five miles north of the town of Bodega.[3]

Counter-Rebuttals

According to Raymond Aker, "Bodega Bay is situated in the midst of moorlike country which extends so far inland and with so gradual a change that it is unlikely that Drake's party could have penetrated far enough to mark the inland as being far different from the shore."[4]

While this would be true for a party hiking inland from the shores of Bodega Bay proper or Tomales Bay, figure 96 makes it abundantly clear that the statement is incorrect for a party heading inland from Bodega Harbor.

---

2. Khlebnikov, *Reports*, p. 126.
3. Ibid., pp. 126–28.
4. Aker, *Report of Findings*, p. 369.

CHAPTER 22
Tenet 10 – The Flora

Botanical observations in "The World Encompassed" include mention of the Indians' decorative use of "a certain down, which groweth up in the country upon an herb, much like our lettuce which exceeds any other down in the world for fineness," and a description of the landscape which showed "trees without leaves, and the ground without greens." Correlate these observations [and other similar observations of the native vegetation] with the flora of the proposed landing site.

Fig. 99. Trees without leaves growing near the site of Drake's former encampment. *Photograph taken at Campbell Cove by the author, August 1994.*

Clues

- During their visit to the upland country, Fletcher observed an herb resembling English lettuce that apparently bore its seeds in late July. The Indians used its delicate white down to decorate their chieftain's headdresses. They collected the seeds and down in baskets and offered them in their sacrifices (*W.E.*, p. 74).

- The Indian women visiting the fort site with the king's retinue on June 26, 1579, brought baskets full of a native herb that they apparently smoked, and a root that they either ate raw or ground and baked. They called the herb *tobah* (tobacco) and the roots *petah* (potato), presumably, bastardized English words picked up from the sailors (*W.E.*, p. 74).

- On June 23 when the Indians made a stand at the top of the hill, at the bottom of which the English had built their fort, Fletcher, looking up from within the fort site, observed the women "cast themselves upon the ground, never respecting whether it were clean or soft, but dashed themselves on hard stones, knobby hillocks, stocks of wood, and prickling bushes" (*W.E.*, pp. 71–72).

- The coastal landscape in the anchorage locale included "trees without leaves" and "ground without greens in those months of June and July" (*W.E.*, p. 65).

Answers

The downy herb resembling lettuce – You will find the herb just as Fletcher described it if you visit the Bodega Bay area in late July or early August. Departing the town of Bodega Bay on Route 1, heading south into the coastal hills, my friend Tony Meyer first spotted the telltale white balls waving in the breeze near the intersection of Bay Hill Road. We also observed it further inland in the grassy meadows along the roadway between the towns of Bodega and Freestone. Earlier in the day, we had observed it growing in abundance along the roadsides further north on Route 1, near Fort Ross. We did not find it along Route 1 as it passes Bodega Harbor or on Bodega Head.[1]

This wildflower is in the thistle family, the Venus Thistle (*Cirsium occidentale v.*) as near as I can tell from the common field guides. It is very similar to its cousin English Lettuce (*Laticus serrianus*), being about the same height (two to five feet) and having prickly leaves that hug the bottom of the stem. It produces its spectacular bloom of deep lavender in May and early June. Except for one radiant flower cluster, all we saw of the flower heads in early August were the very pregnant-looking white balls. The balls were formed of edible black seeds (pinole) with their stems attached to a parachute of delicate, but durable, white filaments that at least appeared well-suited for decorating an Indian chief's hairnet. The plant does not appear to be common in the area, perhaps explaining why the Indians treated it with such reverence.[2]

Fig. 100. The Venus Thistle dots the hillsides along the old Indian trails leading to Drake's fit harbor. We found these native wildflowers growing near the junction of today's Bodega Highway and Bay Hill Road. The white balls consist of delicate white filaments surrounding edible black seeds. *Photograph by the author, August 1994.*

1. The Drake Navigators Guild identified the lettuce-like plant as California plumeseed (*Rafinesquia californica*); Neasham and Power identified it as milkweed (*Asclepias mexicana*). Cited in Holliday, "The Francis Drake Controversy," pp. 233–34.
2. I suspect the king's headdress was actually decorated with milkweed down which is much stickier and softer than thistle down.

Fig. 101. "Prickling bushes" and "knobby hillocks." They can be observed looking down at Drake's Campbell Cove campsite from the edge of the bluff and along the foot of the bluff. *Photograph by the author, August 1994.*

Native tobacco – According to Tom Smith, one of the last full-blooded survivors of the Coast Miwok nation, the Bodega Miwok obtained their tobacco through trade with the Western Pomo: "had to go to Healdsburg for tobacco."[3]

Peter Kostromitinov made the following observations in 1839 for the Bodega Miwok and Pomo Indians then living in the vicinity of Fort Ross (Western Pomo territory): "They also have a special herb resembling tobacco, which largely grows near the rivers in sandy locations."[4]

The roots – It appears that all the California tribes harvested edible roots. Stephen Powers made the following observations in the early 1870s after interviewing the remnants of the Pomo still living

3. Cited by Kelly, in Collier and Thalman, eds., *Interviews*, p. 356. The quote is from her interview with then ninety-five-year-old Tom Smith in 1932 (see chapter 36).
4. Kostromitinov, "Observations," in Wrangell, p. 44.

near Fort Ross: "the Gualala also eat considerable quantity of a wild potato, probably cammas, which they call *hipo*, and which is said to be quite good eating when cooked and peeled."[5]

Ground without greens – I will borrow Raymond Aker's description of the west side of the Point Reyes Peninsula to describe the entire vista that Drake and his company would have seen in June and July 1579 from Bodega Head:

> We are treated to an aspect of rolling, treeless downs, dry in the summer as is most of California. . . . the cold, fog, and wind blowing across the open downs in the spring and summer produce a thin and stunted ground cover in all exposed areas . . . The soil appears dry, dusty and cracked in places where it shows through the thin ground cover; hence the face of the earth can well be said to be unhandsome and deformed, generally squalid and barren.[6]

Trees without leaves – Since the early accounts consistently reveal that the Bodega Harbor environs were nearly devoid of trees, it is likely that Fletcher was referring to the shrubs growing along the bluffs in the immediate vicinity of the encampment. This dense vegetation is mostly comprised of lupines that behave in a very peculiar fashion: "many lupine shrubs die each year on the head [Bodega Head]. The foliage first wilts, then turns gray and brittle, and finally falls off . . . sometimes an acre patch of many shrubs succumb at the same time."[7]

The ecologists have traced this phenomenon to a moth larvae that attacks the root system, not the nipping colds cursed by Fletcher. They reported that it happens at several other localities along the coast, including Half Moon Bay, about a hundred miles to the south, but did not mention the Drakes Bay area or San Francisco Bay.[8]

Fig. 102. Defoliated lupine branches. This blight is caused by Bodega Head's moth larvae, not Fletcher's nipping colds. *Photograph by the author, June 1995.*

5. Powers, *Tribes of California*, p. 189.
6. Aker, *Summation*, p. 29.
7. Barbour et al., *Coastal Ecology of Bodega Head*, pp. 72–73.
8. Allen, "An Examination of the Botanical References," p. 40, also concluded that Fletcher was referring to a defoliated shrub: lichen-hung specimens of the Blue Blossom. Neasham and Power identified the leafless trees as buckeye (Holliday, "The Francis Drake Controversy," pp. 233–34).

CHAPTER 23
Tenet 11 – The Fauna

Identify the fauna described in the Drake sources, especially "The World Encompassed's" observation: "We saw . . . a multitude of a strange kind of conies."

Clues

* Fletcher used the term conies to refer to small burrowing rodents (*W.E.*, p. 80).
* The visiting king wore a coat made of the skins of conies (*W.E.*, p. 74).
* The king's guard wore coats made of the skins of other types of conies (*W.E.*, p. 74).
* Fletcher described the conies he saw on the king's coat as follows: "Their heads and bodies, in which they resemble other conies, are but small; his tail like the tail of a rat, exceedingly long; and his feet like the paws of a want (mole?); under his chin, on either side, he has a bag, into which he gathers his meat when he has filled his belly abroad" (*W.E.*, p. 80).
* The countryside was apparently infested with conies (*W.E.*, p. 80).
* On the visit inland, they saw thousands of very large and fat deer (*W.E.*, pp. 79-80).

Answers

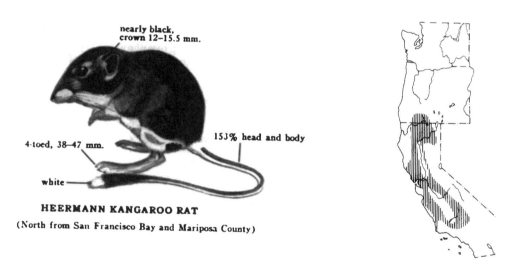

nearly black, crown 12–15.5 mm.

4-toed, 38–47 mm.

white

15J% head and body

HEERMANN KANGAROO RAT

(North from San Francisco Bay and Mariposa County)

Fig. 103. The California kangaroo rat (*left*) and its distribution along the Pacific Coast (*right*). The kangaroo rat (a gerbil) has a small body, an exceedingly long ratlike tail, front feet like the paws of a mole, cheek pouches, and is indeed rather strange looking compared to your ordinary conie. *Reproduced from MAMMALS OF THE PACIFIC STATES by Lloyd Ingles, Stanford University Press, 1965. Courtesy of Stanford University Press, © 1947, 1954 and 1965 by the Board of Trustees of the Leland Stanford Junior University.*

The king's coat – The king's "holiday coat" seems to have been made from the skins of *Dipodomys heermanni*, the Heermann kangaroo rat (also called the California kangaroo rat). Its physical appearance corresponds exactly with Fletcher's descriptions. These small solitary animals can be found throughout California's valley grasslands and foothill woodlands and therefore do not provide a significant clue to the anchorage location. The use of such fur coats by Western Pomo chieftains is documented, although not specifically for this rodent.[1] Aker rejected this identification for

---

1. Bean and Theodoratus, "Western Pomo and Northeastern Pomo," in Heizer, ed., *Handbook*, p. 291.

Fletcher's conies, arguing that the kangaroo rat "does not meet the full criteria of the description given in the accounts," presumably because of its characteristically sparse population density.[2] It is appropriate to consider, however, that the coat was worn for a grand ceremony and not an occasion for a king to be wearing just ordinary skins. Indeed, Fletcher pointed out a clear distinction between the pelts of the king and his guard.

The multitude of conies – During the visit inland, Drake and his men would have found Salmon Creek Valley completely pockmarked with the holes of the boca pocket gopher. Since the gophers never venture out of their holes during daylight hours, Fletcher could not have known what type of rodent was living in them. He must have incorrectly assumed they were the conies he had seen on the king's coat.[3]

The infestation of Salmon Creek Valley by the boca pocket gopher is well-documented for the colonial period.[4] The Russians found them a nuisance: "The plantations here have suffered damage both below ground and on the surface, . . . in the former case from the hamsters that eat the plant's roots and cause great devastation."[5]

Samuel Barrett identified the name of the Indian village associated with the former Khlebnikov/Smith Ranch at today's town of Bodega as Tsüwütena, from "pocket gopher" in the Bodega Miwok dialect.[6] Today along almost any roadside in Bodega there is ample evidence that these destructive rodents continue to plague its residents.

Thousands of elk – There is plenty of documentation that vast herds of deer once roamed southwestern Sonoma County, none any better than the following description that also provides some solid support for the arguments presented for the visit inland under Tenet 9 (*emphasis added*):

> The country in whatever valley we went was an interminable grain field; mile upon mile, and acre after acre of wild oats grew in marvelous profusion, in many places to a prodigious height—one great glorious green of wild waving corn . . . wild flowers of every prismatic shade charmed the eye, while they vied for each other in the gorgeousness of their colors, and blended into dazzling splendor. One breath of wind and the wild emerald expanse rippled itself into space, while with a heavier breeze came a swell whose rolling waves beat against the mountainsides, and, being hurled back, were lost in the faraway horizon; shadow pursued shadow in a long merry chase. The air was filled with the hum of bees, the chirrup of birds, and an overpowering fragrance from the various plants weighted the air. The hillsides, overrun as they were with a dense mass of tangled jungle, were hard to penetrate, while in some portions the deep dark gloom of the forest trees lent relief to the eye. The almost boundless range was intersected throughout with divergent trails, whereby the traveler moved by from point to point, progress being as it were in darkness on account of the height of the oats on either side . . . The hissing of snakes, the frightened rush of lizards, all tended to heighten the sense of danger, while the flight of quail and other birds, the nimble run of the rabbit, and *the stampede of elk and antelope, which abounded in thousands*, added to the charm, causing him, be he whosoever he may, pedestrian or equestrian, to feel the utter insignificance of man, the "noblest work of God."[7]

There are numerous tales of deer hunting and gopher infestation in Stephen Fowler's journal covering his first years at Bodega Corners (today's town of Bodega) in the early 1850s. Some other early accounts also document the fact that "at this season [late July and August] elk are fatter than at any other."[8]

---

2. Aker, *Report of Findings*, p. 311.
3. The Drake Navigators Guild identified Fletcher's conies as the Botta pocket gopher; and Neasham and Power identified them as the California ground squirrel (Holliday, "The Francis Drake Controversy," p. 236).
4. Tikhmenev, cited in Essig and Ogden et al., *Fort Ross*, p. 9; and Gibson, *Imperial Russia in Frontier America*, p. 137.
5. Khlebnikov, *Travel Notes*, p. 57.
6. Barrett, "Ethno-Geography," pp. 304–5.
7. Munro-Fraser, *History of Sonoma County*, pp. 49–50.
8. Duhaut-Cilly, 1827, cited in Smilie, *The Sonoma Mission*, p. 28.

CHAPTER 24
Tenet 12 – The Indians

What is the significance of Drake's contact with and the observations of the people of Nova Albion, description of whom comprises a large portion of the narratives of Drake's visit?

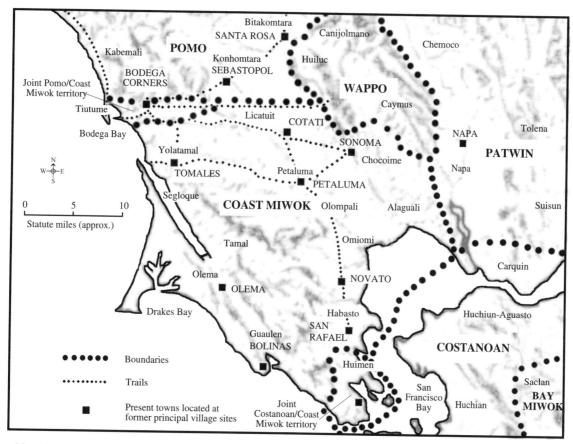

Map 16. Relief map showing approximate boundaries, major tribal regions, and some of the main trade routes of the principal Indian linguistic groups just north of San Francisco Bay circa 1775 (pre-mission era). In all likelihood, the thin triangle of land encompassing Bodega Harbor and Salmon Creek Valley was a no mans land, open to fishing and clamming parties from the Coast Miwok, Pomo, and Wappo nations. Bodega Corners received its name from the fact that it lay at the junction of the well-trodden trails leading into their respective territories. The map is derived largely from Randall Milliken's map in A TIME OF LITTLE CHOICE (1995), and modified on the basis of early Spanish and Russian accounts. Milliken produced his map of tribal regions based on an exhaustive review of mission records and associated military correspondence for the period 1776–1810. Milliken's maps did not delineate trails or designate joint territories.

Clues

The Indians who lived at the harbor and immediate surrounding country:

- To greet the English on June 18, 1579, an Indian paddled a canoe out from shore, made a long, animated speech, and gave them several presents including the following: (1) a feather headdress consisting of neatly arranged black feathers cut to equal length and gathered into a round bundle by a string, and (2) a grass basket filled with an herb resembling tobacco (*W.E.*, pp. 67–68 and *F.V.*).

- The Indians were armed only with bows and arrows (*W.E.*, pp. 68 and 79).
- Within three days of the English arrival, a great number of Indian men and women had assembled bearing the following additional gifts: (1) feathers, (2) netlike caps, (3) quivers of their arrows made of fawn skins, (4) the skins that the women were wearing, and (5) bags of native tobacco (*W.E.*, pp. 68–69).
- They lived in semi-subterranean lodges, covered with earth that from the outside must have looked just like giant anthills; the roofs were cone-shaped and low-vaulted and had a door like the hatch of a ship (*W.E.*, p. 69).
- The men, for the most part, went naked (*W.E.*, p. 70).
- The women wore loose skirts made from tule or deerskins, and deerskin cloaks (*W.E.*, p. 70).
- The women were submissive and respectful of the men and did the bulk of the housekeeping chores (*W.E.*, p. 70).
- Within another three days of the English arrival, more Indians from the surrounding country had assembled bearing similar gifts as before (*W.E.*, p. 71).
- Their leader loudly addressed the English for a long time with violent gestures of his hands and body (*W.E.*, p. 71). The common people often responded to the speaker with a communal "Ohhh" given in a dreamy sort of manner (*W.E.*, p. 71).
- In their rituals, the women scratched their cheeks and breasts to draw blood and hurled themselves, bare-breasted, upon the ground (*W.E.*, pp. 71–72).
- The Indians seemed to think that the English were gods or spirits and for whatever reason made burnt offerings of their belongings as sacrifices (*W.E.*, pp. 69 and 72).
- They were generous in their gift-giving and did not want or accept anything in return (*W.E.*, p. 72).
- They returned every third day to offer sacrifices but did not bring food (*W.E.*, pp. 72 and 78).
- They were nimble, fleet of foot, and almost always on the run (*W.E.*, p. 78).
- They were adept at snatching fish from the surf (*W.E.*, p. 78).
- There was one village on the shores of the harbor, about three-quarters of a mile from the encampment and close by the water's edge. When they journeyed inland, they saw several others (*W.E.*, pp. 70 and 78, and *F.V.*).
- The lodges they saw on the trip inland were constructed like those at the harbor (*W.E.*, p. 78).
- The Indians had a tractable, free, and loving nature, without guile or treachery (*W.E.*, p. 78).

The Indians who came with the king nine days after Drake's arrival:

- These Indians came from an upland country (*W.E.*, pp. 72, 74 and 77).
- The king used an emissary to announce his arrival (*W.E.*, p. 73).
- These Indians may not have spoken the same language as those living at the harbor, for the emissary appeared to use an interpreter to communicate with the local Indians and their Elizabethan guests (*W.E.*, pp. 73 and 75).
- The emissary requested a gift for the king (*W.E.*, p. 73).
- When they arrived, the king and another important leader were leading the train. They were closely followed by the king's guard, consisting of about a hundred tall and warlike men. (*W.E.*, p. 73).
- The headman next to the king carried a four-to-five-foot-long, black wooden stick (royal scepter) with the following gifts attached to it: (1) three long disk-bead necklaces, folded over many times; (2) two netted headdresses artfully decorated with multicolored feathers; and (3) a bag of tobacco (*W.E.*, p. 73).
- The king wore a very fancy, multicolored feather headdress and a coat reaching his waist that was made from the skins of a small, long-tailed rodent (*W.E.*, p. 74).

- The king's guard were similarly dressed but their headdresses had black feathers and some were decorated with the fine, white down of a certain herb. Their coats were made from the skins of other types of rodents (*W.E.*, pp. 67 and 74).
- The king and his guard wore disk-bead necklaces, the number and size of which were accorded by rank. They had their faces painted (*W.E.*, pp. 73–74).
- The common men were naked, had long hair bunched up behind their heads in a plume of feathers, and wore headdresses that had individual feathers protruding out in the front like horns. They also had their faces painted. They all carried gifts (*W.E.*, p. 74).
- The women and children made up the rear of the procession (*W.E.*, p. 74).
- The women all carried one or two very ornate, bowl-shaped baskets that were artfully decorated with tiny red feathers, mother of pearl, and disk-bead chains (*W.E.*, pp. 74–75).
- The baskets the women brought contained bags of tobacco, roots, broiled fishes, and the seeds and down of a certain herb that grew up in the country (*W.E.*, p. 74).
- As they approached the fort, the king and all the men, led by the scepter-bearer, danced, sang, and behaved with great dignity and gravity. The women danced but did not sing (*W.E.*, p. 76).
- When the king confronted Drake, he had him sit down and, with several other headmen, made some rather solemn orations that included the word "hioh." After the orations, the king led the entire group in a ceremonial song, placed disk-bead necklaces around Drake's neck, crowned him with a feathered headdress, and gave him some additional gifts (*W.E.*, p. 76).
- After the coronation ceremony, the common men and women went among the English, looking into their faces as if in search of someone they knew. Finding a suitable face, the women and old men wept and scratched their cheeks as before (*W.E.*, p. 77).
- Their rituals appear to have been similar to those of the Indians that lived at the port, including the three-day sequencing of their visits (*W.E.*, pp. 78–79).
- The Indian men were very strong and could carry huge loads on their backs for great distances (*W.E.*, p. 79).

Answer

The substantial and remarkably detailed ethnological evidence in *The World Encompassed* points directly at Bodega Harbor. It is the only location that fits the demand for an Indian tribe that spoke a Coast Miwok dialect, but demonstrated an overtly Western Pomo culture. It is the only candidate harbor that could have realistically lain within the sphere of influence of a great Pomo (or Wappo) king.

Comments on the Ethnologists' Findings

As discussed in chapter 9, three eminent University of California anthropologists have carefully evaluated the ethnological evidence set forth in *The World Encompassed*: Samuel Barrett, Alfred Kroeber, and Robert Heizer. Based on their findings, they all concluded that Drake and his men encountered Coast Miwok Indians that once inhabited the shores of Marin County.

Without challenging the findings of these scholars, I will take issue with their conclusions. In so doing, I will use their findings to help demonstrate that Drake most likely encountered the Coast Miwok subgroup of Bodega Harbor, not the subgroup of the Drakes Bay area.

Samuel Barrett – Based on the timing, nature and extent of the fieldwork he conducted in 1903, Samuel Barrett may have been the best qualified of the three noted ethnologists to give us an opinion on the matter. But for his apparent deference to the illustrious George Davidson, Barrett might have concluded that Drake landed on the shores of Sonoma County, not Marin. He certainly hinted at such a conclusion:

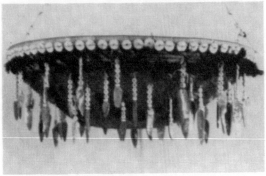

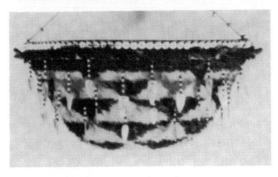

Fig. 104. Three examples of classic Pomo basketry decorated with shell money, mother of pearl, and woodpecker feathers. *Reproduced from FRANCIS DRAKE AND THE CALIFORNIA INDIANS, 1579 by Robert Heizer, University of California Press, Berkeley, California, 1947.*

These facts therefore point further to the tenability of the belief that Drake's landing was somewhere north of San Francisco Bay, perhaps even north of Point Reyes, though Pomo of the Southern or Southwestern dialect area may have journeyed down to Drakes Bay bringing their boat-shaped ornamented baskets with them.[1]

Barrett was obviously convinced that the visiting king and his retinue were Pomo. He pointed out that their boat-shaped baskets "are almost entirely unknown in California except among the Pomo and perhaps certain of their immediate neighbors." He goes on to explain that the ornamentation of these baskets with shell money and a matted down of red feathers "was, in aboriginal times so far as is known, never attempted by any California people other than the Pomo and perhaps certain of their immediate neighbors."[2]

Alfred Kroeber – Professor Kroeber seems to have agreed with Barrett's assessment regarding the visiting king. He also was struck by how closely some of Fletcher's descriptions correlated with Western Pomo culture:

> Absolutely typical Pomo basketry of the ornate type can be recognized . . . Even the net cap filled with eagle down that the Yuki, Pomo and other tribes wore until recently seems to be described . . . The bunch of feathers . . . tallies closely with Pomo and Maidu specimens.[3]

Robert Heizer – Professor Heizer took issue with Barrett's conclusion respecting the Pomo king. He was convinced that the combination of linguistic and ethnological evidence pointed strictly to the Coast Miwok. It is pertinent to point out, however, that Heizer, in presenting his arguments, added the following to Barrett's and Kroeber's lists of the Indian customs specifically identified with the Pomo: the king's use of an emissary, the emissary's request for a gift, the king's guard, the single feather resembling horns worn by the common men who accompanied the king, and the bizarre rituals in which the Indian women scratched their cheeks to draw blood. Heizer was also puzzled by the fact that neither the visiting

1. Barrett, "Ethno-Geography," p. 37.
2. Ibid., pp. 36 and 37. According to Sylvia Thalman of the Miwok Archaeological Preserve of Marin, there are examples of Coast Miwok basketry in Marin County and Russian museums that closely resemble the type of baskets that Barrett attributes exclusively to the Pomo and their immediate neighbors (telephone conversation with the author, Jan. 1997).
3. Kroeber, *Handbook of the Indians*, pp. 276–77.

king nor his retinue wore the types of feather cloaks that characterize Coast Miwok ceremonial occasions recorded for later times.[4]

Heizer justified his position with these three arguments: (1) Coast Miwok and Western Pomo cultures were practically indistinguishable; (2) the word "hioh" for the visiting king is of Coast Miwok derivation; and (3) the Coast Miwok would not likely have permitted a Pomo king to enter their territory.[5]

Rebutting Heizer's Three Points

Similarities in Coast Miwok and Pomo cultures – Dr. Heizer's contention that Coast Miwok culture was identical to Western Pomo culture is logical for the Bodega Miwok tribes but not for the Coast Miwok dwelling in the Point Reyes area. The Coast Miwok tribes living in the vicinity of Bodega Harbor were bordered on two sides by Western Pomo tribes that are known to have visited the harbor on a regular basis. The Coast Miwok tribes dwelling on the remote western side of the Point Reyes Peninsula, on the other hand, were geographically isolated from both the Pomo and Bodega Miwok by two ridges of coastal mountains, Tomales Bay, and greater Bodega Bay. In addition, as noted by Dr. Heizer, they were politically isolated from the Pomo by a large breadth of Coast Miwok territory.

The linguistic correlation – Dr. Heizer argued for a linguistic correlation with Coast Miwok dialect, chiefly on the basis of two words mentioned in *The World Encompassed* and Madox diary: "cheepe" corresponding to the Coast Miwok word *tcipa* (pronounced "cheepa") for bread; and "hioh" corresponding to the Coast Miwok word *hoipu* (pronounced "hoy-puh") for chief.[6]

While one would have to agree that the word "cheepe," for "cheepa," fairly establishes that Drake and his men made contact with the Coast Miwok, phonetically, "hi-oh" for "hoy-puh" seems to be reaching too far.

There is a much more likely explanation for this term. According to Stephen Powers, who conducted extensive ethnological research in the early 1870s, the Southern Pomo exchanged the following words in their ceremonial greetings with important visitors: *a-mi-ka?* (Is that you?) – *hi-o!* (yes!).[7] It is not difficult to deduce how the English and Indians could have misunderstood each other in exchanging these first words, particularly in view of the close resemblance of the term *amika* for the Spanish word *amiga* (friend).

There is more evidence that the visiting king's tribe did not speak the local language. In the following passages, Fletcher seems to be describing how the king's emissary communicated with the local Coast Miwok and their guests through the use of an interpreter:

> They in the delivery of their message, the one spoke with a low soft voice, prompting his fellow; the other pronounced the same, word by word, after him with a voice more audible . . . Whereupon, he who bore the scepter before the king, being prompted by another whom the king assigned to that office, pronounced with an audible and manly voice what the other spake to him in secret.[8]

In concert with this evidence, Dr. Heizer's argument against a Pomo visit is further rebutted by four contemporary maps (French, Dutch, Mellon, and Hondius) that all contain a caption to the effect that the inhabitants of Nova Albion crowned Drake twice on the same day—ostensibly, once by the Coast Miwok, and then again by the visiting Pomo.

The likelihood of a Pomo king visiting Coast Miwok territory – Unlike the Coast Miwok of Marin County, the Bodega Miwok of Sonoma County would most certainly have welcomed a visiting Pomo king; Bodega Harbor was clearly within the sphere of influence of Pomo kings when European

---

4. Heizer, *Elizabethan California,* pp. 69–73 , 80 and note 40 on page 62.
5. Ibid., p. 62.
6. Ibid., p. 98.
7. Powers, *Tribes of California,* p. 176.
8. Drake, *The World Encompassed,* pp. 73 and 75.

explorers first arrived. The well-beaten trail leading from Pomo territory, via Santa Rosa and Sebastopol to the Bodega clam flats, was perhaps the most important Indian trade route in the northwest section of aboriginal California.[9] In fact, as shown in map 16 and discussed in chapter 36, there is some reason to believe that in Drake's day, the Pomo and Coast Miwok boundaries overlapped Bodega Harbor.

Correlation with Descriptions in the Accounts of Other Early Visitors

To complete my discussions on this tenet, I will demonstrate the close correlation between Fletcher's descriptions and the descriptions of native culture found in the accounts of some of Bodega Harbor's early European visitors and settlers (see chapter 35). I will also provide a pertinent observation by Isabel Kelly who, in 1931, conducted interviews with several of the last full-blooded survivors of the Coast Miwok nation (see chapter 36).

James Colnett, September 1790 – The next European vessel known to have anchored off Doran Beach after the *Golden Hind* was the English ship *Argonaut*, under James Colnett. Two natives in a canoe-shaped "rush float" paddled out to the English and gave them some baskets of seeds. When Colnett took the ship's boat into Campbell Cove for water and wood, he found an Indian village close by the water's edge about three-quarters of a mile away (Tókau). The Indians living at the village presented the English with "feather caps" and more baskets full of seeds. They used dip nets to snatch "sardines" from the surf. They wore deerskins and were armed with bows and arrows. At least one Indian was wearing a feathered crown and a disk-bead necklace and bracelet. By signs, they indicated that deer meat could be provided in four days' time.[10]

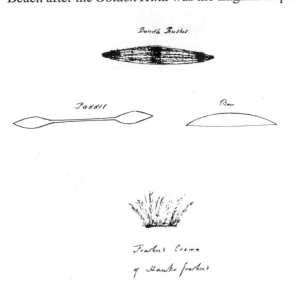

Fig. 105. Implements used by the natives of "Port Sir Francis Drake"- bundle rushes, paddle, bow, and feathers crown of hawks feathers: detail of a sketch drawn at Bodega Harbor by Captain James Colnett, 1790 (photo-edited and partly redrawn). *Courtesy of the Public Record Office, Richmond, England.*

Felipe de Goycoechea, August 1793 – When a Spanish horse-squadron from San Francisco arrived at the east shore of Bodega Harbor to assist the Juan Matute expedition in the construction of the aborted Port Bodega presidio, people from six Indian settlements came to visit them. They wore few clothes and had little to trade. They offered to go fishing and presented two bunches of feathers and some "stones" (clamshell pieces) out of which they made beads. They had nets for fishing.[11]

Archibald Menzies, October 1793 – When a ship of the Vancouver expedition anchored off Doran Beach and dispatched the jolly boat to explore the backwater lagoon, the English found a hut along the west shore and met five women and some children and four naked men armed with bows and arrows.[12] The Indians immediately sat down and shouted out "amika" (the Southern Pomo

9. James Davis, *Trade Routes*, map 1.
10. Colnett, *Journal*, pp. 174–76.
11. Wagner, "Last Spanish Exploration," pp. 342–45.
12. A Spanish expedition under Juan Bautista Matute had attempted to establish a presidio at Campbell Cove in the summer of 1793 (chapter 36).

ceremonial greeting?). About a dozen and a half additional natives "came down from the country." The women wore deerskin wraps and mantles (sleeveless shoulder wraps) made from the skins of small mammals. The English found the Indians to be "remarkably friendly and docile, readily parting with anything they had which they thought would be anywise acceptable to us." The gifts they presented included bows, quivers full of arrows, and a small basket full of fish. The natives had broad flat visages, high cheekbones, depressed noses, and long, straight black hair tied up in front of or behind their heads. They were tattooed with a streak across the chest but wore no ornaments.[13]

Peter Corney, December 1814 – When Peter Corney, second-in-command of the English schooner *Columbia,* anchored off Doran Beach to visit the Russian American Company's new fur hunting outpost, the shore party found an Indian village along the west side of the lagoon. The Indians were living underground, presumably in the same anthill-like dwellings that Fletcher had observed two centuries before. They encountered several women and children and some men. The men hunted by firing the grass and killing the rabbits that fled the flames.[14]

Peter Kostromitinov, manager of the Fort Ross colony from 1831 to 1838 – Kostromitinov made the following observations for the Fort Ross natives who at that time consisted primarily of Coast Miwok tribes that had migrated north from the Bodega region:[15]

Cold-weather lodges – A pit is dug, some vertical poles are driven into the ground with their pointed ends first, and covered with wood bark [presumably clefts of redwood bark], twigs, and grass; an opening is left on the top and side, the former to allow the smoke to escape, the latter to serve as an entrance.[16]

Mourning ceremonies – On the third day, however, they betake themselves to the relatives of the deceased, who await them in their *barabaras* and, after a suitable welcome, commence lamentations together; the old women scratch their faces and strike their chests with stones. The relatives of the deceased positively believe that they are seeing their deceased friends in these actors. During this presentation, the entire settlement exercises great abstinence in matters of nourishment, and meat is not eaten, sometimes for a long time.

Celebration ceremony – Upon arrival of the guests, the host presents them with everything he possesses. . . . when they are all satisfied, they start giving each other good advice . . . this is followed by song and dance; . . . when a song is over they all call out "hoi" and then continue their song.

13. Menzies, in Eastwood, ed., "Archibald Menzies Journal," pp. 302–4.
14. Corney, cited in Lightfoot et al., *Archaeology and Ethnohistory*, pp. 125–26.
15. The Russian accounts suggest that by the mid-1830s, Fort Ross colonists had little positive interaction with Western Pomo tribes and dealt almost exclusively with Bodega Miwok Indians who had remained friendly towards them. Russian sources, as well as Isabel Kelly's notes, reveal that most of the original Bodega Miwok tribes migrated to the Fort Ross area sometime in the 1820s (Kelly, in Collier and Thalman eds., *Interviews*, p. 22; Kostromitinov, in Stross and Heizer, eds., "Ethnographic Observations," p. 7; and Khlebnikov, *Travel Notes*, pp. 193–94).
16. A team of researchers, who recently conducted archaeological and ethnological research at Fort Ross, have reported that during cold weather Western Pomo and Coast Miwok subgroups living at Fort Ross and Bodega Harbor probably dwelled in earth-covered lodges similar to those described by Fletcher, and not the types of structures ascribed by later researchers: redwood bark or grass-thatched lodges. They arrived at this conclusion by reference to the accounts left behind by the earliest visitors (Lightfoot et al., *Archaeology, and Ethnohistory*, p. 143).

Fig. 106. A Southeastern Pomo semi-subterranean, earth-covered lodge. This is actually a small men's sweathouse (sudatory), but according to Samuel Barrett ("Pomo Buildings," p. 42), the cold-weather, earth-covered lodges looked much the same. *Courtesy Phoebe Hearst Museum of Anthropology, University of California at Berkeley—photographed by Pliny Goddard, circa 1904 at Sulphur Bank.*

Demeanor – "Simplicity and good nature are the main traits in their character. Thievery and homicide are practically nonexistent with them.[17]

Anonymous correspondent, "Sonoma County Journal," 1860:

Mourning ceremonies – "Died, November 23, 1860, at the rancheria near the Bodega Ranch House, Juan Grande, for many years a faithful vaquero of the Bodega rancho . . . Juan Grande is dead, and his dusky sons and daughters and gray-haired kindred stand around his lifeless body, tearing their hair, beating their breasts with stones, throwing themselves upon the earth, and weeping and wailing with frantic violence over the lost and gone . . . giving themselves cuts and bruises that must last for weeks."[18]

Maria Copa, Coast Miwok Indian, May 1932 – In reporting to Isabel Kelly what the village chief (at Bodega?) did in the old days:

In the evening the *hoipu* went to the top of the sweathouse and gave advice to the people. He did this almost every evening — not only when they were having a dance. He talked his own language (not any society language), and the people all said "hü" (with rising inflection, like "hooray").[19]

---

17. Kostromitinov, in Stross and Heizer, eds., "Ethnographic Observations," pp. 7–13.
18. Cited by Lewis Publishing Company, *Illustrated History*, 1889, p. 294.
19. Cited by Kelly, in Collier and Thalman, eds., *Interviews*, p. 346. From the manner in which the information is presented, it seems that Maria Copa witnessed this at Tom Smith's village at Bodega. Tom, in correcting Maria's statement, indicated that the *hoipu* (a fellow named "Kupan" or his son "Müska") did harangue every night, but from "any place . . . not necessarily speaks from top of dance house or sweathouse" (ibid., pp. 346 and 348). My interpretation of the way the communal response sounded is "ohh-ee." Fletcher had it as "ohh."

CHAPTER 25
Tenet 13 – Archaeological Evidence

Discuss the significance of the archaeological evidence related to Drake's landing.

Fig. 107. The ruins of an Elizabethan rock fort. San Francisco author, columnist, popular historian, and environmental activist Harold Gilliam predicted that the discovery of such ruins on a lonely California beach might one day solve the mystery of Drake's anchorage. And so it came to be: the half-buried stone wall upon which my friend Teri Fung and I are sitting focused my attention on Campbell Cove when we stumbled across it in the summer of 1989. *Photograph by Teri Fung, November 1996.*

Clues and Considerations

- The California campsite, where Drake and some sixty of his men spent a shivering thirty-six days, was located at the foot of a hill (*W.E.*, p. 71). When they departed, the area at the base of this hill must have been heavily littered with Elizabethan artifacts and the charred residues of their campfires.

- To protect themselves from Indian attack, Drake and his men had entrenched themselves within walls of stone (*W.E.*, p. 70). Based on the descriptions we have for other forts Drake built, and considering the size of his company and cargo, the fort was likely a substantial structure.

- The items the English carried aboard the *Golden Hind* for use in trade with aborigines included "French bracelets of small beads of glass" (Francis Fletcher narrative, in Penzer, ed., p. 116), and "certain glasses, beads, and other trifles" (Edward Cliffe account, in Penzer, ed., p. 194).

- Upon their first encounter with the natives of Nova Albion on June 21, Drake and his company, in making peaceful overtures and entreating them to cover their nakedness, offered the Indians "many good things" and distributed "shirts, linen cloth, etc." (*W.E.* p. 69). Immediately upon returning to their village, located three-quarters of an English mile from the camp, the Indians commenced "a kind of most lamentable weeping and crying out" suggesting that they promptly sacrificed the gifts in one of their ceremonial burnings (*W.E.*, p. 70). Assuming that the Indians did indeed burn the English gifts, and that the gifts did indeed include bracelets and shirts, the ashes of the sacrificial fire would include strings of fused glass beads and probably some metal buttons.

- About seven months before their arrival, the English had plundered 1,770 earthen jugs of "good Chilean wine" at the Spanish port settlement of Valparaíso, Chile (Nuño da Silva's second relation, in Nuttall, p. 261). Among their litter, there was probably an abundance of empty wine jugs.

- Just a few months before their arrival, Drake had pilfered four chests of China dishes from the ship of a Spanish nobleman ("Anonymous Narrative," in Wagner, *Drake's Voyage*, p. 271). Drake might have distributed a few dishes to the various chieftains who arrived.

- During their stay the English had beached the *Golden Hind* close by the encampment (*W.E.*, p. 68). The processes of unloading, cleaning, breaming, caulking, tarring and finally reloading the ship would have resulted in the generation of large amounts of debris. Because they were so heavily laden with their booty, they would have been most anxious to discard anything considered superfluous. They presumably discarded or piled their debris somewhere nearby if they did not toss it in the water.

- Here Drake left behind the small ship (Tello's bark) he had captured from the Spanish. Presumably it was used for firewood (John Drake's second declaration, in Nuttall, p. 51). If so, the campfire residues should contain some of its iron.

- The day of Drake's departure, the Indians lit at least one sacrificial fire near the fort and then ran to the abutting hills and lit fires atop and all around them. In the fire near the fort, they were burning "a chain and a bunch of feathers" (*W.E.*, p. 81).

Answers

To date, there have been no attempts to archaeologically confirm or deny the possibility that Drake encamped at Campbell Cove. In the epilogue, I predict what the archaeologists will find when they come.

Key Points

From what I can tell, there have been only three scientific archaeological investigations conducted on Bodega Head: the first by the University of California Archaeological Survey under R. E. Greengo in the early 1950s; the second under Thomas King and Omar Conger in 1960; and the last in 1962 under state archaeologist Dr. David Fredrickson. When I tried to obtain the Fredrickson report, I found such reports are unavailable to laymen due to understandable concerns about trespassing and looting. I have also seen the picture-catalogue of the late Rose Gaffney's remarkable collection of surface artifacts, a portion of which are now on exhibit at the Bodega Marine Laboratory.[1]

In the absence of any dramatic announcements, we may presume that neither the archaeologists nor Rose Gaffney found anything in the way of Elizabethan California—or did they?

Thomas King's report for his 1960 investigations is entitled: *An Unusual Archaeological Site on Bodega Head, Sonoma County, California*. What were the unusual findings? The archaeologists found evidence of the large-scale ceremonial burning of trade goods—not associated with a burial—that included about fifty clumps of small glass beads among other European items. While an expert reportedly dated the beads and some of the other items to a temporal range of 1830 to 1900,

---

1. I do not think any of the investigations were done at the village site that is referenced in the Drake account. It would be interesting to know if Rose had a separate collection of European/Russian artifacts. I have not seen a record of one.

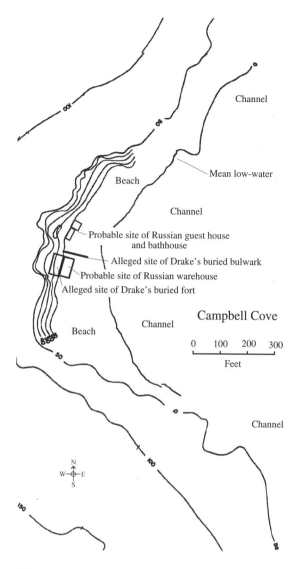

Channel

Mean low-water

Beach

Channel

Probable site of Russian guest house
and bathhouse

Alleged site of Drake's buried bulwark

Probable site of Russian warehouse

Alleged site of Drake's buried fort

Campbell Cove

Beach   Channel

0   100   200   300

Feet

Channel

N
W—⊕—E
S

Fig. 108. The location of Drake's buried fort. *Adapted from a drawing prepared by Dames & Moore, San Francisco, for Pacific Gas & Electric Company, circa 1962, contained in the Public Utility Commission files for Application Number 43808.*

there is some possibility that the clumps of beads—that were found fused in strings of two to fifty—may actually be the remains of sixteenth-century French bracelets.[2]

According to Professor Fredrickson, his research team found "no indication" of any historic structures at Campbell Cove during their 1962 "excavations in advance of Pacific Gas & Electric Company's plant construction," except for what he believed may have been some sort of stone box to collect water from one of the springs. He also told me that they found "no trace of the Russians."[3]

Why no trace of Drake or the Russians? We know for certain that there were three Russian buildings, constructed circa 1812–15, one of which was still standing among the hillocks at the foot of Campbell Cove's rear bluff as late as about 1870. There was a sturdy 70 ft. long by 35 ft. wide warehouse constructed of Redwood planks; a 20 ft. by 20 ft. house; and a bathhouse about 28 ft. long by 14 ft. wide. Records show that these port facilities were in active use first by the Russians, and then by the Americans for the better part of forty years.

As the area in which the Russian structures were located spans about 500 running feet, I can only conclude that Dr. Fredrickson's group, in investigating the hillocks at the foot of the bluff, was working outside of the immediate areas of Drake's fort and the former Russian structures. It may be that they were focusing efforts more on the investigation of Indian village sites than on the search for evidence of former historic structures.

Enter Karl Kortum. First, consider this concise definition from Webster for *retaining wall*: "a wall built to keep a bank of earth from sliding." Now please closely examine figure 109, taken by the late Karl Kortum of San Francisco on October 21, 1962. Kortum's photograph suggests the presence of a buried stone wall running for perhaps sixty or seventy lineal feet along the foot of the bluff. Although

2. The midden is located near the water's edge about a mile away from my alleged fort site. Arthur Woodward, the expert who dated the beads, authored a book entitled *Indian Trade Goods* (1965) that contains a chapter on beads that I reviewed. All things considered, I would encourage the archaeological community to take a close second look at the scorched artifacts. According to King's report, as of 1966 they were in the possession of the Boyd Museum, San Rafael, California.

3. Fredrickson, telephone conversation with author, Dec. 17, 1996. My understanding is that Fredrickson's team investigated two sites along the west shore of the harbor and one within Campbell Cove.

Fig. 109. View of Campbell Cove from the foot of the bluff looking south on the eve of its destruction, October 21, 1962. The child in the foreground is standing at the probable site of the former Russian warehouse. Behind him there appears to be a substantial buried stone wall that by means of its retaining effect, has created a perhaps 70-by-20-foot rectangular indentation among the hillocks lying at the foot of the landslide-prone rear wall of the cove. This is the location where I believe PG&E contractors encountered the buried walls of Drake's fort. *Photograph © by Karl Kortum, October 1962. Courtesy of Jean Kortum, San Francisco, California.*

just one small area of the alleged wall is visible in the photograph (*bottom right*), the retaining effect of the remaining length seems evident.

After Drake and his company departed in 1579, landslides would have eventually filled in the area inside the walls of their fort. We can be confident of this based on documentation that a landslide destroyed the Russian warehouse at Campbell Cove, and that an entire Indian village disappeared into the foot of the bluff: "Bank has dropped and covered site."[4]

---

4. Munro-Fraser, *History of Sonoma County*, p. 183; and Bodega Indian Tom Smith quoted by Kelly, in Collier and Thalman, eds., *Interviews*, p. 68.

Fig. 110. View of Campbell Cove from the bluff looking southeast in the wake of its destruction, October 6, 1963. On what used to be the stretch of beach fronting the cove's former rear bluff, there appears to be a substantial pile of stones and a string of mounds (*boxed area*) lying in front of the area I just pointed to in figure 109. *Detail from a photograph entitled "The Hole in The Head," © 1963 by Karl Kortum. Courtesy of Jean Kortum, San Francisco, California.*

Assuming that landslides did bury the fort, the only visible indications remaining at the time Europeans first arrived at Bodega Harbor, may have been a rectangular indentation at the foot of the bluff caused by the retaining effect of the buried front wall (figure 109) . This indentation would have represented to the Russians an ideal location for constructing their Campbell Cove warehouse.

Enter Pacific Gas & Electric Company (PG&E). While the buried walls of Drake's fort could easily have escaped notice in 1962 by a team of archaeologists, such is not the case for the bulldozer operators who tore into Campbell Cove's rear bluff the year after. If they hit the buried walls of Drake's stone fort and any sizable debris from former Russian structures, the operators would have automatically set at least the larger material aside, if for no other reason than the fact that this is standard practice in sorting excavation spoils for reuse or disposal. If they were saving such stones to preserve the archaeology of the area, they were undoubtedly working under the assumption that the stones were relics of some sort of Russian structure.[5]

When PG&E stopped work on their project in late 1963, its construction contractors had stockpiled about 10,000 cubic yards of soils at the foot of what remained of the south side of Campbell Cove's rear bluff. A photograph taken by Karl Kortum on October 6, 1963 (figure 110) shows this huge stockpile. Lying in front of the area where I contend that Drake built his fort, I can also see what appears to be a substantial pile of stones and a string of small mounds of earth that may or may not be the half-buried stone wall that can be seen in figure 111.[6]

---

5. PG&E had no reason to believe that Drake had built a fort at Campbell Cove. The Drake controversy was raging in the early 1960s, but the focus was entirely on Marin County's candidate anchorage sites.

6. There are several other similar photographs that show the rocks and mounds. See figure 161 of this book. I can not discern the half-buried piles of stone I point to in figure 111 on any of these photos and I am not sure whether the string of mounds correlates

Original foot of the rear bluff

Original shoreline
(mean high water)

Half-buried stone wall
(bulwark?)

Half-buried rock piles
(fort?)

Fig. 111. View of Campbell Cove from the bluff looking north over the site of Drake's former encampment. The half-buried rock piles I point to here lie in the very area where I contend that the Elizabethans built their stone-walled fort. The half-buried stone wall lies in the area where I contend they constructed a bulwark across the beach. *Photograph by the author, 1994.*

Enter the California Department of Parks and Recreation (then Beaches and Parks). Shortly after PG&E sold its Bodega Head holdings to the parks department circa 1973, state contractors removed PG&E's huge stockpile of soils and constructed a parking lot and some recreational facilities. In doing so, however, it seems that either by chance or design they left in place the piles of stones that I believe PG&E encountered in the area where Drake built his fort (figure 111).

In connection with these observations, it is interesting to note that someone filling out a National Park Service information form in the early 1970s reported that "although some historic and prehistoric sites have been damaged or destroyed by road building, plant construction, erosion and vandalism, the majority of prehistoric and historic remains are relatively intact."[7]

---

exactly with the position of the half buried stone wall that can be seen today. This is not to say that the stones are not there, however.
7. The National Park Service Information form (undated) is appended to the following report: *Environmental Impact Statement: Bodega Bay Public Utility District Proposed Sewerage Facilities*, by Jones & Stokes Associates, Sacramento, 1972–74.

CHAPTER 26
Tenet 14 – The Plate of Brass

Before Drake left the bay in which he had anchored, according to "The World Encompassed,"
he "caused to be set up a monument of our being there, . . . namely a plate of brass." This plate
was discovered in 1936 by Beryle Shinn on a hill overlooking San Francisco Bay. After the
discovery was announced, William Caldeira claimed to have previously discovered this same
plate near Drakes Bay in 1933 and to have discarded it near the Greenbrae hillside where Shinn
found it. What is the significance of the plate to the landing site controversy?

Fig. 112. Detail of a radiographic photograph of the Shinn plate revealing the forger's error. Notice how the inscriber attempted to remove the tops and bottoms of his three letter "J's," apparently thinking that he had used the modern form in error. As it turns out, however, Drake may actually have drawn his "J's" in just this fashion, meaning the telltale correction was for naught. *Courtesy The Bancroft Library, Berkeley, California.*

Clues

- The plate was "fast-nailed to a great and firm post whereon is engraven Her Grace's name, and the day and year of our arrival there, and of the free giving up, of the province and kingdom, both by the king and the people, into Her Majesty's hand: together with Her Highness' picture and arms in a piece of sixpence current English money, showing itself by a hole made of purpose through the plate, underneath was likewise engraven the name of our general & c" (*W.E.*, p. 80).
- The plate may have been made of lead ("Anonymous Narrative").

Answers

As discussed in chapter 10, the first team of experts to examine it concluded that the Shinn plate was authentic while the second team concluded it was a hoax. This latter conclusion is expressed in so many words in the plate's display case at the Bancroft Library in Berkeley. Putting aside the work of all the experts, for me, the determining factor is in the lettering as noted by the observant Ethel Sagen of Bakersfield, California. In examining figure 112, it is apparent how the sham artist attempted to hide his telltale error. The three letters "J" each apparently has a cross mark at its top, and a cusp at its bottom that is partially obliterated by peening. It seems that the forger discovered too late that the Elizabethan alphabet may not include the letter "J."

Even in the very unlikely event that the plate is eventually proven authentic, I would not consider the discovery point of such a portable article a definitive clue in pinpointing the location of Drake's anchorage. The site of Shinn's 1936 discovery is not far from Mission San Rafael, the location where many of the Coast Miwok population met their tragic end.

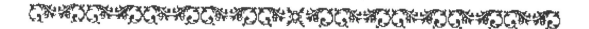

CHAPTER 27
Tenet 15 – The Isles of Saint James

Drake departed from Nova Albion on July 23, 1579, and "The World Encompassed" reported that "not far without this harbor did lie certain islands . . . one of which we fell July 24." What is the significance of the dates and apparent time elapsed on this leg of his journey?

Fig. 113. The "Isles of Saint James." The rocky Farallon Islands are an easy two-day sail from Bodega Harbor, especially with the winds northwest. *Reproduced from "Nova Albion, 1579" by Alexander McAdie, in AMERICAN ANTIQUARIAN SOCIETY PROCEEDINGS, October 16, 1918.*

Clues

- Drake departed his California port on July 23, 1579, and on July 24, "fell" on one of a group of "certain islands" having "great stores of seals and birds" (*W.E.*, pp. 81–82).
- After anchoring off the island, some of the crew went ashore to bolster their supplies of birds and seals (*W.E.*, p. 82).
- They departed the islands and set sail for the Moluccas on July 25, 1579 (*W.E.*, p. 82 and Stow).
- It appears that they named the islands to commemorate the date they departed from them; the Christian church celebrates the vigil of Saint James' feast on July 25 (*W.E.*, p. 81).

Answers

Concerning the dates, the evidence is 100 percent consistent with a July 23 departure from Bodega Harbor. With the winds northwest, the Farallones are an easy one-to-two-day sail from Bodega Harbor, depending on the time of departure, fog, wind and seas. In October 1775, Bodega y Quadra,

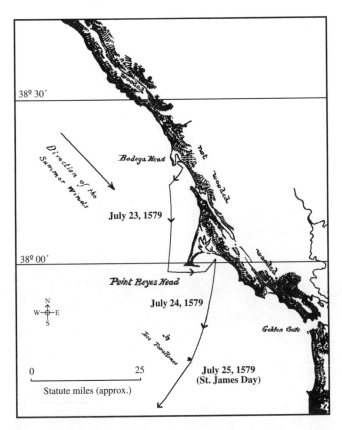

Map 17. Relief map of the San Francisco area showing Drake's likely route of departure. The Elizabethans probably spent the night of July 23, 1579, anchored under the protection of Point Reyes Head. *Redrawn and adapted from Chart no. 1 in IDENTIFICATION OF SIR FRANCIS DRAKE'S ANCHORAGE ON THE COAST OF CALIFORNIA IN THE YEAR 1579 by George Davidson, California Historical Society, San Francisco, 1890.*

in command of the next square-rigger to depart Bodega Bay and sail south, likewise fell upon the Farallones the next day.[1]

Concerning the appellation, the evidence is consistent with the English first landing on the Farallones upon their departure.

Counter-Rebuttals

According to Raymond Aker, the Farallones do not lie "without" Bodega Bay, are screened from sight by Point Reyes, and, given a departure from Bodega Bay, would have been well out of the way on Drake's course for the Moluccas.[2]

First of all, the Farallones do indeed lie "not far without" Bodega Bay, especially when you appropriately consider that the Point Reyes Peninsula forms Bodega Bay's southern headland (see Tenet 4). Secondly, as discussed under Tenet 3, there is every reason to believe that Drake, during the course of his stay at Bodega Harbor, would have conducted enough coastal reconnaissance to determine what lay on the lee side of Point Reyes Head—namely the white cliffs lining the inner shores of the huge headland and the seal-populated Farallones lying due south. Finally, even in the unlikely case that Drake did not have the inclination or occasion to explore the Gulf of the Farallones before departing Bodega Harbor, there is solid evidence in *The World Encompassed* that suggests he did so after. On page 82, the account expressly states that Drake did not decide to give up the search for the Straits of Anian and sail for the Moluccas until July 25, the day the English departed the Isles of Saint James. It necessarily follows, that when he set sail from the Portus two days earlier, on July 23, he was not on a course for the Moluccas. Given the constant north to northwest winds, we can safely assume that he was heading for the lee side of Point Reyes Peninsula, either to explore the area or in search of additional provisions.

---

1. Mourelle, in Russell, ed., *Voyage of the Sonora*, pp. 57 and 73.
2. Aker, *Report of Findings*, p. 369; and conversation with the author, November 1996.

CHAPTER 28
Tenet 16 – The Portus Plan

The Vera Totius Expeditionis Nauticae map of the world by cartographer Jodocus Hondius (London, 1589) contains in its left-hand corner a plan of the Portus Nova Albionis. Discuss the significance of the Portus plan and its correlation with the proposed landing site.

Now we come to the most tantalizing and controversial clue of them all: a contemporary drawing apparently copied from a Drake original showing the lost harbor on the day of departure complete with ship, fort, Indians lighting fires on the hillsides, and a troop of Indians preparing to descend a trail that will take them to the fort site.

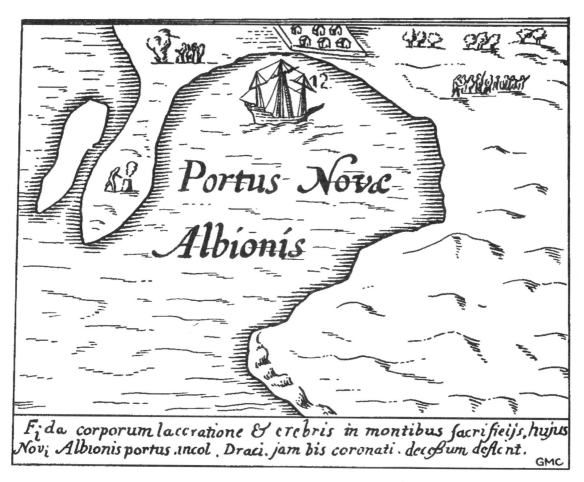

Fig. 114. The Portus Nova Albionis inset from the Hondius Broadside Map (facsimile). The caption (translated from Latin) provides an important clue to the anchorage location: "With appalling lacerations of their bodies and numerous sacrifices <u>in the hills</u>, the inhabitants of this port of New Albion lamented the departure of Drake, whom they had already twice crowned." *Reproduced from LOST HARBOR by Warren Hanna, University of California Press, 1979, courtesy of Mrs. Warren Hanna, Kensington, California, and the University of California Press, Berkeley, California.*

The "Portus Nova Albionis": the name itself seems to cast a spell. Before we start to compare the Portus features with Campbell Cove, let us sound the toll of the victims who have succumbed to its siren's call:

- John Dwinelle, 1878, and Edward Berthoud, 1894, for Bodega Harbor.
- George Davidson, 1889, for the cove under Point Reyes Head (Drakes Bay).
- Henry Wagner, 1926, for Trinidad Bay.
- The Drake Navigators Guild, 1952, for the cove at the mouth of Drakes Estero.
- Douglas Watson, 1937, and Robert Power, 1954, for upper San Francisco Bay.
- Alan Brown, 1960, for Half Moon Bay.
- Robert Becker, 1960, for the cove inside Tom's Point at Tomales Bay.
- Robert Pate, Richard Dobson, and Ethel Sagen, 1967, for Mallagh Landing, San Luis Obispo Bay.
- Don Viles and Wayne Jensen, 1971, for Nehalem Bay, Oregon.
- Aubrey Neasham, 1974, for Bolinas Lagoon.
- Robert Thomas, 1978, for the mouth of the Petaluma River, San Pablo Bay.
- George Epperson, 1980, for Agate Beach, on the north side of Bolinas Bay.
- Robert Ward, 1987, for Whale Cove, Oregon.
- Justin Ruhge, 1991, for the Goleta Slough, Santa Barbara Bay.
- Gerald Weber, 1993, for Point Año Nuevo Cove.

I honestly held out for as long as I could, never counting on the Portus to make my case. It was the half-buried stones that first drew my attention to Campbell Cove, not the Portus. Nevertheless, like Raymond Aker, I had immediately noted some correlation between Campbell Cove's shorelines and "the hard features of the inset," and with my 1958 and 1962 aerial photos tucked away in a drawer, I was looking forward to the day I would start in on this chapter, confident that I could use the inset to some extent as corroborating evidence.[1]

It was only when I started cutting and pasting to make my comparisons that I too was smitten. You can add me to the long list of those who clearly discern his chosen harbor in the amoebic shape of the Portus. But this time, perhaps others will agree that I have finally placed the slipper on the proper foot.

I believe I have stumbled upon the key to unlocking the secrets of the Portus. It seems that Jodocus Hondius, working from what was probably a tattered and faded original, misinterpreted the manner in which Drake (or perhaps it was his young cousin) drew in the shorelines, bluffs and hills of their California port. The horizontal lines they used to depict cliff-banks and hillsides too closely resembled the markings trained cartographers used to show shorelines. Look what happens to the Portus plan when you reinterpret the drawing to account for this transparent error and compare it to a corresponding topographic view of Campbell Cove (figure 115).[2] Bear in mind that in making his drawings, a sixteenth-century "painter" often endeavored to draw a "bird's-eye view" of the coast as seen from the ship, producing a map and view combined.[3]

Assume that Drake sketched the Portus plan while anchored about a half-mile off Doran Beach, directly opposite the entrance to the cove (figure 85). See how the secrets unfold for both the island (he drew in the twin points, the gulch, and the hillside), and the dead end (he could not see the channel as it wound into the lagoon).

---

1. Aker, *Report of Findings*, p. 370.
2. I suspect the error stemmed from Hondius merging the striations Drake used to depict the double-pointed headland with the striations he used to illustrate the elongated hill running along the south side of the headland.
3. Wallis, "The Cartography of Drake's Voyage," in Thrower, ed., pp. 124–25.

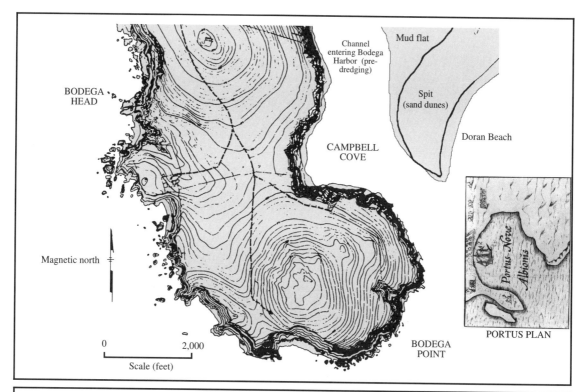

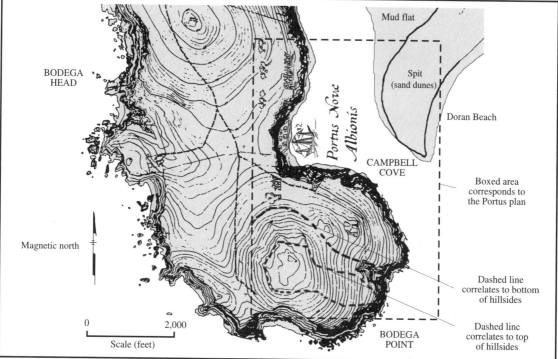

Fig. 115. The true Portus correlation. *Above*, I compare the Portus plan to a topographic map detail of Bodega Head aligned to magnetic north. The correlation is based upon the premise that the so-called Portus island actually represents the hill forming the southern side of Point Bodega and that the Portus, standing on end, is oriented to magnetic north. *Below*, I have marked up the Bodega Head topographic map to further illustrate the correlation. The dashed lines on Bodega Point correspond to the tops and bottoms of hillsides (see figures 116 and 117). Note that the Portus now correctly correlates with its caption that talks of Indians lighting sacrificial fires <u>in the hills</u> to mourn Drake's departure. *The topographic map detail is adapted from a drawing prepared by Dames & Moore, San Francisco, for Pacific Gas & Electric Company, circa 1962, contained in the Public Utility Commission files under Application Number 43808.*

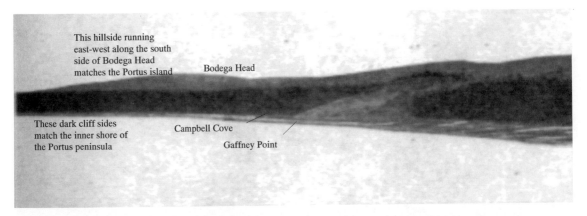

Fig. 116. Profile view of the inner (north) side of Bodega Head looking south across the harbor. Note how well the aspect of the hill rising above the dark cliffs forming the south side of Bodega Head correlates with the Portus island and peninsula. *Reproduced from a photograph in HISTORY OF SONOMA COUNTY by Honoria Tuomey, S. J. Clarke, 1926.*

Fig. 117. View looking south over the hilly, double-pointed headland forming the south side of Campbell Cove. *Reproduced from an aerial photograph taken by William Garnett that appeared in PACIFIC DISCOVERY, March–April 1959. Courtesy of the California Academy of Sciences, San Francisco, California.*

Fig. 118. Cartographic detail showing how early English navigators depicted hillsides and shorelines on their charts and sketches. Look at the way the hillsides are drawn on the islands and then scrutinize the Portus peninsula and island. It is not difficult to discern how a misinterpretation could have occurred. *Reproduced from a chart of the Galapagos drawn in 1793 by Captain James Colnett of the Royal Navy in his journal, VOYAGE TO THE SOUTH ATLANTIC . . . , Da Capo Press, New York, 1968.*

Geologic Considerations

The twin points – The gulch forming the double points at the eastern tip of Bodega Head was probably more pronounced in Drake's day. This gulch seems to lie along a fault line extending about halfway across the head. It seems likely that landslides and erosion have filled in what, in the recent geologic past, was a better-defined cleft. In addition, there has probably been some significant erosion of the more seaward (southerly) of the two points.

Doran Point and its mud flat – The spit forming Doran Beach is obviously a changeable sand shoreline. It could very well have correlated even better with the corresponding Portus shoreline in Drake's day. On figure 115 (*bottom*), I show a mud flat on the north side of the spit at the entrance to Campbell Cove that significantly improves the respective shoreline correlation. This mud flat, originally exposed throughout most of the tidal cycle, was largely eliminated in the early 1940s when the U.S. Army Corps of Engineers opened up the harbor by dredging today's artificial entrance channel. The early charts reveal that the channel originally turned west into Campbell Cove and then hugged the western shoreline (see figure 85).

The cove's eroding back wall – From figure 115 (bottom) we can see that Campbell Cove's bluffs, especially the rear wall, were subject to landslides. While this did not affect my Portus correlation, it helps explain why Drake's fort site, located among the "knobby hillocks" (landslide debris) at the foot of this bluff, may have escaped notice when Europeans and Russians began exploring and settling the area in the late 1700s and early 1800s (see chapter 35).

Key Points

Unlike so many of the other anchorage sleuths, I do not consider the Portus plan the most important clue in identifying Drake's lost harbor. History has shown that the Portus can be successfully correlated to varying degrees with many harbors on the coasts of California and the Pacific Northwest. Of the evidence I have presented, I rate the latitudes and the tentative archaeological findings (half-buried walls and piles of stones) higher. I do consider my Portus correlation to be very powerful corroborating evidence, however, based on the following considerations:

The orientation – Assuming Drake sketched the Portus plan by compass, as is likely, it should align to a magnetic north axis. Assuming the train of Indians shown on the Portus is arriving from the village that *The World Encompassed* describes as lying three-quarters of a mile upwind (north) of the fort site, the north axis, as the Portus lies on the Hondius map, would be horizontal. Bodega Head/Campbell Cove oriented to magnetic north aligns exactly with the Portus plan when it is oriented to its most likely magnetic-north axis.

The seal head – The distinctive seal-head shape of Campbell Cove correlates closely with the peculiar seal-head shape of the Portus—or, if you will, in either cove one can discern "the outline of a knight's helmet with the visor up."[4]

The peninsula and island – The aspect of the hilly, double-pointed peninsula forming the south side of Campbell Cove correlates closely with the Portus peninsula and its adjoining island.

The Indian trail – The old Indian trail cutting down across the face of the bluff on the north side of Campbell Cove (figure 119, left view, upper left) correlates closely with the trail similarly depicted in the Portus plan (figure 114, upper right). This trail once led to the former Coast Miwok/Pomo village area on Gaffney Point, located three-quarters of a mile away in the direction from which the Portus Indians are heading.

The trees – On Bodega's windswept headland, you can find a small grove of gnarled cypress located just south of where indicated on the Portus plan.

---

4. Ruhge, *Drake in Central California*, p. 21.

The sand dunes – The only overt sand dune symbols on the Portus plan correspond to the rolling dunes forming Doran Beach spit.

The notation "12" – Assuming this is a sounding, it correlates well with the navigable depth of Campbell Cove either at its barred entrance at high tide, or at just this point in the channel at low tide. In the case that the notation actually depicts an Indian tending a fire, as contended by the Drake Navigators Guild, it would correlate with an Indian lighting a fire on the beach near a Campbell Cove fort site.

The fort – There are piles of half-buried stones at what was once the foot of Campbell Cove's rear bluff that are located in the area correlating to the Portus fort site.

Fig. 119. Two aerial views showing the location of Drake's former encampment at Campbell Cove: 1962 (*left*), a year prior to PG&E site work showing the proposed access road (*white line*); and 1980 (*right*), pretty much as the area looks today after PG&E aborted its construction project. The boxed area on the two photographs correlates with the position of the fort site on the seal head of the Portus plan. Today this area is still heavily littered with half-buried piles of stones. Note also the very close correlation between the Indian trail on the Portus plan (*figure 114, upper right*) and the trail that once descended from bluff to beach on the north side of Campbell Cove (*left view, upper left*). The 1962 photo is reproduced from an aerial photograph in the the Public Utility Commission files under Application Number 43808; the 1980 photo courtesy of Pacific Aerial Surveys, Oakland, California.

CHAPTER 29
Tenet 17 – The Montanus Illustration

What is the significance of Arnold Montanus' illustration, "The Crowning of Drake," which was published in Die Nieuwe en Onbekonde Weereld of Beschryvning van Americo en 1 + Zuid-Land . . . (Amsterdam, 1671)?

Fig. 120. "The Crowning of Drake": an illustration by Arnold Montanus from the German edition of DIE NIEUWE EN ONBEKONDE . . . , Amsterdam, 1673. It has no value as a clue to the anchorage location. *Reproduced from SEVENTY-FIVE YEARS IN CALIFORNIA by William Heath Davis, John Howell, San Francisco, 1929.*

Answer and Comments

The Arnold Montanus illustration of Drake's coronation by California Indians is from a Dutch world history written in the late seventeenth century. Robert Power introduced it to the anchorage controversy by noting geographic features that bear a "startling resemblance to those of the Point San Quentin area."[1]

Although Montanus engraved the illustration almost a century after the crowning incident took place, Power contended that the illustrator likely used firsthand information in its preparation. Power noted that Montanus, a Flemish craftsman, was related to Jodocus Hondius, presumed author of the Portus inset drawing, and could have inherited any of Drake's original paintings and sketches that Hondius had come to possess.[2]

The Drake Navigators Guild, while at first skeptical of the illustration's value as a clue, eventually came to accept Power's thinking that Montanus had prepared it from an original Drake sketch. The Guild, however, crossed Power up by demonstrating a geographic correlation for their Drake's Cove anchorage site that is every bit as convincing as Power's correlation for Point San Quentin."[3]

Clearly, the Montanus illustration can not be considered an important clue. From the positions of the tents and Indians filing down the face of the bluff, it does seem apparent that the illustrator made use of the Portus plan. To assume that he used something preliminary to the Portus plan, however, is utter and highly dubious speculation.

There are more good reasons for dismissing the evidentiary value of the Montanus illustration. As noted by Professor Robert Heizer, the figure is very careless in its historical content: Drake is beardless; Drake and his crew are dressed in seventeenth-century attire; the *Golden Hind* is under full sail with the captain and his crew still ashore; and the plate of brass is depicted as a banner.[4]

The caption under the Montanus illustration reads as follows:

The King of Nova Albion or New England, so called because the English have become very powerful in this country of America situated in the western portion of Mexico near the kingdoms of Anian, Tolm, Conibas, Totoneac, and New Granada and California. It is separated by a great river from Canada as well as New France. It limits Norumbega to the westward; to the east great rocks surround it and the sea is upon its north. The English do much business there.

Its people are civilized, particularly the sovereign, the grandees, the priests of law and the magistrates. They observe justice carefully. They retain still some of the ancient idolatries. They believe nevertheless in the immortality of the soul. The sovereign and the grandees are curious and magnificent in their clothing made of beautiful skins embroidered with precious stones and fastened with golden threads. These they carry on their shoulders—their heads are ornamented with fine feathers and they deck themselves with necklaces, bracelets of pearls and precious stones.[5]

---

1. Heizer was actually the first anchorage researcher to publish the Montanus illustration in his *Francis Drake and the California Indians, 1579* (1947).
2. Power, "Drake's Landing," p. 119.
3. Aker, *Summation*, pp. 76–78.
4. Heizer, *Elizabethan California*, p. 27.
5. William Heath Davis, *Seventy-Five Years in California*, p. x.

CHAPTER 30
Tenet 18 – The Cermeño Account

What is the significance of the declaration by the Spanish Captain Sebastián Rodríguez Cermeño and others who were shipwrecked in 1595 in the bay presently known as Drakes Bay?

Answers

The Cermeño declaration provides no evidence, pro or con, for Campbell Cove. It does provide some powerful evidence, however, that Drake never set foot on the shores of today's Drakes Bay. The following excerpts from Henry Wagner's translation of the account contain the key statements in this respect (*emphasis added*):[1]

> On the day on which the ship anchored in the bay, about four o'clock in the afternoon, many Indians appeared on the beach and soon one of them got into a small craft which they employ, like a *cacate* of the lake of Mexico. He came off to the ship where he remained quite a time talking in his language, no one understanding what he said. Some cotton cloth and silk things were given him and a red cap. He took them and went back to land. Early the next morning, the seventh, four other crafts like the first came out from land to the ship, and in each one was an Indian. They came alongside where they remained some time talking in their language. Captain Sebastián Rodríguez gave each of them some cotton cloth and taffetas and entertained them the best he could. They went ashore and the captain at once embarked in the ship's boat with twenty-two men, seventeen being harquebusiers with their arms. Accompanying him were Captain Francisco de Chaves and his ensign, the sergeant, and the corporal and three men with shields. These went ashore with the Indians and landed on the beach of the port near some of their underground habitations, in which they live, resembling caves and like those of the Chichimecos Indians of new Spain.[2] They are well-made people, robust and more corpulent than the Spaniards in general. They go naked without covering and with their private parts exposed, but the women cover theirs with straw and skins of animals. Their arms are bows and arrows. They wear their hair long and have no beards; any one who has any removes it. They are painted on the breast and on certain parts of the arms, but the paint is not so decorative as with the Chichimecos.
>
> On the same day the captain went ashore with his men, he asked all to witness that he took possession of the land and port in the name of the king, our master. He gave it the name "La Baya de San Francisco," and the Reverend Father Fray Francisco de la Concepción of the order of the barefoot Franciscans, who comes in the ship, baptized it. The captain, with his ensign, Juan del Rio, carrying the banner, and the sergeant with the men in order went marching to a village which is about a harquebus-shot from the beach. *Here all the Indian men and women, perhaps fifty all told without counting the children, were looking on with great fright in seeing people they had never seen before.* They were all very peaceable and had their arms in their houses, it not being known up to that time that they had any. They produced a seed the shape of an anise seed, only a little thinner, and having the taste of sesame, of which they make the bread they eat. *Their food consists of crabs and wild birds,* which are in great abundance near where they live, and many *deer, as these have been observed going about.* They are beyond comparison the largest that have been seen, as will be apparent from the horns which were found, of which the captain carries a sample . . .
>
> November 30, the ship *San Agustin* having been lost with the food supply in her, the captain left the camp with eleven Spaniards and some Indians and slaves with their arms on an expedition inland to

---

1. Wagner, *Spanish Voyages*, pp. 158–60. This is Wagner's translation of Cermeño's declaration given before Pedro de Lugo, scrivener of the king.
2. According to Wagner (*Spanish Voyages*, p. 370), the Spanish used the term *Chichimecos* generically for all roving Indians of New Spain. In Cermeño's narrative of the voyage, he further describes the lodges as "covered with grass." Collectively, Cermeño's descriptions seem to fit the semi-subterranean, dome-shaped, grass-thatched lodge common among many Californian and Mexican tribes. As pointed out by Robert Thomas in his arguments for a San Pablo Bay anchorage, this description does not correlate very well with the cone-shaped, earth-covered lodges Fletcher described.

Fig. 121. Deer at Point Reyes: engraving by H. Eastman entitled "Sir Francis Drake's Bay or Jack's Harbor," circa 1855. *Reproduced from THE ANNALS OF SAN FRANCISCO by Frank Soule et al., D. Appleton, San Francisco, 1855. Courtesy The Bancroft Library, Berkeley, California.*

hunt for food to sustain those who had survived the wreck, as there was nothing to eat. Having journeyed three leagues they found three villages of Indians of the same class and character as those above described, who were settled somewhat apart from each other on an arm of a river of sweet water on the banks of which there were many trees bearing acorns, hazelnuts of Castile, and other fruit of the country, madrones, thistles, and fragrant herbs. From here they brought that day and on another day, December 2, a quantity of acorns and another round fruit like a hazelnut, and a quantity of thistles with which the men were sustained and which was to be taken in the launch.

Why were the Indians expressing such fright just sixteen years after Drake's visit—and what about those huge deer the Indians were dining upon? [3]

3. Aker, *Report of Findings*, pp. 308 and 310, argued that the Spanish saw the deer after they journeyed inland, citing the declaration of Cermeño's ensign, Juan del Rio, who described sighting "a great quantity of deer horns" during their trip inland to Olema Valley in December 1595. Aker also offered the dubious argument that even if the English saw deer along the shores of the estero, they were "not of sufficient size to occasion any special notice."

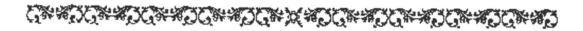

CHAPTER 31
Tenet 19 – The Dudley Maps

Discuss the significance of the cartographic information about the California coast contained in Robert Dudley's 1647 atlas, Arcano del Mare, and manuscript chart.

Fig. 122. California detail from three Robert Dudley maps, drawn circa 1630: the Carta Prima General del India . . . (*left*); the Carta Particolara chart number 33 (*middle*); and unpublished manuscript chart number 85 (*right*). They provide no significant clues to the anchorage.

Answer

I do not think the Dudley maps have any value whatsoever in identifying the anchorage. First of all, in comparing Dudley's charts to a modern atlas one can see that he did not hesitate to use his imagination in depicting uncharted shorelines. Secondly, in his placing of bays and harbors on the California coast, I see nothing to indicate that Dudley used anything more in the way of contemporary information than we have on hand today: specifically "The Famous Voyage," *The World Encompassed*, the Hondius Broadside Map, and some of the accounts and maps of the 1603 Sebastián Vizcaíno expedition. The fact that he provided three substantially different charts of the same area indicates that he was justifiably confused by the information he had on hand.[1]

The Point Reyes Peninsula is the most prominent feature on the coast of Northern California. From the "Isles of St. James" episode in *The World Encompassed*, we know that Drake spent a full day at anchor at the Farallon Islands from which point he had a sweeping view of Point Reyes and the Gulf of the Farallones (fog permitting). On this basis, assuming Drake did indeed anchor in Bodega Bay, the map he presented to Queen Elizabeth upon his return to England in 1580, by all rights, should have depicted a large gulf lying just south of his California landfall.

We have two maps drawn circa 1581–85, both recognized as close representations of the queen's map, that do indeed depict a large gulf lying just south of their respective California landfalls: The French Drake Map (Figure 24) and the Drake-Mellon Map (reproduced inside the rear cover of this book). From figure 123 you will note that both of the early maps show two bays in the area of the California landfall. On the French Drake map, that bears the notation "seen and corrected by Francis

---

1. Henry Wagner, one of the foremost experts on the cartography of the Pacific Coast, shared these views.

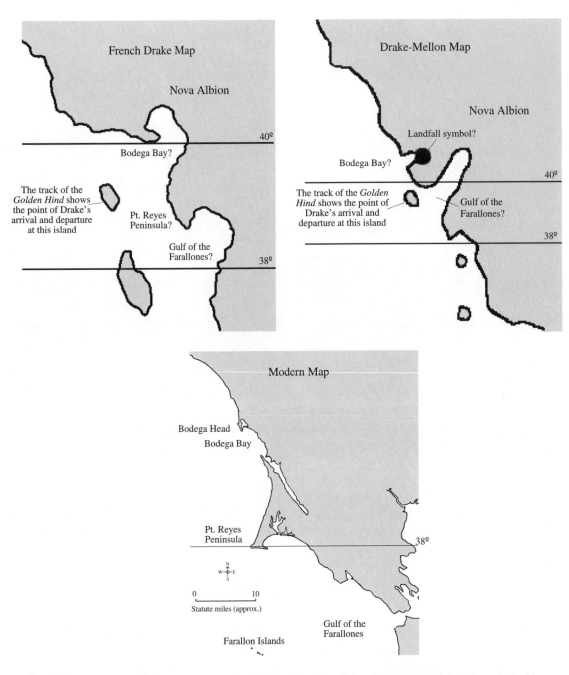

Fig. 123. Correlation of California detail on the French Drake Map (*above left*) and Drake-Mellon Map (*above right*) with coastal features of the greater San Francisco Bay area (*below*). Both early maps—redrawn here in outline—show two bays in the vicinity of their designated California landfalls. In each case, the track of Drake's voyage points to the more northerly bay. In redrawing the early maps, I added the two parallels at 38 and 40 degrees according to their respective latitude scales. Considering the large scale of the maps, it should be noted that it was impossible for the mapmakers to delineate closely adjoining bays at cartographically correct positions. For this reason, we should put no great credence in the latitudes at which the bays appear to fall on the maps.

Drake," the inlets fairly resemble Bodega Bay and the Gulf of the Farallones. On both maps, Drake's track is drawn in such a way as to suggest that his California landfall was in the more northerly bay.

The two maps generally acknowledged to be the earliest and most faithful copies of the queen's map point us to Bodega Bay.

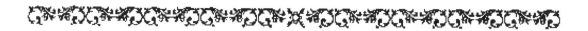

CHAPTER 32
Tenet 20 – The N. de Morena Story

What is the significance of the account of Drake's pilot, N. de Morena, who claimed he had been put ashore at an "arm of the sea" which he believed connected with the Gulf of California and the Atlantic Ocean?

The fascinating story that follows was laid down in 1626 by a Spanish priest named Father Gerónimo de Zárate y Salmerón. He reported obtaining it from a reliable source: Father Antonio de la Ascensión, a friar of the Barefoot Carmelites, who had participated in the 1603 Sebastián Vizcaíno expedition:

> A foreign pilot, named N. de Morena, who steered the Englishman from the sea of the north to that of the south through the Strait of Anian [Magellan], gave his narrative to Captain Rodrigo del Rio, governor that then was of New Spain. When the Captain Francisco Draque returned to his country, this pilot – who had come emerging from the strait in his company – was very sick, and more dead than alive; and to see if the airs of the land would give him life, as a dead thing they put him ashore. The which [the pilot] in a few days recovered health and walked through that land for the space of four years. He came forth to N. M., and from there to the mines of Sombrerete in search of said Rodrigo del Rio. And the said pilot recounted to him the following:

> Having given a long narrative of his much wandering, he told him how the said Englishman, Francis Drake, in the stopping place of the Straits of Anian, had put him ashore for the reason aforesaid, and that after he had recovered health he had traveled through divers lands, through many provinces, more than 500 leagues of mainland, until he came far enough to catch sight of an arm of the sea which divides the lands of New Mexico from another very great land which is on the side of the west. And on the coast of that sea were many great settlements, among the which is a nation of white people, the which are accustomed to go on horseback, and fight with lance and dagger. It is not known what nation this may be. The said Father Fray Antonio says he believes they are Muscovites. I say that when we see them we shall know who they are [i.e., that they were Spanish then settled in Sonora]. This pilot told how this arm of the sea runs from north to south; and that it seemed to him it went on to the northward to connect with the harbor where the Englishman had put him ashore. And that on the sea coast he had seen many and good harbors and great inlets; and that from the point where they put him ashore he would venture to get to Spain in forty days in a good ship's tender, and that he must go to get acquainted with the court of England.[1]

> He offered himself to take the said Rodrigo del Rio to the stopping place of the arm of the sea which he discovered; and said that he could easily cross him over to the other side.

> This arm of the sea is held to be a sure thing. It is that of the Gulf of California, called Mar Rojo; and the land which is on the other side is that of the Californias. As they told me it, so I set it down, without quitting nor adding anything of my own part.

Answer

Since it is obviously at least a thirdhand story, the N. de Morena account can not be considered reliable evidence. Furthermore, even assuming the pilot's story is true and uncorrupted, it does not help in identifying the location of Drake's California anchorage. The account does not make it clear as to whether Drake put de Morena ashore at his first stop on the Pacific Coast near forty-eight degrees, or the second, near thirty-eight degrees. It actually seems to suggest the former as it says

---

1. It is interesting to note that on Drake's last voyage in 1596, he also had aboard a captured Portuguese pilot by the name of Moreno (Andrews, *The Last Voyage*, p. 191). Perhaps N. de Morena did indeed "get acquainted with the court of England."

nothing about an encampment, implies that the pilot was dropped off by himself, and suggests that at the time, Drake still maintained confidence in their search for the strait.

There is also some cartographic evidence that the bay where de Morena convalesced was north of forty-two degrees. Father Ascensión apparently combined the de Morena story with the findings of the Vizcaíno expedition in a map he prepared after the expedition. Although Ascensión's map has not been found, an Englishman, Henry Briggs, wrote a treatise in 1622 in which he claimed to have seen a Dutch map that was evidently derived from it. This map showed California as an island with the Gulf of California stretching from twenty-two to forty-two degrees.[2]

Fig. 124. Detail of North America from the Henry Briggs Map, drawn circa 1622. Henry Briggs' depiction of California as an island may have had its origin in the N. de Morena story. The myth was not dispelled until the Spanish settled upper California a century and a half later. This map also reveals the origin of another California myth; it seems that the Englishman transformed the Spanish name for the bay under Point Reyes located at about thirty-eight degrees from the Spanish "P. S. Francisco," to the English "P. S. Francisco Drake" (*boxed area on map*). The Briggs treatise and the subject map appeared in PURCHAS, HIS PILGRIMS by Samuel Purchas, William Stansby, London 1625.

2. Wagner, *Spanish Voyages*, pp. 386–87.

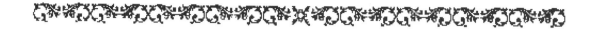

CHAPTER 33
Rebutting Arguments

Without getting into great detail, I will provide here a brief summary of the key arguments against the other candidate anchorage sites.

Tenet 1 – The Approach

To accept the notion that Drake anchored at any of the bays south of Bodega, excepting Tomales, we must assume that either he did not see Bodega Bay or rejected it. Given that vessels from three out of the next four expeditions into the area ended up at anchor in Bodega Bay, we might say the chances are three to one against this. As Raymond Aker put it: since "his purpose was to find a port to careen and refit his ship, logically, he must investigate this haven. He found a suitable cove . . . and so went no farther south."[1]

Tenet 2 – The Latitudes

The latitude evidence eliminates all candidate sites except Bodega Harbor. In order to accept the notion that Drake anchored at any other bay, one must reject the clearcut evidence provided in *The World Encompassed*. One must reject it despite the fact that there is every indication that the compiler (as advertised on the book's cover and introduction) was faithfully reporting from firsthand accounts, and not inventing such details. And one must reject it despite the fact that Hakluyt's "The Famous Voyage" confirms that Drake's California landfall was "within 38 degrees toward the line." In short, one would be rejecting it without a single valid reason.

On the other hand, there is absolutely no good reason for treating "within 38 degrees towards the line" from "The Famous Voyage" as a precise measurement. Even the casual reader can discern that Hakluyt provided such approximations to help identify major turning points in the voyage or important geographic landmarks—not to pinpoint the location of landfalls.

Tenet 3 – The White Cliffs and Terrain

The discriminating clue with regard to the coastal vista is the observation of "low plain land," not white cliffs. At Half Moon Bay, Trinidad Bay and all the Marin County sites, the vistas are marked in some directions by prominent coastal mountain ranges, not low and reasonable plain. For instance, according to Aker this is the view to which Drake and his crew would have been treated as they rounded Point Reyes: "a rugged coast backed for the most part by mountains rising steeply out of the sea just like the miles of unfriendly shoreline passed to northward." San Francisco Bay is largely enclosed by coastal mountains.[2]

In addition, the predominant vistas at these sites do not consistently fit the description of squalid, unhandsome and deformed country. Even at Tomales Bay and Drakes Bay where such vistas prevail, the tree-capped hills of Inverness Ridge can be clearly seen along the coast.

Tenet 4 – The Good Bay

The initial anchorage had to be good enough to allow Drake to feel safe and secure at anchor for three full days while he was making his explorations. It also had to be a great way from shore but within earshot of the Indians standing on it, and on a direct line-of-sight from the inner anchorage and fort site. As such, it is hard to believe Drake would have elected to lie in the open roadstead off the mouth of Drakes Estero when he could have easily moved the *Golden Hind* into the shelter of

---

1. Aker, in Holliday, "The Francis Drake Controversy," p. 271.
2. Aker, *Report of Findings*, p. 265.

Davidson's Cove. The estero would have made a fine harbor, but for reasons discussed below for Tenet 5, it is doubtful Drake could have entered the estero at all, much less fully laden on the day of arrival. If he had, the Indians on the beach would not have been "a great way from shore." It is inconceivable that Drake could have laid at anchor off the mouth of Tomales Bay while a great way from shore. His vessel probably would have been swamped by the huge breakers and turbulence that forms off the mouth during the ebb tide. If he had entered the mouth of Tomales Bay as Bodega y Quadra did in 1775, his initial anchorage would not have been "a great way from shore."

The terms "fair and good bay" and "convenient and fit harbor" do not aptly describe San Francisco Bay, one of the most magnificent natural harbors in the world. As Hubert Howe Bancroft put it:

> That Drake and his men should have spent a month in so large and so peculiar a bay without an exploration extending thirty or forty miles into the interior by water; that notes should be written on the visit without a mention of any exploration, or of the great rivers flowing into the bay, or of its great arms; that Drake's companions should have evaded the questions of such men as Richard Hakluyt, and have died without imparting a word of the information so eagerly sought by so many men is indeed incredible. For sailors in those days to talk of inlets they had never seen was common; to suppress their knowledge of real inlets would indeed have been a marvel. Drake's business in the north Pacific was to find an interoceanic passage; if he abandoned the hope in the far north, one glance at the Golden Gate would have rekindled it; a sight of the far-reaching arms within would have convinced him that the strait was found. . . . That a man like Fletcher, who found scepters and crowns, and kings among the central Californians, who found the likelihood of gold and silver where nothing of the kind ever existed, who was so nearly frozen among the snow-covered California hills in summer, should have called the anchorage under Point Reyes, to say nothing of Bodega, a fine harbor would have been wonderful accuracy and moderation on his part. But supposing San Francisco Bay to have been the subject of his description, let the reader imagine the result. The continent is not broad enough to contain the complication of channels he would have described.[3]

Tenet 5 – The Inner Harbor

While Drake may very well have been able to careen the *Golden Hind* at any of the candidate sites, Campbell Cove is the only place, excepting Brazil Beach at Tomales Bay and the Guild's Drake's Cove (hypothetically) where he could have "graved" her. The proper combination of conditions simply did not exist at the others.

I have found no records to indicate that anything other than low-draft coastal schooners ever entered Drakes Estero or Bolinas Lagoon. All the nautical charts that have been prepared since Davidson first charted the area in the 1860s indicate that their barred mouths were too shallow for even an empty *Golden Hind* to have entered without extreme difficulty. They also show Drake's Cove at the mouth of Drakes Estero as completely shoal.

The accounts of the Cermeño and Vizcaíno coastal reconnaissance expeditions suggest that estero entrance conditions were much the same in the sixteenth century as they are today. They document that neither Cermeño nor his chief pilot Francisco Bolanos considered the estero a suitable port for the *San Agustin*, a vessel not much different in size and draft from the *Golden Hind*. Bolanos, who was very critical of his captain's decision to anchor in the open road off the mouth of the estero, reported only one good anchorage at Point Reyes and that was under Point Reyes Head (i.e., at Davidson's Cove). In addition, Bolanos did not encourage Vizcaíno to enter the estero with the tiny *Santo Tomas* when they rode out a winter storm at Drakes Bay in 1603.[4]

---

3. H. H. Bancroft, *History of California*, vol. 1, pp. 91 and 92.
4. The translation of the "Bolanos-Ascensión Derrotero" is in Wagner, *Spanish Voyages*, p. 435.

Fig. 125. View looking north over Bolinas Lagoon, Olema Valley, Drakes Bay, and Tomales Bay. From this vantage point one can see that Olema Valley is neither an inland nor an upland country that is "far different than the shore." It is, in fact, part of the Point Reyes National Seashore. *Photograph © by Ed Brady, October 1965. Courtesy of Grethe Brady, Aero Photographers, Sausalito, California,*

Tenet 6 – The Victuals

The victualing evidence fits no San Francisco area site except Campbell Cove. Consider the following observations recorded by the first visitors to wander the shores of Drakes Bay:

Sebastián Rodríguez Cermeño, November 1595 – "The Spaniards saw deer walking about, the largest ever found, as could be seen by the antlers, of which the captain carried away a sample." They also observed that the Indians living around Drakes Estero maintained their existence on "crabs, wild birds, and deer."[5]

Felipe de Goycoechea, August 1793 – Looking southwest over the land from Inverness Ridge: "I found a peninsula where there is a very fine site but distant from firewood and timber. There are many deer there."[6]

---

5. Cermeño, in Wagner, *Spanish Voyages*, p. 159.
6. Goycoechea, in Wagner, "Last Spanish Exploration," p. 343.

Fig. 126. "Elk Crossing Carquinez Straits": drawing by E. Wyitenbach circa 1889 under the personal direction of William Heath Davis, to illustrate his story of the vast herds of these now almost extinct animals so plentiful in California before the discovery of gold by Marshall, January 24, 1848. *Reproduced from SEVENTY-FIVE YEARS IN CALIFORNIA by William Heath Davis, John Howell, San Francisco, 1929.*

From the journal of the Boston schooner "Albatross," August 1812 (courtesy The Bancroft Library, as holder) – "The second mate and one man (Jerry Bancroft) went on shore to hunt [deer, presumably]. They came across a grizzly bear, and shot him through the head, but not killing him outright, and the bear being very close to them, he seized Jerry in his hug, and before he expired, bit him severely in a number of places in his left thigh and leg."[7]

Peter Corney, August 1817 – "This part of the country is delightfully pleasant, with many small rivers running through the valleys. While on shore we killed a number of large snakes and adders. And saw many deer and foxes, but they were very shy. We also observed the tracks of bears."[8]

Joseph Warren Revere, August 1846 – During his brief stay with Don Rafael Garcia at his Olema Valley rancho, they went deer hunting on Point Reyes Head and saw not less than "400 head of superb fat animals" and on the return passed many places where "moulding antlers and bones" attested to the wholesale slaughter which had been made in previous years by rancheros of the neighborhood.[9]

Steven Richardson, writing in 1918 – From the son of one of Marin County's original settlers: "California elk must have existed in the millions . . . I think the largest herd in the world roamed over the deep grasslands of Point Reyes . . . I fully believe there were a thousand elk in one herd."[10]

There is no shortage of evidence in the early accounts that the north shores of Tomales Bay and San Francisco Bay similarly abounded with deer, elk and other game.[11] When Bodega y Quadra

---

7. Quoted from the *Albatross*' journal by William Dane Phelps, "Solid Men of Boston," p. 62.
8. In Sullivan, ed., "Russian Settlement," p. v.
9. Revere, *Naval Duty in California*, p. 68.
10. Cited in Mason, *Point Reyes the Solemn Land*, p. 19.
11. For early accounts of deer on the shores of San Francisco Bay, we have the following: Auguste Duhaut-Cilly (1827); Eugène Duflot de Mofras (1841); and Edward Visher (1842). All can be found quoted in Smilie, *The Sonoma Mission*, pp. 26, and 77–79.

anchored the *Sonora* inside Tomales Point in August 1775, he observed deer grazing along the opposite shoreline.[12]

Let us allow San Pablo Bay anchorage enthusiast Robert Thomas to describe "the natural music" of the San Quentin Cove and Petaluma River environs before the white settlers appeared:

> After the night had fallen upon the Marin Peninsula, a bedlam of wild yelping and howling arose on all sides of this cove [the mouth of the Petaluma River] as coyotes greeted the night from every hill and mountainside in a serenade that the Englishmen at first found to be most unpleasant, but which every native of this land found to be the sweetest music known to ears, the personal greeting of an old friend and benefactor. From one end of the peninsula to the other, it rang through the night, interrupted only by the screams of a mountain lion exulting over her kill, or the muffled snarl of a hostile bear.[13]

Simply put, why would Drake and his carnivorous crew have been dining primarily on the likes of mussels and sea lions if they had encamped along the shores of Drakes Bay, San Francisco Bay or any of the other candidate sites? How could the effusive Fletcher have possibly resisted the opportunity to enliven his account with the most vivid descriptions of such fascinating wildlife?

Tenet 7 – The Fort Site

At all the candidate sites except Campbell Cove, the Indians could have approached the encampment simply by walking along the shore; they would not have had to descend a steep hill to make their immediate presence at the fort.

Tenet 8 – The Weather

The evidence, as pointed out by many others, simply does not support a San Francisco Bay or San Pablo Bay locale.

Tenet 9 – The Visit Inland

Olema Valley is neither an inland nor an upland country; it is a relatively narrow, low-lying coastal valley that joins Tomales Bay and Bolinas Lagoon within the area designated today as the "Point Reyes National Seashore" (figure 125). Considering Olema Valley as part of the coastal area, the inland country is not distinctly different from the shore traveling eastward from Drakes Bay or Bolinas Bay. The same can be said for traveling eastward from Tomales Bay, or traveling northward from the north shores of San Francisco Bay.

Tenet 10 – The Flora

I do not consider the flora an important clue and have not conducted sufficient research to comment on the other sites.

Tenet 11 – The Fauna

Drake and his men would not have had to travel inland to catch sight of large herds of deer if they had encamped at any of the candidate sites along the coast of Marin County.

Tenet 12 – The Indians

As the eminent anthropologist Robert Heizer noted, it is very unlikely that a Western Pomo chieftain would have come to visit Drake at any of the candidate sites along the coast of Marin County.

---

12. Bodega y Quadra, "Viaje de 1775," p. 20.
13. Thomas, *Drake at Olompali*, pp. 17 and 18.

Tenet 13 – Archaeological Evidence

Notwithstanding their monumental efforts, the Drake Navigators Guild has not produced any truly conclusive archaeological evidence to support their Drake's Cove theory. Archaeological investigations at Bolinas Lagoon, Davidson's Cove, and San Quentin Cove have also come empty with respect to Drake's visit.[14]

The porcelains of Drakes Estero – Edward Von der Porten and Clarence Shangraw have not provided convincing evidence that Drake camped on the shores of Drakes Bay from their evaluation of porcelain shards recovered from Drakes Estero shellmounds. Their conclusion that Drake's presence is archaeologically confirmed is based on the following findings: (1) a substantial percentage (33 percent) of the 235 "statistically significant" porcelain shards recovered by archaeological investigations at the estero appear to be of both older and better pedigree than the rest of the shards; and (2) a high percentage of these older, higher-quality pieces show no signs of any surf wear and appear to have been purposefully carried ashore, while a high percentage of the others do show surf wear and appear to have been washed ashore from the wrecked *San Agustin*.[15]

It seems to me that when the Spanish set up camp on shore in anticipation of a prolonged stay, the officers' attendants would have brought along a good supply of the *San Agustin*'s eating ware to serve the captain and officers. Given that the ship was a Manila galleon, there is every reason to believe she would have been well stocked for the purpose with Ming porcelain plates, bowls and vases that perhaps dated to the time the ship was first commissioned. After the shipwreck, the officers' fancy china would have been at the very bottom of the priority list when loading the tiny launch for the long and dangerous voyage home. In light of these considerations, it is pertinent to note that the vast majority (sixty-five) of the seventy-seven so-called non-water-worn "Drake shards" were discovered on or near Limantour Spit in the immediate vicinity of where state archaeologists believe Cermeño and his men established their camp. Conversely, none turned up in Guild excavations of the alleged "Drake camp" at Drake's Cove and only five have turned up from all archaeological investigations along the west side of the estero.[16]

Another possibility is that the archaeological findings have done no more than document the fact that the Spanish were successful in salvaging one or more of the more valuable chests of porcelain before the ship completely broke apart in the surf.

The Guild's Raymond Aker, writing in the early 1970s, offered this explanation for the non-water-worn shards in question:

Chests of porcelain which did not break up in the surf zone or become buried in the sands would have tumbled along the bottom until they were cast up on the beach, probably after Cermeño departed. In this circumstance, we may visualize a damaged chest lying for some time in the surf zone with part of its contents exposed to the abrasive action of sand and water; the bowls or dishes being nested and tightly packed in clay so that only the outer surfaces are abraded. Later, they are discovered by the Indians, taken to the village, subsequently broken, and then discarded. On this basis, there is good evidence for attributing much of the porcelains to Cermeño.[17]

---

14. Because I have not seen the archaeological reports, I am going by what Robert Heizer (*Elizabethan California*, pp. 18–19) had to say on the subject circa 1974: "There is no point in summarizing the work done since 1940 at Drakes Bay by various archaeologists, or in the vicinity of San Quentin Cove since the discovery of the brass plate, for the reason that all such rummaging in Indian sites has produced nothing that can be referred to Drake." He points out that a substantial file of this information is in the archaeological research facility at the Department of Anthropology, University of California, Berkeley.
15. Shangraw and Von der Porten, *The Drake and Cermeño Expeditions' Chinese Porcelains*, p. 5. For identification purposes, they only used shards that could be specifically matched with other shards.
16. Notwithstanding the concentration of finds around the presumed Cermeño camp, Shangraw and Von der Porten consider it unlikely that the ware had been brought ashore for the officers' use on the basis that it was too valuable (Ibid., p. 10).
17. Aker, *Report of Findings*, p. 357.

We will never know for sure how this unblemished porcelain came ashore, but, all things considered, Cermeño's Manila galleon is clearly the most likely source.

The Agate Beach finds – I find it well within the realm of possibility that some of the superstructure from Cermeño's galleon, the *San Agustin*, washed up on Agate Beach after it was dashed to pieces on the shore of Drakes Bay in 1595; Agate Beach is directly downwind (southeast) of the presumed site of the wreck, and, due to the peculiar configuration of reefs and currents, the area does seem to attract floating debris. This is the most likely explanation for George Epperson's impressive collection of alleged sixteenth-century finds, some reportedly verified by carbon-dating.

Though there is no evidence to support it in the contemporary accounts, there is at least a possibility that Tello's bark somehow came to rest on Agate Beach, notwithstanding a Campbell Cove encampment. For example, perhaps she struck on Duxbury Reef during a reconnaissance expedition.

Finally, there is documentation of many square-riggers having wrecked on the shores of Drakes Bay and Bolinas Bay, any and all of which could have littered the shores of Agate Beach with some of their bones. According to George Davidson, the steamship *S. S. Lewis* was dashed to pieces on Agate Beach in April 1853 after running aground on Duxbury Reef.[18] Other ships that have come to grief on this reef include the *Duxbury*, 1849; the *Montana*, 1867; the *Prince Alfred*, 1874; the *Western Shore*, 1878; the *Iaqua*, 1901; the *Polaris*, 1914; and the *Hanalei*, 1914. These shipwrecks may account for some of Mr. Epperson's other finds.

In any event, I see no possibility that Drake would have selected Agate Beach as a careening or graving site as it exists today. Furthermore, neither do I see any indications in the very earliest maps and charts to suggest conditions were significantly different in Drake's day, nor do I find any good reason to believe so.

Tenet 14 – The Plate of Brass

I do not consider the plate of brass a valid clue.

Tenet 15 – The Isles of Saint James

Assuming that the Isles of Saint James are indeed the Farallon Islands, the dates of arrival and departure set forth in *The World Encompassed* provide evidence that solidly refutes the greater Drakes Bay and Bolinas Bay anchorage theories.

The World Encompassed provides the following unambiguous evidence relating to the departure sequence: (1) the calendar date of departure from Portus Nova Albion was July 23, 1579; (2) the calendar date of departure from the Farallones was July 25, 1579, the day the Christian church celebrates the vigil and feast of Saint James; (3) the English departed the Farallones the day after they "fell upon" (arrived at) them on calendar day July 24, 1579; (4) at the time of departure, the wind was still blowing from the northwest; and (5) the day they departed, the English sailed out of sight of the Indians who were making fires in the hills around the port. The John Stow account confirms the date of departure as July 25.

This evidence can not be logically applied to support any of the greater Drakes Bay/Bolinas Bay anchorage theories. For example, running before a northwest wind after a hypothetical departure from Drakes Estero, Drake would have easily fallen upon the Southeast Farallon, located about twenty nautical miles due south, in about half a day's sail or less—i.e., on July 23.[19]

Aker's explanation was that in Elizabethan times, navigators dated their logbooks at noon, so that Drake could have departed the morning of July 23, 1579, and arrived the same afternoon on July 24.

18. Davidson, *Pacific Coast Pilot*, (1889) p. 192.
19. In appendix F, I summarize information on the speed early navigators typically made in sailing south along the coast of California: the average rate was 1.25 degrees per day (about seventy-five nautical miles/day). Twenty nautical miles is 0.33 degrees. The Southeast Farallon is by far the largest of the group and the most likely location for Drake's Isles of Saint James anchorage.

Alternatively, he contends that due to light winds or fog, the English may have hove-to en route to the islands.[20]

First of all, Aker's statement about Elizabethan navigators dating their logbooks at noon warrants close scrutiny. In offering the noon dating system argument, Aker cited David Waters' *The Art of Navigation . . .* , that reads as follows (*emphasis added*): "The Elizabethan and Stuart navigator . . . counted his day, *while he was out of sight of land*, from noon to noon. . . . When, however, the navigator was in waters of pilotage he logged his entries in accordance with midnight to midnight date system used by landsmen."[21] Logically, Drake would not have switched to the noon dating system until after departing the Farallones on July 25 at which time, after conferring with his officers and men, he "bent his course directly [into the open sea] to run with the islands of the Moluccas."[22]

In offering the hove-to explanation, Mr. Aker seems to be reaching too far. Given that *The World Encompassed* specifically states that upon their departure the wind was "blowing still (as it did at first) from the northwest," it is unlikely that Drake would have found it necessary to hove to during the short sail from Drakes Estero to the Farallones.[23]

George Davidson, in debunking the San Francisco Bay theory, questions why Drake would have bypassed Seal Rock off Point Lobos if his intent was to bolster the *Golden Hind's* stock of fresh meat. In this respect, he points out that in July the Farallones are a very difficult sail from the Golden Gate due to the prevailing northwest winds.[24]

Tenet 16 – The Portus Plan

History has shown that none of the anchorage investigators have been able to win much support for their various interpretations of the Portus. Noting that proponents of eleven different anchorage sites used the Portus in their arguments, Hanna was unable to score the debaters on this clue. I think the Portus plan correlates far better with Campbell Cove than with any of the other candidate sites, except, perhaps, the mouth of Tomales Bay as first charted in 1793.

Tenet 17 – The Montanus Illustration

I do not consider the Montanus Illustration a valid clue.

Tenet 18 – The Cermeño Account

The accounts of this visit provide compelling evidence that Drake never set foot on the shores of Drakes Bay. Here are the key points: (1) Cermeño reported that the Indians acted as though they had never seen Europeans before; (2) Cermeño did not report finding so much as a trace of Drake's visit; (3) at least some of the Indian lodges were covered with grass, not earth; (4) the Indians happily accepted the Spaniard's gifts; (5) the Indians maintained themselves on crabs, birds, and deer; (6) the chief's ceremonial garb was made from bird skins; and (7) the Spaniards saw evidence of huge deer immediately upon coming to shore.

Tenet 19 – The Dudley Maps

I do not consider the Dudley maps to be valid clues.

Tenet 20 – The N. de Morena Story

I do not consider the N. de Morena story an important clue.

---

20. Aker, *Report of Findings*, pp. 327–28.
21. Waters, *The Art of Navigation*, p. 579. Also cited by Morrison, "An Investigation," p. 103.
22. Drake, *The World Encompassed*, p. 82. The account makes it clear that Drake and company did not decide to abandon the search for the Northwest Passage and set course for the Moluccas until July 25. This evidence refutes Aker's argument that "logically the day of departure from the mainland was the time to set the normal sea routine" (*Report of Findings*, p. 328).
23. Aker, *Report of Findings*, p. 328.
24. Davidson, *Identification*, pp. 5 and 56.

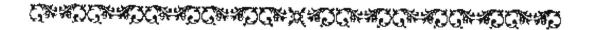

CHAPTER 34
Summary of Findings

In the following table, I take the California Historical Society's twenty tenets and weigh my arguments for Campbell Cove against arguments that others have offered for alternative sites. For the purpose of comparison, I have assigned my arguments a value of ten on a scale of zero-to-ten and weighed the others against this standard. You may need to refer back to the clues for each tenet and to the rebuttals to understand my rationale for the ratings.

While one could subjectively rate the evidence for each tenet somewhat differently, I see no way the final tally could yield a significant difference in the overall rankings; the evidence overwhelmingly favors Bodega Harbor with the mouth of Tomales Bay a solid but distant second.

Table 2 – Summary of Findings							
Tenet	Bodega Harbor	Tomales Bay sites	Davidson's Cove	Drakes Estero	San Francisco Bay sites	Half Moon Bay	Bolinas Lagoon and Agate Beach
Approach	10	10	5	5	0	5	2
Latitudes	10	8	2	2	2	0	2
Terrain	10	8	5	5	4	0	0
Good bay	10	5	10	5	5	10	0
Fit harbor	10	10	5	4	5	5	1
Victuals	10	5	2	2	2	2	2
Fort site	10	8	8	8	8	8	8
Weather	10	10	10	10	0	10	8
Inland	10	2	7	7	5	0	0
Flora	10	10	10	10	10	10	10
Fauna	10	2	2	2	2	2	2
Indians	10	7	5	5	5	0	5
Archaeology	10	0	0	0	0	0	0
Brass plate	NA	NA	NA	NA	NA	NA	NA
Isles of St. James	10	10	5	5	5	0	5
Portus	10	8	7	7	5	0	5
Montanus	NA	NA	NA	NA	NA	NA	NA
Cermeño	10	10	0	0	10	10	10
Dudley	NA	NA	NA	NA	NA	NA	NA
N. de Morena	NA	NA	NA	NA	NA	NA	NA
Tally	160	113	83	77	68	62	60
Rank	1	2	3	4	5	6	7

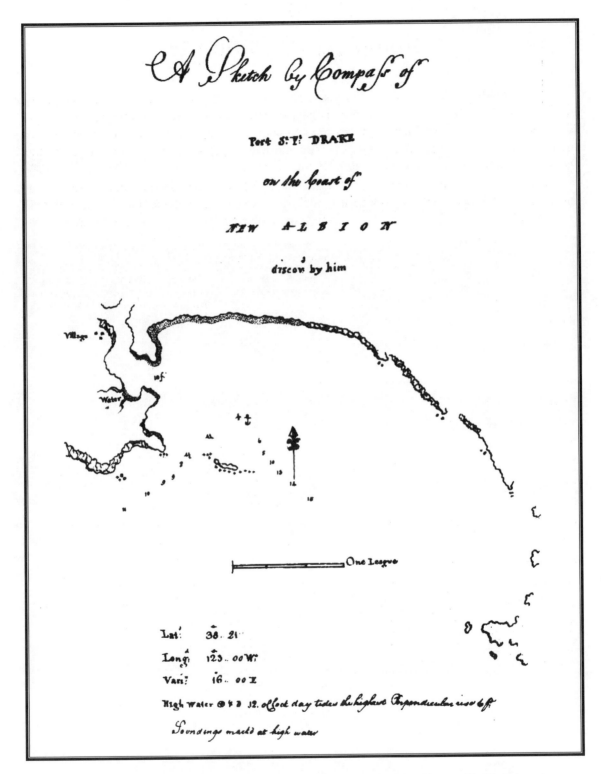

Fig. 127. "A Sketch by Compass of Port Sir Francis Drake on the Coast of New Albion, Discovered by Him": drawn by Captain James Colnett, September 1790 (photo-enhanced and condensed). Colnett, the very next English seafarer after Drake to seek shelter on the shores of Nova Albion, found this haven on the north side of Bodega Bay under circumstances that were virtually identical to Drake's. Note the Indian village located by the water's edge, about three-quarters-of-a-mile from the springs of today's Campbell Cove. *Courtesy of the Public Record Office, Richmond, England.*

PART V

Lost Harbor

Fig. 128. Lost Harbor! View looking northwest over the site of Drake's former encampment at Campbell Cove. If Francis Drake were to rise from his watery grave and look down at the site of his June 1579 encampment, there is little chance he would recognize his Portus Nova Albionis as it appears today. *Photograph by the author, 1995.*

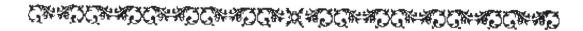

CHAPTER 35

Port Sir Francis Drake

W hen Francis Drake departed Nova Albion in July 1579, he made no attempt to cover his tracks. On the contrary, he went out of his way to stake his claim clearly, setting up a great post and plaque, presumably so located that those aboard the next ship entering the harbor would be sure to see it. The English also left behind a large stone-walled fort, over a month's worth of refuse from some sixty individuals and from the refitting of the *Golden Hind*, and the remnants of Tello's bark. In addition, they left the Nova Albion natives with a variety of souvenirs, at least a few English words, and memories of truly legendary proportions. On top of all this, by the mid-1590s, Richard Hakluyt, the premiere chronicler of the Elizabethan Age, had told the world that Drake had brought the *Golden Hind* into "a fair and good bay within thirty-eight degrees toward the line."

How did the Europeans that first came to explore and settle the coastal region north of San Francisco fail to discern Drake's presence? I often pondered this question after setting forth on my search for the encampment. Somebody should have stumbled upon the fort. Where I grew up in Massachusetts there is no shortage of centuries-old stone walls, and they show no indication of vanishing in the next millennia. The fact is, it was by stumbling upon a misplaced pile of stones that I focused in on Campbell Cove. There is also some reason to believe that the first Europeans to tread upon the shores of Campbell Cove did indeed discern Drake's presence.

How can this be? We will start off by seeing how close the late sixteenth and early seventeenth-century Spanish came to nipping California's great maritime mystery in the bud. We will then follow the remarkable history of Port Sir Francis Drake in exploring how the mystery persisted for so long. It turns out that this remote little haven on the Sonoma, California coast boasts more than an important sixteenth-century landmark. It also bears the distinction of being the most politically coveted locale in the United States. Five sovereign nations and one would-be nation have laid claim to its soil: England, Spain, Russia, Mexico, the Republic of California, and the United States. France lusted for it, and one of the most famous men in California history claimed it for what was in effect the independent province of New Switzerland.

The Aftermath – 1580 to 1590

Following Drake's Pacific Coast raids in 1579, New Spain's viceroy began petitioning King Philip to establish a port settlement in Alta (upper) California. At the time, the most northerly Spanish settlements were in Baja California (northern Mexico) and there had been just one expedition north of today's San Diego: the Juan Rodríguez Cabrillo–Bartolomé Ferrelo expedition of 1542 which explored the coast as far north as the San Francisco Peninsula. The viceroy argued that a northern settlement was needed not only to discourage foreign intrusion, but to provide a much needed way station for the richly laden Manila galleons that ran once a year from Acapulco to the Philippines. Philip, however, was distracted by a more pressing matter: his preparations to invade the British Isles with the so-called invincible Armada.[1]

---

1. Based on his 1542 expedition, Cabrillo is credited with the discovery of Monterey Bay which he named the *Baia de los Pinos* (Bay of Pines). He gave its position as thirty-nine degrees, about two degrees north of its actual position: an error consistent with his other observations. George Davidson also credited Cabrillo with the discovery of Point Reyes/Drakes Bay at thirty-eight degrees and even Cape Mendocino at forty degrees thirty minutes. From the accounts of the voyage, however, it appears very unlikely that Cabrillo and his company saw any part of the shore north of Point Año Nuevo on the southern end of the San Francisco Peninsula.

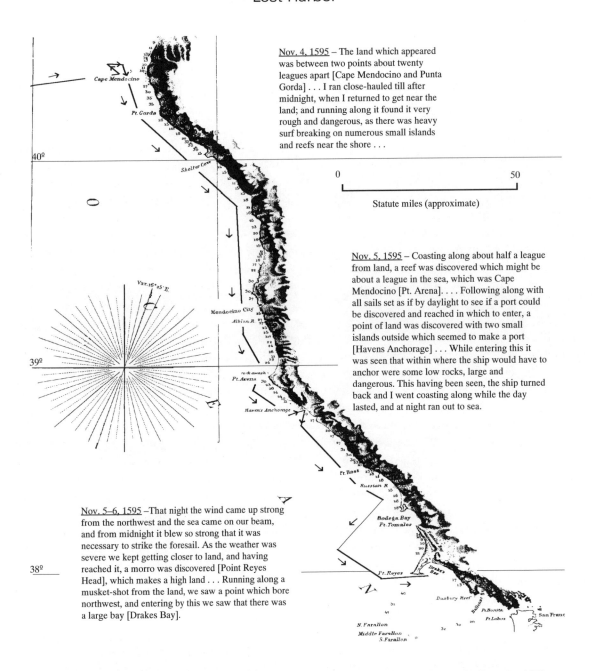

Nov. 4, 1595 – The land which appeared was between two points about twenty leagues apart [Cape Mendocino and Punta Gorda] . . . I ran close-hauled till after midnight, when I returned to get near the land; and running along it found it very rough and dangerous, as there was heavy surf breaking on numerous small islands and reefs near the shore . . .

Nov. 5, 1595 – Coasting along about half a league from land, a reef was discovered which might be about a league in the sea, which was Cape Mendocino [Pt. Arena]. . . . Following along with all sails set as if by daylight to see if a port could be discovered and reached in which to enter, a point of land was discovered with two small islands outside which seemed to make a port [Havens Anchorage] . . . While entering this it was seen that within where the ship would have to anchor were some low rocks, large and dangerous. This having been seen, the ship turned back and I went coasting along while the day lasted, and at night ran out to sea.

Nov. 5–6, 1595 –That night the wind came up strong from the northwest and the sea came on our beam, and from midnight it blew so strong that it was necessary to strike the foresail. As the weather was severe we kept getting closer to land, and having reached it, a morro was discovered [Point Reyes Head], which makes a high land . . . Running along a musket-shot from the land, we saw a point which bore northwest, and entering by this we saw that there was a large bay [Drakes Bay].

Fig. 129. The track of the Manila galleon, *San Agustin*, from Cape Mendocino to today's Drakes Bay in November 1595. *The chart detail is adapted from the 1854 U.S. Coast Survey chart "Reconnaissance of the Western Coast of the United States." The text is from Henry Wagner's translation of Cermeño's journal in "The Voyage to California of Sebastián Rodríguez Cermeño in 1595," CALIFORNIA HISTORICAL SOCIETY QUARTERLY, vol. 3, 1924.*

In July 1586, the viceroy's worst fears were realized. Another heretic Englishman, Thomas Cavendish, slipped through the Strait of Magellan, repeated Drake's depredations along the coasts of Chile and Peru, and captured a returning Manila galleon fully laden with the treasures of the orient.

Cermeño and the Original Port of San Francisco – 1595

Following the disastrous defeat of his Armada in 1588, King Philip authorized the Alta California enterprise. As the first step, he ordered that a returning Manila galleon make a careful exploration of the California coast, with particular attention, it seems, to the area "within thirty-eight degrees toward the line." The expedition's command was awarded to a Portuguese merchant, Sebastián Rodríguez Cermeño.

Cermeño embarked from Manila with the 200-ton *San Agustin* in July 1595 carrying about seventy men and officers and the usual oriental trade goods. After five months at sea, they sighted land between Cape Mendocino and Punta Gorda. Running down the California coast about a mile offshore, they rounded Point Arena at thirty-nine degrees and attempted to enter today's Havens Anchorage. Alarmed at some dangerous-looking rocks, Cermeño turned the *San Agustin* about, coasted a little farther south, and stood off to sea as darkness approached. He had veered off the coast at about thirty-eight and a half degrees latitude, perhaps within sight of Bodega Head.[2]

Late that night a gale came up from the northwest, driving them southeast. The next morning they came in sight of the coast off Point Reyes Head, rounded it, and anchored in today's Drakes Bay. Cermeño called it "el Puerto de San Francisco" (the Harbor of Saint Francis). They had barely missed Bodega Bay, and Drake's former encampment was now approximately thirty miles away by land.

They set up camp on the shores of Limantour Estero where they found several Indian villages and herds of huge deer. The Coast Miwok living along the isolated Point Reyes Peninsula apparently were not among those that had visited Drake's camp sixteen years before; according to the accounts, the Indians acted as though they had never seen Europeans before and had no European articles among them. Hunting parties made several excursions inland noting Tomales Bay and exploring Olema Valley.[3]

Disaster struck on the last day of November when a winter storm sprang up while Cermeño and his party were ashore building a pinnace. The *San Agustin* broke her cables and was driven aground, a total loss, her valuable cargo spilled over the floor of Drakes Bay. Using the dangerously overcrowded pinnace, Cermeño and most of his company barely made it back to Mexico.

Vizcaíno and the Discovery of Tomales Bay – 1603

In the wake of the *San Agustin* disaster, Mexico's new viceroy, the conde de Monterey, changed tactics. In 1599, he convinced the Spanish crown that northern exploration should be undertaken in shallow-draft boats sailing directly from Acapulco. The huge, deep-drafted Manila galleons were ill-suited for exploration and too valuable to risk.

Monterey gave the expedition's command to Sebastián Vizcaíno, a New Spain merchant with conquistadorean ambitions. For this expedition, however, Vizcaíno's instructions were simply to chart and explore the coast from the tip of Baja California to Cape Mendocino, identifying the best ports and harbors.[4]

Vizcaíno left Acapulco in May 1602 with about two hundred men and three ships. Tortuously making their way north against the prevailing winds, they entered and named San Diego Bay and Monterey Bay. Upon departing Monterey, Vizcaíno sent the *San Diego* back to New Spain with many debilitated seamen; scurvy was already becoming a serious problem.

The two remaining ships arrived off Point Reyes in January 1603 where they were separated. Vizcaíno took the *Santo Tomas* under the shelter of Point Reyes Head while Martin de Aguilar sailed

---

2. For the accounts of Cermeño's voyage I refer the reader to Wagner's translations including: Cermeño's journal in "The Voyage to California" (*California Historical Society Quarterly* vol. 3 (1924), pp. 3–23; and Cermeño's sworn declaration in *Spanish Voyages*, pp. 154–67. Wagner had Cermeño first sighting land between Cape Mendocino and Cape St. George, about a degree north of where I do, and thought the small cove was Shelter Cove rather than Havens Anchorage. There are many problems with his interpretation including the descriptions of the southern cape (reef) and small cove, and the sequence of events that Cermeño reports to have taken place after leaving the cove. Excerpts from Wagner's translations are included in chapter 30 of this book.

3. The Cermeño account actually implies that the Indians' arrowheads were forged of iron. Even if so, however, this would be no proof for a Drakes Bay landfall in that any iron arrowheads forged by Drake's men would have likely become a popular trade item among the Coast Miwok—as were obsidian arrowheads they obtained from the Pomo.

4. For the Vizcaíno voyage, I refer the reader to Wagner's translation of the Ascensión journal in *Spanish Voyages*, pp. 249–55. Vizcaíno was also presumably instructed to retrieve what Cermeño and his men had salvaged from the *San Agustin* and left behind on the shores of Drakes Bay, including a quantity of wax and a great many cases of silks (Wagner, *Spanish Voyages*, p. 249).

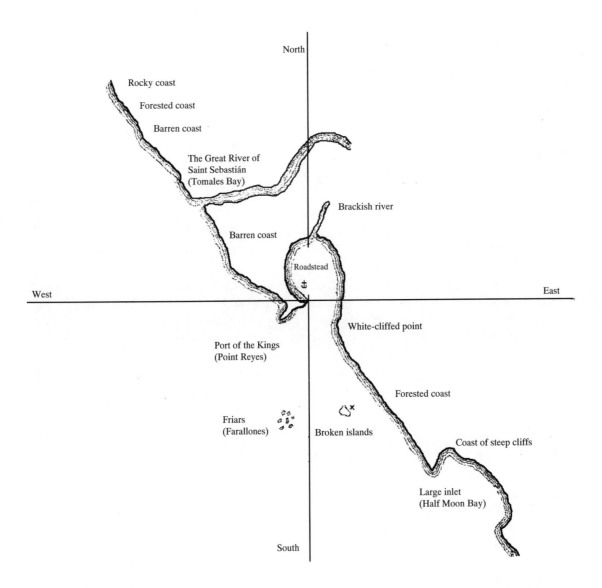

North

Rocky coast

Forested coast

Barren coast

The Great River of
Saint Sebastián
(Tomales Bay)

Brackish river

Barren coast

Roadstead

West

East

White-cliffed point

Port of the Kings
(Point Reyes)

Forested coast

Friars
(Farallones)

Broken islands

Coast of steep cliffs

Large inlet
(Half Moon Bay)

South

Fig. 130. "Puerto de los Reyes": detail from a map drawn by Gerónimo Martín Palacios, circa 1603, based on the results of the Sebastián Vizcaíno expedition (redrawn and translated). The captain and pilot having succumbed to scurvy, the boatswain, Estevan López, took the tiny *Tres Reyes* into the mouth of Tomales Bay, which Palacios named the "Río Grande de San Sebastián." The "white-cliffed point" shown to the south-southeast of Point Reyes probably represents today's Fort Point at the entrance to San Francisco Bay. It is interesting to note that Vizcaíno sailed close enough to the Golden Gate to observe white cliffs near the entrance, but not the huge harbor just within. *The map was redrawn from CALIFORNIA UNDER SPAIN AND MEXICO, 1535–1847 by Irving Richman, Houghton and Mifflin, Boston and New York, 1911. Palacios' derrotero is held in the Archivo General de Indias, Seville: Mexico, Historica, legajo 372, ff. 47–91v.*

on in the *Tres Reyes*. Concerned over the fate of his tender, Vizcaíno departed the next day, narrowly escaping shipwreck in a sudden storm. He took shelter somewhere in the vicinity of Cape Mendocino, his crew ravaged by scurvy. In his desperate flight back down the coast, Vizcaíno either did not see, or more likely, because of all the sickness, did not bother to enter Bodega Bay.

Aguilar, meanwhile, made it a degree or two north of Cape Mendocino in the *Tres Reyes*, discovering Humboldt Bay where both he and the pilot succumbed to scurvy, leaving the boatswain Estevan López in charge. Retreating down the coast, the surviving crew fell upon Bodega Bay, coming to anchor just inside the mouth of Tomales Bay. Given the debilitated state of the crew and with the principal officers dead, there was no question of exploring or charting the area; they immediately departed for New Spain. Had they had enough strength for exploration, there never would have been a Drake's Bay controversy.

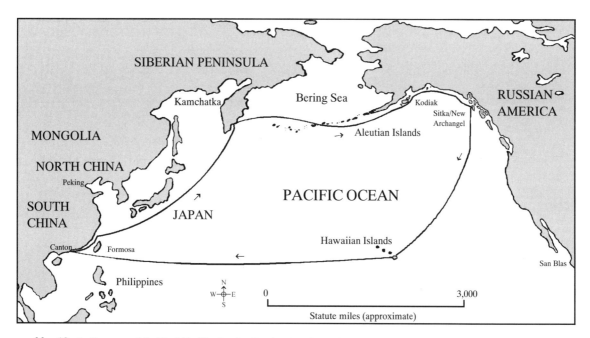

Map 18. Outline map of the North Pacific showing Russian America and the fur-trade route in the late eighteenth century.

The Dust Settles – 1603 to 1769

The 1603 Vizcaíno expedition had satisfied its objectives. It had crudely charted the upper California coast and had identified three potential sites for port settlements: San Diego, Monterey, and Point Reyes. Spain, however, would not follow up on these findings for more than a century and a half. By then all overt signs of Drake's visit were gone, and one can only speculate as to what myths and legends remained among the Indians.

Why the delay? First of all, the Cermeño and Vizcaíno voyages had proven that such expeditions were enormously difficult and expensive; contrary winds, fog and unpredictable weather severely discouraged further exploration. Secondly, the reports brought back by Cermeño and Vizcaíno were completely negative from the conquistadorean viewpoint; there were no economic incentives as the natives had no gold or silver. Finally, there was no compelling foreign danger; Spain was not going to colonize Alta California until a foreign power threatened to encroach. Since the Elizabethan Age, the English had never even attempted to follow up on Drake's claim on the west coast of America; they had their hands full on America's east coast.

It was to be Russia that finally forced Spain's hand.

The Russian Menace – 1727 to 1800

In the 1700s, the coasts of California and the Pacific Northwest (Alaska, Canada, Washington, and Oregon) were teeming with fur seals and sea otters. When the world learned they were there and that the Chinese were willing to pay enormous prices for their pelts, the ships came in droves.

The Russians were the first to come. By the early eighteenth century, Russian cossacks had conquered the Siberian Peninsula, had reached its eastern shores, and were wondering what lay beyond. Imperial Russia's Vitus Bering–Alexeli Chirikov voyages of 1727–42 told them. They discovered and charted the Bering Sea, the Aleutian Islands, and the coast of Alaska. On his last voyage in 1742, Bering perished on a remote Aleutian Island. His surviving crew, however, brought back some sea otter furs they had bartered from the natives. Private Russian fur-hunting ventures into the Aleutians soon followed. By 1784, Spanish fears of encroachment on their northern frontiers were realized when a Russian trading company established a permanent outpost on Kodiak Island

Fig. 131. Juan Francisco de la Bodega y Quadra (1744–94): oil painting by Julio Garcia Condoy. Bodega y Quadra, a Spanish navigator, was commandant of the San Blas, Mexico, naval base from 1789–94. The California anchorage debate began in 1775 with his strongly worded opinion that Drake had harbored in Puerto de la Bodega at the mouth of today's Tomales Bay, and emphatically not in San Francisco Bay. *Courtesy Maritime Museum of British Columbia.*

off the Alaskan coast. In 1799, the tzar granted the Russian-American Company (RAC) exclusive control over the Alaskan otter trade. The RAC immediately founded a settlement off the coast of Southern Alaska, first called Sitka, later New Archangel. It remained the RAC's center of operations until Russia sold Alaska to the United States in 1867.

The Initial Spanish Settlement of Alta California – 1769 to 1776

Responding to the Russian threat, it took Spain very little time to colonize the southern and central California coasts. In one of those peculiar twists of fate, however, their northward expansion ended where Drake had established Nova Albion's southern boundary: at Bodega Bay.

Between 1769 and 1776, Franciscan missionaries supported by Don Gaspar de Portola, governor of Baja California, colonized the coast of California north to today's San Francisco Peninsula. Their basic strategy was to locate the first missions at the principal navigable ports and gradually fill in the gaps with new missions. By the time Junipero Serra, the first president of the California missions, died in 1784, his friars had established nine missions between San Diego and San Francisco. Eventually, there would be twenty-two.

The Bodega and Hezeta Expedition and the Original Port Bodega – 1775

Spain's next great maritime expedition to the Pacific Northwest, after Vizcaíno's in 1603, was dispatched from San Blas, Mexico, in the spring of 1775 by Viceroy Don Antonio María Bucareli y Ursúa (Bucareli). The expedition was timed to coincide with an expedition of land forces under Captain Juan Bautista de Anza who was charged with establishing a mission and presidio on the tip of the San Francisco Peninsula.[5]

---

5. Bucareli was New Spain's viceroy during the period of Spanish colonization of Alta California: 1771–79. The Hezeta expedition was actually the second voyage of exploration that Bucareli dispatched. In the first Bucareli expedition, Don Juan Perez took the *Santiago* out of San Blas in January 1774 and surveyed the coast to about fifty-five degrees, taking advantage of south to southeast winds. He discovered Nootka Sound but did not land or formally claim any territories for Spain. According to Russell, ed., *Voyage of the Sonora*, p. 98, "his observations were not good owing to fogs and bad weather." Perez served as Hezeta's pilot on the second Bucareli expedition. H. H. Bancroft in *History of the Northwest Coast*, pp. 151–56, provides a detailed account of the Perez voyage derived from three of the four diaries still extant.

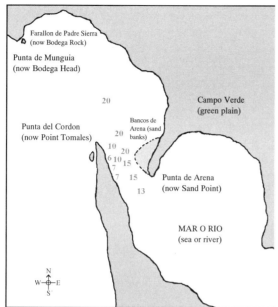

Fig. 132. "Plano de Cap. Bodega . . .": Bodega y Quadra's sketch of cove inside Sand Point at the mouth of Tomales Bay, drawn circa 1775 (redrawn and translated). From the mid to late 1800s, this cove was called "Smith's Landing" after the first European to settle there—a Dutchman who called himself Ralph Smith. The port facilities included a wharf and two ramshackle buildings enclosed by a fence that may have been erected with timbers stripped from the hull of a British ship that wrecked off the mouth of the cove in 1849. Smith's Landing was the first port facility constructed in Tomales Bay and was used by French, English and American vessels to engage in trade with the local Indians. When the Indians died off, Smith took up dairy farming, a land use that presumably started the dramatic erosion of Sugarloaf Hill, the huge sand dune that backed the rear shore of the cove. The north side of the cove is now buried in sand from the continued erosion of Sugarloaf Hill over the last century and a half. On modern charts the cove's inner shore is called Brazil Beach after the Portuguese family that has owned the surrounding land since 1906. *This is one of three versions of Bodega's map held by The Bancroft Library, Berkeley, California. Case XB G4362. B64 1775. B65.*

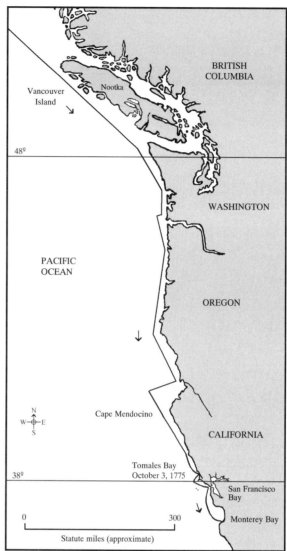

Map 19. Outline map of the Pacific Northwest Coast showing the return voyage of Bodega y Quadra's *Sonora*, in October 1775. Like Drake, Bodega y Quadra, coasting south behind northwest winds, fell upon a fair and good bay as he came within thirty-eight degrees toward the line.

Again, three ships were involved. Juan Manuel de Ayala took the *San Carlos* through the Golden Gate in August 1775, the first European vessel to enter San Francisco Bay. The expected de Anza land expedition did not show up. The two other ships were separated on their way north. Bruno de Hezeta took the *Santiago* just north of forty-nine degrees where he claimed Santa Cruz de Nootka (Nootka Sound) on today's Vancouver Island, British Columbia. His crew decimated by scurvy, Hezeta was unable to chart the coast on his return voyage to Monterey, and missed the entrance to San Francisco Bay due to fog. Meanwhile, Juan Francisco de la Bodega y Quadra in the *Sonora* made it as far north as Alaska before he too was forced to retreat with a sickly crew.

On the return voyage, Bodega y Quadra reached the vicinity of thirty-eight and a half degrees in early October 1775, and took the *Sonora* in towards shore looking for the entrance to San Francisco

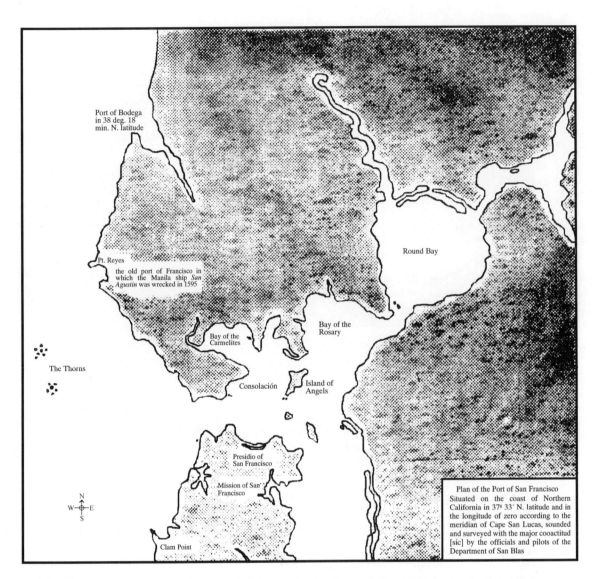

Map 20. The "San Blas Naval Officers' Map" of the San Francisco Bay area, circa 1777, showing the Port of Bodega (redrawn and translated). *The map was redrawn from the copy in CALIFORNIA UNDER SPAIN AND MEXICO, 1535–1847 by Irving Richman, Houghton Mifflin, Boston and New York, 1911.*

Bay. Instead, he fell upon the mouth of Tomales Bay, and mistaking it for the former, tried to enter. Finding many shoals inside the mouth, he anchored off Sand Point, observing a promising cove just inside that was inhabited by many Indians. Seeing no signs of a Spanish settlement, Bodega correctly surmised that it was not San Francisco Bay that he had entered, but some other river or sea to its north. This inlet had been previously discovered by the Vizcaíno expedition, and notwithstanding appearances, Bodega y Quadra did not name the bay for himself. He named the cove just inside its mouth "el Puerto de la Bodega" for the good shelter it affords (the port of the wine cellar). He never saw what today is called Bodega Harbor on the north side of the bay.

In his journal, Bodega y Quadra stated that Puerto de la Bodega (today's Tomales Bay) was undoubtedly the site visited by Drake, presumably based on considerations of latitude and the manner in which he had literally fallen into it on his approach to San Francisco. This was the first educated opinion ever rendered on the matter, and the beginning of more than two centuries of controversy. As we have seen, Bodega y Quadra was not far off.[6]

6. Bodega y Quadra, "Viaje de 1775," p. 21.

San Francisco Bay – 1776 to 1777

California's history could have been much different if San Francisco's Golden Gate were not so hard to spot from an offshore position. Both the Cermeño and Vizcaíno expeditions had failed to observe the huge bay as they sailed by, as did Drake. It was not until November 1769 that Gaspar de Portola's land expedition stumbled upon it during the search for Vizcaíno's Port of Monterey.

Recognizing the bay's strategic importance, Viceroy Bucareli quickly set in motion plans for its settlement. He assigned the task to Juan Bautista de Anza who in 1769 had blazed the trail linking Sonora, Mexico, to San Diego. In 1776 de Anza's company founded Mission San Francisco de Asis and the San Francisco presidio. Within the next two years they established a second mission at Santa Clara and the pueblo of San Jose, and, in 1797, a third mission a little to the north of San Jose.

English Advances – 1778 to 1787

The famous English explorer, Captain James Cook, made his three voyages of discovery in the 1770s. His third (1776–79) plays a significant part in our story. Departing from England in the year of the American Revolution, he took his two ships, the *Resolution* and *Discovery*, around the tip of Africa and crossed the Pacific, discovering the Hawaiian Islands en route. He reached the coast of southern Canada and put into the very same bay (Nootka Sound) that Hezeta had discovered and claimed for Spain two years before. Cook named it King George's Sound and claimed the territory for England.

By the time the *Resolution* and *Discovery* returned to England in 1780, the more enterprising members of the expedition were already planning a return to the area to capitalize on the fur trade. Published accounts of the voyage generated similar mercantile interest in England and other European countries. By 1786, there were seven British sealers on the coast. Although not all these expeditions were profitable, and some ended in disaster, the flood gates had opened.

The Nootka Sound Controversy – 1788 to 1790

We have seen that both Spain, via the Hezeta expedition in 1775, and England, via the Cook expedition in 1778, had laid claim to Nootka Sound–King George's Sound on today's Vancouver Island. The troubles that came of this brought these age-old enemies once again to the brink of war. The fallout also added some interesting twists to our lost harbor story.[7]

Early in 1789, New Spain's viceroy followed up on Hezeta's claim by sending two vessels under naval Lieutenant Estevan Martínez. He had orders to build a small fort and ward off any interlopers. Upon encountering several English brigs, the rash Martínez immediately arrested their captains, releasing the English after confiscating their cargoes.

Enter James Colnett. He began his naval career as a midshipman on the *Resolution* during Cook's second voyage of discovery (1771–75). After fifteen years of service, the admiralty promoted him to first lieutenant and placed him in command of a private fur-trading venture that stopped at Nootka Sound in 1787.

In late 1789, Colnett took command of a second expedition to Nootka of a much more strategic nature. He carried the necessary men and supplies to leave behind a permanent English trading outpost at Nootka Sound that was to be called Fort Pitt. When Colnett arrived at Nootka that summer and arrogantly announced his intentions to the Spanish, Martínez arrested him following a heated argument. After holding the livid Englishman and his crew in confinement for several weeks, Martínez sent his captives and two prizes, the *Argonaut* and the *Princess Royal*, to San Blas, Mexico. The port commandant, Bodega y Quadra, treated Colnett courteously and finally released him with the *Argonaut* after some tense diplomatic exchanges between their respective governments. The

---

7. For the story of Colnett and the Nootka controversy I refer the reader to Howay, ed., Champlain Society, who published Colnett's journal in 1940.

Fig. 133. "Port Sir Francis Drake": detail of the sketch by Captain James Colnett, September 1790. *Courtesy of the Public Record Office, Richmond, England.*

Princess Royal had already been impressed into Spanish service, but Bodega y Quadra agreed that Colnett could return to Nootka to retrieve it.

The curt diplomatic exchange resulting from the Nootka incidents had the English talking war. By October 1790, however, the two governments had signed the so-called Nootka Convention whereby Spain agreed to restore buildings, lands, and properties that had been seized from British subjects, and allow Englishmen to freely navigate, fish, and trade along the Pacific Coast, except within ten leagues of places already occupied by Spain.

Without so intending, Spain had in effect given up its claims to sovereign rights over the unsettled lands between Bodega Bay at thirty-eight degrees and Vancouver Island at forty-eight degrees—the boundaries of Drake's Nova Albion.

James Colnett and Port Sir Francis Drake – 1790

How fitting that the next European to set foot on the shores of Portus Nova Albionis was not just an Englishman, but another Plymouth sea dog born in Devonshire.

Colnett finally secured his liberty in July 1790, after "twelve months and four days' cruelty, robbery, and oppressive treatment by the Spaniards of New Spain." He immediately sailed for Nootka. At this time of year, however, the winds were contrary, so that after two months at sea Colnett was only about halfway along, with water and wood nearly expended.[8]

Approaching shore in the approximate latitude of Point Arena, Colnett headed south following the same course that Cermeño had in January 1595, this time actually putting into Havens Anchorage on September 10, 1790. This allowed him to fall upon Bodega Bay about two o'clock the following day. For reasons he never explained, Colnett was convinced that Drake had harbored at Campbell Cove, and on the sketch he made of the cove and outer anchorage he aptly named it "Port Sir Francis Drake." He stayed only two days securing water, wood and provisions and again set off for Nootka.

His little sketch fell into Spanish hands in April 1791 when Colnett finally caught up with the *Princess Royal* in Hawaii and traded the chart for a Spanish chart of Puget Sound. It must have aroused the immediate attention of Bodega y Quadra when it reached him at San Blas later that year.[9]

Colnett was the first European to explore Campbell Cove after Drake's departure. Did he see anything of the fort? We will never know. Perhaps, like Bodega y Quadra, he simply assumed Drake had entered because he had fallen upon the harbor under such similar circumstances.

8. Colnett, *Journal*, p. 169.
9. Ibid., p. 219. Colnett made this exchange with Captain Manuel Quimper who was under Bodega y Quadra's command and presumably obligated to turn this important information over to him as soon as possible.

Vancouver's Visit to Bodega and the Misnaming of Drakes Bay – 1792

In 1792, Spain and England both sent official expeditions to Nootka Sound to implement the terms of the Nootka Convention. Spain sent Bodega y Quadra with the *Activa, Mexicana* and *Sutil*. England sent George Vancouver with the *Discovery* and the *Chatham*. After additional disputes arose between Vancouver and Bodega y Quadra over the terms of the agreement, the governments amended it in 1794. Under the new terms, the subjects of both nations were allowed to frequent Nootka and erect temporary buildings, but neither could make a permanent settlement in the port or claim the right of sovereignty. Accordingly, the Spanish abandoned their Nootka settlement in 1795.[10]

When Vancouver and Bodega y Quadra left Nootka in the fall of 1792 to continue their negotiations at Monterey, Vancouver made diversions to explore the ports of Bodega, Point Reyes, and San Francisco. Vancouver's account of his expeditions includes a chart of the area (figure 42) that has considerable bearing on the anchorage controversy. On it, he names the bay under Point Reyes "The Bay of Sir Francis Drake." Due to the widespread use of this chart in the early 1800s, it influenced British, American, and Russian thinking on the whereabouts of Drake's landfall at the time the area was first settled.

How did Vancouver arrive at the name? Here is what he says in his journal: "According to the Spanish, this is the bay in which Sir Francis Drake anchored . . ."[11]

Coming when it did, this is a very curious statement. We know that Vancouver paid a ten-day visit to San Francisco Bay in late November 1792 and then spent fifty days at Monterey meeting with Bodega y Quadra. This makes Bodega y Quadra the most likely source of the alleged Spanish opinion that the Spanish believed Drake landed at today's Drakes Bay.[12] Yet we also know that Bodega y Quadra actually believed that Drake's landfall was at the Spanish Puerto de la Bodega (Tomales Bay) and that by November 1792, Colnett's sketch showing a Port Sir Francis Drake at Bodega Harbor was almost certainly in Bodega y Quadra's hands.[13] Finally, we know that at the time of Vancouver's visit to Monterey, the Spanish were very concerned that the English would attempt to draw the Spanish California boundary at Point Reyes by establishing an outpost at Bodega Harbor.[14]

Was Vancouver deliberately misled? Under the terms of the Nootka Convention, the bay under Point Reyes was off limits to the British: it was within ten leagues of an existing Spanish settlement. Bodega Harbor, on the other hand, located twenty leagues away, was open for settlement. Under the circumstances, it is difficult to escape the conclusion that the crafty Spanish commander had purposefully and very successfully deceived Vancouver to divert English attention from Bodega Harbor.[15]

Juan Matute and the Aborted Port Bodega Presidio – 1793

The Spanish were justly alarmed over the turn of events of 1790–92. Smarting from the painful concessions they were compelled to make in resolving the Nootka controversy, they now feared the English would use Colnett's alleged discovery to justify planting a settlement at their doorsteps.

10. For Vancouver's story I refer the reader to Vancouver's own account, *Voyage of Discovery*; Anderson's biography; *Surveyor of the Sea*, and Wilbur's two-volume, *Vancouver in California*, that contains the original account of the California sojourn.
11. Vancouver, *Voyage of Discovery*, p. 701.
12. Anderson, *Surveyor of the Sea*, pp. 119 and 124. Vancouver may have obtained this information during his visit to the port of San Francisco in December 1792 or October 1793 in the course of his meetings with Sr. Heamegildo Sal, the commandant of the presidio.
13. It seems that Colnett's chart was in Viceroy Revilla Gigedo's hands by no later than November 24, 1792, at which time he wrote to California's governor (José Joaquin Arrillaga) ordering him to be on his guard against English ships (cited in H. H. Bancroft, *History of California*, vol. 1, p. 514).
14. Wagner, "Last Spanish Exploration," p. 338. Also in H. H. Bancroft, *The History of California,* vol. 1, pp. 514–15.
15. It is evident from the journal of Archibald Menzies (Menzies, in Eastwood, ed., "Archibald Menzies Journal," p. 301), the expedition's botanist, that by mid-1793 Vancouver had somehow come into possession of Robert Gibson's journal of Colnett's 1790 voyage. Gibson was Colnett's first mate who had departed "Port Sir Francis Drake" in the *Argonaut*'s longboat, making a brief return to repair a leak. Because Vancouver left England before Colnett's return, it is uncertain whether Vancouver had come into possession of Colnett's own journal or chart as well.

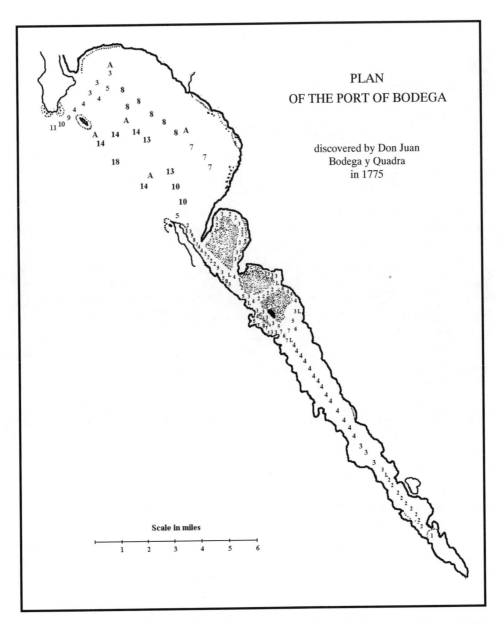

PLAN
OF THE PORT OF BODEGA

discovered by Don Juan
Bodega y Quadra
in 1775

Scale in miles

Fig. 134. "Plano del Puerto de la Bodega": the first detailed Spanish chart of today's Bodega and Tomales Bays prepared from the survey by Juan Martinez y Zayas in September 1793 (redrawn and translated). Redrawn from "The Last Spanish Exploration of the Northwest Coast and the Attempt to Colonize Bodega Bay," by Henry Wagner, in CALIFORNIA HISTORICAL SOCIETY QUARTERLY, vol. 10, 1931. *The original is held by the Archivo Histórico Nacional, Madrid, Spain. A manuscript copy is held by The Bancroft Library, Berkeley, California.*

These fears were greatly heightened in the winter of 1792–93 when the commander of the San Francisco presidio notified Monterey that two English ships were stationed at Bodega Bay and had landed guns.[16] Responding quickly to this threat of invasion, by March 1793, Bodega y Quadra had formulated the following detailed plans for occupying Colnett's Port Sir Francis Drake, a harbor the Spanish had never even seen before:[17] (1) leaving San Blas in April 1793, Juan Bautista Matute, in command of a small contingent of soldiers and priests, would take the *Sutil* into Bodega Harbor,

16. Ogden, *Sea Otter Trade*, p. 156, reports that these two English ships were the *Jackal*, under Captain William Brown, who arrived on March 15, 1793, and the *Prince Lee Boo* under Captain Sharp who stayed from January through March 1793. Spanish records suggest these two ships first arrived at Bodega in October 1792 (H. H. Bancroft, *History of California*, vol. 1, p. 514).
17. Wagner, in "Last Spanish Exploration," covers the aborted Port Bodega settlement, translating the accounts of Matute, Goycoechea, Zayas and Eliza.

make peace with the natives, and erect some buildings—they arrived in June; (2) Salvador Valdés would follow in the *Aranzazu* with the garrisons and artisans needed to form the establishment—they arrived in July; (3) Felipe de Goycoechea, the commandant of the Santa Barbara presidio, would then come with a small land expedition from San Francisco, both to assist in forming the settlement and to identify the route for a new road that would link the two establishments—he arrived in August; and (4) Don Francisco Eliza and Juan Martinez y Zayas in the *Mexicana* and the *Activa*, after making what would be Spain's last exploration of the Pacific Northwest Coast, would stop in at the port on their return to assist as needed with the settlement and chart the port—they arrived in late August and early September 1793 respectively.

Despite the efficient execution of these plans, the would-be Port Bodega presidio never came to fruition. Matute was back in San Francisco in August 1793 after spending some three months encamped along the shores of Campbell Cove. The following excerpts from a letter/report he issued from San Blas later that year explain why he aborted the effort:[18]

> I did not find there any vessel or any foreign establishment, and am of the opinion that there will never be any there, as the port is not capacious enough for larger vessels than the *Sutil* [forty-six tons], and has no timber or wood near the anchorage. As the lack of this made it impossible for me to build a house and live there, the same would be experienced by foreigners. The signs of some foreign vessel having been in the port are reduced to my having seen that the only small trees in the interior of the port had been sawed off. Afterwards I learned that this had been done by the Englishman Colnett, who took refuge there in a storm. This fact is noted on his plan of 1790, which at the present time has served all our vessels sent to that destination.[19]

Matute waited several months at San Francisco for orders to renew the effort, but they never came. His health suffering from the cold and foggy San Francisco climes, he returned to San Blas that fall. After the tensions associated with the Nootka affair had subsided, the Spanish seem to have let down their guard respecting Bodega. They would come to regret this.

Since Matute and his men spent at least three months at Bodega Harbor in close contact with the direct descendants of Drake's Nova Albion Indians, one might suppose they would have noted some evidence of Drake's visit. My suspicions are aroused by the following statement in his one surviving report that was written at San Blas, Mexico, in November 1793:

> On my arrival at this port [San Blas, Mexico] it has been called to my attention that the advises sent from the Port Bodega, July 23, and from San Francisco, August 23, giving the results of the commission with which I have been entrusted, have not been received at the office of the commandant [Bodega y Quadra], although I supposed they all had gone together in the same post, dispatched [from Monterey] by the governor of California August 26.[20]

Were Matute's first two reports really lost?

In the wake of the Nootka Convention, there were overt reasons for the Spanish to have carefully suppressed any indications they found of Drake's visit. If Matute and his men did find any vestiges of Drake's rock fort, they probably would have done their best to hide the evidence. Furthermore, had Matute made any reference to the matter in his reports, the information clearly would have been considered sensitive to national security, and accordingly handled as what would today be referred to as "highly classified" information. Perhaps Matute's first two reports still exist, tucked away in a folio in the archives of Spain.[21]

---

18. From the accounts of Menzies and Goycoechea, it appears Matute's company may have camped near the Indian village of Tókau, just south of Gaffney Point.
19. Matute, in Wagner, "Last Spanish Exploration," p. 339.
20. Ibid., p. 339.
21. Wagner mentioned seeing some "collateral documents of little importance" in the same Mexican archives in which he found the journals of the 1793 expedition (Tomo 70, Cuaderno 23; and Tomo 71, Cuaderno 35). Perhaps there is still something important to be found there. Two copies of Matute's sketch map were found in 1959 at the Archivo Histórico Nacional (Madrid). The Bancroft Library holds a manuscript copy. The sketch is crude and provides little detail.

Menzies at Bodega Harbor – 1793

When Vancouver completed his negotiations with Bodega y Quadra and left Monterey in January 1793, he returned to the Pacific Northwest Coast, via Hawaii, to continue his coastal surveys. On their return to San Francisco that fall, he had Captain Peter Puget take the *Chatham* into Bodega Bay to complete the survey of its two ports, Bodega Harbor and Tomales Bay. Archibald Menzies, the expedition's botanist, left us a journal that describes the visit.[22]

The *Chatham* arrived at Bodega the morning of October 20, 1793, anchoring off Doran Beach, just as Drake had on his day of arrival. Launching the ship's boat, they sounded the channel as they rowed into Campbell Cove:

> The soundings were found to decrease very gradually as we neared the northern shore and the channel leading into this backwater, which might with more propriety be termed a lagoon than a harbor was only about eight feet deep even at high water, and that too very narrow, scarcely half-a-cable's length across and winding round a low sandy point . . .[23]

After exploring the harbor, they beached the boat at Campbell Cove, climbed the steep trail up the bluffs, and did some exploring before casting off later that day. Menzies noted considerable evidence of Matute's visit earlier that year: horse and cattle tracks and many Spanish words in use by the Indians, and one other very interesting item:

> We landed on the west side and ascended the high ground which formed the bluff headland in expectation of a fine prospect which was, however, very limited from a thick fog that enveloped the inland country. Here we found a cross that did not appear to be long erected, it was formed of a piece of the stave of a cask fastened to a pole by a rope yarn.[24]

How close to the location Drake selected for his post did Matute and his friar erect their cross? Based on the premise that both commanders would be seeking the optimal vantage point to stake their respective claims, I suspect the Spanish were right on top of it.

Solid Men of Boston – 1794 to 1806

Once stories began filtering into New England ports of the great riches to be made in the Pacific Northwest, it did not take long for the Americans to arrive on the scene. In 1787, the U.S. Congress sent Bostonian Robert Gray to examine the region and open a fur trade with China. Gray, the first American to circumnavigate the globe, made two voyages, the first in 1788 and the second in 1792, effectively accomplishing his mission. By 1794, the so-called Boston men were arriving at a rate of over a half-dozen ships a year.

The Fur Trade Comes to Bodega – 1807 to 1810

By the turn of the nineteenth century, the combined onslaught of Russian-Aleut hunters and American, British and other European fur traders had greatly reduced the fur seal population throughout the Pacific Northwest. As the profits started to dry up, the British and European traders lost interest and almost disappeared from the scene. Americans, however, were more determined and enterprising. Recognizing that fur seals and otters were still abundant off the coast of California, they started directing their efforts southward.[25]

22. Extracts of the journal covering Menzies' visit to Bodega Harbor were published by Eastwood, ed., "Archibald Menzies Journal," in the *California Historical Society Quarterly* 2, Jan. 1924, pp. 265–340.
23. Menzies, in Eastwood, ed., p. 302. Aker, *Report of Findings*, p. 370, used these soundings to make the argument that Drake could not have brought the *Golden Hind* into Campbell Cove. First of all, Menzies seems to be referring to the channel leading into the inner lagoon, rather than the entrance to Campbell Cove. Secondly, these soundings could have measured the smaller of the day's two high tides, and possibly a neap. Twelve hours later that same day, the English might have measured ten or more feet of water in the channel.
24. Ibid., p. 302.
25. This falloff in the English fur trade explains why the Spanish abandoned their plans to colonize Bodega Harbor and Tomales Bay; the threat of English encroachment had dissipated. It appears that the Spanish may have intended to establish the Bodega settlement as a fur trading outpost (see Ogden, *Sea Otter Trade*, pp. 27–28).

To exploit the California fur seal population, the Americans would need Russian assistance. The California Indians, unlike the tribes further north, had little skill in hunting fur seals (they hunted by land only), and offered no real opportunity for bartering. Furthermore, the Indians were largely under the control of the Spanish who strictly prohibited such trade. In 1803, Boston merchant Joseph O'Cain journeyed to Kodiak Island to meet with the Russian-American Company's (RAC) governor, Alexander Baranov. O'Cain offered to split his pelts fifty-fifty with the RAC, if the Russians would furnish him with a contingent of Aleutian

Fig. 135. Sketch of a Kodiak Indian hunting sea otter off the Sonoma coast: drawing by Honoria Tuomey, circa 1925. *Reproduced from HISTORY OF SONOMA COUNTY by Honoria Tuomey, S. J. Clarke, 1926.*

Indians equipped with bidarkas (ocean kayaks). Baranov entered the first of what was to be a series of similar cooperative ventures. The enterprise was a modest success. When O'Cain returned to Kodiak from the coast of Southern California in June 1804, he turned over half his 1,100 furs.[26]

In 1807 and 1810, similar Russian-American fur seal expeditions touched on the shores of Campbell Cove. Captain Oliver Kimball, O'Cain's brother-in-law, was there from February through May 1807 in the 100-ton *Peacock*. In September 1810, William Davis somehow managed to squeeze the 200-ton *Isabella* over the barred entrance to the cove. He was paid a visit by a Spanish squadron from the San Francisco presidio under Lieutenant Gabriel Moraga. The Americans lived in three huts on the beach while their Aleuts, notwithstanding constant harassment from the presidio soldiers, went to work in San Francisco Bay. After leaving Bodega, Davis and his crew spent a month at anchor under Point Reyes joined by the *O'Cain* and the *Albatross*. While hunting ashore, one of the *Albatross'* crew was attacked by a grizzly bear![27]

The Boston men probably built their Campbell Cove campsites near where Drake and his company had entrenched themselves two centuries before. Did they see the walls of his stone fort protruding from the sandy beach or the hillocks at the base of the bluffs? As previously indicated, during this period both the Boston Men and the Russians were using Vancouver's chart that showed a "Drake's Bay" at Point Reyes. If they did see something curious at Campbell Cove, they probably would have assumed it was left behind by the Spanish.[28]

Russian California – 1809 to 1811

The idea of forming a Russian California colony was conceived by RAC official Nikolai Rezanov. With the Alaskan colonies facing a critical food shortage in 1806, Rezanov took the supply ship *Juno* south to San Francisco in a desperate effort to obtain provisions. On his return, he reported that the Spanish had treated him cordially and expressed a willingness to open up regular trade. He also announced his betrothal to Doña Concepción Arguello, the fifteen-year-old daughter of the port commandant and "the acknowledged beauty of California."[29]

26. I refer the reader to Ogden's, *The California Sea Otter Trade, 1784–1848* (1941).
27. Phelps, "Solid Men of Boston," p. 62.
28. During Colnett's visit to Campbell Cove in 1790, he was under the false impression that Spanish naval vessels stopped into Bodega Harbor once a year "destroying all canoes and habitations on the sea coast, that in case of being visited by foreign nations they may give them no assistance or information" (Colnett, *Journal*, p. 175). He presumably received this information from his Spanish captor, Estevan Martínez, who was then advising his superiors to aggressively wrest the sea otter trade from foreign nations (Ogden, *Sea Otter Trade*, pp. 26–27).
29. Rezanov to Rumiantsev, "Confidential Report," in Dmytryshyn et al., eds., *Russian American Colonies*, pp. 119–31.

Rezanov further alerted Governor Baranov to the advantages of establishing Russian settlements in the unoccupied territory north of San Francisco Bay: "In God's name I beseech you, gracious sir, to look carefully, in a spirit of patriotism, at the circumstances of this territory which offer the fatherland such great commercial opportunities."[30]

In October 1809, Ivan Kuskov, Baranov's able assistant, took the first independent Russian expedition to California waters. He set up temporary buildings on the beach at Campbell Cove—probably used in 1810 by Davis and his company—and possibly a small fort, "Fort Kuskov," some five miles inland near today's town of Bodega. Kuskov left in August, returning to New Archangel with over 2,000 skins.[31]

Kuskov's voyage was more than a simple fur-hunting expedition. Using information the Russians brought back with the *Peacock* the year before, Baranov had ordered him to establish friendly relations with the Indians and to select the most advantageous site to build a stockade and suitable houses for his company. When he finished his hunting, he had the option of staying with some of his men, or leaving and returning later to form a permanent settlement.[32]

Fig. 136. Ivan Kuskov (1765–1823): portrait by an unknown artist, believed to have been painted at Fort Ross in 1813. Kuskov was in charge of the siting, construction, and first ten-years' management of Russia's California colony. *Reproduced from FORT ROSS: INDIANS-RUSSIANS-AMERICANS by Bickford O'Brien et al., Fort Ross Interpretive Association,1981. The original is at the regional studies museum in Tot'ma, Russia.*

Fort Ross – 1811 to 1841

During a second visit to Campbell Cove in 1811, Kuskov had second thoughts about encroaching too close upon occupied Spanish territory and decided that the RAC better locate its first settlement a bit farther north. Accordingly, when he returned to California in the spring of 1812 with twenty-five Russians and eighty Aleut Indians, they commenced work at a strip of bare, hundred-foot-high coastal tableland, located about twenty miles north of Bodega Harbor. They dedicated Fort Ross in September 1812.[33] As of 1820, this remote frontier outpost included thirty-eight Russians, seventeen Creoles, one hundred thirty-three Eskimos, three Aleut men, five Yakuts, four Hawaiians, two Tlingits, and fifty-six California Indians. Of the native Indians, eight were men, all prisoners, and the rest were women who had married Russian men or Aleut Indians.[34]

The Russians, working mostly from their Farallon Island outpost, had largely depleted the California fur seal population by the mid-1820s. To compensate, they expanded the Fort Ross

30. Ibid., pp. 130–31.
31. According to Sonoma historian Honoria Tuomey (1923), Kuskov built his stockade near an Indian village in a narrow, level valley between high, redwood-crowned hills about five miles inland from the port; the site of the latter-day Khlebnikov and still later Smith Ranch, a mile west of the town of Bodega. Within the stockade, he supposedly constructed "a roomy house for himself," some barracks for his men, and the "Chapel of Saint Helen." Tuomey did not indicate where she got this information and I have not found mention of this fort in any Russian or Spanish account. Nevertheless, I suspect she is at least partially correct because it was in Kuskov's instructions to build these facilities and he certainly had the time. Kuskov's own detailed report of this visit has never been found.
32. Baranov to Kuskov, "Instructions," in Dmytryshyn et al., eds., *Russian American Colonies*, pp. 168–73.
33. The fort has been reconstructed in modern times. The visitor's center includes a museum, bookstore and library that has many recently published translations of Russian documents and titles. The Fort Ross Interpretive Association has published a well-illustrated booklet entitled *Outpost of an Empire* (Tomlin and Watrous, 1993), that provides an excellent general summary.
34. Istomin, *Fort Ross Settlement*, tables 1 and 2.

Fig. 137. Fort Ross: drawing by the Frenchman Auguste Duhaut-Cilly, 1828. *Reproduced from OUTPOST OF AN EMPIRE by Kaye Tomlin and Stephen Watrous Fort Ross Interpretive Association, 1993.*

settlement inland to increase agricultural production. In 1831 and 1833 they constructed two ranches in the interior: the Kostromitinov Ranch about two miles inland on the south side of the Russian River, and the Khlebnikov Ranch about five miles inland from Bodega Harbor. In 1838, the Fort's last manager, Alexander Rotschev, established a third ranch about eight miles northeast of Bodega Harbor in today's Coleman Valley. The Chernykh/Gorky Ranch was known for its scenic environs and remarkable vineyards (figure 98).

Although Kuskov had moved the main settlement north, the Russians ended up needing Bodega Bay and Campbell Cove for an anchorage and port, having discovered that the small harbor near Fort Ross was unsuitable for their purposes. The Russians called Campbell Cove, Port Rumiantsev, and referred to the anchorage off Doran Beach as the roadstead. They called Bodega Harbor, "Rumiantsev Bay" or "Little Bodega Bay," and greater Bodega Bay/Tomales Bay, "Big Bodega Bay." They drew the boundary of Spanish California at the "Levantula River," presumably today's Estero Americano about midway along the east shore of Bodega Bay. They referred to the lands north of the estero as Nova Albion and the lands to the south as Northern California.[35]

At the same time they were constructing Fort Ross, or perhaps the year after, the Russians built a sturdy eighty-by-one-hundred-foot warehouse at Port Rumiantsev (Campbell Cove). In all likelihood, they built this warehouse fronting the buried walls of Drake's old fort (see chapter 25).

To complete the port facilities, the Russians also built several houses, a launch, a boat, corrals, and later, a bathhouse. The Indians who lived at Gaffney Point about three-quarters of a mile north of the warehouse, kept an eye on things when the Russians were not occupying the port. As of 1818, their new chief was a brave young man named Valenila who had recently arrived from Marin County.[36]

California Department of Parks and Recreation Collection

Fig. 138. A Russian-American Company house located near Port Bodega: drawing by the Frenchman Eugène Duflot de Mofras, 1840. Reproduced from FORT ROSS: INDIANS-RUSSIANS-AMERICANS by Bickford O'Brien et al., Fort Ross Interpretive Association, 1981.

35. Golovnin, *Around the World*, p. 162; and Khlebnikov, *Reports*, p. 134. By the Levantula River, the Russians might also have been referring to today's Estero San Antonio, a little further south than Estero Americano.

36. Golovnin, *Around the World*, p. 165. Valenila and his people do not appear to have been of the tribes that occupied Bodega Harbor when the Russians first arrived in 1807–11; the Russian accounts suggest they were probably Marin County Coast Miwok that migrated from Olema Valley and the shores of Tomales Bay when the Spanish established Mission San Rafael in 1817.

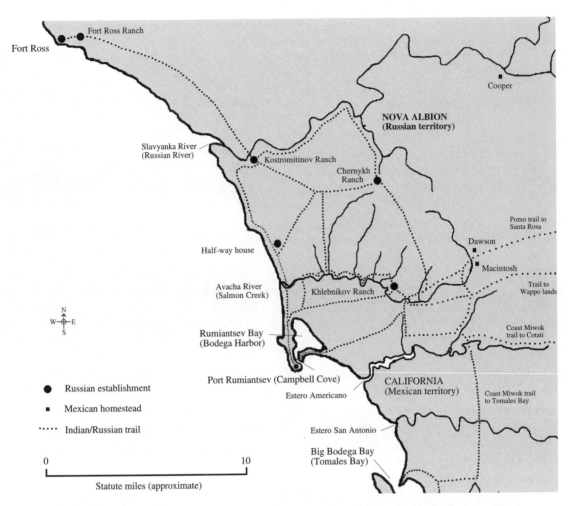

Map 21. Outline map of the area north of San Francisco showing Port Rumiantsev and Russian California circa 1838. The map shows the approximate locations of the Russian-American Company establishments and highways. In placing the ranches and Indian trails/Russian highways, I drew from Honoria Tuomey's descriptions, a sketch by the Frenchman Eugène Duflot de Mofras (1841), A. B. Bowers map of Sonoma County (1863), early USGS topographic sheets, current road maps and correspondence from historian Ruth Burke of Bodega Bay. The three Yankee settlers, John Cooper, James Dawson and Edward Macintosh, arrived at today's Mirabell Heights and Freestone in 1834, representing Mexican efforts to check Russian expansion. The Russians drew the boundary of Mexico's "Northern California" at the Estero Americano.

In their charts and reports, the Russians called the bay under Point Reyes, "Drake's Bay." Since their charts were obviously based on Vancouver's, it seems that the Russians simply assumed that this was where both the Spanish and British had already determined Drake had anchored. Perhaps the Russians never gave a thought to the possibility that Drake had encamped at Campbell Cove; but then again, if the Russians had indeed discovered Drake's fort, what are the chances that they would have announced it to the world?

Ivan Kuskov's invaluable journals and correspondence for 1809–18 appear to have been lost or destroyed.[37] If they could be found, they might very well contain enough documentary evidence to prove unequivocally that Drake built his encampment at Campbell Cove. For example, perhaps Kuskov recorded seeing a half-buried stone wall when they constructed the Campbell Cove warehouse. In addition, Kuskov was on very good terms with the Bodega Indians and became fluent in their language; he may have had a good deal to say about their legends. Who knows what will

---

37. Pierce, ed., *Correspondence of the Governors*, pp. v–vi, reports that in November 1818, almost all the RAC files belonging to the first RAC Governor, Alexander Baranov, were taken from Sitka to St. Petersburg by his replacement, L. A. Hagemeister. "There they remained, until some time after the liquidation of the RAC [1867], when they and other company files were destroyed."

Fig. 139. The last Russian house to stand at today's town of Bodega. According to Sonoma historian Honoria Tuomey, who was born and raised at Bodega in the late 1800s and conducted careful research on the matter, this structure was part of a stockaded presidio that Ivan Kuskov constructed on his initial visits to Bodega Harbor in 1809 and 1811. The records on hand, however, suggest that it was not until 1833 that the Russians constructed the Khlebnikov ranch at today's town of Bodega, including an adobe house, farm buildings, bath, well, and corral; there is no mention of what Ms. Tuomey referred to as "Fort Kuskov" and its "Chapel of Saint Helena." *Reproduced from HISTORY OF SONOMA COUNTY by Honoria Tuomey, S. J. Clarke, 1926.*

eventually emerge from the Russian archives . . . , or for that matter, even our own national archives in Washington, D.C.?[38]

Northern Missions and Mexican Ranchos – 1817 to 1840s

In 1817, the Franciscans transferred 230 converts from Mission San Francisco to the site of a proposed new settlement on the opposite side of the Golden Gate. At first, Mission San Rafael Archangel was intended to be just an extension of the San Francisco mission: a sort of hospital to care for the many sick natives that were suffering from San Francisco's infamous bad weather. By 1822, however, the friars in charge had been so successful in attracting new converts from the local Marin tribes that the authorities granted it independent status. In 1823 the Franciscans completed construction of a second mission in the northern frontier: Mission San Francisco de Solano in the Valley of the Sonomas (Mission Sonoma).

In the meantime, California was no longer a province of Spain. Rebel leaders had proclaimed Mexican independence in February 1821 and by year's end a Mexican government was in control. There was little immediate impact from these political events in the far-off Californias; new flags were not flown at the presidios until the following year.

---

38. It seems that Kuskov's correspondence to Hagemeister, for the period November 1818 to Kuskov's departure from Fort Ross in 1821, may yet be found in our own national archives among some uncalendared materials (over 30,000 documents) received in 1867 when the United States purchased Alaska from Russia (Pierce, ed., *Correspondence of the Governors*, pp. v–vi).

When the Russians began extending their agricultural enterprises inland in the 1830s, the Mexican government became alarmed. In 1834, to satisfy the joint objectives of secularizing the missions and establishing new settlements along the northern frontier, the governor appointed Lieutenant Mariano Vallejo as mission administrator and granted him a huge tract of land, Rancho Petaluma, near the Sonoma mission. By the mid-1840s, Vallejo, by then a general, had succeeded in converting both northern missions into Mexican pueblos, pacifying the frontier tribes, and carving up the former mission and Russian domains into Mexican ranchos.

Johann Sutter's New Helvetia and the Russian Withdrawal – 1841 to 1845

Upon his return from a visit to Fort Ross in 1834, RAC's new governor, Ferdinand Petrovich von Wrangell, recommended expanding the Russian-California colony through the purchase of Missions San Rafael and Sonoma, then being secularized. Towards this end, he journeyed to Mexico in 1835, but failed to win any significant concessions from a Mexican government that was becoming increasingly hostile to the Russian presence. By this time, Vallejo was planting settlers along the colony's boundaries in an attempt to discourage further Russian

Fig. 140. John Bidwell (1819–1900), photographed circa 1850. Bidwell lived in the old Russian guest house at Campbell Cove in 1842–43, during which time he wrote his epic, *A Journey to California*. *Reproduced from SUTTER OF CALIFORNIA by Julian Dana, Halycon House, 1934.*

encroachment. From the financial standpoint, Ross had never been profitable to the RAC, and with the disappointing results of Wrangell's visit, the company decided to pull out.

The RAC first attempted without success to sell their holdings to the Hudson Bay Company, and then to the Mexican government through Vallejo. Frustrated by Vallejo's negotiating tactics, Rotschev sailed up the Sacramento to offer the Russian holdings to an enterprising young Swiss whose name was destined to become famous in California history.[39]

Johann Sutter had arrived in St. Louis in 1834 where he became involved in the Santa Fe Trail covered-wagon trade. En route to California in 1838, he delivered a cargo of provisions to New Archangel where he became well acquainted with the Russians. Upon arriving in California, he applied for Mexican citizenship and a land grant on the northern frontier. He visited Vallejo in July 1839 and Rotschev at Fort Ross shortly thereafter. In late summer he established his prosperous New Helvetia (New Switzerland) colony at the site of today's city of Sacramento.

Sutter arrived at Bodega Harbor in September 1841 and quickly came to terms with Rotschev. He had all the livestock transferred to New Helvetia by month's end. For the sum of thirty thousand dollars, half in cash, half in wheat, he bought the houses, farms, cattle, implements, and vessels. The Russians had departed by year's end.

It was at this time that the French became interested in Bodega. Eugène Duflot de Mofras, who spent five months based at Monterey as part of his diplomatic tour of Mexico, urged his government

39. The first settler that took up stakes in the Bodega area actually preceded Vallejo. In 1827, Irishman John Reed built a tule hut near a Coast Miwok village near what is today's Cotati. He later settled in Sausalito after the Indians drove him away.

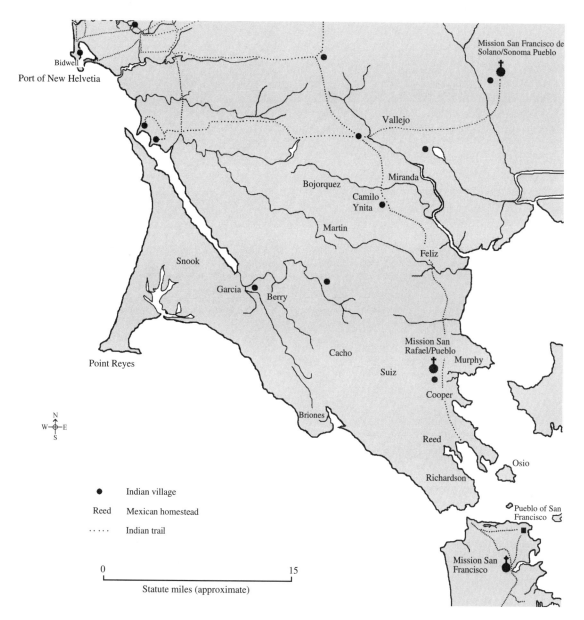

Map 22. Outline map of the area north of San Francisco Bay showing Johann Sutter's port of New Helvetia and Mexico's District of San Francisco circa 1842. This map shows the northern missions, the principal Indian settlements, the main Indian trails leading to Bodega Harbor, and the handful of Mexican homesteads in what was still at the time, a frontier wilderness. For a brief period after the Russians departed, Bodega Harbor remained outside Mexican rule as part of Sutter's independent province of New Switzerland. *Based on the Navare Map of 1830; THE SPANISH AND MEXICAN ADOBE AND OTHER BUILDINGS IN THE NINE SAN FRANCISCO BAY COUNTIES by George Hendry and J. N. Bowman, 1940; and other sources.*

to offer Sutter seventy-five thousand francs for his Russian holdings and then expand the colony south to occupy the northern shores of San Francisco Bay.[40]

Sutter returned to New Helvetia and sent a young Englishman, Robert Ridley, to take possession of his new Russian lands. In 1842, he replaced Ridley with John Bidwell, "prince of California pioneers."[41] Bidwell, who was to become one of early California's most prominent citizens, wrote

---

40. Duflot de Mofras, *Exploration of Oregon and California*, p. 258. Duflot de Mofras' assignment was to assess French trading and colonial opportunities on the Pacific coast of North America. In other words, he was a spy.

41. From the title of R. D. Hunt's biography, *John Bidwell, Prince of California Pioneers.*

Fig. 141. Stephen Smith and his wife, Doña Manuela Torres Smith, photographed circa 1850. *Reproduced from HISTORY OF SONOMA COUNTY by Honoria Tuomey, S. J. Clarke, 1926.*

his epic pioneer diary, *A Journey to California*, while living at the Russian house at Campbell Cove—sitting virtually atop Drake's old encampment.

When Sutter returned home in 1841 after a three-month absence conducting the Fort Ross purchase, he found many new American settlers arriving overland in the Sacramento Valley. In January 1848, one of these, James Marshall, his new sawmill foreman, picked up a few shiny stones unearthed during the construction of a new millrace: gold!

As things turned out, the Gold Rush led to Sutter's downfall. His workmen deserted, his cattle disappeared, and squatters took over his lands. By 1852 he was bankrupt, probably wishing he had not given up his Russian holdings without a fight.

Stephen Smith and his Rancho Bodega – 1843 to 1855

One day in September 1843, a surprised John Bidwell looked out from Campbell Cove to see a large vessel at anchor off Doran Beach. It was the bark *George Henry*, 344-tons, out of Baltimore, Maryland. On board was sixty-one-year-old Stephen Smith, master and owner, Smith's bride, sixteen-year-old Doña Manuela Torres, along with her brother and mother, and a company of some twenty or so men. Stowed below was an assortment of odd piping, boilers, and equipment: the makings of California's first steam mill.[42]

Smith, a Massachusetts seafaring man, had come to Monterey two years before looking for business opportunities. Learning that the Mexican government was seeking settlers for its northern frontiers, Smith persuaded the governor to allow him to build a steam-powered sawmill in exchange for citizenship and a land grant. He then made an expedition north examining all the territories between Monterey and the soon-to-be-vacated Fort Ross settlement. Excited by what he saw at Bodega, he journeyed to Petaluma where he met General Vallejo and broached the idea of building his mill at the abandoned Khlebnikov Ranch near today's town of Bodega. Vallejo favored this plan, and apparently even helped Smith with the arrangements for securing the prerequisite Spanish wife.[43]

42. Munro-Fraser, *History of Sonoma County*, pp. 51–56 and 193–96, covers Smith's days at Bodega in considerable detail.
43. A letter by Smith is reproduced in Lane, *Rancho Bodega*, pp. 9–11.

Fig. 142. Stephen Smith's adobe mansion: drawing by Honoria Tuomey circa 1925. Smith took up residence at the abandoned Khlebnikov Ranch in 1843. He constructed this impressive structure shortly thereafter. *Reproduced from HISTORY OF SONOMA COUNTY by Honoria Tuomey, S. J. Clarke, 1926.*

When Bidwell met Smith and learned of his intentions, he tried to ward him off. Smith, however, knew that Sutter lacked a Mexican title for the land, and ignored the threat, issuing a menacing threat of his own. Bidwell left to apprise Sutter of the turn of events, never to return; Smith went right to work.

Within a few months, Smith completed a sawmill that featured an adjoining gristmill. He invited everyone in the surrounding territory to celebrate the start-up. General Vallejo made the dedication speech, lumber was cut, wheat was ground to flour, and everyone feasted.

Having fulfilled his end of the bargain, Smith filed Mexican citizenship papers and in 1846 was duly granted the 35,487.5-acre Rancho Bodega, apparently including the harbor and old Russian port facilities. Fortunately for Smith, the grant came just before California's Bear Flag Revolt in which a group of Yankee immigrants, under Captain John Charles Frémont, briefly wrested The Republic of California from the Mexican provincial government (July 1846).

Later that month, Smith had to swear allegiance to yet a third flag when Commodore John Drake Sloat sailed the U.S. naval brig *Portsmouth* into Monterey and claimed California for the United States, that had just declared war on Mexico. Historian J. P. Munro-Fraser provides an interesting anecdote of this event, giving the impression that Smith was taking no chances on the outcome of the Bear Flag Revolt.

> Upon receiving the flag [stars and stripes] sent to him from Sonoma, he [Smith] at once proceeded to the woods and selecting a beautiful straight tree about fifty feet high, he cut it and brought it to the top of an eminence near his house. He then fashioned a rude figure of a bear with a star attached to the extremity of its tail. This novel emblem was placed at the top of the flagstaff and reared aloft. The stars and stripes were then run to the top of the staff and unfurled to the breeze for the first time in that section … When the flag had reached the top of the staff there was a curious commingling of the three emblems of liberty, at that time so justly famous, and popular, viz.: the stars and stripes, the bear, and the lone star.[44]

Smith continued to operate the steam mill himself until 1848 at which point he sold the mill and most of the timber on the ranch to Bethuel Philips. From that point, the aging entrepreneur leased the plant to some Yankee settlers and devoted his time to managing his holdings, cattle ranching and building and operating a tannery. The mill building burned to the ground in 1854, by which time some of the boilers had already been moved to Mendocino.

---

44. Munro-Fraser, *History of Sonoma County*, p. 191.

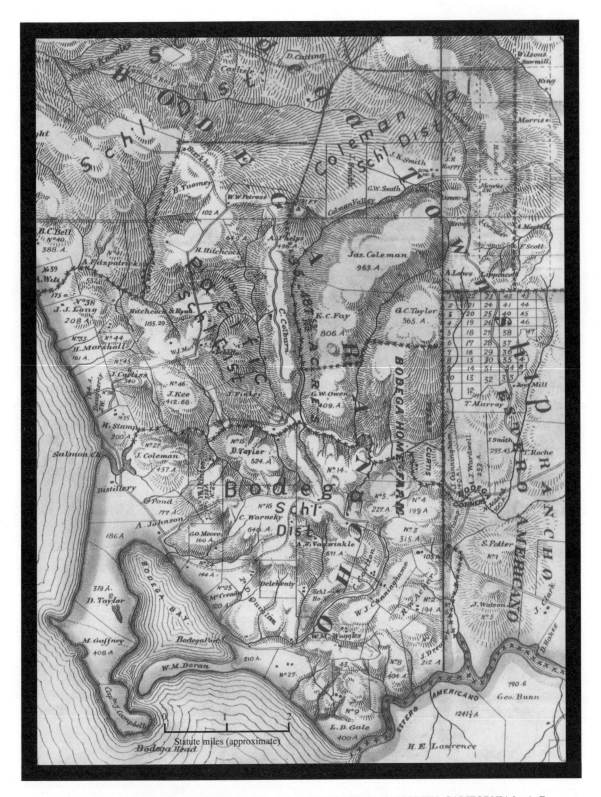

Map 23. Rancho Bodega as of about 1863. Detail from the MAP OF SONOMA COUNTY, CALIFORNIA by A. B. Bowers, San Francisco, 1867. In the nineteenth-century, today's town of Bodega was called Bodega Corners. *Courtesy The Bancroft Library, Berkeley, California.*

The Heydays – 1850 to 1860

Taking advantage of the influx of settlers associated with the Gold Rush, Smith began leasing parcels of Rancho Bodega to some of the more enterprising immigrants. After allowing a pair of squatters named Keyes and Noble to raise potatoes on Bodega Head in 1850, he leased two-hundred acres on the point to Edward Cheney. In 1851, he leased a tract to four farmers who formed a partnership called the Suffolk Company. By the spring of 1852, Smith had rented out the greater portion of land near the port to other farmers, most of whom raised Bodega Red potatoes or grain.

To ship his lumber, Smith built a small warehouse at "Bodega Port" in the southeast corner of the harbor, and used his little schooner, the *Fayaway*, to make regular runs to San Francisco. The Suffolk Company built another schooner, a warehouse and a landing on the north shore of the harbor that later became "Johnson's Port." Quite a number of additional buildings were constructed in the early 1850s, including three large warehouses, the first store, a hotel, and even a bowling alley. In 1852, the federal government made Bodega Harbor a U.S. port of entry.

For a while, the little port was bustling with activity. Smith's former squatter John Keyes purchased a small schooner, the *Spray*, that he ran between Campbell Cove and San Francisco carrying lumber, shingles, potatoes, cattle, sheep and passengers. Between 1852 and 1860, there were several other vessels plying this trade, "there being often as many as six at one time making weekly trips to the city and return." One of these vessels, the *Mary*, was owned by a Captain Tibbey, who came to possess the old Russian port facilities. These he leased to one Captain John Campbell, who eventually purchased all the lands around the cove. Drake's Portus Nova Albionis bears his name today.[45]

The Dog Days – 1860 to 1900

When historian Munro-Fraser observed Bodega Harbor in the 1870s, its brief glory days were already over:

> At the present time there is no more forlorn and dilapidated-looking place in the state. Nearly all the buildings are gone, with one warehouse and a part of another remaining of all the warehouses which have been built there. The hotel and bowling alley still stand, but where, on a Sunday, at least, one hundred men were wont to congregate and pitch fifty-dollar slugs at a peg in front of the barroom door, scarcely a stranger's face is seen once in a fortnight. The few buildings are all going to ruin. Where there was once the bustle and jostle of the great shipping business which loaded a large schooner every day, now the sight of a vessel is something to be wondered at.[46]

Why? According to Munro-Fraser, the port was essentially built around the vitality, enterprise and kindheartedness of Stephen Smith and destroyed by the opposite qualities of his successor. "If a man were poor and appeared to be honest, Smith would advance him the necessary amount of money to put in his crop, without interest."[47]

Smith died in 1855 at the age of seventy-three, leaving his lands to his widow and three children. After the young widow remarried, her new husband, a San Francisco attorney named Tyler Curtis, came to control the entire grant. Soon after Curtis took over, he attempted to oust all the squatters who had been taking advantage of Smith's kindness. The "Bodega War" took place sometime in 1858 or 1859, when Curtis recruited some forty San Francisco roughnecks and brought them up the Petaluma on a steamer. The Rancho Bodega settlers, however, somehow learned of the scheme, joined forces, and confronted Curtis at the blacksmith's shop in Bodega Corners with the local sheriff on hand to keep order. The "bullies" retreated to the city without a shot fired and Curtis was compelled to sell off much of the grant to the disgruntled settlers.[48]

---

45. The local history was compiled from Munro-Fraser, *History of Sonoma County*, pp. 187–88 with input from Ruth Burke, Bodega Bay, correspondence to author, Apr. 9, 1997. Ms. Burke is a local historian who has a book in progress on the history of Bodega.
46. Ibid., pp. 188–89.
47. Ibid., p. 187.
48. Ibid., pp. 196–97; and Burke, Bodega Bay, correspondence to author, Apr. 9, 1997.

Fig. 143. View looking north along the east shore of Bodega Harbor, 1958. *Photographed © by Ed Brady, February 1958. Courtesy of Grethe Brady, Aero Photographers, Sausalito, California.*

It was during this period that the Russian warehouse came tumbling down, burying Drake's fort site in its ruins. According to Munro-Fraser, writing in the late 1870s: "This building was very strong and durable and would probably be standing at the present writing but for the fact that a landslide swept the most of it into the bay some years ago."[49]

Today's Quaint Fishing Port

From David Pesonen, writing in 1963 (courtesy of David Pesonen, Berkeley, California):

The Bodega peninsula resembles an arm raised in defense against the sea. It terminates in a blunt fist of granite called Bodega Head. Curled inside the bend of its elbow is Bodega Harbor, an anchorage for several hundred commercial and sport fishing boats. Commercial seafood landings from this fleet exceed a million dollars a year, twice the value of the catch from San Francisco's more publicized Fisherman's Wharf. Bodega Harbor is considered the safest among the five harbors of refuge along the 300 miles of forbidding coast between San Francisco and Coos Bay, Oregon. Sometimes the winds blow in off the ocean relentlessly for weeks, whipping spindrift from miles out, sending breakers spuming a hundred feet up from the rocks on the seaward side of the head, while the fleet bobs securely in the harbor and the fishermen mend their gear, hoping for a letup.

Looking down on the harbor is the town of Bodega Bay. With nets drying on the wharves and the salty odor of steaming crabs, fresh cod and salmon, the town breathes an atmosphere of old world fishing

49. Ibid., p. 183. After George Davidson visited Campbell Cove in 1853, he reported that "two large wooden houses under the eastern bluff [the Russian warehouse and guest house] . . . were in ruins" (*Pacific Coast Pilot* [1889], p. 254).

Fig. 144. View looking north at the marina on the west shore of Bodega Harbor. *Photograph by the author, 1995.*

ports, like those strung along the Mediterranean from Spain to Yugoslavia. In fact, most of the fishermen who sail from Bodega Harbor, the fish processors, the shipwrights, and the mechanics are immigrants or their descendants from the Mediterranean region. With names like Lazzio, Zankich, Gelardi, their heritage is stamped indelibly on Bodega Bay.[50]

On the surface, not all that much has changed in the thirty-odd years since Pesonen penned these words. By that time there were already five commercial fish-receiving piers, one marine fueling station, one boat repair facility, six small boat landings, and the two breakwaters that were built circa 1940. On the north side of the harbor, we now find an upscale seaside inn and a links-style golf course meandering through a high-priced housing development. There are several new inns and restaurants on the east side and two new marinas on the west. The captains of the fishing fleets do not have the same difficulties entering and leaving the port that Drake and the Russians did. When it installed the breakwaters, the U.S. Army Corps of Engineers dredged a channel that they still maintain.

When Captain George Vancouver's botanist, Archibald Menzies, visited Bodega Harbor in November 1793, he was much impressed with the great numbers of the feathered tribe that inhabited the area: "white and brown pelicans, gulls, plover, and a variety of aquatic fowl . . . eagles, hawks, the red-breasted lark, crows and ravens."[51] Producer/screenwriter Alfred Hitchcock brought Bodega Harbor, its winged residents, the Tides Restaurant, and the Potter schoolhouse at the town of Bodega to international attention in filming one of his best known chillers, *The Birds*.

As you will see, Hitchcock's film was not the only event in the early 1960s that brought national attention to this quaint little fishing port.

---

50. Pesonen, *Visit to the Atomic Park*, p. 3.
51. Vancouver, in Wilbur, ed., *Vancouver in California*, pp. 115–16.

Fig. 145. "Dance of the inhabitants of Mission San Francisco": drawing by Louis Choris, 1816. *Courtesy The Bancroft Library, Berkeley, California.*

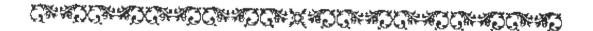

CHAPTER 36
The Indian Story

What short- and long-term effects did Drake's visit have on the local Indians? Were they decimated by European diseases? Were there any pregnancies? Did they assimilate any of the English customs or words? We will never know. We do know what happened to the Indians when Europeans finally returned over two centuries later. It is a tragic story.

Fig. 146. "Natives of Rumiantsev Bay": painting by Mikhail Tikhanov, 1818. There is much more to this relatively unknown and unappreciated Russian painting than meets the casual eye. The scene depicted here is a shaman's last desperate attempt to cure his dying chief, fatally stricken with one of the European diseases ravaging the Indian population at the time. Here is how Tikhanov's commanding officer, Vasili Golovnin, described the heartrending scene: "Sitting over the sick man in the tent, he kept repeating incantations and singing, while waving around a stick with feathers attached to it. The patient's family, who were in the same tent, responded and joined in the singing at certain prescribed times. This went on for over an hour in our presence, and when we left the shaman was still continuing with his cure" (Golovnin, *Around the World*, p. 169). The artist unwittingly captured the tragic fate of the Nova Albion Indians. *Courtesy of the Scientific Research Museum, Academy of Arts, St. Petersburg, Russia.*

A Curious Legend

It seems likely that for many generations, the tribes frequenting Bodega Harbor would have considered Drake's encampment site a hallowed ground. Some shadowy evidence supporting this conjecture appears in a curious legend recorded for the Wappo Indians, who once occupied the upland country some thirty miles due east of Bodega Harbor. At death, they believed the spirit was borne aloft and then flown to a grotto hard by the sea at the end of a peninsula. In this grotto a fire burned without ceasing, that no living being could behold without being instantly stricken blind. The disembodied spirit entered the grotto, hovered over and around this fire for a season, then fluttered forth again, winging its way over the ocean to the "happy western land."[1]

Drake's Missing Men and Another Legend

There has been considerable speculation that Drake left behind some of his men when he departed California. This speculation stems partly from the N. de Morena story (chapter 32), and also from the fact that a credible Spanish source indicates that Drake sailed north from Guatulco with seventy-two in his company, while *The World Encompassed* reports he returned to England with only fifty-eight.[2]

Although none of the contemporary accounts mention anyone left behind, there was a legend among the Coast Miwok that lends some credence to the notion. It was recorded in September 1824 by the Russian sea captain, Otto von Kotzebue. During an overland journey from San Francisco Bay to Bodega Harbor, the Russians stopped at the crest of one of the coast range hills opposite Bodega Harbor. Pointing east, their Coast Miwok Indian guide Marco informed them that the valley they could see extending inland (Salmon Creek Valley) was named "The Valley of the White Men." He explained that there was a tradition among them that the survivors of a shipwreck chose this valley for their residence and had lived there in great harmony with the Indians.[3]

Although Marco told him that the Coast Miwok legend did not record what ultimately became of the white men, Kotzebue speculated that the Indians to whom the legend pertained must have been the ancestors of a particular tribe that dwelled near a tree-covered mountain off in the distance. Marco had explained that this tribe, apparently the Wappo, were "a distinct class from the common race of Indians" set apart by their courage. In numerous battles, they had won the fear and respect of the Spanish, preferring death to the dominion of the missionaries.[4]

Tribal Geography at the Time of European Settlement

Bodega Harbor, from the standpoint of ethnogeography, was unusual. It lay near the border of two major language groups, the Coast Miwok and the Pomo, and was frequented on a regular basis by a third, the Wappo. While the early accounts and later ethnological studies both suggest that Bodega Harbor was in Coast Miwok territory from at least the early stages of Russian occupation,

1. Powers, *Tribes of California*, p. 200. Powers reported that the peninsula was Point Reyes. While it is conceivable that his informant was actually referring to a grotto on Bodega Head (Campbell Cove), the evidence is against it. Isabel Kelly's Coast Miwok informants, Tom Smith and Maria Copa, both mentioned legends associating Point Reyes with the dead (Collier and Thalman, eds., *Interviews*, p. 451). Some Drakes Bay anchorage advocates have cited these legends in support of their case, for example Aker, *Report of Findings*, pp. 325–27. I include the Wappo legend here simply because I find it so interesting. Perhaps it was somehow born of the Cermeño visit in 1595.
2. Those who have discussed the matter include Wagner, *Drake's Voyage*, pp. 148 and 167; Aker, *Report of Findings*, p. 330; and Hanna, *Lost Harbor*, pp. 54–65. Nicolás Jorje, a Spaniard held captive aboard the *Golden Hind* from February 5 to March 5, 1579, surreptitiously counted seventy-one to seventy-two in Drake's company (Nuttall, *New Light on Drake*, p. 137). The discrepancy seems to be at least partly accounted for in the first declaration of John Drake, who reported that Drake lightened his load at Ternate by reducing his company to sixty, including María and the two other Africans Drake left behind shortly thereafter on Crab Island. He also reports that one hand died rounding Cape Horn (Nuttall, *New Light on Drake*, pp. 32–33).
3. From Kotzebue's journal in Sullivan, ed., *Russian Settlement*, pp. 39 and 40.
4. Ibid. Even if the legend were true, and white men did briefly inhabit Salmon Creek Valley, it does not necessarily mean that Drake left any of his crew behind. The legend may have been born simply from Drake's visit inland. There is also the possibility that the legend, as recorded, did precipitate from a latter-day shipwreck. Marco's reference to Wappo battles with the Spanish probably stemmed from the Luis Antonio Argüello expedition of 1821.

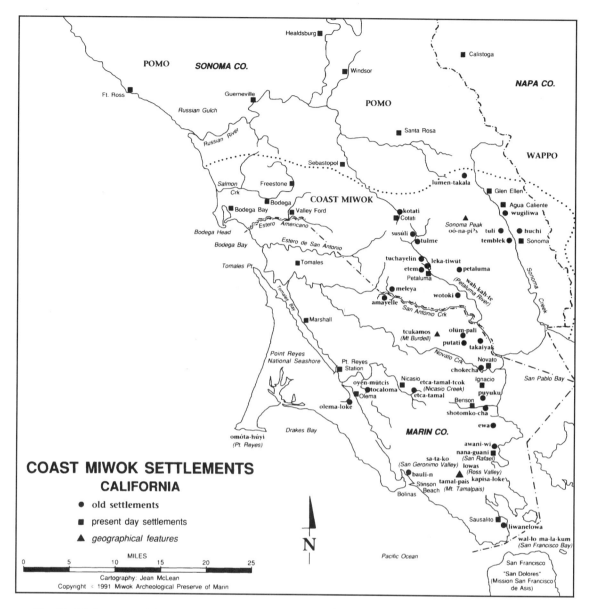

Map 24. Modern map of Coast Miwok territory and settlements in the late nineteenth century: drawn by Jean McLean, 1991. *Courtesy of the Miwok Archaeological Preserve of Marin, San Rafael, California.*

there is some reason to believe that this was not the case beforehand. The Russian accounts reveal a northward migration of Marin County Coast Miwok into Russian territory with the founding of Mission San Rafael in 1817. They reveal a similar flight of Sonoma County Coast Miwok with the founding of Mission Sonoma in 1822. They also document the flight of Sonoma County Pomo out of Russian California starting in the late 1820s when the Russians expanded their agricultural enterprises inland from Fort Ross and Bodega Harbor.[5]

5. Kostromitinov (in Stross and Heizer, eds., "Ethnographic Observations," p. 7) reported that by the early 1820s many of the original inhabitants from the area of the little and big Bodega Bays had migrated to Fort Ross as a result of the founding of the new missions at San Rafael and Sonoma. Further documentation of the migration of Marin's Tomales Bay tribes to Sonoma County is found in Fyodor Matiushkin's *A Journal of a Round-the-World Voyage on the Sloop Kamchatka under the command of Vasili Golovnin*, recently translated from Russian by Watrous. According to Matiushkin: "The Spanish [after founding Mission San Rafael in 1817] had expanded their hunt for people to Tomales Bay itself. By now all the bands fled for safety under the guns of Fort Ross or to Port Rumiantsev. Last year [1817] when a large number of people gathered at Fort Ross and asked for his protection, Kuskov persuaded them to settle in the forests and mountain gorges . . . The savages followed his advice and settled in the forests that are visible from

According to the late Rose Gaffney who grew up on Bodega Head in the early 1900s, "there was a triangle of land stretching from six miles along the ocean to Freestone where anyone was welcome, and could travel in safety. The Coast Miwok and Pomos made one stipulation, that no one try to take possession of it." This statement, presumably passed down from the original American settlers who interacted with the tribes of Tókau on Gaffney Point, fits the geographic setting as well as the ethnological evidence; it rationally explains a well-documented pattern of subsistence in which the Coast Miwok, Pomo and Wappo all harvested the same Bodega clam flats in sustaining their respective shell-money economies (see map 16).[6]

The Coast Miwok's southern boundary also appears to have included a joint linguistic group border zone. The very earliest Spanish accounts and mission records suggest that a mixture of bilingual Costanoan and Coast Miwok people occupied the area around Richardson Bay at the tips of the Sausalito and Tiburon peninsulas. The same may also have been the case with the eastern border due to the intermarriage of neighboring Coast Miwok, Wappo, and Patwin tribes.[7]

According to the Russian accounts, as of 1818 the Indians called Bodega Harbor *Chokliva* (creek water) and Bodega Head *Tiu-tuiya*. According to Mariano Vallejo, at the time of his visit to Bodega Harbor in May 1833, the Indian village at Bodega Harbor was named *Tiutume*. George Davidson, after visiting Bodega in 1852, reported that the resident Indians called Bodega Harbor *Tuliataelivo*, and Bodega Head *Tiu-tuiya*. Tom Smith, perhaps the last full-blooded (or near full-blooded) survivor of the Coast Miwok race, who was born and raised in the Bodega area in the late 1800s, called Bodega Harbor *Koyo-liwa-puluk* (salt water pond) and Bodega Head *Nai-utci*. He said the word *tókau*, then applied to the village at Gaffney Point, meant a kind of *hena* (spirit), like the wind as opposed to a dead man. It would seem that Vallejo's appellation is the most appropriate village name for the Bodega Miwok at the time of first European contact: the Tiutumes.[8]

The Coast Miwok

As with the other linguistic groups living in the San Francisco Bay area, the Coast Miwok territory was occupied by scores of small tribal groups, each of which held territories of roughly one hundred square miles. The tribal groups typically consisted of approximately two hundred to four hundred members of intermarried families, loosely united under a common headman, or chief (hoipu). Based on mission records, it appears that there were about a dozen principal tribal groups in Coast Miwok territory, some of which contained splinter groups (map 16). Some of the tribes may have shared a

Port Rumiantsev [presumably today's Salmon Creek Valley and Coleman Valley]."

6. Cited in "Bodega Head, PG&E, and Rose Gaffney," *Sonoma County Bugle*, Dec. 10, 1970, and Jan. 6, 1971. According to Driver (1936), Bodega Bay was common territory for Pomo, Wappo, and Coast Miwok clam digging (cited in Collier and Thalman, eds., *Interviews*, p. 204). Some of the first ethnological researchers, Powell and Powers, actually had Bodega Harbor in Pomo territory. Barrett drew the border at Salmon Creek based on his studies of the remnants of post-mission tribes in the early 1900s. It would be interesting to know what Matute's friars observed in 1793 or to study Kuskov's missing accounts for his visits in 1808 and 1811.

7. Evidence for Costanoan occupation of the southern shoreline of Marin County is found the diaries of Father Vicente María Ayala (1775) and Father Francisco Palóu (1776). During the first Spanish entry into San Francisco Bay aboard the *San Carlos* in August–September 1775, Ayala found the native population of the Richardson Bay environs conversant in the Costanoan language and listed eight individuals, seven of whom, by mission records, are identifiable as Costanoan visitors of the Huchiun tribal group (Milliken, *Time of Little Choice*, p. 131). Palóu, involved in founding Mission San Francisco the next year (August 1776), watched the Costanoan tribes occupying the tip of the San Francisco Peninsula flee to the north and east sides of the bay after an attack by their warring neighbors. The records for Mission San Francisco reveal that the exodus of Huimen tribes from the north shores of San Francisco Bay coincides with that of Costanoan Huchiun tribes occupying the adjacent East Bay shores of today's Richmond, Berkeley and Oakland and that the two groups were interrelated by marriage. The Costanoan exodus from across the bay (1785–1805) began substantially earlier than the Coast Miwok exodus (1800–1817). Heizer, ed., *Handbook*, pp. 4–5, on one side, comments on the apparent early removal (to Mission San Francisco) of Costanoans living north of the Golden Gate and general uncertainties about tribal boundaries. Milliken, *Time of Little Choice*, p. 244, on the other side, concludes that Huimens of Richardson Bay were bilingual Coast Miwok. Milliken based his conclusion on an intensive evaluation of mission records, noting three Miwok village names ostensibly located on the shores of Richardson Bay and that "the Huimen vocabulary that Father Arroya de la Cuesta recorded in 1821 [at Mission San Rafael] is definitely a Coast Miwok word list."

8. Lütke, "Observations on California," in Dmytryshyn et al., eds., *Russian American Colonies*, p. 279; Farris, "Bodega Miwok," p. 4; Davidson, *Pacific Coast Pilot* (1889), pp. 252–54; and Kelly, in Collier and Thalman, eds., *Interviews*, pp. 4 and 107. Appellations such as Olementke (Powell), Olamentko (Merriam) and Yolatamal (Milliken) assigned by modern anthropologists appear to reflect the northward flight of Marin County's Coast Miwok into Russian territory in the early nineteenth century.

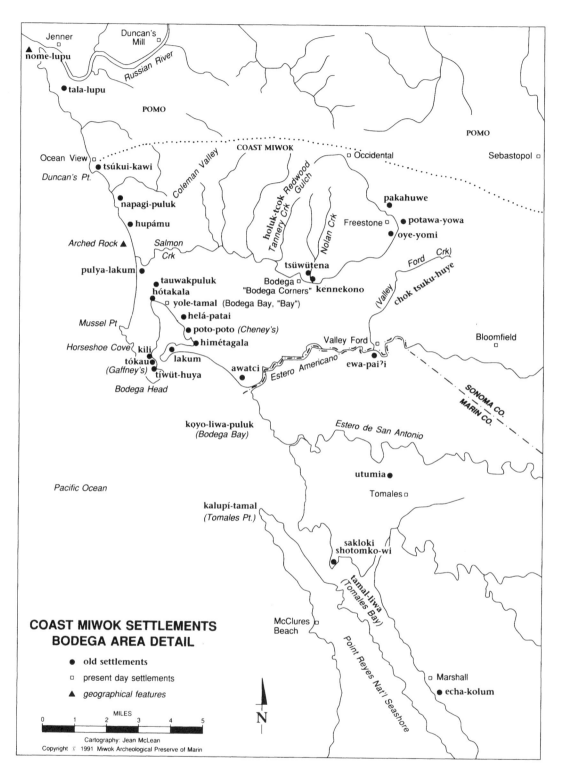

Map 25. Modern map of late nineteenth-century Coast Miwok settlements in the greater Bodega Bay area: drawn by Jean McLean, 1991. *Courtesy of the Miwok Archaeological Preserve of Marin, San Rafael, California.*

Fig. 147. "New neophytes at Mission San Francisco": including from left to right: Ululato (Patwin); Numpali (Wappo); Suisun (Patwin); Olompali (Coast Miwok); and Cholvon (Yukot): drawing by Luis Choris, 1816. *Courtesy The Bancroft Library, Berkeley, California.*

single central village site for at least part of the year, but, in general, the natives lived in a dispersed pattern within their territory and when necessary moved their settlements seasonally as dictated by their hunting and gathering mode of subsistence:[9]

> The season dictates the place they must visit to find sustenance. In spring, they live near rivers and other places where water is abundant in order to catch fish and gather roots and herbs; the summer brings them to the forests and steppes, where they collect berries and the seeds of wild plants; in autumn they lay in stores of acorns, wild chestnuts and other nuts, and kill bison and wild goats [deer] with their arrows. . . . When this is finished they go without lose of time to collect the seeds of a plant that grows abundantly on the plain.[10]

Neighboring tribes were tied together by trade bonds and intermarriage, and occasionally visited each other en masse for resource-specific trade feasts and ceremonial dances. In addition to foodstuffs and tobacco, the major items of trade were basket materials, obsidian for arrowheads, feathers, and shell beads. There was occasional intertribal feuding, with territorial disputes and women-stealing the most commonly documented source of hostilities.[11]

Modern anthropologists have differentiated three Coast Miwok subgroups: the Western or Bodega (the Bodega Harbor area), the Southern (coastal Marin County), and the Valley (the area between Freestone and Petaluma). The Western group had a slightly different dialect than the other two, perhaps reflecting the influence of the closely surrounding Pomo.[12]

Based on local geography, one suspects that Marin's Olema Valley and Point Reyes Peninsula tribes would have had little if any interaction with the tribes inhabiting Bodega Harbor; they were separated by Tomales Bay, greater Bodega Bay, and the coastal mountain range. Furthermore, they had no good reason to travel to Bodega since their territory had similar clam flats and fishing grounds. The Point Reyes tribes were apparently so isolated that University of California researchers were not able to confirm they existed from their interviews with the post-mission period Coast Miwok.[13]

---

9. Milliken, *Time of Little Choice*, pp. 21–24 and 228–61.
10. Kostromitinov, "Observations," in Wrangell, *Statistical and Ethnographic Information*, pp. 42–43.
11. Milliken, *Time of Little Choice*, pp. 22–23.
12. Kelly, "Coast Miwok," in Heizer, ed., *Handbook*, p. 414. The boundaries that researchers have assigned to the three Coast Miwok subgroups roughly coincide with the territorial boundaries of the two northern missions and the Fort Ross settlement.
13. For example, Barrett, "Ethno-Geography," p. 307; and Kroeber, *Handbook of the Indians*, p. 273. King and Upson, "Protohistory on Limantour Sandspit," in Schenk, ed., *Archaeology of Point Reyes*, pp. 172–73, discuss archaeological evidence of this isolation.

This may explain why the Indians Cermeño encountered at Drakes Bay in 1595 showed no indication of having seen Drake's party just sixteen years before.

The Valley and Western Coast Miwok, on the other hand, were connected by well-worn Indian trails. The main trail ran north from today's San Rafael, through Novato, Petaluma and Cotati, and then west along the north bank of the Estero Americano to Bodega where it intersected a trail leading to Freestone and Sebastopol in Southern Pomo territory. The Valley Coast Miwok tribes presumably frequented Bodega Harbor to harvest the clam flats and to trade venison and deerskins for Pomo tobacco and obsidian. It seems likely that the large contingent of natives who visited the English the sixth day after their arrival were Valley Coast Miwok.[14]

From *The World Encompassed* we are given the impression that the Elizabethans encountered many thousands of Indians during their stay. By the time the great king paid his visit nine days after their arrival, Fletcher observed "the greatest number of people which we could reasonably imagine to dwell within any convenient distance round about." Modern researchers, however, have estimated the entire Coast Miwok population in aboriginal times at only fifteen hundred to two thousand. Based on numbers alone, it seems that Drake encountered other tribes in addition to the Coast Miwok.[15]

The Pomo

More is known of the distribution of Sonoma County's Pomo because all except for the most southern tribes were beyond the reach of the missions. Researchers have identified seven known Pomo dialect groups: Southern, Southwestern, Southeastern, Eastern, Central, Northern, and Northeastern. Their culture was very similar to Coast Miwok although they seemed to have had more political organization. Within villages, there were chiefs and subchiefs and the villages sometimes united under a common chief. Early estimates of Pomo population ranged from about three thousand to twelve thousand. Both estimates, however, were provided after severe epidemics had ravaged the population.[16]

All seven Pomo subgroups may have been seasonal residents of Bodega Harbor. The inhabitants of at least some villages belonging to these tribes would pack their belongings into beautifully decorated rush baskets and make coastal expeditions to fish, clam, and trade with the Coast Miwok. The Pomo prized the harbor's mud flats for harvesting clams, especially the huge Washington clam. They used its shells to make disk-bead shell money, the principal unit of trade in central California. The Pomo were the chief agents of an extensive trade in this media that linked all the principal tribes of northwestern California. The Central Pomo tribes, as middlemen in the exchange, benefited most, with some of the chiefs attaining considerable wealth and status.[17]

The Wappo

As conveyed in the Indian legends previously related, there is a remote possibility of a connection between Drake's visit and this rather unusual tribal group. The Wappo dwelled along the scenic mountain range separating Sonoma and Napa Valleys, surrounded by Pomo, Coast Miwok and Patwin tribes. Under a great chief named Colorado (Spanish name), Wappo warriors bravely battled the succession of Spanish, Mexican, and American intruders who attempted to subjugate them. According to Stephen Powers who observed them in the early 1880s:[18]

14. Teather, without citing her reference, reports that John Reed provoked the Cotati Indians in 1825 by building a tule hut near the main Indian trail to the Bodega clam flats (cited by Adair Heig, ed., *History of Petaluma*, p. 9).

15. Kroeber, *Handbook of the Indians*, p. 275; and Cook cited by Kelly, "Coast Miwok," in Heizer, ed., *Handbook*, p. 414. It is difficult to believe that these estimates, based on mission baptismal counts, accurately account for the large losses to European diseases in the early to mid-nineteenth century. My impression from the early accounts is that the northern missions baptized a lower percentage of the Coast Miwok population that existed circa 1800 than is generally believed. I think the population could have easily been twice the estimate: 3,000 to 4,000.

16. McLendon and Oswalt, "Pomo: Introduction," in Heizer, ed., *Handbook*, p. 274; Bean and Theodoratus, "Western Pomo and Northeastern Pomo," in Heizer, ed., *Handbook*, pp. 293–94; and Barrett, "Ethno-Geography," pp. 15, 20 and 42.

17. Bean and Theodoratus, "Western Pomo and Northeastern Pomo," in Heizer, ed., *Handbook*, p. 298.

The Wappo presents a finer physique than the lowland Gallinomero (Southern Pomo). He is shaded perceptibly lighter; has a more even and well-rounded head, though it is large like the Yuki head; less angularity and coarseness of feature; a much more prominent chin; a brighter eye; less protuberance of belly.[19]

Could it be that chief Colorado was a descendant of the Indian king who visited Drake at Bodega Harbor? Considering the legends, the lighter, more-European features, the distinctive language, the contempt for the Spaniards, and the fact that Fletcher did say the king was from the upland country . . . who knows?

The Spanish Mission Period

According to baptismal records, Coast Miwok began to appear at Mission San Francisco in the mid-1780s. As of 1799, the mission register included about twenty Coast Miwok names, excluding the bilingual Huimens. The Franciscans may have started recruiting from the Bodega Harbor region in 1793 with Matute's aborted attempt to establish an outpost there but there is nothing in the mission records to confirm this. The next Spanish military expedition into Marin territory was not until 1810, and this was simply to check on an American fur-hunting party encamped at Campbell Cove. In terms of physically entering their territory, it seems that the missionaries and presidio soldiers had little interaction with the Coast Miwok until they established the San Rafael and Sonoma missions in 1817 and 1823 respectively.[20]

Nevertheless, between 1800 and 1818, Christian Indians whom the missionaries sent into the northern frontier on recruiting expeditions induced a substantial percentage of Marin County's native population to abandon their homelands. The exodus to Mission San Francisco included 800 to 900 natives from the areas of today's Bolinas and Tomales Bays, Olema, Novato, and the Point Reyes Peninsula (Southern Coast Miwok), some of whom returned with the founding of Mission San Rafael. Approximately 200 to 300 Valley Coast Miwok from the areas of today's Petaluma and Sonoma went to Mission San Jose in 1815–18, some of whom returned with the founding of Mission Sonoma in 1823.[21]

With its founding in 1817, Mission San Rafael appears to have rapidly assimilated most of Marin County's remaining native population. As late as 1824, however, the priests were still constantly on guard for Indian attacks, and bands of Coast Miwok warriors from the eastern shore of Tomales Bay were at large, within a day's journey northwest. According to mission records, the mission fathers baptized 916 Coast Miwok Indians at San Rafael and another 48 at Sonoma over the twenty-odd years they were operating. The Sonoma mission appears to have recruited mostly from Valley Coast Miwok, Southern Pomo, Wappo, and Wintun tribes. From Russian records it appears that many Bodega Miwok remained in Russian-California territory and refused to join the missions.[22]

The missionaries relied on incentives and word of mouth to recruit new converts, not force. In addition to spiritual salvation, the missionaries offered the Indians food, lodging, medical attention, warm blankets, and protection from their enemies.

Although the priests did not force them to come, once the Indians were baptized, the monks would not allow them to leave the mission fold. This is when the presidio soldiers became involved:

---

18. Powers, *Tribes of California*, pp. 197–98. Lieutenant Warren Revere mentions an encounter with *Colorado* in 1846 (Revere, *Naval Duty in California*, pp. 114–16).
19. Ibid., p. 198.
20. Aside from the aborted attempt to occupy Bodega Bay in 1793, neither H. H. Bancroft or Milliken found records of any Spanish military expeditions to Coast Miwok territory before Gabriel Moraga's trips to Bodega Bay in 1810, 1811 and 1812, and these expeditions were to visit the Russians, not to capture or recruit mission Indians. Milliken (*Time of Little Choice*, pp. 202 and 234) reports that proselytizing Franciscans visited a Huimen village in October 1809 and another village near the mouth of Sonoma Creek in August 1811.
21. Milliken, *Time of Little Choice*, p. 271. In his thoroughly researched and well-written book, Milliken provides keen insight into the complex circumstances that compelled the native tribes to voluntarily abandon their homelands and freedom for the disciplined life of the missions. In short, he concludes, the mission era was "a time of little choice."
22. Colley, "Missionization," p. 156; and Kotzebue, in Sullivan, ed., *Russian Settlement*, p. 37.

The soldiers capture the Indians by means of lassos made of horsehair. Galloping up to an Indian at full speed, the soldier throws a lasso tied to his saddle and, pulling him off his feet, drags him for some distance along the ground until he is powerless; then the soldier binds the Indian and leaves him while he goes off to capture another one; after capturing the desired number of Indians, he ties their hands and drives them to the presidio.[23]

Under the Spanish frontier settlement scheme, the missions were only intended to be temporary. In theory, their function was to Christianize and civilize the native tribes, converting the Indians to God-fearing and productive citizens within ten years of their baptisms. When they completed the process in one area, they were to move on to the next, in this manner peacefully subjugating the entire Alta California territory.

In reality, however, this did not take place, even after New Spain's viceroy issued a law in 1813 that specifically required that all missions older than ten years be secularized. The mission priests refused to comply, retorting that their California converts still needed their guidance and protection and were neither ready nor willing to govern themselves in what the Spanish authorities would consider a civilized or Christian manner. They justly feared that the consequences to the Indians would be disastrous.[24]

The Russian Interaction

When Russian settlers first arrived in California in 1812, they made it a point to establish friendly relations with the native inhabitants, in part to justify their existence on California shores:

The friendly relations of these people [Indians] with the Russians, which continue to this day [1818], clearly prove that the Russians did not seize this land by force. Singly or in pairs, Russian promyshlenniks [RAC employees] go hunting for wild goats in the woods, frequently spend nights with the Indians, and return safely without being injured or accosted by them. In contrast, the Spaniards do not dare appear in small numbers or unarmed for fear of being killed. These Indians willingly give their daughters in marriage to the Russians and Aleuts, and there are many Indian wives in Fort Ross. This establishes not only friendly but family ties.[25]

The climax in goodwill occurred in 1817, when Captain-Lieutenant Ludwig von Hagemeister entered into a formal treaty with three headmen from local Pomo tribes. During the ceremony, the captain thanked them for "ceding" their lands to Russia and gave Chu-gu-an, the principal chief, a medal ornamented with the Imperial Russian seal. The treaty entitled the Indians to receive the Russian's respect, but required of them their loyalty and assistance if the need arose. Chu-gu-an reportedly replied, "We are very satisfied with the occupation of this place by the Russians, because we now live in safety from other Indians, who formerly would attack us and this security began only from the time of the Russian settlement." On their departure from the fort, the chiefs were honored with a salute.[26]

From the beginning, the Russians seemed to have established a special relationship with the Coast Miwok tribes that frequented Bodega Harbor. When Vasili Golovnin visited the port in September 1818, he reported the following:

The chief of the people living next to Port Rumiantsev came to see me when my sloop was anchored there. He brought gifts consisting of various parts of their regalia, arrows, household items, and asked to be taken under Russian protection. An Aleut who had lived over a year among these people acted as interpreter. This chief, called Valenila, definitely wanted more Russians to settle among them in order to protect them from Spanish oppression. He begged me for a Russian flag, explaining that he wanted to raise it as a sign of friendship and peace whenever Russian ships should appear on the shore.[27]

23. Golovnin, *Around the World*, p. 167.
24. Engelhardt, *Missions and Missionaries*, pp. 636–40.
25. Golovnin, *Around the World*, p. 163.
26. Cited in Dmytryshyn et al., eds., *Russian American Colonies*, p. 296.
27. Golovnin, *Around the World*, p. 165. Golovnin's observations about the Coast Miwok chief Valenila make it clear that as of 1818 Coast Miwok tribes were dwelling at Bodega Harbor. Valenila's complaint of Spanish oppression, however, strongly suggests that he and his people had recently migrated north from Marin County when the Spanish founded Mission San Rafael the year before:

While Russian relations with the Bodega Miwok may have been improving with time, such was not the case with the Pomo. When the Russians increased the intensity of their agricultural enterprises and expanded inland in the late 1820s and early 1830s, the Pomo men simply stopped coming to the settlement. In response, the Russians resorted to force-recruiting tactics similar to the atrocities Mexican rancheros and American settlers would commit in the years to come. There was one raid, sometime around 1833, in which the Russians drove Pomo men, women and children over forty miles to Fort Ross with their hands tied. When the chief manager learned of the practices, however, he initiated new policies intended to stop the exploitation and bring the workers back through better treatment.[28]

The Frenchman, Cyrille LaPlace, who visited Fort Ross and a nearby settlement of several hundred Indians in August 1839, provided these observations from an interview he had with the Fort's last manager:

> Mr. Rotschev, seeing my astonishment that the contact with the Russians had not modified more the ways and habits of the natives, assured me that these people . . . obstinately refuse to exchange their customs for ours. However, he added, thanks to a lot of perseverance and enticements, I have succeeded in diminishing a little this adverse sentiment to whites, among the natives of the tribes which frequent Bodega Bay; several chiefs and a good number of young people, encouraged by the bounty and generosity with which they were treated by the Russian agents . . . remain near the fort during the bad season, working with our colonists and are nourished like them. . . . they return each spring in greater numbers than the year before, to cultivate our fields.[29]

The Mexican Rancho Period

Under Mexican rule there was increasing pressure on the government to secularize the Alta California missions. There were reports that the mission priests mistreated their converts, reaped undue profits at the Indians' expense, or deprived new settlers the opportunity to occupy frontier lands. In January 1833, the new governor, José Figuero, made a trip along El Camino Real, closely examining the mission chain to ascertain the truth of these complaints. Although his reports largely exonerated the priests, by April 1834 the Mexican government had passed a series of laws that required complete secularization within four months' time.[30]

When Mariano Vallejo secularized Mission San Rafael and Mission Sonoma in 1835, he duly divided up most of the huge mission landholdings and livestock among the Indians. As the priests had warned, however, the converts, for the most part, were not ready or willing to adopt Spanish customs or conform to the pueblo laws. It did not take long for the mission lands to fall into the hands of the Vallejo family and a new wave of Mexican settlers.

At the same time he was secularizing the missions, Vallejo was dividing up the Indians' outlying homelands among new settlers the Mexican government attracted to the frontier with the promise of land grants. When these rancheros moved into their allotted tracts, they typically allowed the natives to maintain villages on their ranchos, engaging them for fruit picking, reaping and other labor. Thus each rancho had its associated Indian rancheria, that probably correspond to the old settlements shown in maps 24 and 25. With secularization and the ultimate closure of Missions San Rafael and Sonoma in the early 1840s, many of the surviving Coast Miwok moved to these rancherias.

---

they may have been members of the Yolatamal tribe that were occupying the mouth of Tomales Bay at the time of Gabriel Moraga's land expedition of October 1810. Valenila appears to have been the son of the Tamal chieftain Yolo/Ióllo who, based on an entry in the diary of the mission father Mariano Payeras, appears to have entered into a treaty with the Russians circa 1810 when they first started seal hunting operations in the greater Bodega Bay area (discussed by Farris in "The Story of the Purchase of Fort Ross and Payment for Bodega Bay by the Russian Promyshlennik, Tarakanov").

28. Wrangell, *Statistical and Ethnographic Information*, pp. 210–11.

29. Cited in Lightfoot et al., *Archaeology and Ethnohistory*, p. 132.

30. There has been much written about the Indians' treatment at the hands of the Spanish missionaries and presidios. For the most comprehensive work on the subject from the missionaries point of view, I direct the reader to Zephyrin Engelhardt's monumental *Missions and Missionaries of California* (1913). For a more critical perspective, I direct the reader to the works of modern anthropologists such as Cook, Colley and Milliken (see bibliography of this book).

The Decimation of the Coast Miwok Population

The Russian accounts provide grim evidence of the tragic fate of the Nova Albion tribes. They reported unspecified pestilences that "largely exterminated" the population between 1815 and 1822. They also reported measles epidemics in 1828 and 1833, and varied epidemics of measles, chicken pox, whooping cough, and smallpox between 1836 and 1838.[31]

The Russians themselves appear to have unwittingly delivered the single most catastrophic blow. According to Mexican records, in 1838, a smallpox epidemic that started at Fort Ross killed between sixty thousand and seventy thousand Indians in the northern frontier.[32]

The American Settlement Period

By the time of the Gold Rush, there were few Coast Miwok remaining. From J. P. Munro-Fraser we learn that in the early 1850s, Stephen Smith and his wife were treating the Indians living at Bodega Corners with kindness, but that one of the tribe was hanged for an alleged murder. We also learn that when Smith's Rancho Bodega passed into the hands of the next owner, Tyler Curtis, he forced all the Indians to move to a reservation near Healdsburg. As of 1880, there were only a very few left who would still occasionally return to Bodega to visit the scenes of their childhood.[33]

Thanks to one of those who did return, and the efforts of a young woman with a strong interest in native cultures, we now have a wealth of fascinating details about the last of the Bodega Miwok Indians.

Tom Smith and Isabel Kelly

While a graduate student at the University of California, Isabel Kelly, working under the direction of Alfred Kroeber, made two visits to Bodega Bay in December 1931 looking for surviving members of the Coast Miwok tribe. There she found two Indians living on a farm near the old Smith Ranch at Bodega: Tom Smith, perhaps one of the few full-blooded Coast Miwok Indians then alive, and his younger, half-white brother Bill. From her lengthy interviews with Tom, she put together extensive notes that have only recently been organized and published.[34]

Tom Smith was born circa 1845 at a village in Coleman Valley about five miles north of today's town of Bodega (see map 23). His paternal grandmother was from the village of Tókau at Gaffney Point on the west shore of Bodega Harbor. His paternal grandfather, presumably also from the Bodega area, was among those who migrated to Fort Ross during the Russian period. His mother was from Petaluma, and was half Tomales (Southern Coast Miwok) and half Petaluma (Valley Coast Miwok). His mother seems to have been an ex-mission Indian, for Tom was baptized at birth. Tom's parents were members of the tribes associated with Stephen Smith's Rancho Bodega, where his father drove oxen. Tom's Bodega Miwok father left him at a young age and moved back to Fort Ross.[35]

Tom Smith married a number of times, but did not leave behind any full-blooded children. He may have been the last surviving Bodega Miwok descendant of Drake's Nova Albion Indians. He died in 1932.

31. Lightfoot et al., *Archaeology and Ethnohistory*, p. 26.
32. Smilie, *The Sonoma Mission*, p. 67.
33. Munro-Fraser, *History of Sonoma County*, pp. 193–94. Barrett, "Ethno-Geography," pp. 304–6, reported that in 1904, four or five full-blooded Bodegas were living at the mouth of the Russian River. He indicated that the post-mission survivors of Marin's Coast Miwok tribes had first settled about the town of Marshall on Tomales Bay, but that at least some had moved to a village near Bodega in the late 1800s (including Tom Smith's mother?). By 1904, the six remaining full-blooded Marin Coast Miwok were all residing near Windsor in the Russian River Valley, presumably the Healdsburg reservation mentioned by Munro-Fraser.
34. Kelly's notes were compiled posthumously by Collier and Thalman (1991). According to Thalman, Tom Smith's brother Bill recorded Tom as three-quarters Bodega Bay in a 1933 California Indian census. She believes Tom Smith's father, Com-cha-tal, was half Russian and half Pomo or Coast Miwok. Her genealogical research also suggests that there may have been other full-blooded Coast Miwok living in Tom's day (Thalman, telephone conversation with the author, Dec. 1996).
35. Smith's father may have been among the Bodega Miwok Barrett reported to be living at the mouth of the Russian River in 1904.

Fig. 148. Tom Smith of Bodega Bay (circa 1845–1932). He may have been the last full-blooded (or almost full-blooded) survivor of the Bodega Miwok tribes. *Courtesy of David Peri, Guerneville, California.*

There may be alive today some direct descendants of the Pomo or Wappo Indians who visited Drake's encampment. Members of these tribes survived Spanish missionaries, Mexican rancheros, and epidemics of European diseases, only to face the endless onslaught of American settlers. Theirs is a story unto its own.[36]

---

36. See Barrett, *Ethno-Geography*; or Bean and Theodoratus, "Western Pomo and Southeastern Pomo," in Heizer, ed., *Handbook*, pp. 299–305.

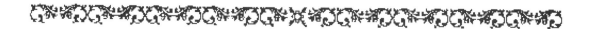

CHAPTER 37
The Hole in The Head

Fig. 149. Portus Nova Albionis/Campbell Cove, October 1962. *Photograph © 1962 by Karl Kortum. Courtesy of Jean Kortum, San Francisco, California.*

January 7, 1963. The dawn breaks as it has more than 140,000 times since Drake bid farewell to the brokenhearted natives of Nova Albion. As the gloom begins to lighten, a gull's sharp cries enter the cove's timeless symphony: a melody of gurgling springs, hissing sands, and whispering leaves playing to the faint drum of waves pounding the seaward shores.

Across the beach under the dark cliff banks surrounding the cove, there is no sign of the human saga that has unfolded within these insular walls. All one can see at the foot of the bluffs are Fletcher's knobby hillocks, densely overgrown with a tangled mass of lupine. But even without any visible signs, I wonder how many others sensed something extraordinary when they first found their way into this remote little haven.

What fabulous array of artifacts lay hidden among those fertile mounds at the foot of the bluffs? Spanning the Stone Age to the Nuclear Age, there must have been thousands of them. The Indian artifacts would have covered their entire history on the Central California coast: granite mortars, bird bone whistles, obsidian arrowheads, stone tools, disk-beads. Forgetting Elizabethan California for the moment, the rich storehouse of other relics would have covered the Spanish mission period, the fur-hunting era, the colonial Russian era, Rancho Bodega, and the Gold Rush days.

How much of Elizabethan California had survived until that day? I expect we would have found Drake's fort largely intact, buried at the foot of the cliffs by centuries of landslides. Within the buried earthen floor of the fort, we would have found myriad Elizabethan artifacts, countless fragments of 400-year-old Spanish wine jugs, and the ashes of many large campfires, some bearing the iron of a small Spanish bark. And what about that fabulous horde: did Drake's men spill a few gold doubloons or pieces-of-eight in the shuffle; did one of his company hide a little cache in the hillside anticipating a return?

As the morning wears on, coarse new sounds interrupt the cove's gentle ones: voices, shouts, the clanking of metal on metal, and then suddenly a deafening roar . . .

Thus begins the saga of Atomic Park and the Hole in the Head. It is the story of Pacific Gas & Electric Company's (PG&E) monumental attempt to construct the world's then-largest atomic power plant at the very location where Francis Drake had encamped almost four centuries before. It is one of California's saddest and most remarkable stories: the unfortunate destruction of the state's most important historic landmark under the noses of a team of archaeologists and a host of officials representing the following public agencies: the California Department of Beaches and Parks, the University of California, the Sonoma County Planning and Zoning Departments, the Sonoma County Harbor Commission, the Sonoma County Board of Supervisors, the state Public Utilities Commission, the U.S. Army Corps of Engineers, the Atomic Energy Commission and the National Park Service.

PG&E and the Atomic Energy Commission

The U.S. Congress established the Atomic Energy Commission (AEC) in 1951, in part, to draw up regulations concerning government sponsorship and control over peaceful uses of the atom. The AEC's efforts bore fruit in 1954 when Congress enacted the Atomic Energy Act with its provisions for stimulating private development of nuclear power plants.

PG&E was among the first private utility companies the AEC approached in its effort to help commercialize nuclear power. In the early 1950s, the AEC awarded huge research and development contracts to the utility, giving them a running start when the Atomic Energy Act went into effect. By the late 1950s, PG&E had sited, self-designed, and supervised the construction of two nuclear power stations in California: a small experimental unit at Vallecitos, and a small (60-megawatt) commercial unit at Humboldt Bay near Eureka.

By 1961, PG&E had invested almost $100 million in its nuclear power program and was anxious to demonstrate that full-scale atomic power plants could be economically competitive with conventional fossil fuel plants. In planning its first large commercial atomic plant for a projected start-up date of 1965, the utility was poised to spend upwards of $61 million of its own money, using it own engineering staff to design the plant and supervise its construction.

The installation would be powered by a boiling-water nuclear reactor utilizing "pressure suppression containment, developed by PG&E, that would allow the reactor to be installed below ground." The plant would also feature sea water evaporators, pioneered by PG&E, to provide the plant with fresh water for the reactor and steam system, and auxiliary uses. Atomic Park was to have an initial generating capacity of 325 megawatts, enough to serve a city of half a million people. The

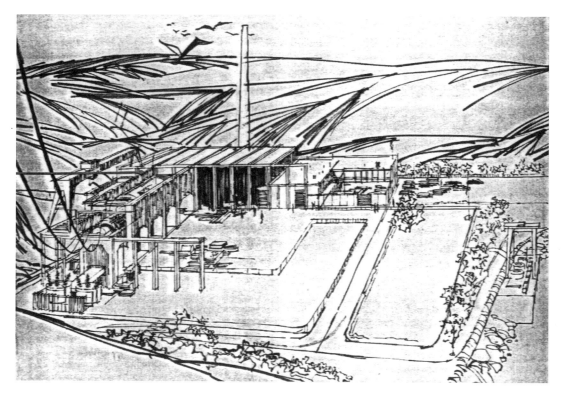

Fig. 150. Artist's rendition of PG&E's proposed Atomic Park. *Reproduced from the Pacific Gas and Electric Company 1961 Annual Report. Courtesy of PG&E, San Francisco, California.*

facility layout would allow for eventually tripling the capacity through the installation of additional reactors.[1]

The Interest in Bodega Head

PG&E developed its interest in constructing a power plant at Bodega Head soon after the Atomic Energy Act was passed; by 1957 the utility was already well along in its preliminary planning. There were a number of factors that made the Bodega Head site appear a "natural" for the construction of a power plant. Although located just fifty miles north of San Francisco as the crow flies, it was still comparatively remote from population centers, and, having no public access road, was not being used by the public. The site also offered an unlimited cooling water source, seemingly good foundation conditions, and adequate harbor facilities for receiving and shipping heavy loads. This combination of siting and physical conditions was not to be found at other potential reactor sites then under consideration. What's more, in the late 1950s the Sonoma County Board of Supervisors had embarked upon a national campaign to promote more industrial development to balance out their agricultural and recreational economy, and were actively encouraging the utility to construct a big power plant.[2]

An Array of Obstacles

Though economics and some of the siting criteria may have appeared favorable, PG&E was facing some formidable obstacles when it first started laying plans for Atomic Park in the late 1950s. In

1. PG&E News Bureau press release, June 28, 1961.
2. Sidney Allen, "Battle of Bodega Head," p. 38.

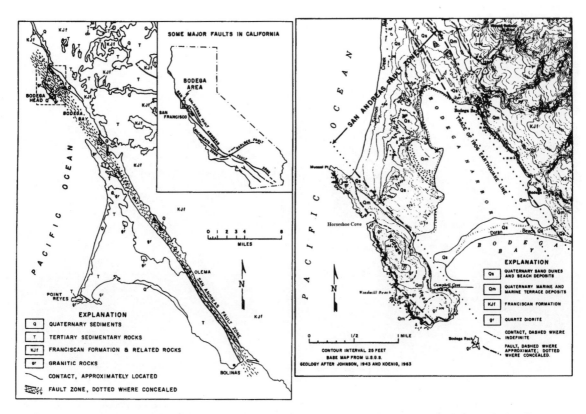

Fig. 151. The location of the San Andreas Fault zone at Bodega Harbor (*right*) and in the greater Point Reyes/Bodega Bay area (*left*). *Reproduced from "The Geologic Setting of Bodega Head." Sacramento: California Division of Mines and Geology, "Mineral Information Service,"* July 1963.

addition to the normal problems of acquiring land from the current owners and overcoming the inevitable staunch local opposition, there were some special difficulties at Bodega.

Conflicting public interest – In 1955, the National Park Service had recommended public acquisition of Bodega Head "to preserve its beauty, history, remarkable geology, and unduplicated marine environment."[3] Following up on this recommendation, the 1956 state legislature appropriated $350,000 to add Bodega Head to the adjacent Sonoma coast state park system. The California Department of Beaches and Parks (park agency) started the necessary procedures to acquire the headlands from private landowners by year's end. In addition, in late 1956, the University of California initiated procedures to acquire part of the headlands for constructing a marine laboratory.

Seismic concerns – For its proposed reactor site, PG&E originally selected Horseshoe Cove, on the ocean side of Bodega Head, despite the fact that the cove is located alongside California's most infamous and dangerous earthquake fault, the San Andreas. Utility officials were either oblivious to the problem in the early stages of the project or simply underrated its magnitude. In 1959, however, the AEC issued draft siting rules that would prohibit nuclear power plant construction on an active earthquake fault. These proposed rules caused PG&E to make its fateful and very dubious decision to move the plant site to Campbell Cove, just a quarter mile from the main San Andreas fault line and within hundreds of yards of a known related fracture.

Private landowners – In 1957, the Bodega headlands belonged to two families whose ownership went back to the 1800s. The Stroh-Campbell family owned Campbell Cove and the southern end of the head. The Rose Gaffney family owned the rest of the head and a stretch of the sand dune peninsula

3. Pesonen, *Visit to the Atomic Park*, p. 3, citing a 1955 National Park Service survey for the Pacific Coast.

connecting it to the mainland. PG&E officials would come to feel the sharp thorns of this particular "rose" before all was said and done.

Archaeology – It was not until after PG&E switched the site from Horseshoe Cove to Campbell Cove that the utility encountered this problem. Campbell Cove has since been proposed as a state historic landmark because of its very important and unique position in California's native American and colonial eras.

Lining up Local Government Support

PG&E's actions must be judged in the context of the times; its officers were following the company's mandate to site, design and build economical new electrical generating capacity to support northern California's burgeoning population. The record is clear that Sonoma County officials were in favor of a large power plant at Bodega Head and were doing everything they could to encourage PG&E to build one there. As you will see, local and county officials share the responsibility for the hole in their head; they were the ones who took liberties with the existing environmental and land use codes.

When the project was first proposed, Sonoma County officials, much of its citizenry, and its press were, in general, very enthusiastic supporters of the Atomic Park. In the 1950s Sonoma County was almost completely rural and sparsely populated. Its principal industries were poultry and dairy farming. With the San Francisco Bay communities to the south rapidly developing into some of the nation's foremost urban-industrial areas, PG&E was offering Sonoma County the opportunity to take part in the boom. In a February 1960 editorial, a leading county newspaper offered the following points in favor of the plant:

> (1) Sonoma County and all the North Bay Region need the electric power that will be generated at Bodega Bay. A plant there will assure abundant energy to plan and meet the residential and industrial growth that is our inevitable destiny; (2) The investment by PG&E of $60 million or more will provide substantial new tax revenues to the county government, to schools and to the several districts involved. This can provide services to our citizens literally impossible except through this or similar industrial development; (3) Money that will be spent here during two or three years of construction will exert a tremendous upward impact upon our economy. Hundreds of skilled craftsmen will be working at Bodega Bay, earning and spending hundreds of thousands of dollars; (4) When the plant is completed it will be staffed by engineers and technicians who, with their families, will make useful and valuable residents of Sonoma County. The plant's permanent payroll will have a salutary influence upon our economy; (5) If the plant is powered by nuclear fuels, a good possibility now, it will become an attraction to scientists and engineers from all over the world and to citizens all over the nation. Its presence will add measurably to Sonoma County's prestige and fame. . . . Let's get on with the job![4]

There was one crucial detail that PG&E may not have provided to the county commissioners when they initially broached the idea of a Bodega Head power plant: the matter of the atom. When PG&E first went public with its plans in May 1958, it was only in "preliminary purchasing negotiations" for a site to build a "steam-electric generating plant." Biding its time, the utility would not actually confirm the persistent rumors that the plant was to be atomic until the summer of 1961. In the meantime, the county had granted PG&E three use permits: the first to construct overhead transmission lines across Doran Park, the second to construct a steam-electric power plant at Campbell Cove, and the third to construct an access road (the so-called tidelands road) along the west shoreline of the harbor. In each case, despite vehement protests from the residents of Bodega Bay, the county officials determined that public hearings were not necessary.[5]

In its eagerness to enhance the tax base, county officials gave the utility unconditional approval to design and build whatever size and type of plant it saw fit.

---

4. Editorial, "Bay Plant Should be Built," *Santa Rosa Press-Democrat (SRPD)*, Feb. 23, 1960, p. 4.
5. Lora, "Bay Site of PG&E A-Plant," *SRPD*, May 23, 1958, p.1.

Winning Over the Department of Beaches and Parks

Odd as it may seem, the state's proposal to add Bodega Head to the park system may have contributed heavily to PG&E's success in winning such wholehearted support from the county officials. PG&E representatives needed only point out the obvious; if the county allowed the state to take Bodega Head, the area would never provide any significant tax base; if the county allowed PG&E to take the head, it could count on a dramatic and sustained increase in its tax base, in the order of a million dollars a year.[6]

In the summer of 1957, the park agency abruptly ceased its efforts to acquire Bodega Head. How the utility was able to "push aside the needs of the state" was a riddle to those who looked into the matter at the time. Some of the more vocal critics suspected some sort of collusion or political influence from the governor's office.[7]

What I have seen of the record suggests to me that the county and PG&E simply formed a united front in presenting a reasonable alternative to the park agency's plans that were dormant at the time due to funding constraints. First, the county made it clear that it firmly opposed the concept of the state's removing 947 acres from its tax base. PG&E then offered a sensible compromise. The utility would develop just a portion of the head for its power plant, conceal the installation behind the sand dunes as much as possible, lease or sell to the park agency any portions of the head it did not need, and build and maintain an access road to the area along with recreational facilities. Recognizing they would have a poor chance of defeating this formidable alliance, it is not surprising that park agency officials looked very favorably upon PG&E's alternative proposal.

The Troubled University of California

What a delicate situation the University of California's administrators encountered in the summer of 1957 when PG&E won park agency support for its plans. At this time, the university was receiving over fifty percent of its operating budget from the Atomic Energy Commission: over 250 million dollars alone in 1962. As much as the administrators may have wanted the Bodega Head marine laboratory, they certainly had nothing to gain by interfering with PG&E's plans for Atomic Park. In this respect it is pertinent to note that the university chancellor, who withdrew plans for the Bodega Marine Laboratory in the fall of 1957, was appointed director of the Atomic Energy Commission in 1960.[8]

Under the circumstances, one can reasonably assume that, when university officials backed off in 1957, it was without undo prompting from PG&E representatives or political pressure from the governor's office. It is also clear that in its negotiations with the university, PG&E had left the door open for compromise. Shortly after PG&E announced it was moving the reactor site from Horseshoe Cove to Campbell Cove in 1959, the university announced it would construct the marine lab after all. Notwithstanding the protests of some of the faculty, university officials sensibly agreed to share Bodega Head with PG&E and the park agency.

Wresting Away the Stroh-Campbell Property

In October 1958, when PG&E was still planning to build the reactor at Horseshoe Cove, it filed suit in Sonoma County Superior Court to take over 400 acres of land owned by the widowed Rose Gaffney. When the utility switched the reactor site to Campbell Cove a year later, it immediately turned its attention to the 160-acre, Stroh-Campbell property at the tip of the head (figure 155).

6. Thurber, "County Officials Unmoved by Letters Attacking PG&E," *SRPD*, Apr. ?, 1959.
7. Dusheck, "Deal Will Put Power Plant in Beauty Spot—Park, Marine Lab Get Ax in Sonoma," *San Francisco News*, Sept. 12, 1958.
8. Perlman, "Huge UC Contract for Atomic Work," *San Francisco Chronicle (SFC)*, Aug. 25, 1962, p. 1. Most of the AEC grants covered development of atomic weapons and nuclear propulsion systems at two U.C. laboratories: Lawrence-Livermore/Berkeley and Los Alamos, New Mexico.

The Campbell family had owned Campbell Cove and the adjacent headlands for over a century. Much as they wished to keep it, the family members were not willing to sustain a long condemnation battle that they were likely to lose in the end. "We don't want to sell, but we don't want to fight either, so we hope this will be for the greatest good of the greatest number." The deed transferring the property to PG&E was filed by year's end.[9]

The Thorns of a Rose

The 400-acre Gaffney estate included part of the headlands and some of the dune-studded peninsula connecting it to the mainland. The property had been in the family since 1864, acquired from Tyler Curtis. The Gaffney ranch house abutted the Indian village of Tókau.

Rose Gaffney had lived on Bodega Head since emigrating from Poland in 1913. She was outraged when PG&E informed her they needed her precious land for a power plant. During four decades tending sheep and cattle

Fig. 152. Rose Gaffney (1895–1974). *Photographer unknown, circa 1962. Courtesy of Joel Hedgpeth, Santa Rosa, California.*

over its wind-swept contours, she had catalogued its flora and gathered a massive collection of Indian artifacts. Like the Indians before her, Rose knew and loved every inch of this land.[10]

She waged a furious fight, forcing PG&E to file a condemnation suit in 1958. When the utility turned its sights to Campbell Cove and left the bulk of her property for the university and the park agency to divide up, she bravely fought all three of them. A true conservationist, she futilely offered to let her land go to the park agency and university if they would only agree to keep PG&E out.

Even after the state court took away her lands, Rose battled on. At the Public Utilities Commission (PUC) hearings, PG&E had no more determined an opponent. In the PUC files, one can still find the colored slides she presented showing Campbell Cove and the Bodega headland in all its varied moods. One can also find reference to her heated tirades and the disparaging remarks she directed at PG&E and county officials as they attempted to justify their proposals.

Mitigating the Archaeological Impacts

It would seem that PG&E at least tried to do the right thing. In November 1961, they contracted with a group of professional archaeologists to conduct investigations at Bodega Head in advance of the plant construction. Throughout the winter of 1961 and early spring of 1962, these experts collected artifacts and catalogued some of the areas where known Indian encampments had existed:

> Artifacts are now being catalogued and a report is being prepared in order that the early history may not be lost. Other work remains to be done in the area, including an investigation of the supposed Russian fort site, which will be left for the future when there is more time and when new methods of investigation may be discovered. We believe all the available history is being preserved for the future.[11]

---

9. Alexander, "PG&E Can Buy Bay Plant Site," *SRPD*, Oct. 18, 1959, p. A3.
10. Part of Rose Gaffney's collection is now on exhibit at the Bodega Marine Laboratory.
11. PUC exhibit No. 27, May 21, 1962 in application no. 43808. I find this reference to a Russian fort intriguing. From what I can tell there is no record of the Russians building any stone bastions or stone structures at Bodega Head. Juan Bautista Matute's report says nothing about the Spanish building fortifications in the summer of 1793, nor did Archibald Menzies mention any when he explored

HAROLD GILLIAM *Atom vs. Nature at Bodega*

CURVING into the ocean some, 45 miles north of the Golden Gate is the magnificent granite promontory of Bodega Head. It bounds Bodega Bay on the north and encloses the picturesque little fishing port of Bodega Harbor

Granite is rare along this section of the California

Fig. 153. Clip from Harold Gilliam's article in the *San Francisco Sunday Chronicle*, February 11, 1962.

Showdown at the PUC Hearings

With its county use permits in hand, PG&E still needed the approval of the PUC before it could begin plant construction. In October 1961, the utility submitted its application for a certificate of public convenience and necessity to operate Unit No. 1 of the Atomic Park.

The PUC hearings on the application opened in March 1962. At first it looked like a sure thing. There was very little opposition and county officials lined up to present their enthusiastic support.

Fig. 154. "Karl Kortum at Bodega Bay" October 7, 1962. Kortum, who passed away in September 1996, was known throughout the world for his dedication to historic ships and the preservation of maritime history. In galvanizing public attention at the most critical point in the process, Kortum played perhaps the key role in the battle of Bodega Head. Among his most important contributions to the preservation of maritime history could be the outstanding photography Kortum shot at Campbell Cove in October of 1962 (see figures 109 and 149). *Photograph © 1962 Karl Kortum Courtesy of Jean Kortum, San Francisco, California.*

The park agency and University of California were absent. Anticipating prompt PUC approval, PG&E prepared to start site work.

Shortly after the hearing closed, however, the *San Francisco Chronicle* published a strongly worded letter from Karl Kortum, director of the San Francisco Maritime Museum and a prominent figure in the San Francisco political scene, rallying public opposition. The flood of letters the PUC received in response compelled them to set the application aside for another round of hearings.[12]

Both sides were better prepared for the May–June 1962 hearings. The opposition included the Sierra Club, numerous conservation-minded citizens living in the greater San Francisco Bay area, the indomitable Rose Gaffney, and many other concerned residents from Bodega Bay and nearby townships. Offering support for the application were PG&E officers and technical staff, expert witnesses, and a host of Sonoma County officials, business owners and citizens. Both the University of California and the park agency sent representatives to answer questions, but not to take a pro or con position.

In short, PG&E officials and experts were successful in diffusing each and every concern

the area that fall (chapter 35). I have seen reference to an 1862 U.S. Geodetic Survey map purportedly showing some bastions on the headland at the point overlooking the narrows (Jones & Stokes Assoc., "Environmental Impact Statement," p. C-3). Such bastions and any stone-walled structures PG&E stumbled across at Campbell Cove were most likely left behind by Drake, or (for the bastions only) possibly the English who holed up there in early 1793 (see chapter 35 of this book). My 1862 USGS map shows no such bastions.

12. Kortum, letter to editor, *SFC*, Mar. 14, 1962, p. 34.

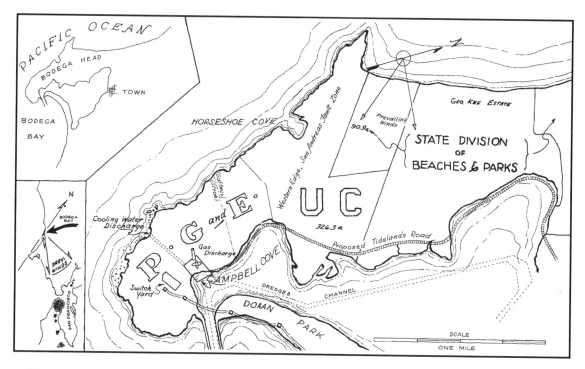

Fig. 155. PG&E's proposed Atomic Park tidelands road and overhead transmission lines. The Gaffney estate originally included all of the lot labeled "UC" (University of California) and part of the lot labeled "State Division of Beaches and Parks." *Obtained from the Public Utility Commission files for Application Number 43808.*

raised at the hearings. Responding to concerns regarding public safety from accidents and/or radioactive fallout, PG&E engineers provided detailed information on design safeguards. Responding to concerns about earthquakes, PG&E engineers claimed that the reactor would be located over a mile from the San Andreas fault, bedded on granite, surrounded by concrete, and conservatively designed and constructed in accordance with all AEC safety criteria. Responding to the environmentalists, they brought in experts in marine biology to allay concerns and claims that the cooling water or radioisotope emissions would have any significant effect on the coastal ecology. Responding to concerns by nature lovers and conservationists, PG&E officials pointed out that they were going to be sharing the headlands with the park agency and University of California's Bodega Marine Laboratory. They also announced plans to develop and enhance the recreational value of the head by reconstructing a "Russian fort" at Bodega Head, constructing nature trails and possibly even a museum to exhibit Russian and Indian artifacts.[13]

The testimony that park agency and university officials gave at the hearing was intended to be neutral. In both cases, however, by simply stating that they did not oppose PG&E's plans, or consider them to be conflicting with their own plans for Bodega Head, they provided some of PG&E's most powerful supporting arguments.

In November 1962, relieved PG&E officials had reason to celebrate; the PUC granted a conditional interim permit for Atomic Park. The conditions required them to perform diffusion studies to determine the impacts of their wastewater and cooling water discharges, fund a two-year study to establish background radiation levels, and provide public access to as much of the headlands and tidelands as safely possible. The granting of a final permit, however, was contingent upon PG&E securing all necessary AEC approvals for the facility design.

13. Staff writer, "PG&E Would Open Bodega Site to Public," *SFC*, May 20, 1962.

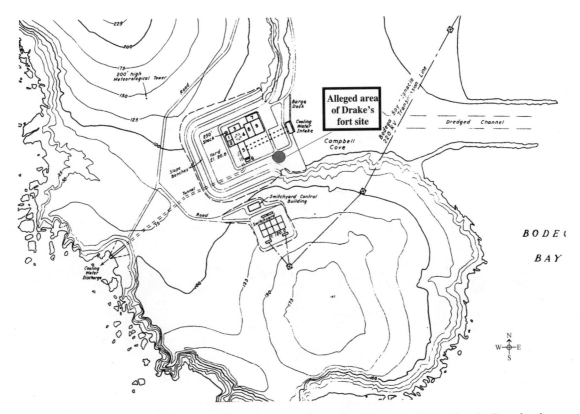

Fig. 156. Site plan for PG&E's proposed Atomic Park with a notation showing the location of Drake's fort site. *Reproduced from the Public Utility Commission files for Application Number 43808.*

The Tidelands Road Battle

The most controversial issue that came up in the PUC hearing, and the one that ended up having the most bearing on the saga of Drake's lost harbor, involved the proposed access road for the plant. In this regard, it seems appropriate that the following statement was entered into the record by one of the PUC commissioners who approved the Atomic Park permit application:

> In my opinion, among the ancillary aspects of this application, the location of the access road will have the most indelible effect upon the Bodega Bay area; it more than anything will change its complexion. This view is attested to by the extent of attention focused on the problem during the hearings . . . I find nothing in the arguments for the tidelands road which suggests it is necessary for it to be so located in order to effectively serve the plant site; admittedly, an upland road would accomplish this purpose equally well. Nor do I find in these arguments anything which suggests that such a road would be convenient, for the record is replete with reasons suggesting substantial detriment to the public resulting therefrom.[14]

The state commissioner wanted to make the record clear that, if not for the legal tactics county officials had used to block a PUC decision on the access road issue, the commissioners would have required PG&E to use an uplands road route to mitigate legitimate concerns the public had raised over the destruction of the tidelands. The travesty is that without the tidelands road, there may not have been a hole in the head.[15]

---

14. PUC decision no. 64537, Nov. 8, 1962, appendix A: concurring opinion of Commissioner Frederick Holoboff.

15. Except for the purpose of providing a mountain of earth fill for building the county's tideland road, PG&E had no reason or justification for starting plant construction at Campbell Cove prior to receiving its AEC approvals. I am assuming that if the PUC had mandated an upland route over the dunes, PG&E would not have commenced the excavation work when they did, and that the concerns over earthquake hazards would have ultimately prevented PG&E from going forward with the project. The weakness in this assumption is that the true magnitude of the earthquake hazard was in large part confirmed by the excavation process.

It is unclear why Sonoma County officials wanted the controversial tidelands road so badly. In 1959, during the course of considering PG&E's use permit application, the Sonoma County Harbor Commission drew up a new plan for eventual development of Bodega Harbor that included a heavy-duty access road to Campbell Cove. Rather than following the existing route across the dunes and headlands, this road was to run along the tidal flats.[16]

At the PUC hearings, the county's counsel argued that, by taking this action, the harbor commission had triggered a legislative trust that granted local agencies absolute and irrevocable title to tideland areas if they improved them "for the promotion and accommodation of commerce and navigation." The conservationists countered that needlessly destroying the tidelands to provide access to a power plant was not the use the state legislature had in mind when they offered the county this carrot. The county prevailed, however, by emphasizing that they planned to build the road whether PG&E constructed the plant or not.[17]

From the start, Bodega Bay residents, conservationists, and University of California faculty offered staunch, well-founded opposition to the tidelands road based on the following concerns: it would erase some of the rich clamming grounds inside the harbor; it would cut off access to the harbor for west shore property owners; it would endanger, if not completely obliterate, many plants and animals shoreward of the road; it would grossly conflict with the University of California's objectives in siting a marine laboratory at Bodega Head; it would preempt an area that could later be used for harbor expansion by converting the muddy shoreline to a hard roadside abutment; and it would create a hazard to the boats moored in the harbor, particularly during storms.[18]

In addition, the record is clear that a route over the headlands was a completely viable alternative that was acceptable to PG&E officials. The only drawback from PG&E's standpoint was that they would have to ocean dispose of all the excavation spoils they created in constructing the reactor site at a significant incremental cost.

Despite the vociferous local opposition, in May 1960, county officials granted PG&E a use permit for constructing the tidelands road and, in December 1961, entered into an agreement whereby PG&E would construct the road and the county would own and maintain it.[19]

Since the road potentially affected navigation in the harbor, the U.S. Army Corps of Engineers (the Corps) would have to review the plans and issue a permit before road construction could begin. When county officials learned that the Corps planned to conduct a public hearing, they were so incensed they released the following statement through the local press: "We'll not spend another five cents at Bodega Bay until the Corps approves the application. The Corps might as well know that if they don't cooperate with us . . . they better forget future development at the bay." The Corps issued their permit in March 1962, explaining that conservation, aesthetics, and marine biology were outside its scope.[20]

Under the terms of its PUC permit, PG&E could not legally begin plant construction until its plans had been approved by the Atomic Energy Commission. The record suggests that county officials prompted PG&E to circumvent this requirement. Under the guise of starting work on the county's new tidelands road, bulldozers tore into Campbell Cove's rear bluff on the morning of

16. Pesonen, *Visit to the Atomic Park*, p. 13. Pesonen refers here to the "Sarles plan" for residential and commercial development of the harbor area.
17. PUC Decision No. 64537, Nov. 8, 1962, pp. 22–23.
18. Pesonen, *Visit to the Atomic Park*, pp. 13–14. The "Emerson Report" prepared by University of California faculty in November 1960 concluded that the Bodega area would be reduced from a unique Class-A site to a Class-B site for a marine research center. The greatest concerns were over the access road. Cited in NCATPBH "Petition to Invoke Certain Sections of the PUC Code," Jan. 15, 1963, in PUC Application No. 43808, .
19. Pesonen, *Visit to the Atomic Park*, p. 14. In May 1963, county commissioners, by rezoning Bodega Head to an agricultural zone, successfully thwarted Northern California Association to Preserve Bodega Head and Harbor's (NCATPBH) legal challenge to the use permit.
20. Staff writer, "Action by Board of Supervisors," *SRPD*, Jan. 17, 1962, p. 2.

Fig. 157. "Sierra Club meeting, David Pesonen presiding," June 1962. Future members of the Northern California Association to Preserve Bodega Head and Harbor include clockwise from left: Joel Hedgpeth, Philip Flint, Harold Gilliam, David Pesonen, Ray and Marian Ruebell, Doug Hill, unidentified, and Sam Rodgers. *Photograph © 1962 by Karl Kortum. Courtesy of Jean Kortum, San Francisco, California.*

January 7, 1963. Over the next ten months PG&E's construction contractors succeeded in filling in most of the tidelands area along the west shore of the harbor with the mountain of soil they removed from the proposed Campbell Cove reactor site.

The Final Insurmountable Hurdle

Throughout PG&E's long and expensive battle to construct Atomic Park, the AEC had not only encouraged the giant utility to proceed with the project, but had pressured them to do so. By the early 1960s, the fledgling atomic power industry was rapidly losing momentum and the AEC desperately needed a success story. Atomic Park appears to have been one of the few bright spots on the horizon. PG&E officials probably could not have conceived that the AEC would disapprove their plans.[21]

The AEC's Safeguards Committee announced conditional approval of the plans in May 1963. The approval had the following caveat, however: "tentative exploration indicates that the reactor and turbine buildings will not be located on an active fault zone. If this point is established, the design criteria for the plant are adequate."[22]

The trouble started when David Pesonen, executive director of the Northern California Association to Preserve Bodega Head and Harbor (the Association) got an internationally renowned expert seismologist, Dr. Pierre Saint-Amand, interested in the project. After reviewing the reports of the utility's seismic experts and inspecting the site in April 1963, Saint-Amand prepared his own report, expressing surprise that PG&E was preceding with construction in view of the severe earthquake hazards: "A worse foundation condition would be difficult to envision." During his site visit, he observed a "spectacular fault" exposed high on the northwest face of the reactor excavation, and evidence of other faults that intercepted the area where PG&E planned to build its cooling water discharge tunnel.[23]

With Armand's preliminary findings in hand, the Association first heightened public concern over radioactive fallout during a Memorial Day protest rally at Doran Beach Park, directly across from PG&E's Bodega Head construction site. Toward the end of the rally, they released fifteen hundred helium-filled balloons to each of which was tied a returnable post card reading:

WARNING! This balloon could represent a radioactive molecule of strontium 90 or iodine 131. It was released from Bodega Head on Memorial Day 1963. PG&E hopes to build a nuclear reactor at this spot, close to the world's biggest active earthquake fault.[24]

With the balloons dispersed by the prevailing northwest winds, the majority of the returned cards were from the most densely populated areas of Marin County, between Novato and Sausalito.

Ten days later, the Association filed a memorandum with the PUC, challenging the Safeguard Committee's decision with the following assertions: (1) the reactor was within a quarter of a mile

---

21. Pesonen, *Visit to the Atomic Park*, p. 31, referencing Congressional hearings on the AEC reactor development program.
22. Cited in Barbour et al., *Coastal Ecology of Bodega Head*, p. 234.
23. Saint-Amand, *Geologic and Seismologic Study*, pp. 7, 19 and 20.
24. Conzett, *Battle of Bodega Bay*, p. 7.

of the San Andreas fault zone, not a mile away as PG&E claimed; (2) there was "concrete evidence" that an earthquake fault ran right through the proposed reactor site; (3) PG&E studies had revealed that the reactor site was underlain by badly fractured quartz diorite with the stability of clay, not hard granite rock as PG&E had claimed in its application; and (4) the summary report that PG&E had given to the AEC had repressed or altered initial reports by PG&E staff that showed the unsuitability of the location.[25]

In response to these serious allegations and an intensive lobby effort in Washington led by Jean Kortum and Harold Gilliam, U.S. Interior Department Secretary Stewart Udall wrote to AEC Chairman Glenn Seaborg expressing "grave concern" about earthquake hazards, and offered the services of the U.S. Geological Survey (USGS) to conduct a thorough investigation of the hazards.[26]

After examining the excavation in August and September 1963, the USGS team confirmed that a fault did indeed run through the reactor site. Although PG&E's own seismic experts provided strong arguments that the fault was minor and not likely to be active, the USGS seismologists were not convinced; they concluded that the site was not an adequately safe location for a nuclear power plant and further contended that "acceptance of the Bodega Head as a safe reactor site will establish a precedent that will make it exceedingly difficult to reject any proposed future site on the grounds of extreme earthquake risk."[27]

In an attempt to mitigate these concerns, PG&E engineers revised the plant design to provide additional earthquake safeguards. After reviewing these plans and all the reports, two distinguished University of California geologists, Dr. Garniss Curtis and Dr. Jack Evernden, in an unsolicited letter to PG&E, indicated that despite their initial bias against the plant, now saw no objection. They noted that "no major displacement has ever taken place within the site area and none is expected in the near future." With respect to the minor displacements in evidence on the head, they concluded that they "have not occurred within more than 40,000 years . . . and would not dangerously affect the reactor if they did occur."[28]

In September 1964, the AEC released two reports. In one report the AEC's Safeguards Committee concluded that the plant could be built "without undue hazard to the health and safety of the public." But in the other report the AEC's Division of Licensing concluded that "Bodega Head is not a suitable location for the proposed nuclear power plant at the present state of our knowledge."[29]

PG&E had an opportunity to appeal the decision by the AEC's licensing, but company officials, following a direct public appeal from the governor's office among other factors, elected not to: "We would be the last to desire to build a plant with any substantial doubt existing as to public safety." In October 1964, after seven years and some four million dollars of effort, PG&E's president announced the utility had formally withdrawn the application.[30]

The Activists

Without the efforts of some very determined individuals by the names of Karl and Jean Kortum, David Pesonen, Rose Gaffney, Joel Hedgpeth, Harold Gilliam, Doris Sloan, Hazel Mitchell, J. B.

---

25. *Sebastopol Times*, May 9, 1963. Cited in Barbour et al., *Coastal Ecology of Bodega Head*, p. 234–35.
26. Staff writer, "A-Plant Quake Fear Pointed Out by Udall: Interior Secretary in Unusual Warning," *SRPD*, May 22, 1963, p.1. According to Wellock (doctoral thesis), Jean Kortum and Gilliam played the key roles in bringing about Udall's interaction.
27. PG&E retained the services of two highly qualified experts to address seismic concerns: Dr. Elmer Marliave and Dr. Donald Tocher Their internal report to PG&E is titled "Geologic and Seismic Investigation of the Site for a Nuclear Electric Power Plant on Bodega Head, California," (1964) and can be found at the Bancroft Library among several boxes of Marliave's papers that also include maps, photographs and correspondence. The 1963 USGS report authored by Schlocker and Bonilla is titled *Engineering Geology of the Proposed Nuclear Power Plant Site on Bodega Head, Sonoma County, California*.
28. Staff writer, "Bodega Head is Unsafe, U.S. Expert Finds," *San Francisco News Call Bulletin*, Oct. 4, 1963, p. 1; *PG&E Progress*, Nov. 1963; *PG&E Progress*, Apr. 1964; and Sidney Allen, "Battle of Bodega Head," p. 40.
29. The 1964 AEC report can be found at the Bancroft Library and is titled "Summary Analysis by the Division of Reactor Licensing in the Matter of Pacific Gas & Electric Company, Bodega Head Nuclear Power Plant."
30. *PG&E Progress*, Dec. 1964; and staff writer, "PG&E Gives Up on Bodega," *San Francisco News Call Bulletin*, Oct. 30, 1964, p. 1.

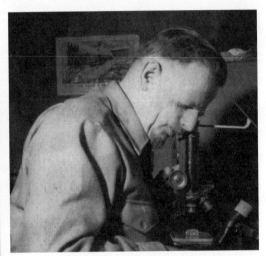

Fig. 158. Environmental activists Hazel Mitchell and David Pesonen at Bodega Head (*left*), October 1964, the day after PG&E had canceled its plans to build the plant. In 1973 PG&E sold its Bodega Head holdings, including Campbell Cove, to the state Department of Parks and Recreation (then Beaches and Parks). The area is now part of the state park system. The tidelands road that PG&E built to access the atomic plant now provides access to the Bodega Head state park and the University of California's Bodega Marine Laboratory, completed in 1966. *Courtesy of David Pesonen, Berkeley, California.*

Fig. 159. Joel Hedgpeth at work, as director of the College of the Pacific's marine research station at Dillon Beach, Marin County, California, 1959. Dr. Hedgpeth went on to become the first director of the Bodega Marine Laboratory. Of all the activists, Joel Hedgpeth fought the longest, the hardest, and the most passionately. The Bancroft Library has two linear feet of his letters and papers documenting the struggle. *Photograph by Don Kelley, then editor for "Pacific Discovery." Courtesy of the California Academy of Sciences, San Francisco, California.*

Fig. 160. Jean Kortum preparing a mailing to the State Assembly Rules Committee, May 1962. *Photograph © 1962 by Karl Kortum. Courtesy of Jean Kortum, San Francisco, California.*

Neilands and the many other members of the Northern California Association to Preserve Bodega Head and Harbor, it is very likely that the residents of Sonoma, Marin and San Francisco counties would be living today under the ominous cloud of a first-generation atomic power plant constructed perilously close to the San Andreas fault system. In the fascinating story of their unrelenting battle to preserve Bodega Head, we can witness the birth of the environmental activism movement in this country.[31]

The Association was founded in early 1962 by Karl Kortum, Harold Gilliam and J. B. Neilands of the University of California, Berkeley. David Pesonen, though not among the founders, emerged as the Association's leader and principal spokesman. Putting his fledgling legal career on hold to devote two years of full time effort to the cause, Pesonen "transformed the association from a disorganized band of about a dozen individuals into a force of nearly two thousand members with a budget of ten thousand dollars by the end of 1963." Jean Kortum also played an instrumental role in this respect, recruiting and organizing grass-roots opposition on the local political front and spearheading lobbying efforts in Sacramento and Washington.[32]

The Hole in the Head

It looks today as if someone dropped an atomic bomb on the cove, blowing away an acre of Bodega Head and perhaps, unwittingly, destroying one of America's most important historical landmarks. If Drake's fort was indeed buried at the foot of the bluffs when the bulldozers arrived at Campbell Cove in January 1963, let us hope that they did indeed leave behind something more than a crater 147 feet wide and 73 feet deep.[33]

---

31. The story is presented in Wellock's "The Battle for Bodega Bay: the Sierra Club and Nuclear Power, 1958–1964" (1993) and Wellock's doctoral thesis: "Critical Masses: Opposition to Nuclear Power in California, 1958–78" (University of California, Berkeley,1996).
32. Wellock, "Battle for Bodega," p. 205.
33. I compiled the Hole in the Head chapter from a number of sources including the Joel Hedgpeth papers (Bancroft Library); the Elmer Marliave papers (University of California, Berkeley, Water Resources Center Archives); PUC files (Sacramento); PG&E files (San Francisco); newspaper clippings; and David Pesonen's *A Visit to the Atomic Park*.

Fig. 161. The Hole in the Head, October 12, 1963. Photograph by PG&E staff or contractors. Courtesy of the University of California Department of Water Resources, as holder.

PART VI

Drake's American Legacy

Fig. 162. The Prayer Book Cross at Golden Gate Park, San Francisco, California. This fifty-seven foot sandstone monument, unveiled January 1, 1894, was donated by Philadelphia philanthropist George W. Childs. It commemorates the Anglican prayer service Francis Fletcher held when the English first came ashore in Nova Albion on June 24, 1579. *Photograph by the author, February 1997.*

CHAPTER 38

The Significance of Drake's Visit

irate – freebooter – master thief of the unknown world – greedy filibuster – pirate – religious zealot – object of wrath – base rascal – devil – buccaneer – irresponsible and impetuous – vigorous social climber – dragon – slaver . . .

Founder of British maritime greatness – one of the greatest of all Englishmen – discoverer – explorer – national hero – statesman – patriot – folk hero – greatest sailor of all time – gallant gentleman – a symbol of resolution in the face of adversity . . .

As can be deduced from these widely divergent descriptions that have been applied to Francis Drake, the whereabouts of his 1579 California landfall is not the only controversy surrounding the man. The very significance of this landfall is a controversy even more compelling in many respects.

What role did Francis Drake play in founding British Colonial America? Was his visit to California an impromptu stop on a voyage of plunder, or an epic event in America's history? Unlike the puzzle over the location of the anchorage, the correct answer to this question is not one that an archaeologist can confirm with the turn of a spade. The subject will always be a matter of opinion.

Let us begin by noting these three bare facts: (1) Drake's 1579 proclamation represents England's first formal claim to American soil;[1] (2) the claim, as far as Elizabethan England was concerned, extended over most of the landmass that now comprises the continental United States (map 27); and (3) the site of Drake's California fort is the British Empire's first Colonial Era landmark in America.[2]

Traditional Opinions

Notwithstanding the above facts, America's historians have traditionally denied, downplayed, or ignored Drake's role in America's founding. Our schoolbooks teach us that Captain John Smith and the Pilgrims were the nation's founding fathers. We visit Plymouth, Massachusetts; Jamestown, Virginia; and Roanoke Island, North Carolina to see the United States' original colonial-period landmarks.

Why? Most historians have portrayed Drake as a glorified pirate. They have indeed painted his California visit in the *Golden Hind* as an unplanned stop on a voyage of plunder. While they recognize Drake as one of the world's greatest mariners, they seem to believe that his so-called buccaneering denies him a place with England's great patriots or the great explorers and discoverers of the age.

Here is a sample of how some of Western America's most influential nineteenth and early twentieth-century historians and writers viewed the man:

George Bancroft, 1852 – This part of his career [the circumnavigation] was but a splendid piracy against a nation with which his sovereign and his country professed to be at peace. . . . The exploits of Drake, except as they nourished a love for maritime affairs, were injurious to commerce; the minds of the sailors were debauched by a passion for sudden acquisitions.

Hubert Howe Bancroft, 1884 – What territorial rights do the discoveries of a privateer or outlaw convey upon his nation?[3]

---

1. By America, I refer here to the continental United States. I discount John Cabot's claims in Nova Scotia and Newfoundland during his first voyage of 1497 because they were made in Canada. I also discount the very shadowy claims to the whole coast of North America associated with Cabot's mysterious second voyage of 1498 during which he and his four ships seem to have disappeared without a trace (see Morison, *The Northern Voyages*, chapter 6).
2. I place the start of England's Colonial Era with Drake's Nova Albion claim. Although the English did not establish the first colony for another five years (Sir Walter Raleigh's 1584 Virginia colony), Drake's circumnavigation, as pointed out by Andrews in *Drake's Voyages*, p. 85, clearly precipitated the event.
3. H. H. Bancroft, *History of the Northwest Coast*, p. 145.

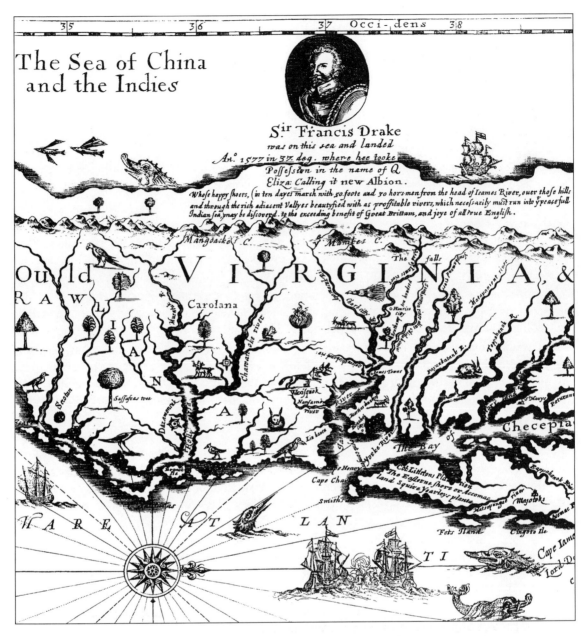

Fig. 163. Colonial Virginia, fifth state, circa 1667: detail of a map first drawn by John Farrer, circa 1651 (west is to the map's top). This detail is from the sixth edition, published by Farrer's daughter, circa 1667. The map lacks geographic accuracy, but not political; it well illustrates the instrumental role Drake played in founding British Colonial America. *Courtesy of the Library of Congress, Washington D.C.: L.C., G3880 1667, F3 vault.*

George Davidson, 1908 – Drake was no novice in unlawful acts at sea . . . Nor was Drake a "discoverer" or "explorer." He had fitted out his vessels for plunder only: the plunder of treasure ships, of towns, estates, individuals, and the altars of the churches. He plundered enough gold and silver to ballast his vessel of one hundred tons. . . . a rover of the ocean who lived by plundering at sea and on land, and who died ingloriously and was dropped to the depths of the sea in a leaden coffin . . . When one studies the history of Spain and England in those times of English slave traders and pirates who sacked the towns of the Spanish colonies on the Atlantic and the Pacific when no war existed, it is easy to comprehend why Spain undertook the invasion of England in 1588.[4]

───────

4. Davidson, *Francis Drake on the Northwest Coast*, pp. 42–53.

Fig. 164. "Queen Elizabeth Knighting Drake on Board the *Golden Hind* at Deptford, April, 4, 1581." A drawing by Sir John Gilbert. *Courtesy of the Astor, Lenox & Tilden Foundations, New York Public Library.*

Henry Wagner, 1926 – The "English Hero," today, in his character of freebooter, no longer appears so noble; times have changed . . . The nineteenth century produced a new sense of morality . . . The examination of his career brings no conviction that he was statesman, explorer or colonizer. He should properly be called a captain of industry. The particular one to which he devoted his attention and which may be said to have been the most popular, if not the most remunerative, pursued in England during the reign of Queen Elizabeth, was plundering the Spanish king or his subjects . . . an unscrupulous seeker after gold . . . in all his enterprises, booty seems to have been somewhere in sight.[5]

Modern Opinions

With good reason, America's modern-day scholars and historians have considerably softened their opinions of Drake, some hesitantly recognizing the significance of his achievements, a few offering outright praise:

Samuel Eliot Morison, 1964 – [Drake] took possession of that country for Queen Elizabeth I, naming it Nova Albion – New England. The profits of this voyage were almost $9 million in gold, and the queen was so pleased with her share that she knighted Drake on the deck of the *Golden Hind*, which was as good as telling the king of Spain, "North America belongs to England – hands off." The efforts of blithe, lusty spirits like Drake, Gilbert, and Raleigh, under that great queen they called Gloriana, blazed the way for the United States of America and the British Commonwealth.[6]

Robert Power, 1974 – I maintain that the historians Bancroft, Davidson, and Hittell were wrong, although they formed the hard core of the consensus history that continues to this day in California textbooks . . . consensus history wrongly states that the British Empire was founded when Sir Humphrey Gilbert laid claim to Newfoundland, in 1583. The honor rightfully belongs to California, where

---

5. Wagner, *Drake's Voyage*, pp. 17, 211 and 212.
6. Morison, *Oxford History*, pp. 79–83. Morison's writings clearly improved Drake's reputation in 1970s America. Sugden writing in 1990 (*Sir Francis Drake*, pp. 321–22) points out that such was not the case in England.

sovereignty over America was first claimed by the Elizabethan nation and the first non-Europeans accepted the authority of the Crown of England's representative, Captain Francis Drake. In a very real sense the plate of brass opens the British epic in North America, and in a symbolic sense it represents the physical founding of the British Empire.[7]

Notwithstanding the opinions of such a highly respected historian as Morison, it is clear that today's historians have not yet removed the tarnish their predecessors heaped on Drake's name. According to a piece in the January 1994 issue of *Discovery*, the magazine of the Discovery Channel: "For Drake's privateering, Spain put a price on his head; England knighted him. Ever since Egyptian ships were attacked in the Mediterranean three thousand years ago, pirates have plied their old profession on the high seas . . . My pirate of choice was a man to be reckoned with . . . a true pirate with a heart of gold."[8]

Patriot or Pirate?

Even in his day, many of his own countrymen angered and offended Drake by questioning his motives and integrity. He arrogantly dismissed them with such colorful expressions as "featherbed milksops." Here is what Drake himself had to say about the motives for his enterprises in verses he penned in 1583:

> Who seeks by worthy deeds to gain renown for hire,
> Whose heart, whose hand, whose purse is pressed to purchase his desire,
> If any such there be, that thirsteth after fame,
> Lo! Hear a means to win himself an everlasting name.
> Who seeks by gain and wealth to advance his house and blood,
> Whose care is great, whose toil no less, whose hope is all for good;
> If any one there be that covets such a trade,
> Lo! Hear the plan for commonwealth, and private gain is made.
> He that for virtue's sake will venture far and near,
> Whose zeal is strong, whose practice true, whose faith is void of fear;
> If any such there be, inflamed with holy care,
> Here may he find a ready means his purpose to declare;
> So that for each degree this treatise doth unfold,
> The path to fame, the proof of zeal, and way to purchase gold.[9]

Thus did Drake at the pinnacle of his career call forth to his countrymen to stir up their heroic spirits, in this case for England's first colonial enterprise in America. In this call we clearly see the potent forces that inspired Drake and sent his countrymen flocking after him: heroism, patriotism, religious zeal, and the common good promoted hand-in-hand with the quests for fame and wealth.

For the English, the Elizabethan Age was a time when these virtuous and base ambitions were joined in a desperate fight for survival. From the English viewpoint, mighty Spain, under Philip II, was plundering New World gold and silver and using it to attack reformed England and her Protestant allies across the English Channel. Because Spain denied impoverished England the opportunity to share in the New World wealth, the only way the English could possibly stop Philip was to rob him of his treasure and use it against him. As aptly put by a contemporary poet lamenting Drake's death:

---

7. Power, *Francis Drake*, pp. 2 and 25. Power reinforced this argument by reference to the verses of a contemporary poet, Stephan Parmenius, which Hakluyt published in his *Principal Navigations*, vol. 3, 1600.
8. Belleville, "A Dead Man's Tale," pp. 20 and 25.
9. Drake composed this sonnet entitled "Sir Francis Drake in Commendation of this Treatise," as the preface to a rare book written by Sir George Peckham in 1583 entitled *A True Report of Late Discoveries, and Possession Taken in the Right of the Crown of England, of the New Found Lands: by that Valiant and Worthy Gentleman, Sir Humfrey Gilbert, Knight*. Peckham appears to have started the book with the idea of reporting on the triumphant results of Gilbert's expedition to plant a colony in Newfoundland, only to find out later that the effort had failed and that Gilbert's ship had been lost on the return with all hands. In his verses, it is clear that Drake is endeavoring to inspire his fellow Englishmen to take up the colonial cause. The Huntington Library, San Marino, California, has a copy of the Peckham treatise.

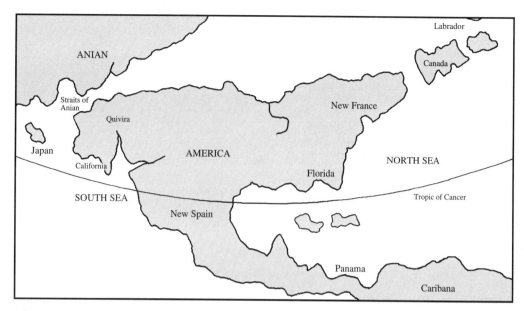

Map 26. Sketch map of North America for 1576 showing England's political and geographic perceptions on the eve of Drake's circumnavigation. The English flag is nowhere to be found. *Derived from the map accompanying A DISCOURSE OF A DISCOVERY FOR A NEW PASSAGE TO CATHAY by Sir Humphrey Gilbert, London, 1576.*

> Whether to win from Spain that which was not Spain's,
> Or to acquit us of sustained wrong,
> Or intercept their Indian-hoped gains,
> Thereby to weaken them, and make us strong[10]

Pirate? – outlaw? – an unscrupulous seeker after gold? Drake was a product of the Elizabethan Age and to judge him fairly one must evaluate his career within the context of his times. This I endeavor to do in the ensuing chapter, where you will find a biographical sketch highlighting the instrumental role Drake played in America's founding. Here let us simply examine the political face of North America before and after Francis Drake arrived on the scene.

The Political Face of America Before and After Drake

Map 26 shows North America derived from several world maps published in England and Western Europe in the late 1570s on the eve of the circumnavigation. As of early 1579, there was absolutely no English presence in what was to become the continental United States.

Map 27 is derived from a world map published in Antwerp in the mid-1580s, some two decades before Captain John Smith founded Jamestown, Virginia. Note the strong British presence in these three respects: (1) the route of Drake's 1577–80 voyage with his Nova Albion claim extending prophetically across the country; (2) Sir Walter Raleigh's 1585 Virginia colony at the mouth of Chesapeake Bay which was sparked by the success of Drake's circumnavigation; and (3) the route of Drake's 1585–86 campaign in which he stopped in to reinforce Raleigh's colony after sacking Santo Domingo, Spain's Caribbean capital; Cartagena, her most important port on the Spanish Main; and Saint Augustine, her frontier outpost in Florida. Witness tiny England demonstrating her determination and ability to implement and aggressively defend her colonial ambitions against the mighty Spanish empire!

---

10. Cited by Michael Allen, in "Charles Fitzgeffrey's Commendatory Lamentation," Thrower, ed., p. 104.

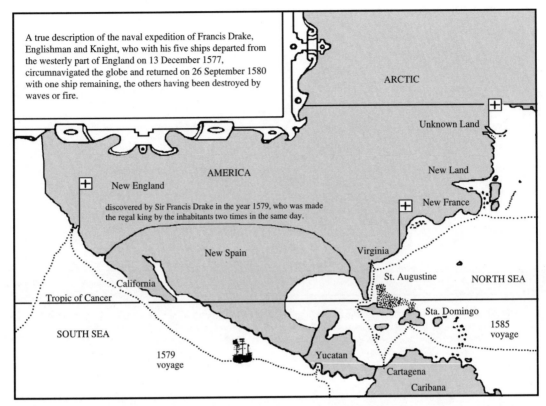

A true description of the naval expedition of Francis Drake, Englishman and Knight, who with his five ships departed from the westerly part of England on 13 December 1577, circumnavigated the globe and returned on 26 September 1580 with one ship remaining, the others having been destroyed by waves or fire.

Map 27. Sketch map of North America for 1586 showing England's political and geographic perceptions of America in the wake of Drake's 1579 and 1585 voyages. Drake is completely dominating the scene and there are now two English flags within the boundaries of today's United States. Drake planted the first one in 1579; in 1585 he staunchly defended the second. *Based on the Drake-Mellon Map, drawn circa 1581–87 (reproduced inside the rear cover of this book).*

What happened in the decade separating the publication of these two world maps that so dramatically altered the political face of North America? How much credit does Drake deserve for these changes?

Drake's Part

With Drake's Nova Albion proclamation, we can precisely date not only England's first formal claim to American soil, but more significantly, her commitment to the enterprise, her writ of manifest destiny, if you will. Putting aside all the learned opinions on the matter, the Drake-Mellon Map, the Hondius Broadside Map, and several similar period maps vividly portray the simple truth that England's commitment to the colonial enterprise was born of Drake's heroic exploits. As you will see in the chapter that follows, this premise is borne out by a review of the period's history. Clearly, it was Drake's astonishing accomplishments that gave Elizabethan England the will and confidence to challenge and defeat the mighty Spanish empire.

It is only fitting that Francis Drake was the Englishman who first stood on American soil and roared out England's claim, for Drake, far more than anyone, made sure his country made good on it. If not for Drake, the United States would have begun as part of Latin or French America.[11]

In the half-buried rocks at Campbell Cove, Americans can look upon the very birthstones of their country.

---

11. In the words of Samuel Morison (*Northern Voyages*, p. 631): "If Spain had defeated England in 1588–1590, she would certainly have planted a chain of garrisoned forts along the North American coast, and it is doubtful whether a humbled and subjected England could have fought her way in. The ultimate fate of Virginia depended on the Armada's victory or defeat."

CHAPTER 39
The Queen's Corsair

A biographical sketch highlighting Drake's part in founding the British Colonies

He was more skillful in points of navigation than any that ever was before his time, in his time, or since his death. He was also of perfect memory, great observation, eloquent by nature, skillful in artillery, expert and apt to let blood, and give physique unto his people according to the climates. He was low of stature, of strong limbs, broad-breasted, round-headed, brown-haired, full-bearded, his eyes round, large and clear, well favored, fair, and of a cheerful countenance.

Fig. 165. Sir Francis Drake: painting by Crispin van de Passe, 1598. *By permission of the British Library, London.*

His name was a terror to the French, Spaniard, Portugal, and Indians. Many princes of Italy, Germany, and others, enemies as well as friends desired his picture. He was the second ever went round about the world: he was married unto two wives both young, yet he himself and ten of his brothers died without issue. He made his younger brother Thomas heir who was with him in most and chiefest of his employments. In brief he was as famous in Europe and America, as Tamberlaine in Asia and Africa.

John Stow, circa 1600.[1]

The Elizabethan War

Francis Drake's career is a microcosm of the Elizabethan War. In relating the chronicles of Drake's voyages, one recounts the history of this war. There were just a few events in the war in which Drake or his mentor and kinsman John Hawkins were not the key actors on the English side. Between them they precipitated the war, masterminded the battle plans, built up and modernized the Royal Navy and merchant marine, set up the syndicates that financed the expeditions, assembled and fitted out the fleets, put together the officers and the crews, organized the victualing, and coerced Queen Elizabeth I and her Privy Council into supporting them. Drake usually led the charge, pitching himself headlong into the heat of battle. The ingenious Hawkins was the man behind the scenes. What they accomplished is astonishing!

The Elizabethan War began in 1568 when a New Spain treasure fleet attacked a Hawkins-syndicate merchant fleet in the West Indies after Hawkins deliberately provoked the Spanish beyond their diplomatic limits. Its climax occurred twenty years later when England defeated Spain's so-called invincible Armada. While the war officially ended in 1604 with Elizabeth's death and the treaty of London, it actually petered out soon after Drake and Hawkins both died in 1596.[2]

Because of its timing, the Elizabethan War has been viewed as part of sixteenth-century Europe's monumental Reformation/Counter-Reformation struggles. In reality, however, it was not a religious war; it is more aptly viewed as Spanish and English merchants squaring off to decide England's right and might to expand oceanic trade and colonize America. The ultimate spoils of England's victorious merchants were the British East India Company and, most importantly for us, the British colonies in America.[3]

Prelude to the Age – 1540 to 1557

To understand Drake's actions and appreciate the magnitude of his achievements, one must have a basic understanding of the Elizabethan Age and the events that led to its momentous war. Before we look at his deeds, let us examine the troubled times into which Francis was born.

Sixteenth-century America – In the first half of the sixteenth century, Spain and Portugal had dazzled Europe with a series of spectacular discoveries and New World conquests. By 1579, Spain held the Caribbean (West Indies), Central America (New Spain), much of South America (Terra Firma), and had a trading outpost in the Philippines (Manila). Portugal held western Africa and Brazil, and had trading outposts in India and the Moluccas (East Indies).

North America, on the other hand, had hardly been touched. As of 1578, there was only one tiny European outpost that the Spanish had founded in 1565: Saint Augustine, Florida. The French had substantial land claims in Quebec and the Saint Lawrence River area based on the expeditions of Jacques Cartier, but they had no settlements in those territories.

Notwithstanding the French pretensions, Spain claimed all of North America based on the explorations of Juan Rodríguez Cabrillo, Francisco de Coronado, Juan Ponce de Leon, and others.

---

1. From Stow, *The Chronicles of England,* cited in Wagner, *Drake's Voyage,* p. 307.
2. While not all historians share this view on when the war started, Morison is an example of one who does (*Northern Voyages,* p. 486). Many historians regard 1585 as the starting date of what I am calling the Elizabethan War.
3. Historians and Drake biographers have not clearly distinguished the Elizabethan War from Western Europe's wars of religion. Today's historians might challenge such a crisp distinction while acknowledging the basic concept that Elizabeth's war had more to do with control of the seas than reforming the church.

The papal bulls of 1493 and 1494, that divided the unknown world between Spain and Portugal, reinforced these claims. With the exception of allowing English, French, Dutch and other European fishing fleets to frequent the coasts of Newfoundland and Labrador, Spain strictly prohibited foreigners from these lands. In fact, just before establishing Saint Augustine, the Spanish massacred a group of French Huguenots who had the audacity to plant a small settlement in northern Florida. In 1567, the Spanish attempted to plant a chain of Jesuit missions from Florida to Virginia to discourage further such encroachments, but they had abandoned them all by 1576.

Tudor England and the Reformation movement – By the time Henry VIII ascended the English throne in 1509, the Church of Rome had amassed huge monastic landholdings and was wielding enormous power through a plethora of church officials. As the Catholic clergy grew in power, the laity became increasingly unhappy with their oppressive rule and the English Crown became increasingly unhappy over the loss of power and tax base. Reformists appeared espousing humanism and lay piety over traditional ritual and scholastic theology. In the 1520s, German reformist Martin Luther published a reformist doctrine, two catechisms and a German translation of the Bible. The Protestant Reformation started when Luther won the support of many princes in Germany, Denmark and Sweden.[4]

When Henry VIII decided to adopt some of the Lutheran reforms in the 1530s, the changes he had in mind for the Catholic church were not simply ceremonial. He began a process of secularizing the church's vast landholdings, stripping the clergy of its power, and throwing off Rome's yoke.[5] Henry's reformation not only threw his own country into turmoil, it also sent shock waves across Europe. His actions infuriated Rome and in 1545 caused a temporary breach in the long alliance between England and Spain. In addition, the windfalls from the sale of church property helped Henry build England's first formidable standing navy and pursue his imperial designs on Scotland and France.[6]

During this turmoil, circa 1542, Francis Drake, the first of a brood of twelve brothers, was born at Tavistock, a tiny town in Devonshire near Plymouth Harbor. His father Edmond came from a long line of tenant farmers and the Drakes were in good standing in the community. Francis' godfather and namesake was Francis Russell, son of the Lord Russell who had gained control of Tavistock Abbey when Henry wrested its lands from the Benedictine monks.[7]

Counter-Reformation struggles and the flight from Devon – Henry's Reformation movement almost fell apart after his death in 1547. By the end of his reign, the wars with Scotland and France had left the English economy in ruins. Under the rule of Henry's ten-year-old son Edward I and his protectorates, the earl of Somerset and the earl of Warwick, there were significant outbreaks of domestic violence as living standards continued to decline due to the combined effects of the Reformation's sudden redistribution of wealth, war with France, and a pronounced slump in the woolens trade. The populace became so disenchanted with the reforms of the era, that in 1549, Cornish farmers marched on Plymouth calling for the repeal of the new Protestant prayer book.

During the Cornish uprising, Edmond Drake, who had enthusiastically adopted Henry's Protestant reforms, was forced to flee Devon with his family. The Drakes ended up living in the hulk of a ship near Chatham Dockyard, England's principal naval base located on the Medway River just outside of London. Here Francis grew up among the ships and seamen of the Royal Navy, while his father, now a Protestant chaplain, read prayers to the sailors. Edmond found Francis a position as an

---

4. The Tudor period starts with the reign of Henry VII (1485–1509) and covers the reigns of Henry VIII (1509–47), Edward I (1547–53), Mary I (1553–58) and Elizabeth I (1558–1603).
5. Henry's breach with Rome began when the Pope would not allow him to divorce his first wife, Queen Katherine of Aragon (Mary's mother), to marry his second wife, Anne Boleyn (Elizabeth's mother).
6. The rise of English sea power in Henry's reign is also attributable to the dynamic growth in the export of English woolens, particularly to Antwerp.
7. Contemporary sources conflict on the date of Drake's birth, ranging from 1540–45.

apprentice to the owner of a small bark, and he learned seamanship, nuances of weather, currents and trade on the English Channel. The bark's owner left it to Francis when he died.[8]

The Bloody reign of Philip and Mary – Upon Edward's death in 1553, Henry's staunchly Catholic daughter, Mary Tudor, ascended the throne. She immediately repealed all of Henry's reformation acts and reenacted the dreaded heresy laws that allowed for the prosecution and execution of heretics. To ensure England's commitment to Catholicism, Mary married Prince Philip, whose father, Charles V, was not only the king of Spain, but the Holy Roman Emperor as well. During their bloody reign, the Catholic crown put hundreds of Protestant leaders to the stake in subduing violent uprisings among the populace.

Francis Drake, like many other English youth of his era, grew up with a deep-seated hatred for Catholics and especially the villainous Spanish prince who had martyred their comrades and kin. Next to the Bible, they revered John Foxe's *Book of Martyrs*, published in 1563, with chilling illustrations of the gruesome burnings.[9]

The Dawn of the Elizabethan Age – 1558 to 1567

Ascension of Elizabeth I – Elizabeth Tudor, Henry VIII's second daughter, was a steadfast Protestant who barely survived the bloody reign of her half sister by feigning a spiritual enlightenment. When she ascended the throne upon Mary's death in 1558, Elizabeth rejected the widowed Philip's marriage proposal and promptly restored her father's Church of England. By 1563, Elizabeth had reinstated much of her father's religious legislation and became England's zealous new champion of reformation.

It was a most daring move. In picking up her father's cause Elizabeth alienated and infuriated not only her erstwhile suitor, now the powerful king of Spain, but also almost half of England's nobility; Francis II, king of France and his wife, Mary Queen of Scots; and the Pope. Her most immediate threat, the French-Scottish alliance, dissipated in 1559 when Francis II died and Scottish Protestants revolted against their Catholic queen. With the ensuing declines of Scotland and France into civil unrest and wars of religion, Spain and Rome became Elizabeth's most serious and lasting threats. It was a miracle she survived them.

Elizabeth probably would not have survived them if sometime around 1563, a very zealous twenty-year-old Protestant had not sold his bark, packed up his *Book of Martyrs* with the rest of his belongings, and headed for Plymouth and Tudor England's famous family of merchants, the Hawkinses.

Seeds of conflict – Spain and Portugal had been gorging themselves on New World treasure since the middle of the sixteenth century. Each winter, Spain would dispatch a huge convoy called the plate fleet to reap the harvest of its new world colonies. After arriving in the Caribbean, the plate fleet would divide into two convoys, the New Spain flota and the Terra Firma flota. The New Spain flota went to Mexico, exchanging colonial supplies for Aztec gold, silver, pearls, emeralds, dyes, and hides. The Terra Firma flota went first to Venezuela, and then to Panama where the Spanish picked up goods that had been transported by mule train across the isthmus: Incan silver and gold from Chile and Peru; and silks, porcelain, jewels, and ivory from the Manila trade. The two flotas would rendezvous in Havana, pick up more treasure, and finally complete the transatlantic loop. Portuguese carracks laden with spices and the treasures of the orient plied an eastern trade route around Africa. Backed by Rome and its famous fifteenth-century papal decrees, the two Iberian countries strictly excluded all other European nations from this extremely lucrative trade.

---

8. Sugden, *Sir Francis Drake*, p. 3, has disclosed some contemporary documents that reveal Edmond Drake was arrested around this time for petty theft. He speculates that Edmond may have "bolted" from Tavistock to escape the law in 1548 rather than fleeing from the inflamed Catholics the year after.

9. Zelia Nuttall, *New Light on Drake*, (pp. 355–57) deduced that Drake was carrying John Foxe's *Book of Martyrs* aboard the *Golden Hind* from the accounts of Francisco Gómez Rengifo, factor of the Port of Guatulco.

Fig. 166. Queen Elizabeth I: painting by an unknown artist, 1590. If Sir Francis Drake is to be considered a founding father, Queen Elizabeth I has to be considered America's founding mother. Drake, by his own assessment, was God's instrument of her noble cause. *Courtesy of the National Portrait Gallery, London.*

As Spain and Portugal grew ever richer on the wealth they reaped from their New World conquests, English merchants became increasingly envious and resentful. William Hawkins, a prominent Plymouth merchant, was among the first to break through the Iberian New World trade monopoly. With the assistance of Henry VIII's powerful navy, he successfully challenged the Portuguese along the coasts of Africa and Brazil, openly smuggling trade goods across the Atlantic. Spurred by Hawkins' successes, other English merchants soon became involved in the illicit trade. During Edward's reign, the English made a series of successful expeditions to Morocco and the Gold Coast (Guinea), ignoring vigorous protests by the Portuguese government. Despite King Philip's outright ban on the Gold Coast trade during Mary's reign, Hawkins and others continued the expeditions with the open support of the English navy.

Storm clouds were forming over the English Channel.

Elizabethan aggression – Except for the temporary breach at the end of Henry VIII's reign, sixteenth-century England and Spain had been strong allies. In addition to a formidable common enemy in France, the two countries shared strong trade interests. Spain's European alliance, that then included the Netherlands, Austria, Naples, and Sicily, was an important market and distributor for British woolens. Spanish, Flemish, Austrian, and Sicilian merchants not only called at English ports, but also relied on the English to keep the English Channel open for their commerce with the rest of Europe. The Anglo-Spanish alliance reached its peak in 1554 with the marriage of Philip and Mary.

Notwithstanding the long alliance and the strong trade interests, Elizabeth's stubborn Protestantism and her aggressive new policy of expanding trade made lasting peace impossible. As far as the queen and her advisors were concerned, "the Pope had no right to partition the world and to give and take kingdoms to whomever he pleased." She limited her recognition of Portuguese and Spanish claims in Africa and America to just those territories under their dominion and tribute.[10]

The rise of English sea power – In order to understand how this bankrupt queen eventually emerged victorious in her war with the mighty Spanish empire, we need to understand how Elizabeth gained her military strength.

At her 1558 coronation, Elizabeth inherited some twenty-six neglected ships from her father's once proud navy. In forming his naval board, Henry VIII had appointed prominent London merchants, a practice that continued under the reigns of Edward and Mary. When Elizabeth took over, some of London's most powerful merchants held three of the five seats on the board: Benjamin Gonson, treasurer; William Winter, ordnance master and surveyor; and George Winter, clerk of the ships. This explains how the English navy was able to participate in trading expeditions to the Gold Coast during Mary's reign despite Philip's ban. It also shows how far private citizens had come in gaining control of British foreign policy.

Elizabeth took this government-merchant alliance to new heights. She not only incited her merchants by defying the papal decrees, she enthusiastically joined them in a reckless and aggressive policy of commercial expansion. On an increasing scale, the queen, her Privy Council members, and her wealthy naval board members formed joint-stock ventures with private companies that financed the Gold Coast expeditions. Usually the Crown provided ships, crew, equipment, and some cash for victualing. The merchants supplied additional ships, paid the crews, provided the cargoes, and bore the cost of necessary repairs. The proceeds were shared proportionate to the investment, with the Crown often receiving as much as a third of the profits.

With these profits Elizabeth was able to refit her old ships and build new ones without taxing her impoverished subjects. Just as importantly, when the need arose, she could count on her merchants to bolster her naval forces with large private fleets of armed merchantmen and supply vessels. It turned out to be a very powerful alliance, indeed.

The Hawkins' slaving voyages – When the elder William Hawkins retired to more statesmanlike pursuits, his sons William and John took over his substantial fleet of merchantmen. William managed the Plymouth headquarters while John commanded the ships and handled foreign affairs. John developed direct inroads with the naval board when he married Benjamin Gonson's daughter. He would come to control the board and was ultimately responsible for refurbishing the Royal Fleet.

With Elizabeth's relations with Philip rapidly deteriorating, John Hawkins devised a rather devious plan to break into the Spanish West Indies trade monopoly under the guise of building on common trade interests. Learning that Spain's New World colonies were facing a critical shortage of forced labor to work plantations and mines, he gambled that Philip would allow the English to trade African slaves for Spanish silver. After all, Spain had long been using the Portuguese and Genoese to provide its New World slaves, and the English could provide this service more to Spain's advantage. They could transport the slaves in heavily armed naval fleets that would protect Spanish Caribbean shipping from the constant menace posed by French corsairs. The English apparently received enough official or unofficial encouragement to put the plan to a test.

Between 1562 and 1565, Hawkins commanded two joint-stock syndicate voyages that amply demonstrated the mutual benefits of the slave trade arrangement. The second voyage was on a much larger scale than the first and was sponsored principally by the Crown. The queen provided a 700-ton warship, and Hawkins sailed under Her Majesty's orders as a British naval officer. He conducted the transactions at gunpoint while making sure the dealings, at least at the Spanish end, were undertaken at fair market value. At first the Spanish colonists were happy with the arrangement.

10. Williamson, *The Age of Drake*, p. 33. He referenced a letter the Spanish ambassador wrote in 1561 that attributes the statement to Lord Cecil, Elizabeth's chief minister.

Fig. 167. John Hawkins: painting by an unknown artist, circa 1581. *Courtesy of the National Maritime Museum, Greenwich, England.*

New Spain merchants at Rio de la Hacha gave Hawkins an order for another consignment. At home, however, Philip fumed at the audacity of the heretic English.

After the second voyage, Philip took strong measures toward denying Hawkins' overtures. He expressly ordered Elizabeth to keep Hawkins and his ships out of the Caribbean and strictly forbade his colonial governors to allow any more English incursions under the penalty of imprisonment. In addition, he significantly strengthened his Caribbean naval defenses by providing a new squadron of twelve, heavily armed war galleons called the Indian Guard.

Hawkins was not easily discouraged. He circumvented the queen's agreement with Philip by sending out a private expedition in 1567 under Captain John Lovell. Unlike the resourceful Hawkins, however, Lovell could not compel the Spanish to trade. He was humiliated at the Port of Rio de la Hacha where, desperately short on provisions, he resorted to unloading his cargo of slaves without payment.

Francis Drake, on his first voyage to the West Indies, shared Lovell's humiliation as one of the fleet's minor officers.

The Opening Salvos of the Elizabethan War – 1568 to 1575

Spanish treachery at Vera Cruz – Elizabeth and her merchants were infuriated by Lovell's rude treatment at Rio de la Hacha. In response, Hawkins returned to the Spanish Main in early 1568 with a fleet of six heavily armed ships carrying some four hundred crew and five hundred slaves. Two of the ships were the queen's: the *Minion* and Hawkins' flagship, the *Jesus of Lubeck*. Hawkins appointed Drake, now in his twenties, as one of the flagship's principal officers. Later in the voyage, he transferred Drake to the fifty-ton bark the *Judith*, giving Francis his first naval command. It was to be an eventful one.

The voyage had been troublesome from the beginning. On the African side, Hawkins was not able to procure the slaves from the Portuguese as he had before. They were forced to do their own dirty work in capturing their unfortunate human cargo. On the West Indies side, they ran into more difficulties. Because of Philip's new policy, it was only with extreme difficulty that Hawkins succeeded in compelling the Spanish to effect a fair trade. At Rio de la Hacha he was forced to capture and ransom the town.

Storms forced Hawkins' returning fleet into the Gulf of Mexico and damaged the queen's galleon. To repair the leaking ship and reprovision, he was compelled to take refuge at San Juan de Ulua, the port for the city of Veracruz, Mexico, from which New Spain shipped most of its goods and treasure back to Spain. It so happened that Philip's plate fleet arrived several days later. Aboard the flagship was Don Martin Enriquez, the colony's newly appointed viceroy. Following good-faith negotiations with the viceroy, Hawkins agreed to let the fleet enter. Enriquez, however, reneged on his agreement and launched a treacherous surprise attack.

Drake and Hawkins barely escaped the onslaught with the *Minion* and *Judith*, leaving behind the queen's flagship and many of their hapless comrades. By the time he limped back to Plymouth, the

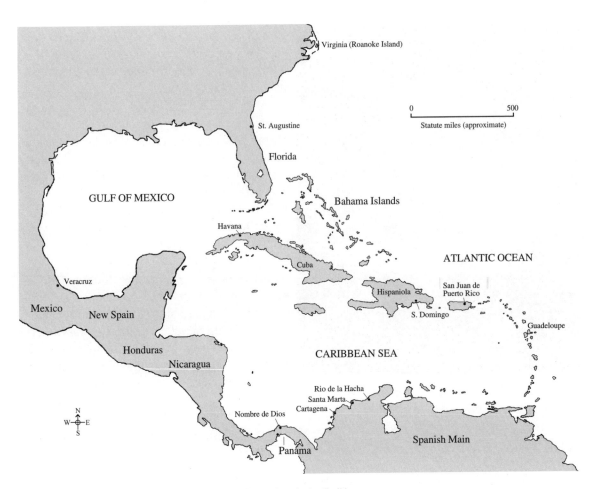

Map 28. Outline map showing Drake's theater of operations in the Caribbean.

half-dead Hawkins carried only fifteen survivors on his ship. Drake's first command ended up as a mortifying defeat at the hands of the treacherous Spanish viceroy.

This event represents the opening salvo in the Elizabethan War.[11] Philip and his new viceroy may have won the first battle, but never could they have dreamed what formidable enemies they had created in Hawkins and his angry young protégé. As Drake would soon prove to the Spanish in no uncertain terms: "If there be cause, we will be devils rather than men!"[12]

The crisis of 1568 – At the very time the Hawkins' fiasco was unfolding in the West Indies, equally foreboding events were taking place in Europe. To the dismay of Spain and Rome, the Reformation movement had swept across Europe, precipitating civil wars in Scotland, France, and the Netherlands. The Scottish Protestants, who had revolted in 1559, finally succeeded in unseating Mary Stuart, Queen Elizabeth's cousin. She fled to England in 1568, finding asylum among England's disgruntled Catholic nobility. The Spanish Netherlands revolted in 1564, and two years later Philip sent the duke of Alva with three brigades of elite Spanish soldiers to suppress another revolt. The French Huguenots created an international crisis in 1568 when they took the important port city of La Rochelle. From there, Dutch sea beggars joined French Huguenot corsairs in attacking papist shipping in the English Channel. Elizabeth responded to Alva's intervention in the Netherlands by harboring these Dutch and French privateers in English ports.[13]

---

11. Williamson, *The Age of Drake*, p. 93, is another historian who has the war starting in 1568. He has it resuming in earnest in 1585, "after a near decade of quasi-peace."
12. From the "note of the English corsairs," presumably written by Drake in February 1571 (cited in Sugden, *Sir Francis Drake*, p. 50).
13. Sea beggars was the contemporary term applied to the Dutch corsairs.

The crisis deepened when French corsairs chased a fleet bearing Philip's payroll for Alva's army into English ports, compelling Elizabeth to seize the fleet's treasure for safekeeping. She feared that Philip was planning to turn Alva's army on her once it had defeated the Protestants on the other side of the Channel. Alva responded by seizing all English merchant shipping in Dutch ports while the Spanish ambassador covertly supported an attempted coup involving Elizabeth's cousin, the recently deposed Mary Queen of Scots.

Tensions between England and Spain were already running high when Drake arrived in London bearing Hawkins' demands for restitution.

Open conflict in the English Channel – Prompted by Hawkins and her other angry merchants, Elizabeth decided to take the offensive. She seized all Spanish shipping in English ports and unleashed her sea dogs (privateers). After appointing Hawkins to take charge of her navy, she sent his fleet to La Rochelle, furnishing the French and Dutch corsairs with money, provisions, munitions, and men. By 1569, English privateers, some operating under French letters of commission, joined these corsairs in attacking Spanish shipping in the English Channel. They almost completely severed Spain's sea routes to the Netherlands and its European markets. Captured goods poured into English ports.[14]

Philip suspended trade, confiscated all English ships in Spanish ports, arrested the heretic crews, and subjected them to the terrors of the Inquisition. He also backed another coup attempt involving Mary Stuart, the so-called Ridolfi plot, this time to be supported by a surprise seaborne invasion from Spain. In 1570, Pope Pious V excommunicated Elizabeth and openly called for her ouster.[15]

Drake in the Caribbean – In 1569, Hawkins had almost convinced Elizabeth to let him go after Philip's returning plate fleet, not simply for restitution but to cripple Spain's capacity to sustain war in the Netherlands or invade England. Since she opted to keep him close to home, Hawkins, it seems, assigned Drake to sever this vital cord.

In 1570, while Hawkins and Elizabeth's other mariners were busy harassing Spanish shipping in the English Channel, Drake took two small ships, the *Swan* and the *Dragon*, into the Caribbean with about fifty men. Little is known of their doings. They may have taken three Spanish ships and attempted two landings. If so, this was the first English privateering raid in these waters. The voyage appears, however, to have been principally one of reconnaissance.

Taking advantage of the information gleaned the year before, Drake took the twenty-ton *Swan* with twenty-five crew back to the West Indies in 1571. For five months his tiny force wrecked havoc along the coast of Panama, capturing some twenty small frigates and barks, several of which he took away as prizes. Before he returned, he had confiscated or destroyed merchandise the offended Spaniards valued at over 400,000 pesos, a king's ransom.

Curiously, many historians and biographers, including his most recent, have labeled Drake a pirate for these raids, notwithstanding the events at Veracruz, the conflict in the Channel, and a political situation amounting to undeclared war. Although there is no direct documentary evidence, the circumstantial evidence strongly suggests that Drake's operations were financed by the naval board, by then heavily under the influence of John Hawkins, and sanctioned by the queen. This would make him a patriot and privateer, not a pirate . . . with or without formal letters of commission.[16]

Nombre de Dios and the capture of the treasure train – So far in this undeclared war, Philip was getting the worst of it. Elizabeth's privateers were feasting on Spanish shipping while English

---

14. Williamson, *The Age of Drake*, pp. 99–107.
15. Corbett, *Drake and the Tudor Navy*, vol. 1, pp. 155–56.
16. Williamson, *The Age of Drake*, pp. 103, 117–21 and 260. Sugden, *Sir Francis Drake*, p. 44, reflecting one school of thought, refers to Drake at this stage of his career as a man "bent on piracy" and this phase of the Elizabethan War as "Drake's War." Williamson, p. 103, puts forth another view that I believe is the more accurate: "the danger at home, the Protestant wars on the continent, and the operations of the English sea captains in the Channel and on the oceans, are all one story, from the crisis of 1568–69 to the end of the age."

merchants were finding new outlets for their goods in Germany. Without pay, Alva's Netherlands army resorted to looting the countryside, inflaming the enemy. Matters were about to get even worse.

By mid-1572, the emboldened Drake, with some seventy young men and boys, was back in the Caribbean with two ships. This time the exploits of his tiny fleet electrified England and sent shock waves through the Spanish empire. He overran and briefly held two of the three towns lying along New Spain's principal treasure route across the Isthmus of Panama: Venta Cruz and Nombre de Dios. He twice ambushed and finally captured the mule train transporting treasure across this route. He entered, briefly blockaded, and otherwise thoroughly harassed Cartagena, the strongest and most important port on the Spanish Main. He captured and looted hundreds of Spanish ships, the choicest of which he brought back to England as prizes. He formed a threatening alliance with a group of French Huguenot privateers. He formed an even more threatening alliance with the dreaded Cimarrones, bands of vengeance-bent African slaves who had escaped the Spanish yoke to live in Panama's tropical jungles.

For Drake personally, the victories were bittersweet. He had lost a good many comrades and two brothers. He barely escaped with his life and would carry a Spanish ball in his painful leg for the rest of his days. On the other hand, he returned a relatively wealthy man and a national hero.

For Elizabeth's England, Drake's triumph at Nombre de Dios was a monumentally successful climax to this phase of the war. Drake's tiny fleet, while only slightly disrupting the flow of treasure that Philip was using to finance his Counter-Reformation campaign, had dealt his powerful regime a crushing blow in confidence and prestige. It left Spanish America stunned, the colonies' weakness and vulnerability to attack laid bare. The English countryside was buzzing with the name of Drake. English unity of purpose, nationalistic pride and confidence replaced the gloom and despair of the 1560s.

"El Draque" had arrived.[17]

As was her practice, Elizabeth officially disclaimed any part in the undertaking. Drake himself, however, tells us he conducted this voyage in her service.[18]

A Nervous Truce

Although he was incensed by Drake's depredations, Philip called off his invasion and coup intrigues when he learned Elizabeth was aware of his plans. The time was not ripe for invading England. Things were still not going well for Alva in the Netherlands, and Philip's navy was tied up in two other campaigns in the Mediterranean.

Elizabeth recalled Hawkins and suspended all types of government sponsored privateering in the English Channel and Caribbean. She not only closed her ports to the Dutch and French corsairs, but used her sea dogs to chase them from the Channel. Drake ended up in Ireland where he put his Spanish prizes to the Crown's service supporting English campaigns to occupy and pacify Ulster. By 1575, Elizabeth had negotiated a nervous truce with Philip and reopened trade. Philip replaced Alva with the duke of Parma, promising to ease up in the Netherlands and respect the nation's autonomy.

The Push for Westward Expansion – 1576 to 1578

Throughout the sixteenth century, the English had, for the most part, groveled before the might of Spain, offering no significant challenge to its overseas empire. The only exceptions were the

---

17. Given that there is no letter "k" in the Spanish alphabet, the English word "Drake" becomes "Draque." In the Elizabethan Age, the name "el Draque" struck terror throughout the Spanish empire.
18. Williamson, *The Age of Drake*, pp. 121–22, points out that in the early 1570s, Drake was one of many captains working under Hawkins and Winter who were conducting privateering raids under the general permission of the queen. This view is borne out by circumstantial evidence. When Drake fitted out his expedition in the spring of 1572, relations with Spain were at their worst due to the Ridolfi Plot and political events in France and the Netherlands. In a January 1592 letter to Elizabeth transmitting his own account of this expedition, Drake clearly states that he performed these services on Her Majesty's behalf and warns her that his reputation was bound to suffer if this information was not disclosed.

shadowy voyages of the Genoese explorers John and Sebastian Cabot (1497–1508). The Cabots had discovered and claimed Newfoundland and Labrador on behalf of Henry VII while seeking the Northwest Passage. The sole attempt to follow up on these voyages was in 1527, when Henry VIII sent Captain John Rut on another futile attempt to locate the Northwest Passage.

Since the days of the Cabots, all of Europe had been intrigued with the concept that North America and Europe could be rounded by sail: that the northern continents had passages like those around the tips of the southern continents. For England, located high in the northern latitudes, the allure of a northwest or northeast passage to the rich markets of Marco Polo's Cathay (China) and Cimpango (Japan) was especially great. Toward this end, in mid-century, a group of English merchants formed a joint-stock company that obtained an exclusive patent for discovering these passages and opening up trade with new lands. Due largely to concerns over antagonizing Spain, this company directed its expeditions exclusively east into Russia, thereby becoming known as the Muscovy Company.

As relations with Spain steadily deteriorated during Elizabeth's reign, the English frantically began to search for alternative trade partners. In the mid-1560s, certain members of the Muscovy Company, in particular Humphrey Gilbert, began looking west. Others looked to the Mediterranean and Baltic Seas.

The nationalistic fervor inspired by Drake's Caribbean exploits and the crisis in the Netherlands set things in motion. During the nervous truce of the mid-1570s, grand schemers stepped up promoting westward expansion as a means out of the current dilemma. Gilbert, John Dee, Martin Frobisher and Michael Lok pointed north, convinced not only of an easy route to Cathay via the mythical Northwest Passage, but also of vast mineral riches to be exploited in North America. Drake, Hawkins and Richard Grenville pointed south, convinced they could wrest from Portugal and Spain, England's fair share of New World wealth. Their dream was to establish outposts and colonies in unoccupied sections of both coasts of South America and develop their own spice trade alliance in the East Indies. They were also intrigued by the prospects of Terra Australis Incognita, the mythical continent then believed to extend north from Antarctica and far into the southern Pacific.

In an impressive three-year spurt of imperialistic energy starting in 1576, Drake and two of Elizabeth's other great mariners, Frobisher and Gilbert, made five voyages of discovery to carry out these grand designs. The Muscovy Company also renewed its efforts to locate the Northeast Passage, sending out an expedition in 1580.

Frobisher's enterprises were disastrous. Between 1576 and 1578, he took three fruitless expeditions into northern Canada on behalf of the short-lived Cathay Company. The investors in these voyages suffered huge losses while the officers and crews suffered miserably from cold and scurvy. Many ships and men were lost. Frobisher brought home two hundred tons of what he thought was gold ore on his second voyage, only to find out on the return of his third expedition with another 1,300 tons, that it was worthless rock. Elizabeth and her merchants were rapidly losing their enthusiasm for westward expansion.

During their 1572 adventures in Panama, Drake and his cohort John Oxenham had climbed a tall tree to look out on the Pacific. As he gazed out over the western horizon, Drake prayed aloud that God might grant him life and leave to sail once on that sea in an English ship. Oxenham beat him to it. In 1574 he took a small company of Englishmen back to Panama, crossed the Isthmus, and looted a good deal of west coast shipping. The Spanish, however, captured Oxenham and his company during their desperate retreat, subjecting them to the terrors of the Inquisition before hanging them.

God ultimately answered Drake's prayer to sail on the Pacific, though in 1576 it looked as if the queen would assign command of the southern expedition to another officer, Sir Richard Grenville. Philip changed her mind, however. He dashed Elizabeth's hopes for a lasting peace that year by ordering his new governor in the Netherlands to batter the rebelling Dutch into submission. With rumors in the air of a new plot to overthrow her, Elizabeth, originally worried about unduly provoking the Spanish with the southern voyage, now became aggressive. She withdrew Grenville's command,

and let loose two of her most trusted and worthy mariners, Drake and Gilbert. Their missions appear to have been twofold: first, to seek out opportunities for overseas empire, colonization and trade, Gilbert on the east coast of North America, Drake in the Pacific; and second, to "annoy the king of Spain."[19] Before the voyage, she took Drake aside and bluntly told him "Drake, so it is that I would gladly be revenged on the king of Spain for the divers injuries I have received."[20]

Gilbert's 1578–79 expedition to the east coast of North America was a dismal failure. Most of his captains abandoned the enterprise before it left port, and a Spanish squadron trounced some of the marauding fleet during a brief engagement in the Caribbean. Drake's mission, however, succeeded beyond Elizabeth's wildest dreams.

As disappointed as was Elizabethan England with the results of the voyages of Frobisher and Gilbert, so much was it exhilarated, exalted and astonished by the accomplishments of Drake.

The Circumnavigation – 1577 to 1580

See chapter 1.

Political Aftermath of the Circumnavigation

Drake's immediate concern for the well being of his queen when he sailed into Plymouth Harbor in September 1580 was a legitimate one. By late 1580, the ominous forces of Counter-Reformation Europe were massing against her.

Mighty Spain had grown mightier still. When Portugal's King Henry died early in 1580 with no direct heirs, Philip sent in his army and seized the Portuguese throne. Along with the vast new territories this put under his control, it greatly bolstered the size of his navy. The Spanish Empire was now at its peak, the most far-reaching and mightiest by far the world had yet known.

Across the English Channel, the Catholics were taking control. In the Netherlands, Philip's troops took one city after another until only the Dutch provinces held out. In France, the Catholic league, with financial assistance from Spain, was getting the upper hand on the Huguenots.

The church of Rome, ever more determined to overthrow Elizabeth, embarked on the *empresa*, the enterprise of England. The *empresa* coordinated the efforts of Jesuit missionaries with Counter-Reformation forces in Spain, France, Ireland, and England. Starting about 1579, Pope Gregory XIII began sending pseudo-missionary forces into England and Ireland to foment a Catholic uprising. When Elizabeth's troops put the Irish rebellion down and massacred the papist contingent, Rome called for her assassination. In the ensuing years, *empresa* forces attempted several coups and assassinations, in one case with the covert assistance of the Spanish ambassador.

Knighthood and Adulation

Disdaining Spain's claims for restitution, in 1580 Elizabeth knighted Drake for his accomplishments in a grand public ceremony aboard the *Golden Hind*. Then she put the proceeds of his voyage to work paying off the large debt she had incurred in putting down the Irish rebellion, assisting her beleaguered Protestant allies in France and Holland, and shoring up the country's defenses. She also richly rewarded her new knight and his crew, and paid off the investors.

Drake's exploits won him lasting fame across Europe. In England he was exalted and glorified. Wherever he went, the crowds thronged about him. His presence was constantly demanded at court.

---

19. Drake's detractors have argued that his written instructions covered only the first objective and that his raids along the Spanish coast were an act of piracy. As discussed by Sugden, *Sir Francis Drake*, pp. 97–98, there is ample contemporary evidence to support Drake's own statement that the queen had specifically told him that she desired retaliation for injuries received. In addition, according to Williamson, *The Age of Drake*, pp. 198–200, in November 1577, Gilbert had presented the crown with "A Discourse how Her Majesty may annoy the king of Spain." Elizabeth granted Gilbert his famous patent the next year authorizing him to plant a colony anywhere north of Florida and attack any foreign ships he encountered along the coast.
20. John Cooke narrative, in Vaux, ed., *The World Encompassed*, p. 216.

In its hour of need, this low man from Devon had exploded onto the scene as England's champion of freedom over despotism. In the years soon to come, Drake would meet the challenge of this heroic role in dramatic fashion.

War Preparations and the Colonial Enterprise – 1581 to 1585

By 1581, everyone knew war was coming; the question was, what to do to prepare. The queen entered into a formal alliance with the French who were then at war with Spain. Drake, Hawkins and some of the queen's Privy Council wanted to take the offensive. In 1581, they put together an enterprise to take the Azores, hoping for assistance from the French. The idea was to install the exiled Dom Antonio, a legitimate pretender to the Portuguese throne. From the Azores, the English could harry Philip's returning plate fleets, cutting off the life blood for his troops. Philip learned of the plot as the fleet was being fitted out. When he threatened war, Elizabeth pulled out, and the French failed miserably when they tried to execute the plan by themselves.

These troubled times also gave rise to England's strengthening its commitment to expand west and establish an American colony. The most ardent promoters of westward expansion were Drake, Humphrey Gilbert and Richard Hakluyt.

Drake was eager to follow up on the spice trade agreement he had effected in Ternate. After putting together a syndicate involving the Muscovy Company and court nobles, he fitted out a four-ship fleet under Edward Fenton that set sail for the Moluccas in 1582. Drake, at the time, was slated to command the Azores expedition. Fenton's ships got no further than the east coast of South America. Many ships and men were lost including Drake's twenty-year-old cousin John who had accompanied him on the circumnavigation. John was shipwrecked on the coast of Brazil and captured by the Spanish.

Several years later, Drake fitted out a much larger expedition for the Moluccas. Queen Elizabeth, however, compelled him to divert this fleet to more urgent tasks: the conquest of the Caribbean and support of England's first American colony.

In 1578, Elizabeth had granted Gilbert a six-year patent to "discover, search, find out, and view such remote and barbarous lands, countries, and territories not actually possessed of any Christian prince . . . and . . . make a settlement with any English subjects who shall willingly accompany him."[21]

As we have seen, Gilbert's expedition that year seems to have been more concerned with annoying the king of Spain than colonial enterprise. In 1582, however, he embarked on England's first serious attempt to colonize North America, taking a fleet of five ships to plant a settlement in Newfoundland. The expedition ended disastrously. The English failed to plant a settlement and on the frantic return voyage Gilbert's ship foundered in the open ocean with all hands.

The next year, Hakluyt wrote his famous "discourse on western planting," a direct appeal to the queen to plant an American colony. In this treatise, that he wrote on behalf of Sir Walter Raleigh, Gilbert's half brother, Hakluyt cited the following objectives: (1) to extend the reformed religion; (2) to replace other English trades that have grown beggarly or dangerous; (3) to supply England's wants from her own dominions instead of from foreign countries; (4) to employ numbers of idle men; (5) to provide overseas bases in the event of war with Spain; (6) to enlarge the queen's revenues; (7) to increase the Royal Navy; and (8) to discover the Northwest Passage.

On the basis of this plea, Elizabeth, without committing the Crown's funds, granted Raleigh a patent similar to Gilbert's. By 1585, Raleigh, working largely with his own capital, had planted England's first tenuous foothold on American soil: a tiny settlement lead by Ralph Lane on Roanoke Island, North Carolina. He named the surrounding territory Virginia on behalf of his unmarried queen.[22]

---

21. Cited in Morison, *Northern Voyages*, p. 566.
22. For the saga of Raleigh's Virginia Colony, I refer the reader to the following books by David Beers Quinn, ed.: *The Roanoke Voyages:*

Although Drake would play an important part in the events of 1585, it was the impact of his earlier exploits, in particular the circumnavigation, that bore the greater significance with respect to his role in the first colonial enterprise. As the renowned Elizabethan scholar Professor Kenneth Andrews put it:

> . . . the main effect of the voyage [circumnavigation] was to inflate the national ego, helping more than any other single achievement to create a new mood of confidence in maritime endeavor. It appeared to open up the oceans to English enterprise and thus stimulated the extraordinary proliferation of projects for trade, plunder, and colonization that marked the period from 1578 to 1584."[23]

Outbreak of War and the Caribbean Campaign – 1585 to 1586

Elizabeth's French alliance fell apart when Philip's troops drove Huguenot forces from the Netherlands in 1584. Early in 1585, Philip entered into a formal alliance with the French Catholic League aiming to wipe out, once and for all, the remaining Protestant resistance. Philip's Netherlands army then took the port city of Antwerp and began a drive to mop up the remaining Dutch Protestants. Elizabeth feared that once Philip finished off the Dutch he would turn his troops on England. When she formally pledged military aid to the Dutch rebels and dispatched a small army, Philip seized British shipping and started making plans to invade the British Isles.

By 1585, the Drake/Hawkins' syndicate had already fitted out an imposing fleet for the Moluccas enterprise. When Elizabeth learned of Philip's invasion plans, she diverted the expedition to the Caribbean and enlarged the fleet. It was another government-promoted, joint-stock enterprise financed by the queen and the usual Drake/Hawkins' syndicate members. The expanded fleet included two British naval vessels and twenty-six other large and small vessels belonging to the merchants of London and Plymouth. It carried about 2,300 men including twelve companies of soldiers.

Drake was assigned command of both land and sea forces with Martin Frobisher, his second. The objectives were to raid and conquer the Spanish colonies and intercept the plate fleet, thereby crippling Philip's capacity to wage war. Such was Drake's fear that Elizabeth would change her mind as she had with the Azores project, that he left port before his fleet was fully provisioned. Here is what Drake accomplished after he hastily departed in September 1585:

Vigo Bay and the Cape Verde Islands – Drake landed troops at Vigo Bay and Bayona where he insulted the Spanish Crown by completing his provisioning at Spain's expense. Here he learned Philip had just lifted the embargo on English shipping and that the plate fleet had already arrived in Spain. Proceeding to the Cape Verde Islands, he sacked the towns of Santiago and Porto Praya. While ashore, his troops were inflicted with a contagious fever that caused three hundred deaths during the transatlantic voyage.

Santo Domingo – With little bloodshed, Drake took Santo Domingo, the capital city of Hispaniola and the administrative center of the West Indies. After destroying nearly a third of the city's beautiful houses and stripping bare the governor's house and the cathedral, Drake raised a significant ransom by threatening to reduce the entire city to rubble. His men plundered all the armament, provisions and hides they could find, liberated the slaves, and destroyed the shipping, including a galley and about twenty barks.

Cartagena – Drake took the well-defended Cartagena, the most important port city on the Spanish Main, with the loss of only twenty-eight men. The citizenry had fled the city with almost all the valuables, but after destroying about 250 homes and threatening the rest, Drake again succeeded in

1584–1590 (1955), which includes all the contemporary accounts; and the easy reading *Set Fair for Roanoke* (1985).

23. Andrews, *Drake's Voyages*, p. 85. Morison, *Southern Voyages*, p. 689, points out that the Drake-Mellon World Map (depicted inside rear cover of this book) supports the notion that Gilbert's voyage was linked to Drake's Nova Albion claim, referencing Robert Power's theory that both the 1579 and 1583 voyages were "part of an imperial plan to link up the two coasts into a northern viceroyalty to stop the Spaniard's advance."

raising a substantial ransom. His men plundered provisions and all the armament including over sixty brass ordnance, destroyed a naval galley and two other vessels, and liberated all the slaves and prisoners including numerous Frenchmen and Turks. Drake devised and brilliantly coordinated the attack, sparing many lives on both sides. Again the crew was stricken with fever. Many either died or were mentally or physically debilitated.

Attempts to protect and fortify the Virginia colony – Proceeding north, Drake sacked the small outpost at Saint Augustine, taking provisions, cannon, and the garrison's payroll. He completely destroyed the town and fort in order to eliminate the potential threat they posed to Raleigh's Virginia colony.

He then visited Lane's fledgling English colony on Roanoke Island to determine if they needed assistance. He found the colonists hard pressed by hostile Indians and severe shortages of provisions and clothing. To tide them over until the next supply ships arrived, Drake provided Lane with a four-month supply of food for a hundred men, arms, and clothing. He also offered them a small fleet suitable for exploring the coast and transporting the colony to a better location. The squadron included Drake's own vessel, the seventy-ton *Francis*, two pinnaces, four boats and enough seamen and officers to handle them. Lane accepted the offer, but when a fierce storm scattered Drake's entire fleet and the *Francis* disappeared with some of Lane's men aboard, the colonists lost heart. Fearful that the colonists would be wiped out by Indians, the compassionate Drake took the lot of them back to England. They reached Portsmouth in July 1586.

Outcome – In terms of financial return to the investors, the expedition proved moderately unsuccessful. Drake declined to claim his own considerable investment in the enterprise so the crews could be paid in full and the investors could receive a maximum return. The British lost approximately 750 men, mostly to disease.

Militarily, on the other hand, the expedition was an immense success. The English had inflicted enormous commercial ruin on Philip's colonies and severely damaged his credit with the European bankers who were financing his war effort. The Bank of Seville broke.

Immeasurable, but far more significant, was the dismay the expedition invoked in the enemy versus the enthusiasm it inspired at home. Drake once again had left the Spanish colonies in a state of shock, demonstrating to the world how weak and vulnerable were its principal cities and how fragile its treasure flow. Philip's mighty empire had been dealt another savage blow to morale and prestige, while in England spirits soared. Drake's name was now ringing all across Europe. Protestants in England, France and the Netherlands were inflamed with a new sense of purpose and determination. Thanks to Drake, the tide was already turning in England's favor.

The Preemptive Strike – 1587

By the time Drake returned in 1586, Philip was already planning England's demise. Using Portuguese galleons and his Indian Guard as its nucleus, the marquis of Santa Cruz, his top Portuguese naval commander, would take a huge armada up the English Channel to wrest the seas from Drake. With Drake beaten, the fleet would escort Parma's invading army across the Channel from the Netherlands. The Pope joined the crusade, pledging a million crowns when Parma's army landed. By early 1587, dockyards all along the coast of Spain throbbed with activity as Santa Cruz began assembling his great fleet.[24]

Unlike his fretful queen, Drake was eager for action. At his continued prodding, Elizabeth finally turned him loose for a preemptive strike with the following objectives: distress the ships in their ports; impeach the junction of Spanish fleets; attack, harass and cut off supply lines; and capture the returning West or East Indies treasure fleets.

---

24. Such was Drake's reputation among the Spanish, that his name came to represent the entire English fleet.

Map 29. Outline map showing Drake's theater of operations on the coast of Western Europe.

Drake set sail in April 1587, with a fleet of twenty-three ships and 2,200 men, the largest the English had ever assembled. The fleet included four royal galleons, including his flagship the *Elizabeth Bonaventure*, 400 tons. The rest were armed merchantmen, contributed once again by London and Plymouth merchants like Drake and Hawkins. Drake commanded, representing government and Plymouth interests. William Borough, second-in-command, represented the regular navy. Rear Admiral Robert Flick, third-in-command, represented London interests.

Here are the sensations the intrepid commander accomplished this time:

The razing of Cadiz – "The wind commands me away" is the famous line Drake penned as he hustled his partially provisioned fleet out of Plymouth, again fearful that Elizabeth would call the venture off at the last minute . . ."pray unto God for us that He will direct us the right way."[25]

Impassioned with this spirit of righteous determination, Drake sailed in such haste that he reached the south coast of Spain with most of his fleet trailing far behind. There he promptly pulled off a lightning strike that goes down as one of the greatest individual feats in naval history. Without hesitation and ignoring the vehement protests of the astonished Borough, he took the flagship directly into one of Spain's largest and best fortified ports, Cadiz. Borough had no choice but to follow with whatever vessels he could muster. When a pair of oared galleys came out to query this curious intruder, he greeted them with the roar of his cannons, sinking one and putting the other to flight.

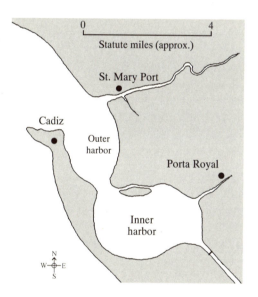

Fig. 168. Cadiz Harbor, Spain, as depicted by William Borough in 1587. *Based on Borough's sketch in Julian Corbett's "Drake and the Tudor Navy," Longmans Greene, London, 1898.*

Eight more galleys sprang to action, but Drake's ordnance was too much for them; they fled to the protection of the shore batteries.

There were about sixty ships anchored in the outer harbor including many destined for the Armada and the next plate fleet. All except one big Genoese galleon of about 1,000 tons lay defenseless. Unable to effect the galleon's surrender after a heated battle, Drake sent the great ship and her valuable cargo to the bottom. He now turned his attention to the rest of the hapless shipping. Notwithstanding the continuous thunder of the shore batteries, the relentless attack of the galleys and fire ships, and the pleas of Borough to retreat, that night the outer harbor was aglow in an eerie light as Drake's crew boarded one ship after another, plundering their cargo before putting them to the torch. One Spanish report said that the flames rose like "a huge volcano, or something out of hell."[26]

In the early morning hours the next day, Drake discovered that a huge Portuguese galleon, belonging to Santa Cruz himself, lay in the inner harbor. He was determined to have her. While the now frantic Borough withdrew six ships to a safer point near the mouth of the outer harbor, Drake boldly led a flotilla of small craft into the fury of the inner harbor. Beating off repeated attacks from fire ships, galleys and the troops amassing along the shores, Drake torched the great galleon and destroyed everything else in reach.

The destruction of shipping complete, Drake tried to leave but was becalmed in the outer harbor. For fourteen long hours he held out against the galleys. He made his escape when the wind finally freshened the next morning.

---

25. Drake to Walsingham, April 2, 1587, cited in Sugden, *Sir Francis Drake*, p. 206.
26. Anonymous eyewitness, cited in Sugden, *Sir Francis Drake*, p. 210.

Cutting off supply lines at Cape St. Vincent – At Cadiz, Drake correctly discerned that much of the Armada's shipping and supplies had not yet arrived from the Mediterranean. He therefore proceeded northwest and lay off Cape St. Vincent looking for prey. With an urgent need for a watering and provisioning place, Drake decided to take the heavily fortified port town protecting the anchorage that lay under the cape. Borough objected, intervening with a written protest that such an attempt would be "vainglorious folly." The angry Drake stripped him of his command and placed him under arrest aboard Borough's own ship, the *Golden Lion*.

Drake then took personal command of a shore party that made a desperate attack on the seemingly impregnable castle of Sagres. Drake himself lit the faggots at the castle's gate, and when it crumbled in flames, the terrified garrison surrendered with three surrounding forts. Drake, to the great indignity of the Spanish, was now in firm command of an anchorage on their Portuguese turf. He quickly put it to England's gain and Philip's detriment by using small craft to capture and loot the passing fishing and merchant fleets that were carrying victuals and supplies to provision the fleet at Lisbon.

The blockade of Lisbon – Drake then further antagonized the enemy by taking part of his fleet north along the Portuguese coast, anchoring off Lisbon. There Santa Cruz was still in the process of arming and fitting out the Portuguese war galleons with which he would lead the Armada. The entrance to Lisbon, unlike that of Cadiz, was too heavily fortified even for Drake to dare entry. Sitting outside the range of the batteries, he blockaded the port and taunted Santa Cruz to come out and fight. The furious marquis would undoubtedly have obliged had he the armament and men to do so.

Capture of the India Galleon – By May, the English were dangerously low on provisions so Drake sent some ships home and took the rest to the Azores. Borough's ship, the *Golden Lion*, deserted the fleet and returned to England, prompting Drake to issue a court-martial. In the Azores, Drake hoped to take a returning East Indies' treasure ship to pay off investors and replenish the queen's much-depleted war reserves. They were in luck. Just after the rest of the fleet arrived, so did the huge Portuguese galleon, the *San Felipe*, crammed full of East Indies' trade goods including spices, silks, china, jewelry, precious stones, gold, and silver. Drake took her after a few hours of fighting and after safely discharging the Portuguese crew, shepherded the great ship back to England.

Outcome – The voyage was an immense military and economic triumph. Drake, as he aptly put it himself, had "singed the king of Spain's beard."[27] In Cadiz he destroyed some thirty to forty vessels of 200 to 1,500 tons each. The final tally for the treasure lifted off the *San Felipe* was staggering. It took seventeen ships to carry the loot from Plymouth to London. Most significantly, Drake's exploits postponed the Armada's departure for a year. His raid on the Azores compelled Santa Cruz to fit out and command a time-consuming expedition to ensure the safety of the remaining treasure fleets. Elizabeth gained valuable time in marshaling her defenses, while Philip's army and navy sat at port rapidly consuming his treasure. The tormented Santa Cruz died early in 1588.

The following dispatch from Rome accurately portrays the profound negative effect on Spanish morale:

> The king goes trifling with this armada of his, but the queen acts in earnest. Were she only a Catholic, she would be our beloved, for she is of great worth! Just look at Drake! Who is he? What forces has he? And yet he burned twenty-five of the king's ships at Gibraltar, and as many again at Lisbon. He has robbed the flotilla and sacked San Domingo. His reputation is so great that his countrymen flock to him to share his booty. We are sorry to say it, but we have a poor opinion of this Spanish Armada, and fear some disaster.[28]

One can well imagine the opposite psychological effect Drake's exploit had at home—national confidence soared!

27. Cited by Sugden, *Sir Francis Drake*, without reference, p. 210.
28. Giovanni Gritti to the Doge and Senate, August 20, 1588. Cited in Sugden, *Sir Francis Drake*, p. 211.

Defeat of the Great Armada – 1588

The moment of truth had arrived. At stake was not only the fate of reformed Europe and an autonomous England, but that of colonial America as well. Would it be Spain or England who settled North America's shores in the seventeenth century?

The telling blow – When Drake returned in June 1587, it was with plans to reprovision, add reinforcements and return to the coast of Spain. To his chagrin, however, the queen received him coldly. As far as she and many of her advisers were concerned, this time Drake had gone too far. Elizabeth still held out hope that Philip might call off the invasion and feared Drake's "bearding" had pushed him beyond his diplomatic limits. Her fears were justified because Philip's resolve was now so great that nothing could steer him from his course, not even the pleas of his commanders and advisors who were frantically trying to convince him that the timing was not right and that his invasion plans were lunacy. What Elizabeth did not realize was that Drake had sealed the Armada's fate by taunting Philip into hurried actions he never should have taken. The Armada was doomed before it set sail and its commanders knew it.

The Armada – What an invincible looking armada it was that left the Spanish port of Corunna and sighted the southernmost tip of England some two weeks later on July 20, 1588. There were about 125 sails, forming a front almost two miles across. The duke of Medina Sidonia's lead squadron of ten big war galleons protruded headlike from the center with his 1,000-ton flagship, the *San Martin*, leading the way. Two other squadrons of war galleons, some twenty ships strong, protruded from each end, protecting the flanks. These squadrons were commanded by two worthy admirals, Juan Martinez de Recalde on the right flank and Alonso Martinez de Leiva on the left. The vessels in between the three fighting squadrons consisted principally of armed transports, dispatch boats, and supply ships. The convoy carried some 19,000 soldiers armed with muskets, pikes and arquebuses, and about 8,000 sailors.

Philip's plan was as follows: Medina Sidonia would take the Armada north to the mouth of the English Channel, engage Drake's western squadron and defeat him. The fleet would then sail east up the Channel, dispatch Howard's eastern squadron, and seize a landing place on the English coast, preferably near Margate. Medina Sidonia would then use his galleons to escort Parma's army of some 17,000 soldiers across the channel using a flotilla of shallow-draught transports.

Neither Medina Sidonia nor Parma favored this plan. Sidonia had personally witnessed Drake in action at Cadiz and had grave doubts about the ability of his bulky galleons to defeat Drake's sleek ships at sea. Parma recognized that his transports were not robust enough to survive bad weather in the crossing, and that they would be extremely vulnerable to Dutch attack when they first set out. There were no good ports in the area and his transports would be unguarded until they reached the Armada ships anchored in the open roadstead off the coast.

With a following wind the Armada entered the Channel on July 20 and sailed due east past Lizard Point, approaching Plymouth Sound as night fell. As they followed the coast, they could see the beacon fires burning brightly along the distant hilltops, spreading word of their arrival across the realm.

Where was El Draque?

The English fleet – Originally the English fleet was divided in half. Lord Admiral Charles Howard, the executive officer in the queen's regular navy, commanded the eastern squadron and had a force of about 4,000 men at his disposal. His fleet consisted of sixteen of the queen's powerful war ships and was based in London. Drake commanded the western squadron that consisted of five of the queen's galleons, two of her pinnaces, and some twenty-three armed merchantmen. He had a force of about 2,800 men. Drake was assigned to protect the south coast of England and Ireland and represented the first line of defense. Howard's fleet represented the second line and guarded the Strait of Dover, the narrow part of the Channel across which Parma was expected to come.

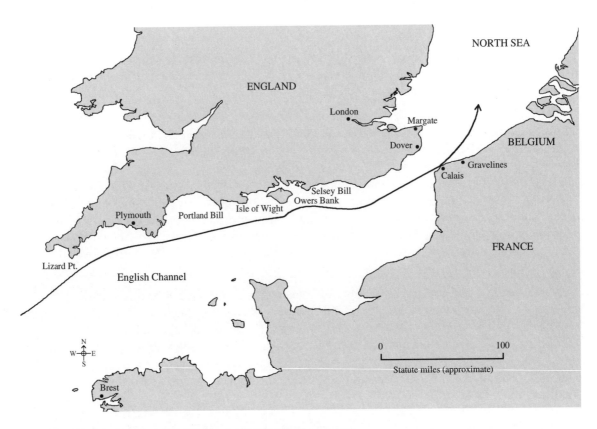

Map 30. The course of the Spanish Armada in the English Channel, 1588.

Drake was opposed to this defensive strategy from the start. Since his return from the Cadiz expedition, he had been pleading with Elizabeth to let him take an enlarged fleet back to Spain to finish the work he had begun. She gave in to his appeals in March 1588, only when she conceded that Philip was not going to deviate from his invasion plans. She surprised Drake, however, by not merely sending most of Howard's ships, but by also sending the Lord Admiral himself to take over the command of the expedition. Drake, who despised the conservative tactics of the naval commanders, must have been very upset. Nonetheless, he accepted this slight gracefully. When Howard arrived, Drake greeted him with the utmost goodwill and respect, demonstrating that he placed the common cause above his wounded pride.

The combined English forces, while not as visually impressive as the Armada's, were just as formidable. There were fifty-one fighting ships of over 200-tons each and hundreds of smaller supply and dispatch vessels that Plymouth merchants had volunteered to the cause. Lord Admiral Howard commanded the 600-ton *Arc Royal*, Admiral Sir Francis Drake the 450-ton *Revenge*, Rear Admiral Sir John Hawkins the *Victory*, and Vice Admiral Martin Frobisher the *Triumph*, the largest ship in the fleet. More than half the larger ships were privately owned and many had been built with the proceeds of privateering raids on Spanish shipping and ports, including two of Drake's ships, the 60-ton *Elizabeth Drake* and the 200-ton *Thomas Drake*. Thanks to John Hawkins, unlike Philip's huge lumbering galleons, the English ships were built for speed and maneuverability. These attributes would provide a decisive advantage in the battles to come.

Victualing requirements and bad weather delayed the English fleet's departure for the counteroffensive. They finally got under way in late June, but contrary winds forced them back to Plymouth. On July 19, 1588, as the great Armada entered the Channel, the English were busy reprovisioning the fleet for a second attempt to launch the offensive. Medina Sidonia had caught

Fig. 169. The Game of Bowls: a nineteenth-century painting by Seymour Lucas. The painting's theme is the popular anecdote/legend that in bringing news of the Armada's arrival on the coast of England, the messenger found the commanders of the English fleet playing a game of bowls on Plymouth Hoe; the news set all the captains scurrying for their ships except for Drake who calmly replied: "There is plenty of time to finish the game and beat the Spaniards too." *Courtesy of the National Maritime Museum, Greenwich, England.*

them off guard. Now the English would have to break up the Armada and prevent it from seizing an English port.

Although Drake was not in overall command of the British fleet that would fight five epic battles with the great Armada, he had more to do with their outcomes than Lord Howard or any other British commander.

The battle off Plymouth – During the night of the twentieth, Drake and Howard were able to move fifty-four ships out of Plymouth Harbor and into the English Channel. In a brilliant feat of seamanship, Drake's squadron discreetly worked their way around and behind the Armada. What a shock it must have been to the Spanish watch as dawn broke the next morning; there was a huge English fleet upwind and behind them, bearing down for an attack. While Medina Sidonia's squadron took off after some stragglers, the English swooped in to engage the rear guard. Howard took on Leiva's squadron while Drake, Hawkins, and Frobisher engaged Recalde's. When the smoke cleared about four hours later, lives had been lost on both sides but none of the vessels had incurred much damage. The English refused to come close enough for the Spanish to grapple and board, and from the distance the English cannon shot had little effect on the tough hulls of the big wooden galleons.

The day belonged to the Armada. It had easily repulsed the English attack. Drake spoiled it for them, however. Late that night, he captured a huge armed merchant ship, the *Nuestra Señora del Rosario*. The *Rosario* had been damaged in a collision with another Spanish ship, and had fallen behind as Medina Sidonia took the Armada further up the channel. She was a supply ship that also carried almost a third of the Armada's payroll and some silver and jewels. It was England's first and richest prize of the campaign and when the news reached London its citizens lit bonfires in celebration.

The battle off Portland Bill – As dawn broke on the twenty-third, the wind had turned and the English found themselves positioned dangerously downwind of the Armada. Howard led his fleet

Fig. 170. Drake captures the *Rosario*: an eighteenth-century engraving by J. Pine. *Courtesy of the National Maritime Museum, Greenwich, England.*

northwesterly to outflank the Spanish and regain the attacking position. Medina Sidonia, however, cut him off, trapping a detachment of ships inshore including Frobisher's *Triumph* and Flick's *Merchant Royal*. Sidonia's squadron launched a furious attack threatening to send them to the bottom.

Drake came to the rescue, attacking the Armada's southern flank with a force of about fifty ships. With Spanish attention diverted southward to Drake, Howard took the rest of the fleet directly after Medina Sidonia's flagship. One English ship after another sailed by the great Spanish galleon, exchanging broadsides with her. Then Drake suddenly appeared with the *Revenge*. His squadron entered the fray, riddling Sidonia's flagship with no less than five hundred shot.

By midday the action was over as the exhausted English were running out of ammunition. Remarkably, despite the violent exchanges, none of the ships, Spanish or English, received much damage. Most importantly, the Armada had repulsed another English attack with minimal losses. Notwithstanding Drake's heroics, the day again went to the Spanish.

The first battle for the Isle of Wight – The twenty-fourth dawned with Drake launching a furious attack on the Armada's rearguard. Singling out the flagship of one of the support squadrons, Drake's fleet surrounded and for two hours attacked it like a pack of hungry wolves. They failed to sink her, however, and, running low on shot, were forced to fall back when Medina Sidonia's squadron approached.

Drake was justly concerned as he watched the Armada pass the Isle of Wight and come into position to enter the channel that separated it from the mainland. It appeared that Medina Sidonia's intention was to enter the channel and seize the island. The well-protected anchorage on the island's lee side was large enough to accommodate his entire fleet and it would provide a more than suitable base for supporting Parma's amphibious assault. The French had used this same strategy during their attempted invasion late in Henry VIII's reign.

The second battle for the Isle of Wight – By the twenty-fifth, Howard had divided his fleet into four squadrons, one under his command, and the others under Drake, Hawkins, and Frobisher. Drake took the seaward wing, Frobisher the landward, with Hawkins and Howard in between. The day started poorly for the English. With Howard and Hawkins attacking some stragglers, Frobisher tried to outflank the Armada on its landward side. He soon found himself pinned between the Spanish fleet and the island with a number of great galleons bearing down on him. In the meantime, however, Drake, who had outflanked the Armada to seaward, suddenly launched a fierce assault that had the net effect of crowding the Spanish northeasterly toward Selsey Bill. Medina Sidonia discovered the significance of Drake's maneuver when the lookouts suddenly sang out "shoal water ahead." Drake was running the Armada aground on the treacherous Owers Bank. To save his fleet, Medina Sidonia had no choice but to put the Isle of Wight astern and veer south-southeast with all speed.

Thanks to Drake's brilliant maneuver, the day belonged to the English.

The fire ships at Calais – By July 27, Medina Sidonia had brought the fleet further up the Channel, crossed over to the coast of northern France, and into the neutral French port of Calais. Here he counted on Parma to supply ammunition and provisions before he made his next assault on the English fleet. On the night of the twenty-eighth, however, the English ruined those plans. About midnight, the silhouettes of eight large ships appeared in a line at the mouth of the harbor, sails fully set. As they came forward borne by wind and current, one after another suddenly burst into flames: "Fire ships"!

As the flaming ships bore down on the anchored fleet, a series of deafening explosions filled the night air as flames touched off loaded ordnance. The Spanish captains panicked, cutting cables, desperately struggling to get to sea. The next morning, Medina Sidonia's galleon, his four consorts, and the galleon *San Lorenzo*, that had run hard aground, were the only vessels remaining at port. The rest of the fleet was scattered in the channel.

The gallant Drake had sacrificed the 200-ton *Thomas Drake* to the cause.

The victory at Gravelines – While Howard's twenty-ship squadron occupied itself capturing the beached *San Lorenzo*, Drake led the rest of the fleet to the important business at hand, attacking the Armada before it could re-form. Medina Sidonia, however, sailed out to intercept him, bravely offering his small but powerful squadron to the entire approaching English fleet. Drake did not hesitate to oblige. The action that ensued was furious. Drake squared off with Medina Sidonia, directing the *Revenge*'s blazing cannon directly into the *San Martin's* hull, this time at point-blank range. Taking Drake's lead, the encircling English closed in, firing volley after volley while the galleons pounded back. When Spanish reinforcements arrived, the English engaged them too. According to an eyewitness account, "Francis Drake's ship was pierced with shot above forty times and his very cabin was shot twice through."[29]

The great battle ended late in the day with the English running out of ammunition and the Armada approaching the perilous Zeeland shoals. The Spanish had lost only four ships in the battle but many others, including Medina Sidonia's flagship, were seriously damaged. Thousands had been killed or wounded. Barely skirting the shallows, he headed the crippled fleet northward up the Channel, putting their sterns to Drake, Elizabeth's England, Parma's troops, and Philip's dreams. What the British did not know at that time was all too clear to Medina Sidonia: "We are lost!"[30]

The terrible retreat – By the time the scattered fleet had returned to Spain, almost half of the 130 ships that set out were lost along with some 15,000 men. During the two months it took to round Scotland and Ireland, storms, the wounds sustained in the battles, sickness, and starvation combined to take a horrible toll. Many of the dispirited survivors who did make it home died shortly afterward. Some ships were simply swallowed up by the cold gray seas. Others were wrecked on the coast of Ireland where the merciless Irish murdered any survivors who managed to make it ashore.

29. Corbett, *Drake and the Tudor Navy*, vol. 2, p. 292. Citing Petruccio Ubaldino, he attributes the statement to Meteren.
30. From the Calderon narrative, cited by Sugden, *Sir Francis Drake*, p. 258.

The Portugal and Azores Projects – 1589 to 1594

The victorious English did not lose a single ship in the fighting and less than a hundred men. The toll lost to disease, however, was quite another matter. A typhus-like epidemic, ever a threat in the cramped unsanitary conditions aboard ship, had broken out even as the fighting was in progress. Thousands died in the returning ships and ports where disembarked seamen littered the streets. To make matters worse, Elizabeth had no money to pay these patriots. Drake, Hawkins, and Howard pitched in to the extent that they could to see their men were properly discharged.

In the winter of 1588, the burning question was how to exploit the defeat of the Armada. The queen wanted simply to destroy what was left of the Armada by attacking Spanish ports. Hawkins wanted to maintain a continuous blockade between the Azores and the coast of Spain in order to intercept the flow of Spain's New World and East Indies' treasure. Drake and most of the queen's council argued for landing an army in Lisbon under the Portuguese pretender Dom Antonio. They reasoned that such an action would set off a rebellion among the many dissident Portuguese, put Dom Antonio on the throne, and transform Portugal into England's strategic ally. Elizabeth opted for Drake's plan over Hawkins', but this time she probably would have been much better served by listening to Hawkins. As it turned out, Dom Antonio had very few supporters among the mainland Portuguese.

Drake took what amounted to an English counterarmada out of Plymouth in April 1589. Once again it was financed by a joint-stock syndicate involving the queen, Drake and other prominent navy officials and merchants. There were six important navy ships, including the *Revenge* as Drake's flagship, and over a hundred private merchantmen. The fleet may have carried as many as 4,000 seamen and 13,000 soldiers. Few, however, had formal training or experience. Drake commanded the fleet while Sir John Norris commanded the army. The mission had the following military and naval objectives: destroy the Armada's surviving ships in the ports of northern Spain; attack Lisbon, land an army to test the strength of Dom Antonio's support; and if unsuccessful at Lisbon, proceed to and attack the Azores, placing Dom Antonio in power over at least some of the islands.

When the great fleet finally got under way in April 1589, it was dangerously underprovisioned and lacked the siege guns the queen had promised to provide for attacking Lisbon.

Most of the Armada's surviving ships were at the port of Santander, but Drake had been informed that there was also a large contingent at Corunna. He easily took this port but was disappointed in finding only seven of the Armada's fleet. After securing the port and destroying buildings and shipping, Norris led his army inland but was rebuffed when he attempted to capture the adjacent town. His one moment of glory came when the English routed a Spanish relief force, killing thousands of enemy soldiers and ravaging and pillaging the countryside. The looting solved the victualing problem, but only temporarily; when bad weather delayed their departure for several weeks, the huge, undisciplined army quickly consumed almost everything. Worse still, a deadly epidemic broke out.

When they were finally able to get underway again, Drake and Norris opted to proceed to Lisbon rather than Santander. They had a number of valid reasons for this choice. Santander was heavily fortified, provisions were in short supply, and the epidemic was raging. Nevertheless, based on its disappointing results, Elizabeth came to view the decision with great displeasure.

Drake landed Dom Antonio with Norris' army about fifty miles outside Lisbon and proceeded to the mouth of the Tagus River. There he assaulted the castle of Cascaes and took some sixty merchant ships. Norris, meanwhile, was to march on Lisbon, picking up provisions and reinforcements from the hordes of Portuguese supporters that Dom Antonio promised would be waiting en route. He would send a dispatch to Drake when he arrived at Lisbon, at which point Drake would take the fleet up the Tagus to reinforce the attack.

Things did not turn out as planned. Norris marched unopposed to the gates of Lisbon behind which the Spanish and Portuguese forces were firmly entrenched. He found almost no rebel support for the invasion, and without Drake's fleet on hand he had no artillery with which to breach the city walls. When the dispatch arrived in Cascaes with Drake assembling for the attack, it was with the disappointing news that the Portuguese had not rebelled and that Norris' disease-stricken army was in desperate retreat.

Despite the setback at Lisbon, the lack of provisions, and the epidemic, Drake was intent on taking the remnants of the fleet to the Azores. Bad weather and contrary winds completely frustrated the attempt. Drake's scattered fleet had all retreated to England by the end of June, having satisfied none of the expedition's objectives. The English may have lost as many as 10,000 men in this ill-fated expedition, mostly to disease. In addition, Drake, the queen and other syndicate investors suffered huge financial losses. Such was Drake's chagrin over the calamity that he declined to take to sea again for almost five years. For the first time in his naval career he had experienced the bitter taste of defeat.

The war went on, however. With the failure of Drake's Portugal enterprise, Elizabeth attempted to execute the Hawkins' plan to blockade the coast of Spain and capture the plate fleet. Frobisher and Hawkins took the first squadrons out in 1590, forcing Philip to postpone the departure of his Indies' fleet. In 1591, she sent Lord Howard to the Azores with a fleet of six of her best ships, this time placing Sir Richard Grenville in charge of Drake's former flagship, the *Revenge*. In the battle of Flores, the plate fleet's powerful consorts captured Grenville and the *Revenge* and put the rest of Howard's squadron to flight. Frustrated by the loss of the *Revenge* and the continued failures of the expensive blockading tactic, by 1594 Elizabeth had abandoned the project as a naval enterprise.

It was high time to call back her master mariner.

The Final Voyage – 1595 to 1596

Philip was not deterred by the disastrous defeat of his Armada in 1588. His plate fleets and Indies' galleons continued to unload their immense treasures at Spain's docks. Spanish morale improved with the defeat of Norris at Lisbon and especially with the ensuing defeat of Howard in the Azores. Philip rebuilt Spain's navy, shored up his port defenses on the continent and in the West Indies, and resumed his Counter-Reformation contests in France and the Netherlands. In 1590 and 1592, he sent Parma's army into France and another into Brittany where the Spanish captured the strategic port city of Brest. Elizabeth responded in 1594 by sending Frobisher and Norris. They routed the Spanish, but Frobisher was fatally wounded in the action.

Drake and Hawkins, despite their advancing years, had both been petitioning Elizabeth for a commission. Hawkins, now over sixty years of age, had long dreamed of taking Philip's plate fleet and Drake, in his early fifties, was convinced he could repeat his triumphant expeditions to the Caribbean. In December 1594, with rumors in the air of Philip's assembling another armada, Elizabeth appointed them joint commanders of what was to be their final expedition. She placed Sir Thomas Baskerville in command of the land forces.

The queen provided six war ships and Plymouth merchants provided another twenty or so. Drake and Hawkins engaged their own men and each provisioned his own half of the fleet. They departed Plymouth in August 1595, promising to return within six months. Their first objective was to capture San Juan de Puerto Rico where the English had learned a richly laden treasure galleon lay under repair. The second was to capture Nombre de Dios, Panama, and the treasure train.

Drake's contingent, as usual, was not adequately provisioned when they departed, and he was hard pressed to convince the disgruntled Hawkins to make a victuals raid on Spain's Canary Islands. The diversion proved fruitless and costly. They found the Canaries too well defended when they arrived, and Spanish dispatch boats sped to the Caribbean spreading the word ahead of them. They had lost the critically important advantage of surprise.

The scattered fleet assembled off Guadeloupe as the first wave of Philip's West Indies' relief forces arrived: a squadron of five heavily armed galleons. The impetuous Drake was all for engaging them when they were sighted, but the cautious Hawkins insisted on delay to get the fleet in fighting condition. Hawkins' hesitation proved disastrous as the Spanish squadron proceeded directly to Puerto Rico, providing a decisive edge to the defenders. Meanwhile, many English, including Hawkins, were stricken with tropical fevers.

Just as the fleet anchored off San Juan for the attack, Hawkins died, a terrible blow to the English. Following the solemn funeral, Drake launched an attack that clearly lacked the passion, brilliance and tenacity of those he led in younger days. The stout Spanish defenses frustrated his forces and inflicted significant casualties. Counting on easier pickings, Drake gave up the siege and headed for the familiar Spanish Main. Hawkins' death must have added keenly to the bitterness of the defeat.

If Drake had taken the fleet directly to Panama, the expedition probably would have succeeded. Instead, however, he went south, a diversion that gave the Spanish just enough time to place a weak defense network across the isthmus. It seems that an older, more calculating Drake had lost the instinctive military genius of his youth.

The English sacked Rio de la Hacha and Santa Marta but obtained only a little booty for their trouble. They found Cartagena well defended and Drake declined to attack. When they finally arrived at Nombre de Dios it was virtually abandoned. Four days after Baskerville had set off to capture Panama, he was back at port, his forces repelled by a small but strongly entrenched Spanish garrison. The deeply disappointed but still resolute admiral took the fleet out of Nombre de Dios, his sights now set on the coasts of Nicaragua and Honduras.

By the time they set sail, the fleet was ravaged by another epidemic. Drake had miraculously escaped these pestilences throughout his career, but not this time. The stricken commander directed his forces back toward Panama probably hoping for a chance to make one last onslaught on the treasure train. It was not to happen. On the night of January 27, 1596, the delirious Drake rose from his bed demanding his armor. The great warrior died a few hours later as the fleet lay off Porto Bello.

Twilight of the Elizabethan Age – 1596 to 1604

While England mourned the passing of its two great naval heroes, Philip, of course, was jubilant. His Catholic subjects "illuminated" Seville to celebrate the event. Flushed with renewed optimism, Philip set the Spanish back to work assembling another great armada.

Drake may have been gone, but in 1596 his fighting spirit clearly lived on in Elizabeth's England. In a fitting tribute to the brilliance of his 1587 expedition, Lord Howard joined by Sir Walter Raleigh led a lightning-strike on Cadiz, burning the shipping and sacking the port. Notwithstanding the attack, Philip's armada set out later that year and again in 1597. In both cases it was turned back by foul weather. When Philip passed away in 1598, it must have been with a sense of bitter failure.

The last years of Elizabeth's reign were rather uneventful. The great Elizabethans and their antagonists were almost all gone. English privateers of colonial period fame like Bartholomew Gosnold, Christopher Newport, and Captain John Smith continued to torment Spanish shipping, but the 1596 raid on Cadiz was to be the last significant battle. When Elizabeth died in 1603, the king of Scotland ascended the English throne. In 1604, King James I signed the Treaty of London and the Elizabethan War was officially over.

Services Ashore

Of the thirty years between Drake's first voyage to the Caribbean and his death there in 1596, he spent the better part of twenty at sea. The only sustained intervals he spent ashore were the two five-year periods after his circumnavigation (1580 to 1585) and the Portuguese expedition (1590 to 1595). What he accomplished in these interludes is very revealing about the man's character.

Drake met his first wife Mary Newman, a local girl of humble background, just before or after Hawkins' third slaving voyage. They were married in 1569, and settled in Plymouth, owning a modest

Fig. 171. The "Jewel portrait" of Sir Francis Drake: painting attributed to Marcus Gheeraerts, circa 1591. *Courtesy of the National Maritime Museum, Greenwich, England.*

property probably secured with part of the proceeds of his early Caribbean raids. Mary, who must have seen very little of her wandering husband, died in 1583.

Drake's knighthood ceremony in 1581 marked a substantial rise in his family's status and financial standing. The queen allowed Drake to keep 10,000 pounds, a substantial fortune, from the proceeds of the voyage, gave him a coat of arms, and granted him title to a number of valuable properties the Crown had previously seized from the Church.

We have seen that Drake was busy in the five years after the return from his circumnavigation planning and fitting out new enterprises. He was also busy with more statesmanlike services. He was elected mayor of Plymouth for the year beginning September 1581, during which time he passed a regulation controlling the local pilchard trade. Afterwards he continued to serve as a justice. In 1581 and then again in 1584, Sir Francis was elected to seats in Parliament, presumably through the efforts of his godfather, Francis Russell, then earl of Bedford. This brought him to London where he sat on numerous committees mostly concerned with pending legislation governing maritime affairs. The most famous of these bills concerned the issuing of Raleigh's license to establish an American colony, a cause Sir Francis enthusiastically supported. He earned the reputation of an eloquent and bold speaker.

Drake further advanced his social position in 1585 when he married Elizabeth Sydenham, the "beauteous and virtuous" twenty-year-old heiress of an extremely wealthy nobleman.[31]

Drake was remarkably active after his return from the Portuguese expedition. He was heavily involved in a project to fortify Plymouth, including the construction of a wall, trench, and fort. He was also instrumental in the construction of an aqueduct and reservoir system that brought critically needed fresh water to the city. Demonstrating his entrepreneurial and business management skills, he built six profitable gristmills along the new aqueduct and ran these businesses while managing his merchant shipping business and substantial landholdings. In the early 1590s, Drake must have spent considerable time supervising the compilation of the narratives of his voyages. As discussed in chapter 5, he presented a sampling of his writings to the queen in January 1592.

There was more public service. The queen's Privy Council appointed Drake deputy lord lieutenant and justice for the Admiralty Court. As lord deputy, Drake supervised the local justices of the peace. This involved him in presiding over numerous legal suits, including a famous murder case. In those days English corsairs were preying heavily on Spanish shipping. When the ships returned to port, the plunder had to be divided fairly among the ships' owners, crews, investors, and the Crown. Drake's duties for the Admiralty Court were primarily those of prize commissioner. The magnitude of this task can be appreciated when one considers that from 1589 to 1591 alone, English corsairs took some three hundred Spanish prizes. His duties included recovering, accounting, and securing the cargoes, and in trying and disciplining any mariners found guilty of pilfering them. By far his biggest challenge in this role was sorting out the booty taken from the *Madre de Dios*, an immensely rich Portuguese carrack taken off the Azores in 1592 by a swarm of English privateers sponsored by various syndicates.

In 1593, Drake managed to get himself elected to another seat in Parliament and spent the better part of that year in London voicing his opinions on various committees and petitioning the queen for his next commission.

The Compassionate One

The following statement relating to Drake's standard of military conduct is attributed to Commandante Don Alonso de Stotmayor, whom the Spanish viceroy sent from Peru to Panama to do battle with Drake in 1595.

> One of the most famous men of his profession who have existed in the world, very courteous and honorable with those who have surrendered, of great humanity and gentleness; virtues which must be praised even in an enemy.[32]

Sixteenth-century Europeans were, in general, merciless to their captives. During the queen's Irish and Spanish campaigns, Norris' blood-thirsty soldiers almost always put the vanquished to the sword. The French were well known for their cruelties to their Spanish captives. The Spanish were considered the worst of all. They would first submit their captives to the horrors of the Inquisition

---

31. From Charles Fitzgeffrey's contemporary poem, cited in Sugden, *Sir Francis Drake*, p. 173.
32. Sugden, *Sir Francis Drake*, p. 315, citing Jameson, "Some New Spanish Documents Dealing with Drake."

before condemning them to slavery or some horrible form of execution. Such was the lot of many of Drake's personal friends or crew including those he lost at Nombre de Dios in 1567, at Panama in 1576, and at Chile in 1577. What must have been most painful of all to Drake was the uncertain fate of his young cousin John, whom the Spanish captured in 1582.

Drake's record is remarkable not only in its stark contrast to the norm, but particularly in consideration of the horrible treatment so many of his companions received at the hands of his enemy. Of the thousands of prisoners Drake took, the records show only one instance in which he intentionally took human life, and that was under extenuating circumstances.[33] Though there were a few instances where Drake tortured his prisoners to obtain information, the means, for the time, were very mild. For each case one can find to accuse Drake of cruelty, there are many more examples of his compassion to go along with his standard practice of treating his prisoners courteously and releasing them without harm as soon as the opportunity arose. Here are a few examples. Despite their protests, he refused to let his Cimarron or French allies harm any of his Spanish captives during his 1572 Panama raids. At Mucho Island in 1579, he declined to punish the Indians who had treacherously attacked his shore party, almost killing Drake and his companions. He reasoned that these Indians had suffered enough at the hands of the Spanish. In 1590, he, John Hawkins and Lord Howard founded the first English naval charity.

Resurrection of the Virginia Colony

What became of Raleigh's Virginia colony? Only weeks after Drake picked up Lane's group in 1586, Grenville arrived with Raleigh's relief forces. Finding the colony deserted, he left fifteen men to maintain possession and returned to England. In 1587, Raleigh leased the colony's rights to a group of investors. They sent out an expedition under John White that reached Roanoke Island in July. Despite finding that Grenville's men had all been killed by Indians, they left behind almost two hundred determined settlers, including White's daughter and her husband.

By the time White had fitted out the next relief force, the war of the Armada had started and his ships and men were impressed. Raleigh complicated matters in 1589 by selling off all the rights of his patent to yet another group of Plymouth investors that included Richard Hakluyt and Sir Thomas Smyth. By the time the anxious White finally made his way back to Roanoke Island, it was 1590. By signs left behind, he discerned that the settlers had moved to nearby Croatan Island where the Indians were less hostile. En route to this island, storms battered the small fleet and the ships were nearly wrecked. White could not oblige the commander to resume the search for the colonists.

Incredibly, it was over ten years later that Raleigh, in 1602, finally sent an expedition under his nephew Bartholomew Gilbert to determine the fate of the colony. The expedition was unsuccessful in locating the settlers and Gilbert was killed by the Indians. During the long interlude, English adventurers and their financiers had been too busy with the war with Spain to pay any attention to colonization efforts. It was only when the Treaty of London eliminated the opportunities for lucrative privateering raids that the colonial enterprise again came to the forefront.[34]

The new promoters of colonization, Bartholomew Gosnold and Richard Hakluyt, appear to have initiated the program to resurrect Raleigh's Virginia colony as soon as the treaty was signed. By mid-1605 they had recruited Captain John Smith and others. King James I signed the articles of the Virginia Company in November 1606.

---

33. Ibid., pp. 189–90. Sugden discusses this incident in some detail. It occurred in 1585 during Drake's raid on Santo Domingo. Drake hanged two Dominican Friars when the Spanish murdered a messenger boy he had dispatched under a flag of truce.

34. There has been much written about the so-called Lost Colony. In a recent book, *A Witness for Eleanor Dare*, Robert White contends that the colonists survived long enough to construct a brick church, still standing, in the Tidewater section of Virginia.

Chronology of the Elizabethan War: Birth of British Colonial America

1490s and early 1500s: Spain and Portugal dazzle Europe with spectacular New World discoveries and conquests in Africa, the West Indies (Caribbean), America, and the Orient. John and Sebastian Cabot discover and claim Newfoundland for Henry VII of England.

1520s on: The Iberian crowns and their merchants net huge profits as Spanish and Portuguese plate fleets return home with New World treasure. England and other European nations are completely excluded from this trade.

1530s: Religious reformation under Henry VIII leads to civil unrest. The profits the Crown reaps from the sale of church properties allow Henry to put together England's first impressive standing navy and pursue his imperial designs on France and Scotland.

Early 1540s: Henry's policies create a temporary breach in the alliance with Spain. War breaks out with France and Scotland.

Late 1540s to early 1550s: William Hawkins and other English merchants challenge Portuguese trade monopolies on the coasts of Africa and Brazil with the support of the Royal Navy (the Gold Coast expeditions). Under Edward I, the ongoing wars with Scotland and France severely depress the English economy leading to Counter-Reformation uprisings.

Mid-1550s: Edward's successor, Mary I, weds Philip of Spain and repeals Henry's Protestant reforms; the papist crown puts hundreds of Protestant leaders to the stake.

Late 1550s to early 1560s: Philip II and Elizabeth I ascend the thrones of Spain and England, respectively. Elizabeth reinstates Protestant reforms and refuses to respect Iberian claims to unoccupied lands in Africa and America. She begins to rebuild the royal fleet from the proceeds of increasingly aggressive Gold Coast expeditions.

Mid-1560s: John Hawkins attempts to break into Spain's West Indies trade monopoly by force-trading African slaves at West Indies ports. Francis Drake participates in the third voyage as low-ranking officer.

Late 1560s: Religious movements in Scotland, France, and the Netherlands plunge Western Europe into civil wars. Philip responds by sending troops into the Spanish Netherlands.

1568: The opening salvos of the Elizabethan War – under orders of the Mexican viceroy, a Spanish squadron treacherously attacks the fourth Hawkins slaving expedition after Hawkins' fleet was forced by weather to put into a Mexican port. Drake and Hawkins barely escape, leaving behind many men and one of the queen's ships.

1568: Elizabeth confiscates a Spanish ship carrying payroll for Philip's Netherlands troops; the Spanish retaliate by seizing all English shipping in their Netherlands' ports.

1569–71: Undeclared war – both sides seize enemy shipping in their respective ports. The Spanish ambassador is implicated in an attempted coup. English privateers join French and Dutch corsairs in attacking papist shipping in the English Channel. Elizabeth is excommunicated. John Hawkins is placed in charge of rebuilding and modernizing the Royal Navy. Drake makes two privateering voyages to the West Indies.

1572–73: Loot from plundered Spanish merchant ships pours into English ports. The Spanish ambassador is implicated in another attempt to oust Elizabeth, this time supported by a seaborne invasion from the Spanish Netherlands. Drake again attacks the West Indies, blockades Cartagena, captures a treasure train in Panama, and plunders Spanish shipping.

1574–76: A nervous truce – ports are reopened; Elizabeth suspends privateering raids; and Philip eases up in the Netherlands. English merchants promote westward expansion. Martin Frobisher makes two futile voyages to Canada in search of the Northwest Passage. Richard Grenville plans an aborted voyage south to open trade in the South Pacific and East Indies.

1576–77: Renewed hostilities – Philip sends a new army to the Netherlands and the Spanish are implicated in another attempt to depose Elizabeth; John Oxenham attacks Panama. Elizabeth embarks Humphrey Gilbert and Drake on New World trading voyages partly intended to annoy the king of Spain.

1577–80: The circumnavigation – the key event of the Elizabethan Age and birth of the British colonial era. Drake attacks and plunders Spanish and Portuguese ports and shipping on the coasts of Portuguese Africa and the Spanish Pacific; he plants English flags in Tierra del Fuego and California, claiming North and South American territories for the English Crown; and he strikes a trade alliance in the Moluccas, laying the foundation for the British East India Company. Humphrey Gilbert's expedition falls apart and Martin Frobisher makes another failed attempt to find the Northwest Passage.

1579–82: Rome and Spain embark on the *empresa*, a coordinated attempt to depose Elizabeth and create Catholic uprisings in Britain and Ireland. Portugal falls to Spain. Elizabeth assigns Drake command of an expedition to attack and capture the Portuguese Azores, but changes her mind just before the fleet sets sail.

1582–83: Gilbert commands a failed expedition to plant an English colony in Newfoundland and perishes on the return. Drake sends a failed trading/colonial expedition under Edward Fenton to the South Seas to honor the terms of his spice trade agreement. Richard Hakluyt authors his "Discourse on Western Planting" on behalf of Sir Walter Raleigh.

1584–85: The first English colony – Sir Walter Raleigh sends several expeditions to North America and plants a small English outpost on Roanoke Island, North Carolina; he names it Virginia. Drake and Hawkins outfit another expedition to the Moluccas. Spanish troops take the offensive in the Netherlands and capture Antwerp. Elizabeth sends troops to Holland. Philip seizes English shipping and puts in motion a plan to invade England with a huge armada.

1585–86: Drake diverts his Moluccas enterprise and commands the English royal fleet in attacking the West Indies: he captures and ransoms Santo Domingo and Cartagena and levels Saint Augustine. He attempts to reinforce Raleigh's Virginia colony, but is compelled to return the disheartened colonists to England.

1587–88: Raleigh leases his colonial rights and the new investors dispatch additional colonists to Virginia under John White. Drake commands the royal fleet in attacking the coast of Spain and Portugal: he sacks Cadiz, destroying scores of ships destined for the Armada; he takes Cape Saint Vincent and blockades Lisbon, plundering hundreds of ships attempting to provision the Armada; and he captures a huge Portuguese East India galleon fully laden with oriental trade goods. Drake's attack on Spain effects a telling one-year delay in the Armada's departure.

1588: Climax of the Elizabethan War – defeat of the invincible Armada. Drake, though second in command of the English fleet, is again England's hero: he devises the key strategies and plays the key role in the epic sea battles.

1589: The Portugal enterprise – Drake commands largely unsuccessful attacks on the coasts of Spain and Portugal.

1590–94: Frobisher and Hawkins make failed attempts to blockade the Iberian coast and capture a returning plate fleet. Grenville loses Drake's former flagship the *Revenge* at the battle of Flores. Philip sends troops into France and Brittany, briefly taking Brest. English troops recapture Brest. White takes an expedition to Virginia, but is unsuccessful in locating the lost colony.

1595–96: Drake and Hawkins jointly command a failed attack on the Spanish West Indies. Both commanders succumb to tropical fevers.

1596–97: Lord Admiral Charles Howard and Sir Walter Raleigh sack Cadiz. Philip is thwarted by weather in two attempts to launch seaborne attacks on England.

1598–1603: Philip and Elizabeth die. English privateers continue to plunder Spanish shipping.

1604: The Treaty of London ends the war and lucrative English privateering.

1604–5: Bartholomew Gosnold and Richard Hakluyt initiate a venture to resurrect Raleigh's Virginia colony.

1606: The key outcome of the Elizabethan War – King James I grants two colonial charters – one for the Virginia Company and the other for the New England Company.

Fig. 172. A Spanish galleon of the Elizabethan War: detail of an early seventeenth-century engraving by C. J. Visscher. *Courtesy of the National Maritime Museum, Greenwich, England.*

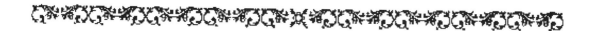

Epilogue

The Mystery Resolved

On June 17, 1579, Captain-General Francis Drake anchored the *Golden Hind* off what is now called Doran Beach at the northern end of Bodega Bay in Sonoma County, California. Four days later, with the tide cresting at approximately eight o'clock in the evening, Drake brought the ship through a winding channel on the lee side of Bodega Head and grounded her on the shore of the snug little harbor we now call Campbell Cove. There the Elizabethans constructed a stone-walled fort at the foot of a steep bluff and spent the next month repairing the ship, interacting with the native Indians and exploring the hinterland in the area of today's Salmon Creek Valley.

Every shred of credible evidence in the large body of contemporary accounts and maps relating the events of Drake's circumnavigation and California sojourn supports these conclusions.

Why is it then that a mystery and controversy surrounding the location of Drake's California anchorage has persisted for more than four centuries? To some degree, the fascinating series of events that took place at the encampment site as Spaniards, Russians, Mexicans and Americans explored and settled the San Francisco area help explain how the ruins of the Elizabethan rock fort escaped discovery. In truth, however, I believe it was pioneer scientist George Davidson—the West Coast's first and most renowned expert in the fields of cartography, navigation and maritime history—who was largely responsible for the Drake conundrum. If Davidson had conducted a more careful and objective analysis of the latitude readings in the contemporary accounts, California's Drake anchorage controversy probably would have ended before the turn of the nineteenth century.

Instead, one anchorage researcher after another came to overlook/misconstrue the key piece of evidence in the contemporary accounts: the precise latitude of the California anchorage reported in what is by far the most accurate and detailed contemporary account of the voyage, *The World Encompassed* by Sir Francis Drake, published by one of Drake's heirs in 1628.

To appreciate the strength of this evidence, check the latitude readings Drake provided in *The World Encompassed* against a modern world atlas. Except where the positions are approximated, and in several cases where discrepancies are attributable to reporting errors, not faulty measurement, you can trust Drake's navigational data to identify the forty-odd bays and islands he visited in making his way around the world, including: the farthest point south (Cape Horn), the farthest point north (the coast of Washington), and the "Island of Thieves" (one of the Palaus).

There is additional documentary and circumstantial evidence that foils the proponents of other candidate anchorage sites. Those who theorize that Drake anchored at Drakes Bay, for instance, put Drake in uncharted waters, desperately in need of water and provisions, and then have him sail his leaking ship right past a perfectly acceptable harbor located at the very latitude *The World Encompassed* reports he came to anchor. They have the expedition's chronicler, Francis Fletcher, describing Inverness Ridge—the tree-capped coastal mountain range forming the eastern shore of Drakes Bay—as "low plain land." They have Drake and his men dining exclusively on clams, sea lions and the like, while the area of their proposed anchorage abounds in deer, grizzly bears and other game. They have Drake sailing two days to reach an island he could have reached easily in about half a day. They have a Spanish explorer visiting the site of Drake's encampment sixteen years after the fact without seeing so much as a trace of the Elizabethans. In addition to most of these flaws, the Bolinas Lagoon and Agate Beach advocates have Drake, in one case, carelessly taking the *Golden Hind* into a barred channel where she surely would have grounded; and in the other, onto an exposed, rocky patch of shore where she likely would have wrecked. The San Francisco Bay advocates have

Drake entering one of the largest and most stupendous natural harbors in the world and describing it simply as a "fair and good bay" and "convenient and fit" harbor.

On top of the documentary evidence, there is the cartographic evidence. We have the Portus Nova Albionis, a drawing of a peculiar seal-head shaped cove the famed sixteenth-century cartographer Jodocus Hondius copied from a sketch or painting either Drake or his nephew created when they first came to anchor in California. When you reinterpret the drawing to account for a simple, understandable error Hondius made in redrawing it from the original, Bodega Head and the seal-head shape of Campbell Cove emerge clearly into view.

The Portus shows the stone-walled fort that Drake and his men built at the foot of a steep hill. Today one can go to the corresponding location at Campbell Cove and observe piles of half-buried stones that Pacific Gas & Electric Company (PG&E) construction crews appear to have left behind in 1963 after their bulldozers had torn apart the landslide debris at the foot of the rear bluff. A photograph taken by the late Karl Kortum in 1962 provides evidence of a substantial buried wall at the foot of the bluffs located and oriented in a position that correlates closely with the Portus fort.

We also have two sixteenth-century world maps that appear to be reasonably accurate and authentic facsimiles of a large map of the world Drake presented Queen Elizabeth I when he returned from the circumnavigation. Both maps depict Drake's California landfall in an inlet that correlates with Bodega Bay in terms of its relationship with the Point Reyes Peninsula, the most prominent feature on the Northern California Coast.

The Missing Pieces

In attempting to solve this historical jigsaw puzzle, I have assembled all the pieces I could find. Though I see a clear picture through the gaps, others perhaps will not. The single biggest piece of the puzzle still missing is some solid physical evidence that unequivocally places the Elizabethans at the scene in 1579. Despite what happened at Campbell Cove in 1963, I am confident that qualified archaeologists will still be able to produce this evidence. For example:

The stockpiles – On the south side of the Campbell Cove parking lot, PG&E and parks department construction crews left behind some large stockpiles of soil. I suspect that these piles may be some of the landslide debris PG&E contractors excavated from the foot of the cliffs. If so, when the archaeologists come, I predict they will find a wealth of Elizabethan artifacts when they screen these piles.

The half-buried rock piles at the foot of the former bluff – From what can be seen on the shores of Campbell Cove today, it appears to me that in the summer of 1963 PG&E's contractors unearthed the stones of Drake's buried fort and, under the assumption that they were the relics of Russian structures, purposefully, or by chance, left at least some of them behind. Although the destruction of Campbell Cove took place more than thirty years ago, I presume that there are individuals who took part in construction activities that may be able to shed some light on the half-buried stones and the stockpiles as well. Furthermore, notwithstanding the efficient work of the bulldozers, some archaeological trenching in the area of the alleged fort site may yet reveal a thin, highly compacted strata representing the heavily littered earthen floor of an Elizabethan fort.

Indian village site – As far as I can tell, the Indian village noted in *The World Encompassed* has not been throughly investigated by archaeologists. In fact, it is possible that it has never been investigated. Given the continuous occupation of this native site for three centuries after Drake departed, it may yield a wealth of Elizabethan artifacts from relatively deep, well-protected strata.

There is also the possibility of some additional documentary evidence falling into place, including Spanish and Russian documents for the period they first explored and settled Bodega Bay. In particular, the missing accounts of the 1793 Juan Bautista Matute expedition and the 1809 and 1811 Ivan Kuskov expeditions are conspicuous by their absence—they may contain references to the Elizabethan fort.

Other Latent Controversies

The matter of Drake's lost California harbor is not the only controversy surrounding our famous Elizabethan hero. I see a reputation that has suffered heavily at the hands of nineteenth and twentieth-century moralists. In closely examining Sir Francis Drake's career in the context of his times, the objective observer will see not only England's all-time greatest patriot, but one of British Colonial America's true founding fathers—its founding grandfather, if you will.

In addition, I do not think historians have given Sir Francis his just due as an author. Elizabethan scholars, Drake biographers and anchorage sleuths alike have historically credited Drake's nephew, namesake and heir, Sir Francis Drake, the first baronet, as the effective author of *"The World Encompassed* by Sir Francis Drake"—a 108-page account of Drake's circumnavigation, published in 1628, that is brimming with navigational details and colorful eye-witness commentary. The evidence, however, suggests that the great admiral himself co-wrote the account with Francis Fletcher, the expedition's minister and official chronicler—making full use of the original journal and logbook they respectively kept during the voyage.

America's Birthplace?

"The State of Virginia on Thursday unveiled the nation's 'most important' archaeological discovery—the exact location and actual foundation of America's first fort." So read an Associated Press article for September 13, 1996, as news flashed across the country that archaeologists had unearthed the long-sought remains of the original fort that Captain John Smith and some one hundred colonists constructed in 1607 on the shores of the James River, Virginia. In making the official announcement, Virginia Governor George Allen proclaimed, "We have discovered America's birthplace—the original fort."

America's birthplace and first fort? . . . I trust by now you know where to look if you wish to view the true birthstones of this country.

Reflections

Notwithstanding the acclaim that would be attached to the successful identification of Drake's lost harbor, in a curious way, I hope that the 1963 destruction of Campbell Cove was complete and that the archaeologist's shovels come up empty once again. Despite their frustrations and disappointments in not being able to confirm their own Drake's anchorage theories with physical proof, perhaps fellow anchorage-sleuths Raymond Aker, Aubrey Neasham, and Robert Power felt somewhat the same way. For more than two centuries, inquisitive minds have sifted through the evidence and puzzled over the mystery of Drake's lost harbor. You can tell from the passion of their writings how much they all enjoyed the hunt. Should the archaeologists bring it to an end, they will be depriving the anchorage sleuths yet to come the excitement of pursuing the elusive Drake down California's thundering shores and across the pages of history.

Fig. 173. Thirty-foot-high steel statue of Sir Francis Drake. Crafted in 1990 by Marin sculptor, Dennis Patton, Sir Francis stands at the entrance to Remillard Park near the ferry landing in Larkspur, California. *Photograph by the author, February 1997.*

Appendix A

Excerpt from "The Famous Voyage" 1589 Edition

I have produced the following excerpt verbatim from the copy of the account reproduced by Hans P. Kraus in Sir Francis Drake, A Pictorial Biography *(1970). It covers the California sojourn.*

This Pilot brought us to the hauen of Guatulca, the towne whereof as he told us, had but 17. Spaniards in it. Assoone as we were entred this hauen wee landed, and went presently to the towne, and to the Towne house, where we found a Judge sitting in judgement, he being associate with three other officers, vpon three Negroes that had conspired the burning of the Towne: both which Judges, and prisoners we tooke, and brought them a shipboord, and caused the chiefe Judge to write his letter to the Towne, to command all the Townesmen to auoid, that we might safely water there. Which being done, and they departed, we ransaked the Towne, and in one house we found a pot of the quantitie of a bushell, full of royals of plate, which we brought to our shippe. [margin: Guatulca.]

And here one Thomas Moone one of our companie, tooke a Spanish Gentleman as he was flying out of the towne, and searching him, he found a chaine of golde about him; and other jewels, which he tooke, and so let him goe.

At this place our Generall among other Spaniards, set a shoare his Portingall Pilot, which he tooke at the Islands of Cape Verde, out of a shippe of S. Marie porte of Portingal, and hauing set them a shoare, we departed hence, and sailed to the Island of "Canon, where our Generall landed, and brought to shoare his owne ship, and discharged her, mended, and graued her, and furnished our shippe with water and wood sufficiently. [margin: The Portingall Pilot set on land.] [margin: The Island of Cockles.]

And whiles we were here, we espied a shippe, and set saile after her, and tooke her, and founde in her two Pilots, and a Spanish Gouernour, going for the Islands of the Philippinas: we searched the shippe, and took some of her marchandizes, and so let her goe. Our Generall at this place, and time, thinking himselfe both in respect of his priuate iniuries receiued from the Spaniards, as also of their contempts and indignities offered to our countrey and Prince in generall, sufficiently satisfied, and revenged: and supposing that her Maiestie at his returne would rest contented with this seruice, purposed to continue no longer vpon the Spanish coasts, but began to consider and to consult of the best way for his Countrey. [margin: A ship with a gouernour for the Islands of Philippinas.]

He thought it not good to returne by the Streights, for two speciall causes: the one, least the Spaniards should there waite, and attend for him in great number and strength, whose hands he being left but one shippe, could not possibly escape. The other cause was the dangerous situation of the mouth of the Streights in the south side, where continuall stormes raining and blustering, as he found by experience, besides the shoales, and sands vpon the coast, he thought it not a good course to aduenture that way; he resolued therefore to auoide these hazards, to goe forward to the Islands of the Moluccaes, and therehence to saile the course of the Portingals by the Cape of Bona Speranza. [margin: A purpose in Sir Francis to returne by the Northwest passage.]

Upon this resolution, he began to thinke of his best way to the Moluccaes, and finding himselfe where he nowe was becalmed, he sawe, that of necessitie he must be forced to take a Spanish course, namely to sail somewhat Northerly to get a wind. We therefore set saile, and sailed in longitude 600. leagues at the least for a good winde, and thus much we sailed from the 16. of Aprill, till the 3. of June.

The 5. day of June, being in 42. degrees towards the pole Arctike, we found the aire so colde, that our men being greeuiously pinched with the same, complained of the extremitie thereof, and the further we went, the more the cold increased vpon vs. Whereupon we thought it best for that time to seeke the land, and did so, finding it not mountanous, but lowe plaine land, & clad, and couered ouer with snowe, so that we drewe backe againe without landing, till we came within 38. degrees towards the line. In which heighth it pleased God to send vs into a faire and good Baye, with a good winde to enter the same.

In this Baye we ankered, and the people of the Countrey, hauing their houses close by the waters side, shewed themseues vnto vs, and sent a present to our Generall.

When they came vnto vs, they greatly wondred at the things that we brought, but our Generall (according to his naturall and accustomed humanitie) curteously intreated them, and liberally bestowed on them necessarie thing to couer their nakedness, whereupon they supposed us to be gods, and would not be perswaded to the contrarie; the presents which they sent to our Generall were feathers, and cals of networke.

Their houses are digged round about with earth, and haue from the uttermost brimmes of the circle, clifts of wood set vpon them, ioyning close together at the toppe like a spire steeple, which by reason of that closenes are very warme.

Their beds is the ground with rushes strowed on it, and lying about the house, haue the fire in the middest. The men goe naked, the women take bulrushes, and kembe them after the manner of hempe, and thereof make their loose garments, which being knit about their middles, hang downe about their hippes, hauing also about their shoulders, a skinne of Deere, with the haire upon it. These women are very obedient and seruiceable to their husbands.

After they were departed from vs, they came and visited vs the second time, and brought with them feathers and bags of Tabacco for presents: And when they came to the top of the hill (at the bottome whereof we had pitched our tents) they staied themselues: where one appointed for speaker, wearied himselfe with making a long oration, which done, they left their bowes vpon the hill, and came downe with their presents.

In the meane time, the women remaining on the hill, tormented themselues lamentably, tearing their flesh from their cheekes, whereby we perceiued that they were about a sacrafice. In the meane time, our Generall, with his companie, went to praier, and to reading of the Scriptures, at which exercise they were attentiue, & seemed greatly to be affected with it: but when they were come vnto vs, they restored againe vnto vs those things which before we bestowed vpon them.

The newes of our being there, being spread through the Countrey, the people that inhabited round about came downe, and amongst them the King himselfe, a man of goodly stature, & comely personage, with many other tall, and warlike men: before whose comming were sent two Ambassadors to our Generall, to signifie that their King was coming, in doing of which message, their speech was continued about halfe an howre. This ended, they by signes requested our Generall to send some thing by their hand to their King, who marched to vs with a princely maiestie, the

June.
Sir Francis Drake sailed on the backe side of America to 42 deg. of Northerly latitude.

38. Degrees.

A description of the people and Countrey of Noua Albion.

A long oration

people crying continually after their manner, and as they drewe neere vnto vs, so did they striue to behaue themselues in their actions with comelines.

In the fore front was a man of goodly personage, who bare the scepter, or mace before the King, whereupon hanged two crownes, a lesse and a bigger, with three chaines of a maruelous length: the crownes were made of knit worke wrought artificially with fethers of diuers colours: the chaines were made of a bonie substance, and few be the persons among them that are admitted to weare them: and of that number also the persons are stinted, as some ten, some 12. etc. Next vnto him which bare the scepter, was the King himselfe, with his Garde about his person, clad with Conie skins, & other skins: after them followed the naked comon sort of people, euery one hauing his face painted, some with white, some with blacke, and other colours, & hauing in their handes one thing or another for a present, not so much as their childre, but they also brought their presents.

In the meane time, our Generall gathered his men together, and marched within his fenced place, making against their approching, a very warlike shrewe. They being trooped together in their order, and a general salutation being made, there was presently a generall silence. Then he that bare the scepter before the King, being informed by another, whome they assigned to that office, with a manly and loftie voice, proclaimed that which the other spake to him in secret, continuing halfe an howre: which ended, and a generall Amen as it were giuen, the King with the whole number of men, and women, (the children excepted) came downe without any weapon, who descending to the foote of the hill, set themselues in order.

In comming towards our bulwarks and tents, the scepter bearer began a song, obseruing his measures in a daunce, and that with a stately countenance, whom the King with his Garde, and euery degree of persons following, did in like manner sing and daunce, sauing only the women which danced, & kept silence. The Generall permitted them to enter within our bulwarks, where they continued their song and daunce a reasonable time. When they had satisfied themselues, they made signes to our General to sit downe, to whom the King, and divers others made seueral orations, or rather supplications, that he would take their prouince & kingdome into his hand, and become their King, making signes that they would resigne vnto him their right and title of the whole land, and become his subjects. In which to perswade vs the better, the King and the rest, with one consent, and with great reuerence, joyfully singing a song, did set the crowne vpon his head, inriched his necke with all their chaines, and offred vnto him many other things, honouring him by the name of Hioh, adding therunto as it seemed, a signe of triumph: which thing our Generall thought not meete to reiect, because he knewe not what honour and profite it might be to our Countrey. Wherefore in the name, and to the vse of her Maiestie, he tooke the scepter, crowne, and dignitie of the said Countrey into his hands, wishing that the riches & treasure thereof might so conueniently be transported to the inriching of her kingdome at home, as it aboundeth in ye same.

The king resignes his crowne and kingdome to Sir Francis Drake.

The common sorte of people leauing the King, and his Garde with our Generall, scattered themselues together with their sacrifices among our people, taking a diligent viewe of euery person: and such as pleased their fancie, (which were the yongest) they inclosing them about offered their sacrifices vnto them with lamentable weeping, scratching, and tearing the flesh from their faces with their nailes, whereof issued abundance of bloode. But wee vsed signes to them of disliking this, and staied their hands from force, and directed them vpwards to the liuing God, whome onely they ought to worshippe. They shewed vnto vs their wounds, and craued helpe of them at our hands, whereupon wee gaue them lotions, plaisters, and ointments agreeing to the state of their griefes, beseeching God to cure their diseases. Euery thirde day they brought their sacrifices vnto vs, vntill they vnderstoode our meaning, that we had no pleasure in them: yet they could not be long absent from us, but daily frequented our companie to the houre of our departure, which departure seemed so greeuous vnto them, that their ioy was turned to sorrow. They intreated vs, that being absent we would remember them, and by stelth prouided a sacrifice, which we misliked.

Our necessarie busines being ended, our Generall with his companie trauiled vp into the Countrey to their villages, where we found heardes of Deere by 1000. in a companie, being most large, and fat of bodie. *Great hearbs of Deers.*

We found the whole Countrey to be a warren of a strange kinde of Connies, their bodies in bignes as be the Barbarie Connies, their heads as the heads of ours, the feete of a Want, and the taile of a Rat being of great length: vnder her chinne on either side a bagge, into the which she gathereth her meate, when she hath filled her bellie abroad. The people eate their bodies, and make great accompt of their skinnes, for their Kings coate was made of them. *Abundance of strange conies*

Our Generall called this Countrey, Noua Albion, and that for two causes: the one in respect of the white bankes and cliffes, which lie towards the sea: and the other, because it might haue some affinitie with our Countrey in name, which sometime was so called. *Nona Albion.*

There is no part of earth here to be taken vp, wherein there is not a reasonable quantitie of gold or siluer. *Gold and silver in the earth of Nona Albion.*

At our departure hence our General set vp a monument of our being there, as also of her Maiesties right and title to the same, namely a plate, nailed vpon a faire great poste, whereupon was ingrauen her Maiesties name, the day and yeare of our arriuall there, with the free giuing vp of the prouince and people into her Maiesties hands, together with her highnes picture and armes, in a peece of six pence of currant English money vnder the plate, where vnder was also written the name of our Generall.

It seemeth that the Spaniards hitherto had neuer bene in this part of the Countrey, neither did euer discouer the lande by many degrees, to the Southwards of this place.

Appendix B

Excerpt from "The World Encompassed"

I have produced the following excerpt verbatim from the copy of Sir Francis Drake's account reproduced by Readex Microprint Corporation (1966). It covers the California sojourn.

From Guatulco we departed the day following, viz. Aprill 16. setting our course directly into the sea: whereon we sayled 500. leagues in longitude, to get a winde: and betweene that and Iune 3. 1400 leagues in all, till we came into 42. deg. of North latitude, where in the night following we found such alteration of heate, into extreame and nipping cold, that our men in generall did grieuously complaine thereof; some of them feeling their healths much impaired thereby, neither was it, that this chanced in the night alone, but the day following carried with it, not onely the markes, but the stings and force of the night going before; to the great admiration of vs all, for besides that the pinching and biting aire, was nothing altered; the very roapes of our ship were stiffe, and the raine which fell, was an vnnatural congealed and frozen substance, so that we seemed rather to be in the frozen Zone, than any way so neere vnto the sun, or these hotter climates.

Neither did this happen for the time onely, or by some sudden accident, but rather seemes indeed, to proceed from some ordinary cause, against the which the heate of the sun preuailes not, for it came to that extremity, in sayling but 2. deg. farther to the Northward in our course: that though sea-men lack not good stomaches, yet it seemed a question to many amongst vs, whether their hands should feed the mouths. or rather keep themselves within their couerts, from the pinching cold that did benumme them. Neither could we impute it to the tendernesse of our bodies, though we came lately from the extremitie of heate, by reason whereof we might be more sensible of the present cold: insomuch as the dead and senceless creatures, were as well effected with it as ourselues, our meate as soone as it was remooued from the fire, would presently in a manner be frozen vp; and our ropes and tackling, in a few days were growne to that stiffnesse, that what 3. men afore were able to with them to performe, now 6. men, with their best strength, and vttermost endeauour, were hardly able to accomplish: whereby a sudden and great discouragement seased vpon the mindes of our men, and they were possessed with a great mislike, and doubting of any good to be done that way, yet would not our general be discouraged, but as wel by comfortable speeches, of the diuine prouidence, and of God's louing care ouer his children, out of the Scriptures; as also by other good and profitable perswasions, adding thereto his own cheerfull example, he so stirred them vp, to put on a good courage, and to quite themselues like men, to indure some short extremity, to haue the speedier comfort, and a little trouble, to obtaine the greater glory; that euery man was throughly armed with willingnesse, and resolued to see the uttermost, if it were possible, of what good was to be done that way.

The land in that part of America, bearing farther out into the West, than we before imagined, we were neerer on it then wee were aware; and yet the neerer still wee came vnto it, the more extremitie of cold did sease vpon vs. The 5. day of Iune, wee were forced by contrary windes, to runne in with the shoare, which we then first descried; and to cast anchor in a bad

1578.

Apr. 16.

June 5.

bay, the best roade we could for the present meete with: where wee were not without some danger, by reason of the many extreme gusts, and flawes that beate vpon vs; which if they ceased and were still at any time, immediately upon their intermission, there followed most uile, thicke, and stinking fogges; against which the sea preuailed nothing, till the gusts of wind againe remoued them, which brought with them, such extremity and violence when they came, that there was no dealing or resisting against them.

In this place was no abiding for vs; and to go further North, the extremity of the cold (which had now vtterly discouraged our men) would not permit vs; and the winds directly bent against vs, hauing once gotten vs under sayle againe, commanded vs to the Southward whether we would or no.

From the height of 48. deg., in which now we were, to 38. we found the land, by coasting alongst it, to bee but low and reasonable plaine: euery hill (whereof we saw many, but none verie high), though it were in Iune, and the Sunne in his neerest approch vnto them being couered with snow.

In 38. deg. 30. min. we fell with a conuenient and fit harborough, and Iune 17. came to anchor therein: where we continued till the 23. day of Iuly following, During all which time, notwithstanding it was in the height of Summer, and so neere the Sunne; yet were wee continually visited with like nipping colds, as we had felt before: insomuch that if violent exercises of our bodies, and busie employment about our necessarie labours, had not sometimes compeld us to the contrary, we could very well haue been contented to haue kept about us still our Winter clothes; yea (had our necessities suffered vs) to haue kept our beds; neither could we at any time in whole fourteene dayes together, find the aire so cleare as to be able to take the height of Sunne or starre.

And here hauing so fit occasion (notwithstanding it may seeme to be besides the purpose of writing the history of this our voyage), we will a little more diligently inquire into the causes of the continuance of the extreame cold in these parts: as also into the probabilities or vnlikelihoods of a passage to be found that way. Neither was it (as hath formerly been touched) the tenderness of our bodies, comming so lately out of the heate, whereby the poores were opened, that made vs so sensible of the colds we here felt: in this respect, as in many others, we found our God a prouident father, and carefull Physitian for vs. We lacked no outward helpes nor inward comforts, to restore and fortifie nature, had it beene decayed or weakened in vs; neither was there wanting to vs the great experience of our Generall, who had often himselfe proued the force of the burning Zone; whose aduice always preuailed much to the preseruing of a moderate temper in our constitutions: so that euen after our departure from the heate wee always found our bodies, not as sponges, but strong and hardned, more able to beare out cold, though we came out of excesse of heate, then a number of chamber champions could haue beene, who lye on their feather beds till they go to sea, or rather, whose teeth in a temperate aire do beate in their heads, at a cup of cold Sack and sugar by the fire.

And that it was not our tendernes, but the very extremitie of the cold itselfe, that caused this sensiblenes in vs, may the rather appeare in that the naturall inhabitants of the place (with whom we had for a long season familiar intercourse, as is to be related) who had neuer beene acquainted with such heate; to whome the countrey, ayre, and climate was proper; and in whome custome of colde was as it were a second nature: yet vsed to come shiuering to vs in their warme furres; crowding close together body to body, to receiue heate one of another; and sheltring themselues vnder a lee bancke, if it were possible; and as often as they could, labouring to shroude themselues vnder our garments also, to keepe them warme, Besides how vnhandsome and deformed appeared the face of the earth itselfe! shewing trees without leaues, and the ground without greennes in those moneths of Iune and Iuly. The poore birds and foules not daring (as we had great experience to obserue it), not daring so much as once to arise from their nests, after the first egge layed, till it with all the rest be hatched, and brought to some strength of nature, able to helpe itselfe. Onely this recompence hath nature affoorded them, that the heate of their owne bodies being exceeding great, it perfecteth the creature with greater expedition, and in shorter time then is to be found in many places.

As for the causes of this extremeity, they seeme not to be so deeply hidden, but that they may at least in part be guessed at: The chiefest of which we conceiue to be the large spreading of the Asian and American continent, which (somewhat Northward of these parts) if they be not fully ioyned, yet seeme they to come very neere one to the other. From whose high and snow-couered mountaines, the North and North west winds (the constant visitants of those coasts) send abroad their frozen nimphes, to the infecting the whole aire with this insufferable sharpnesse: not permitting the Sunne, no not in the pride of his heate, to dissolue that congealed matter and snow, which they haue breathed out so nigh the Sunne, and so many degrees distant from themselves. And that the North and North-west winds are here contant in Iune and Iuly, as the North wind alone is in August and September; we not onely found it by our owne experience, but were fully confirmed in the opinion thereof, by the continued obseruations of the Spaniards. Hence comes the generall sqaulidnesse and barrennesse of the countrie; hence comes it, that in the middest of their Summer, the snow hardly departeth euen from their very doores, but is neuer taken away from their hils at all; hence come those thicke mists and most stinking fogges, which increase so much the more, by how much higher the pole is raised: wherin a blind pilot is as good as the best director of a course. For the Sunne striuing to performe his naturall office, in eleuating the vapors out of these inferior bodies; draweth necessarily abundance of moisture out of the sea: but the nipping cold (from the former causes) meeting and opposing the Sunnes indeuour, forces him to giue ouer his worke imperfect: and instead of higher eleuation, to leaue in the lowest region, wandring vpon the face of the earth and waters, as it were a second sea: through which its owne beames cannot possibly pierce, vnlesse sometimes when the sudden violence of the winds doth helpe to scatter and breake through it, which thing happeneth very seldome, and when it happeneth is of no continuance. Some of our mariners in this voyage had formerly beene at Wardhouse, in 72. deg. of North latitude: who yet affirmed, that they felt no such nipping cold there in the end of the Summer, when they departed thence, as they did here in those hottest moneths of Iune and Iuly.

And also from these reasons we coniecture; that either there is no passage at all through these Northerne coasts (which is most likely) or if there be, that it is vnnauigable. Adde hereunto, that though we searched the coast diligently, even vnto the 48. deg. yet found we not the land, to trend so much as one point in any place towards the East, but rather running on continually Northwest, as if it went directly to meet with Asia;

and euen in that height when we had a franke wind, to haue carried vs through, had there beene a passage, yet we had a smooth and calme sea, with ordinary flowing and reflowing, which could not haue been, had there been a frete: of which we rather infallibly concluded then coniectured, that there was none. But to returne.

The next day after our comming to anchor in the aforesaid harbour, the people of the countrey shewed themselues; sending off a man with great expedition to vs in a canow. Who being yet but a little from the shoare, and a great way from our ship, spake to vs continually as he came rowing on. And at last at a reasonable distance staying himselfe, he began more solemnely a long and tedious oration, after his manner: vsing in the deliuerie therof many gestures and signes, mouing his hands, turning his head and body many wayes; and after his oration ended, with great shew of reuerence and submission, returned backe to shoare again. He shortly came againe the second time in like manner, and so the third time: When he brought with him (as a present from the rest) a bunch of feathers, much like the feathers of a blacke crow, very neatly and artificially gathered vpon a string, and drawne together into a round bundle; being verie cleane and finely cut, and bearing in length an equall proportion one with another; a speciall cognizance (as wee afterwards obserued) which they that guard their kings person, weare on their heads. With this also he brought a little basket made of rushes, and filled with an herbe which they called *Tabáh*. Both which being tyed to a short rodde, he cast into our boate. Our Generall intended to haue recompenced him immediately with many good things, he would haue bestowed on him: but entring into the boate to deliuer the same, he could not be drawne to receiue them by any meanes: saue one hat, which being cast into the water out of the ship, he tooke vp (refusing vtterly to meddle with any other thing, though it were vpon a board put off vnto him) and so presently made his returne. After which time, our boate could row no way, but wondring at vs as at gods, they would follow the same with admiration.

The 3. day following, viz. the 21. our ship hauing receiued a leake at sea, was brought to anchor nearer the shoare, that, her goods being landed, she might be repaired: but for that we were to preuent any danger, that might chance against our safety, our Generall first of all landed his men, with all necessary prouision, to build tents and make a fort for the defence of our selues and goods: and that wee might vnder the shelter of it, with more safety (what euer should befall) end our businesse; which when the people of the countrey perceiued vs doing, as men set on fire to war, in defence of their countrie, in great hast and companies, with such weapons as they had, they came downe vnto vs; and yet with no hostile meaning, or intent to hurt vs: standing, when they drew neere, as men rauished in their mindes, with the sight of such things as they neuer had seene, or heard of before that time: their errand being rather with submission and feare to worship vs as Gods, then to haue any warre with vs as with mortall men. Which thing, as it did partly shew itselfe at that instant, so did it more and more manifest itself afterwards, during the whole time of our abode amongst them. At this time, being willed by signes to lay from them their bowes and arrowes, they did as they were directed, and so did all the rest, as they came more and more by companies vnto them, growing in a little while, to a great number, both of men and women.

To the intent, therefore, that this peace which they themselues so willingly sought, might without any cause of the breach thereof, on our part given, be continued; and that wee might with more safety and expedition, end our businesses in quiet; our Generall, with all his company, vsed all meanes possible, gently to intreate them, bestowing vpon each of them liberally, good and necessary thing to couer their nakedness; withall signifying vnto them, we were no Gods but men, and had need of such things to couer our owne shame; teaching them to vse them to the same ends: for which cause also wee did eate and drinke in their presence, giuing them to vnderstand, that without that wee could not liue, and therefore were but men as well as they.

Notwithstanding nothing could perswade them, nor remoue that opinion, which they had conceiued of vs, that wee should be Gods.

In recompencer of those things which they had receiued of vs, as shirts, linnen cloth, etc. they bestowed vpon our generall, and diuerse of our company, diuerse things, as feathers, cawles of networke, and quiuers of their arrrowes, made of fawne-skins, and the very skins of beasts that their women wore vpon their bodies. Hauing thus had their fill of this times visiting and beholding of vs, they departed with ioy to their houses, which houses are digged round within the earth, and haue from the vppermost brimmes of the circle, clefts of wood set vp, and ioyned close together at the top, like our spires on the steeple of a Church: which being couered with earth, suffer no water to enter, and are very warme, the doore in the most part of them, performes the office also of a chimney, to let out the smoake: it made in bignesse and fashion, like to an ordinary scuttle in a ship, and standing slopewise: their beds are the hard ground, onely with rushes strewed vpon it, and lying round about the house, haue their fire in the middest, which by reason that the house is but low vaulted, round and close, giueth a maruelous reflexion to their bodies to heate the same.

Their men for the most part goe naked; the women take a kinde of bulrushes, and kembling it after the manner of hemp, make themselues thereof a loose garment, which being knitte about their middles, hanges downe about their hippes, and so affordes to them a couering of that, which nature teaches should be hidden: about their shoulders, they weare also the skin of a deere, with the haire vpon it. They are very obedient to their husbands, and exceedingly ready in all seruices: yet of themselues offring to do nothing, without the consents, or being called of the men.

As soone as they were returned to their houses, they began amongst themselues a kind of most lamentable weeping & crying out; which they continued also a great while together, in such sort, that in the place where they left vs (being neere about 3. quarters of an English mile distant from them) we very plainely, with wonder and admiration did heare the same: the women especially, extending their voices, in a most miserable and dolefull manner of shreeking.

Notwithstanding this humble manner of presenting themselues, and awful demeanour vsed towards vs, we thought it no wisedome too farre to trust them (our experience of former Infidels dealing with vs before, made vs carefull to prouide against an alteration of their affections, or breach of peace if it should happen) and therefore with all expedition we set vp our tents, and intrenched ourselues with walls of stone: that so being fortified within ourselues, we might be able to keepe off the enemie (if they should so proue) from comming amongst vs without our good wills: this being quickly finished, we went the more cheerefully and securely afterward, about our other businesse.

Against the end of two daies (during which time they had not againe been with vs), there was gathered together a great assembly of men, women, and children (inuited by the report of them which first saw vs, who as it seems, had in that time, of purpose dispersed themselues into the country, to make knowne the newes), who came now the second time vnto vs, bringing with them as before had beene done, feathers and bagges of *Tabáh* for presents, or rather indeed for sacrifices, vpon this perswasion that we were gods.

When they came to the top of the hill, at the bottome whereof wee had built our fort, they made a stand; where one (appointed as their chiefe speaker) wearied both vs his hearers, and himselfe too, with a long and tedious oration: deliuered with strange and violent gestures, his voice being extended to the vttermost strength of nature, and his wordes falling so thicke one in the necke of another, that he could hardly fetch his breath againe: as soon as he had concluded, all the rest, with a reuerend bowing of their bodies (in a dreaming manner, and long producing of the same) cryed *Oh:* thereby giuing their consents, that all was very true which he had spoken, and that they had vttered their minde by his mouth vnto vs; which done, the men laying downe their bowes vpon the hill, and leauing their women and children behinde them, came downe with their presents; in such sort, as if they had appeared before a God indeed: thinking themselues happy, that they might haue accesse vnto our generall, but much more happy, when they sawe that he would receiue at their hands, those things which they so willingly had presented: and no doubt, they thought themselues neerest vnto God, when they sate or stood next to him: In the meanetime the women, as if they had been desperate, vsed vnnatural violence against themselues, crying and shrieking piteously, tearing their flesh with their nailes from their cheekes, in a monstrous manner, the blood streaming downe along their brests; besides despoiling the vpper parts of their bodies, of those single coeuerings they formerly had, and holding their hands aboue their heads, that they might not rescue their brests from harme, they would with furie cast themselues vpon the ground, neuer respecting whether it were cleane or soft, but dashed themselues in this manner on hard stones, knobby, hillocks, stocks of wood, and pricking bushes, or whateuer else lay in their way; itterating the same course againe and againe: yea women great with child, some nine or ten times each, and others holding out till 15. or 16. times (till their strengths failed them) exercised this cruelty against themselues: A thing more grieuous for vs to see, or suffer could we haue holpe it, then trouble to them (as it seemed) to do it.

This bloudie sacrifice (against our wils) being thus performed, our Generall, with his companie, in the presence of those strangers fell to prayers: and by signes in lifting vp our eyes and hands to heauen, signified vnto them, that that God whom we did serue, and whom they ought to worship, was aboue: Beseeching God if it were his good pleasure to open by some meanes their blinded eyes; that they might in due time be called to the knowledge of him the true and euerliuing God, and of Jesus Christ whom he hath sent, the saluation of the Gentiles. In the time of which prayers, singing of Psalmes, and reading of certaine Chapters in the Bible, they sate very attentiuely: and obseruing the end at euery pause, with one voice still cried, *Oh*, greatly reioycing in our exercises. Yea they tooke such pleasure in our singing of Psalmes, that whensoeuer they resorted to

1579.

vs, their first request was commonly this, *Gnaáh*, by which they intreated that we would sing.

Our Generall hauing now bestowed vpon them diuers things, at their departure they restored them all againe; none carrying with him anything of whatsoeuer hee had receiued, thinking themselues sufficiently enriched and happie, that they had found so free accesse to see vs.

Against the end of three daies more (the newes hauing the while spread itselfe farther, and as it seemed a great way vp into the countrie), were assembled the greatest number of people, which wee could reasonably imagine, to dwell within any conuenient distance round about. Amongst the rest, the king himselfe, a man of goodly stature and comely personage, attended with his guard, of about 100. tall and warlike men, this day, viz. June 26. came downe to see vs.

Before his comming, were sent two Embassadors or messengers to our Generall, to signifie that their *Hióh*, that is, their king, was comming and at hand. They in the deliuery of their messege, the one spake with a soft and low voice, prompting his fellow; the other pronounced the same word by word after him, with a voice more audible: continuing their proclamation (for such it was) about halfe an houre. Which being ended, they by signes made request to our Generall, to send something by their hands to the *Hióh* or king, as a token that his comming might be in peace. Our Generall willingly satisfied their desire; and they, glad men, made speedy returne to their *Hióh*. Neither was it long before their king (making as princely a shew as possibly he could) with all his traine came forward.

In their comming forwards they cryed continually after a singing manner with a lustie courage. And as they drew neerer and neerer towards vs, so did they more and more striue to behaue themselues with a certaine comelinesse and grauity in all their actions.

In the forefront came a man of a large body and goodly aspect, bearing the Septer or royall mace (made of a certaine kind of blacke wood, and in length about a yard and a halfe) before the king. Whereupon hanged two crownes, a bigger and a lesse, with three chaines of a maruellous length, and often doubled; besides a bagge of the herb *Tabáh*. The crownes were made of knitworke, wrought vpon most curiously with feathers of diuers colours, very artificially placed, and of a formall fashion: The chaines seemed of a bony substance: euery linke or part thereof being very little, thinne, most finely burnished, with a hole pierced through the middest. The number of linkes going to make one chaine, is in a manner infinite: but of such estimation it is amongst them, that few be the persons that are admitted to weare the same: and euen they to whom its lawfull to use them, yet are stinted what number they shall vse; as some ten, some twelue, some twentie, and as they exceed in number of chaines, so are they thereby knowne to be the more honorable personages.

Next vnto him that bare this Scepter, was the king himselfe with his guard about him: His attire vpon his head was a cawle of knitworke, wrought vpon somewhat like the crownes, but differing much both in fashion and prefectnesse of work; vpon his shoulders he had on a coate of the skins of conies, reaching to his wast: His guard also had each coats of the same shape, but of other skins: some hauing cawles likewise stucke with feathers, or couered ouer with a certaine downe, which groweth vp in the countrey vpon an herbe much like our lectuce; which exceeds any other downe in the world for finesse, and beeing layed vpon their cawles by no winds can be remoued: Of such estimation is this herbe amongst them, that the downe thereof is not lawfull to be worne, but of such persons as are about the king (to whom also it is permitted to weare a plume of feathers on their heads, in signe of honour), and the seeds are not vsed but onely in sacrifice to their gods. After these in their order, did follow the naked sort of common people; whose haire being long, was gathered into a bunch behind, in which stucke plumes of feathers; but in the forepart onely single feathers like hornes, euery one pleasing himselfe in his owne deuise.

This one thing was obserued to bee generall amongst them all; that euery one had his face painted, some with white, some blacke, and some with other colours, euery man also bringing in his hand one thing or other for a gift or present: Their traine or last part of their company consisted of women and children, each woman bearing anginst her breast a round basket or two, hauing within them diuers things, as bagges of *Tobâh*, a roote which they call *Petáh*, whereof they make a kind of meale, and either bake it into bread, or eate it raw; broyled fishes, like a pilchard; the seede and downe aforenamed, with such like:

Their baskets were made in fashion like a deep boale, and though the matter were rushes, or such other kind of stuffe, yet was it so cunningly handled, that the most part of them would hold water; about the brimmes they were hanged with peeces of the shels of pearles, and in some places with two or three linkes at a place, of the chaines forenamed: thereby signifying, that they were vessels wholly dedicated to the onely vse of the gods they worshipped: and besides this, they were wrought vpon with the matted downe of red feathers, distinguished into diuers workes and formes.

In the meane time, our Generall hauing assembled his men together (as forecasting the danger, and worst that might fall out) prepared himselfe to stand vpon sure ground, that wee might at all times be ready in our owne defence, if any thing should chance otherwise than was looked for or expected.

Wherefore euery man being in a warlike readinesse, he marched within his fenced place, making against their approach a most warlike shew (as he did also at all other times of their resort) whereby if they had been desperate enemies, they could not haue chosen but haue conceiued terrour and fear, with discouragement to attempt anything against vs, in beholding of the same.

When they were come somewhat neere vnto vs, trooping together, they gaue vs a common or generall salutation: obseruing in the meane time a generall silence. Whereupon he who bare the Scepter before the king, being prompted by another whom the king asigned to that office, pronounced with an audible and manly voice, what the other spake to him in secret: continuing, whether it were his oration or proclamation, at the least halfe an houre. At the close whereof, there was a common *Amen*, in signe of approbation giuen by euery person: And the king himselfe with the whole number of men and women (the little children onely remaining behind) came further downe the hill, and as they came set themselues againe in their former order.

And being now come to the foot of the hill and neere our fort, the Scepter bearer with a composed countenance and stately carriage began a song, and answerable thereunto, obserued a kind of measures in a dance: whom the king with his guard and euery other sort of person following, did in like manner sing and daunce, sauing onely the women who danced

1579.

but kept silence. As they danced they still came on: and our Generall perceiuing their plaine and simple meaning, gaue order that they might freely enter without interruption within our bulwarke: Where, after they had entred, they yet continued their song and dance a reasonable time: their women also following them with their wassaile boales in their hands, their bodies bruised, their faces torne, their dugges, breasts, and other parts bespotted with bloud, trickling downe from the wounds, which with their nailes they had made before their comming.

After that they had satisfied, or rather tired themselues in this manner, they made signes to our Generall to haue him sit down; Unto whom both the king and diuers others made seuerall orations, or rather indeed if wee had vnderstood them, supplications, that hee would take the Prouince and kingdome into his hand, and become their king and patron: making signes that they would resigne vnto him their right and title in the whole land, and become his vassals in themselues and their posterities: Which that they might make vs indeed beleeue that it was their true meaning and intent; the king himselfe with all the rest with one consent, and with great reuerence, ioyfully singing a song, set the crowne vpon his head; inriched his necke with all their chaines; and offering vnto him many other things, honoured him by the name of *Hyóh*. Adding thereunto (as it might seeme) a song and dance of triumph: because they were not onely visited of the gods (for so they still iudged vs to be) but the great and chiefe god was now become their god, their king and patron, and themselues were become the onely happie and blessed people in the world.

These things being so freely offered, our Generall thought not meet to reiect or refuse the same: both for that he would not giue them any cause of mistrust, or disliking of him (that being the onely place, wherein at this present, we were of necessitie inforced to seeke reliefe of many things) and chiefely, for that he knew not to what good end God had brought this to passe, or what honour and profit it might bring to our countrie in time to come.

Wherefore, in the name and to the vse of her most excellent maiesty, he tooke the scepter crowne and dignity, of the sayd countrie into his hand; wishing nothing more, than that it had layen so fitly for her maiesty to enjoy, as it was now her proper owne, and that the riches and treasures thereof (wherewith in the vpland countries it abounds) might with as great conueniency be transported, to the enriching of her kingdome here at home, as it is in plenty to be attained there: and especially, that so tractable and louing a people, as they shewed themselues to be, might haue meanes to haue manifested their most willing obedience the more vnto her, and by her meanes, as a mother and nurse of the Church of *Christ*, might by the preaching of the Gospell, be brought to the right knowledge, and obedience of the true and everliving God.

The ceremonies of this resigning, and receiuing of the kingdome being thus performed, the common sort both of men and women, leaving the king and his guard about him, with our generall, dispersed themselues among our people, taking a diligent view or suruey of euery man; and finding such as pleased their fancies (which commonly were the youngest of vs), they presently enclosing them about, offred their sacrifices vnto them, crying out with lamentable shreekes and moanes, weeping, and scratching, and tearing their very flesh off their faces with their nailes, neither were it the women alone which did this, but euen old men, roaring and crying out, were as violent as the women were.

We groaned in spirit to see the power of Sathan so farre preuaile, in seducing these so harmlesse soules, and laboured by all meanes, both by shewing our great dislike, and when that serued not, by violent withholding of their hands from that madnesse, directing them (by our eyes and hands lift vp towards heauen) to the liuing God whom they ought to serue: but so mad were they vpon their Idolatry, that forcible withholding them would not prevaile (for as soone as they could get liberty to their hands againe, they would be as violent as they were before) till such time, as they whom they worshipped, were conueyed from them into the tents, whom yet as men besides themselues, they would with fury and outrage seeke to haue againe.

After that time had a little qualified their madnes, they then began to shew & make knowne vnto us their griefes and diseases which they carried about them, some of them hauing old aches, some shruncke sinewes, some old soares and cankered vlcers, some wounds more lately receiued, and the like, in most lamentable manner crauing helpe and cure thereof from vs: making signes, that if we did but blowe vpon their griefes, or but touched the diseased places, they would be whole.

Their griefes we could not but take pitty on them, and to our power desire to helpe them: but that (if it pleased God to open their eyes) they might vnderstand we were but men and no gods, we vsed ordinary meanes, as lotions, emplaisters, and vnguents most fitly (as farre as our skills could guesse) agreeing to the natures of their griefes, beseeching God, if it make for his glory, to giue cure to their diseases by these meanes. The like we did from time to time as they resorted to vs.

Few were the dayes, wherein they were absent from vs, during the whole time of our abode in that place: and ordinarily euery third day, they brought their sacrifices, till such time, as they certainely vnderstood our meaning, that we tooke no pleasure, but were displeased with them: whereupon their zeale abated, and their sacrificing, for a season, to our good liking ceased; notwithstanding they continued still to make their resort vnto vs in great abundance, and in such sort, that they oft-times forgate, to prouide meate for their owne sustenance; so that our generall (of whom they made account as of a father) was faine to performe the office of a father to them, relieuing them with such victualls, as we had prouided for our selues, as, Muscles, Seales, and such like, wherein they tooke exceeding much content; and seeing that their sacrifices were displeasing to vs, yet (hating ingratitude) they sought to recompence vs, with such things as they had, which they willingly inforced vpon vs, though it were neuer so necessarie or needfull for themselues to keepe.

They are a people of a tractable, free, and louing nature, without guile or treachery; their bowes and arrowes (their only weapons, and almost all their wealth) they vse very skillfully, but yet not to do any great harme with them, being by reason of their weakenesse, more fit for children than for men, sending the arrowes neither farre off, nor with any great force: and yet are the men commonly so strong of body, that that, which 2. or 3. of our men could hardly beare, one of them would take vpon his backe, and without grudging carrie it easily away, vp hill and downe hill an English mile together: they are also exceeding swift in running, and of long continuance; the vse whereof is so familiar with them, that they seldom goe, but for the most part runne. One thing we obserued in them with admiration: that if at any time, they chanced to see a fish, so neere

the shoare, that they might reach the place without swimming, they would   1579.
neuer, or very seldome misse to take it.

After that our necessary buinesses were well dispatched, our generall
with his gentlemen, and many of his company, made a journy vp into the
land, to see the manner of their dwelling, and to be better acquainted, with
the nature and commodities of the country. Their houses were all such as
we haue formerly described, and being many of them in one place, made
seuerall villages here and there. The inland we found to be farre different
from the shoare, a goodly country, and fruitful soyle, stored with many
blessings fit for the vse of man: infinite was the company of very large
and fat Deere, which there we sawe by thousands, as we supposed, in a
heard: besides a multitude of a strange kind of Conies, by farre exceeding
them in number: their heads and bodies, in which they resemble other
Conies, are but small; his tayle, like the tayle of a Rat, exceeding long;
and his feet like the pawes of a Want or Moale; vnder his chinne, on either
side, he hath a bagge, into which he gathered his meate, when he hath
filled his belly abroade, that he may with it, either feed his young, or feed
himselfe, when he lists not to trauaile from his burrough: the people eate
their bodies, and make great account of their skinnes, for their kings
holidaies coate was made of them.

This country our Generall named *Albion*, and that for two causes; the
one in respect of the bancks and cliffes, which lie toward the sea: the other,
that it might haue some affinity, euen in name also, with our own country,
which was sometime so called.

Before we went from thence, our Generall caused to be set vp, a
monument of our being there; as also of her maiesties, and successors right
and title to that kingdome, namely, a plate of brasse, fast nailed to a great
and firme post; whereon is engrauen her graces name, and the day and
yeare of our arriuall there, and of the free giuing vp, of the prouince and
kingdome, both by the king and people, into her maiesties hands: together
with her highnesse picture, and armes in a piece of sixpence currant
English monie, showing itselfe by a hole made of purpose through the
plate: vnderneath was likewise engrauen the name of our Generall & c.

The Spaniards neuer had any dealing, or so much as set a foote in this
country; the vtmost of their discoueries, reaching onely to many degrees
Southward of this place.

And now, as the time of our departure was perceiued by them to draw
nigh, so did the sorrowes and miseries of this people, seeme to themselues
to increase vpon them; and the more certaine they were of our going away,
the more doubtfull they shewed themselues, what they might doe; so that
we might easily iudge that that ioy (being exceeding great) wherewith
they receiued vs at our first arriuall, was cleane drowned in their excessiue
sorrow for our departing: For they did not onely loose on a sudden all
mirth, ioy, glad countenance, pleasant speeches, agility of body, familiar
reioycing one with another, and all pleasure what euer flesh and blood
might bee delighted in, but with sighes and sorrowings, with heauy hearts
and grieued minds, they powred out wofull complaints and moanes, with
bitter teares and wringing of their hands, tormenting themselues. And as
men refusing all comfort, they onely accounted themselues as
cast-awayes, and those whom the gods were about to forsake: So that
nothing we could say or do, was able to ease them of their so heauy a
burthen, or to deliuer them from so desperate a straite, as our leauing of
them did seeme to them that it would cast them into.

Howbeit seeing they could not still enjoy our presence, they (supposing vs to be gods indeed) thought it their duties to intreate vs that being absent, we would yet be mindfull of them, and making signes of their desires, that in time to come wee would see them againe, they stole vpon vs a sacrifice, and set it on fire erre we were aware; buring therein a chaine and a bunch of feathers. We laboured by all meanes possible to withhold or withdraw them but could not preuaile, till at last we fell to prayers and singing of Psalmes, whereby they were allured immediately to forget their folly, and leaue their sacrifice vnconsumed, suffering the fire to go out, and imitating vs in all our actions; they fell a lifting of their eyes and hands to heauen, as they saw vs do.

The 23. of Iuly they tooke a sorrowfull farewell of vs, but being loath to leaue vs, they presently ranne to the top of the hils to keepe vs in their sight as long as they could, making fires before and behind, and on each side of them, burning therein (as is to be supposed) sacrifices at our departure.

Not farre without this harborough did lye certaine Ilands (we called them the Ilands of Saint *James*), hauing on them plentifull and great store of Seales and birds, with one of which wee fell Iuly 24. whereon we found such prouision as might competently serue our turne for a while. We departed againe the day next following, viz. Iuly 25. And our Generall now considering, that the extremity of the cold not only continued but increased, the Sunne being gone farther from vs, and that the wind blowing still (as it did at first) from the Northwest, cut off all hope of finding a passage through these Northerne parts, thought it necessarie to loose no time; and therefore with generall consent of all, bent his course directly to runne with the Ilands of the Moluccas. And so hauing nothing in our view but aire and sea, without sight of any land for the space of full 68. dayes together, wee continued our course through the maine Ocean, till September 30. following, on which day we fell in kenne of certaine Ilands, lying about eight degrees to the Northward of the line.

1579.

July 23.

July 24.

July 25.

Sept. 30.

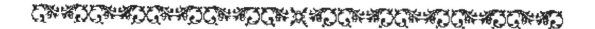

Appendix C

Excerpts from Sloane MS No. 61 entitled:

THE FIRST PART OF THE SECOND VOIAGE ABOUT THE WORLD ATTEMPTED CONTINUED AND HAPPILY ACCOMPLISHED WITHIN THE TYME OF 3 YEARES BY Mr FFRANCIS DRAKE, AT HER HIGHNESS COMMAUND & HIS COMPANY WRITTEN & FAITHFULLY LAYED DOWNE BY FFRANCIS FFLETCHER MINISTER OF CHRIST, & PREACHER OF THE GOSPELL ADVENTURER & TRAUELER IN THE SAME VOYAGE.

As discussed in chapter 5, I believe that circa 1589, in the aftermath of the defeat of the Spanish Armada, Sir Francis Drake succeeded in recovering from his grateful queen, the on-board records of his two most controversial voyages: Nombre de Dios and the circumnavigation. Having secured the records at Buckland Abbey, Drake employed the seafaring preachers, Philip Nichols and Francis Fletcher, to prepare manuscript accounts of the respective voyages. I believe that circa 1591, Drake took both draft manuscripts and edited and enlarged the texts "by divers notes, with his own hand here and there inserted." On January 1, 1592, I believe Drake submitted the "first fruits of his pen" for the queen's review, including his edited versions of the Nichols' account and the first half of the Fletcher narrative. I believe the Sloane manuscript is a copy of the heavily edited Fletcher manuscript that Drake submitted to the queen in January 1592, or perhaps an earlier draft.

This appendix, by permission of the British Library, includes excerpts of the so-called Fletcher notes transcribed in 1677 by one John Conyers, citizen of London (Sloane MS No. 61). I have *highlighted in italics* the passages that I think Drake added to Fletcher's rambling narrative. I also believe that Drake edited-in the illustrations.

Concerning the fertility and resources of the island of Mayo in the Cape Verdes

We found in the Iland though but little yet most sweet water bearing the tast of milke: w**ch** I conceiued to be quallified from the rootes of the muskadine vines from whence it did spring w**ch** bore the fairest & most pleasant grapes that I hadd seen in all my former trauaile in any kingdom. We mett allso wth 2 sorts of most Rare fruits to grow in our parts of the world, The one named Cocos wch is the same wee call nux indica; and the other Palntanos, *the Portraiture herafter you shall see in their Places trees & fruit as they grow as neare as I could describe them*, & his is allso to be noted that all the trees in this Iland as well these as others now being the dead of winter in England were florishing green wth blossoms & green fruit & ripe fruit vpon them at once & at one tyme: that as the ripe fruit is allwayes Ready to be gatherd so the green fruit cometh forward the blossoms turne to green fruit & new blossoms budd out: & this is the state & course of all trees betwixt the 2 troppicks where the Sonn is Zenith East & West in the compass of the whole earth by reason of the nearnes & the presence of the sonn. *wch many of or fetherbedd milkesopps boasting of the deep judgmt in cosmographie at a smyths forge hamering out a globe to make a childish bragg amongst simple poople. doe laugh & mock at & say it is a lye to Report such thing of Gods great & marvailous workes, & further to confirme their blind error & Ignorance Their owne new forges Globe must be shewed whether it be so or no the glory wherof when it is seen standeth in our starr in the heauens & that is venus, one fish in the sea & that is a pearch & the earth a chaos. & yet they will lye & Cogg yea wth brasen faces slander the truthe deliuered by them wch haue seene such the wondrfulle workes of God* About this Iland vnder the bankes & between them and the high water markes do lye huge heapes of salt like drifts of snow & most fine & perfcct in nature, the aboundance wherof is such & they daily increase so exceeding great that they serue all countryes and Ilands about them & is impossible to be consumed *but this is another wonder to our greenheads wch will belieue nothing but that they paint themselues vpon their Iron hoopes this salt whereof I haue spoken cometh groweth,*

& increseth continually without art skill industrye or labour of man, for the verry motion of the flowing of the sea yeilding to the shoare thinn vapers of Saltish matter were taken by the heat of the sonn as they did fall vpon the bankes of salt. & were presently kerned to the infinit increasing of the same, so that it may seeme to be one of the wonders of the world. & so much of the Iland of Maro one of the Ilands of Cape Verde the description wherof followeth. it is a thing to be lamented that so sweet fruitfull & profitable a land should either be possessed by so vngratefull vngratious a poople as are the Portugalls: or be so subject to such caterpillars of euery kingdom. & nation as are pirates and theeues of the sea, but that it should bee but that it should be [sic] inhabited by a poople feareing the Lord to praise him for his benefitts wch he plentifully hath bestowed vpon it to set forth his glory & for the good of his people.

[Illustration]

Concerning the malevolence of Thomas Doughty

Into this shipp the Generall sent on Tho: Dubty Gentilman to be captayne, whoe not long after his enteryng into this charge was charged & accused by John Brewer Edward Bright & som others of their freinds. to haue purloyned to his propper vse to deceaue the voyage som things of gret valew; *& therfore was not to be putt in trust anny longer least hee might robb the Voyage & depriue the company of their hope & her Majesty & other adventurers of their benefitt to inrich himselfe & make himselfe greater to the ouerthrow of all others.* In Regard whereof the Generall speedily went on board the prise to examen the matter who finding certain pares of Portugall Gloves, som few peeces of mony of a strange coine & a small Ringe all w**ch** one of the Portugalls gaue him out of his cheist in hope of fauor all of them being not worth the speaking of.

Concerning the acquiescence of Nuño da Silva

Now the Portugalls of the shipp being discharged & sett freely at liberty as hath been sayed see reserued to our owne seruice only one of theyre company one Syluester their Pilott a man well trauailled both in Basilia & most part of India one this syde of the Land. who when he hard that our trauail was into Mare del Zur that is the Sowth Sea hee of himselfe was most willing to goe with vs

Concerning the fertility of the tropics and contemporary doubt thereof

. . . but all things reuiue flourish & bring forth according to their natures by the presence of the sonn & as the sonns presence in our partes of the world is the cause of increase & fruitfullnes for the tyme that it remainith in power w**th** vs so is it true that where the heat & power euer worketh as between the Tropicks it maketh a perpetuall greenes florishing fruitfullnes and increase as it doth w**th** vs for the tyme *howsoeuer arrogant companions will haue nothing to bee true but lyes which they themselves cannot gather out of a frozen mountaine or from Isacles hanging at their noses in winter.*

Concerning fishes that fly and contemporary doubt thereof

The Dolphin and Bonetta being fresh are most wholsom norishment & fitt for Kings, but being salted & dryed are most dangerous & contagious to breed the Leprosye or at the least pestilent feauers as by Experience I prooued in som w**ch** [I] did eate w**ch** I note for a warneing to to [sic] Trauellers in Hott countryes And because I haue made mencion of flying fishes w**ch** to the most part of men may seem an absurd thing *& as our green headed carpers saye a Lye!* I haue thought good here to sett downe the stoary, & true report of them to the greater glory of that God w**ch** made

Concerning the voyage across the Atlantic and a few points Fletcher failed to note

Now yet these 2 things I thought necessary to be added wch fell out in this Journey, that sometymes the water wch fell out of the ayer when we came neare to the Equator was so quallified of the heat

of the sonn. that it falleing vpon the cloathes of our men wch laye in heapes or folded vpp. if they were not within short tyme washed & hanged abroad they were so burned that they would moulder in peices as a peice of loouse earth.

And the 2 when wee were within one degree of the Æquinoctiall line or thereabouts wee lost the sight of the north pole & not afore: & allso being so much to the southward of the line wee had in sight the South pole. between the guardey [?] wherof wee found great difference: at what tyme the sonn was our Zenith, our bodyes hadd or gaue no shaddow at all but right downe to our feet

Wherevnto allso lett me putt the thirde a thinge worthey the noteing that in our passing from our country being winter lice increased infinitely in the cloathes of our men & were a great plague to many: but no sooner were wee com within the burneing Zone but they all dyed & consumed away of themselues, so that till wee came beyond the southerly tropick to Brasilia there was not one to be found among vs.

Concerning the stay at Port St. Julian and a few thoughts about Magellan

The figure of the Cockel you shall find in the end of this stoary of port St Julian other things in this place we found worthy of remembrance saue onely that wheras Magilanus performeing the first Voyage about the world falling wth this port as wee did did [sic] first name it port St Julian & makeing som aboad there had a Mutiny against him by som of his company for the which he Executed diuers of them vpon a Gibbet close by the sea vpon the maine land ouer against the Iland port of which Gibbet (being of firrwood) wee found there sound & whole of wch Gibbet being 50 years at the least before our tyme of our coming thither of wch wood our Cooper made Tankards or Canns for such of the Company as would drinke in them Wherof for my owne part I had no great likeing seeing there was no such necessity.

I will end this stoary wth Answer to that wch was then proposed to me & often since wch is what might be the reason. that the Giants of this place should degenerate so farr from all others whom we had found so kinde, loueing, and harmless in other partes of the country & that the rest should do vs good & these should deale so bloodily with vs. for Answere (in my owne Opinion) the reason wherof was not difference of Nature nor climate as som would haue it but an accident chanceing to them more then to others. Which if it had fallen our to the Rest as it did to them, it might be as Easily to haue wrought the same in them as in these. for when Magilanus was there he injuriously tooke from this poople 2 of their men. & that with violence, to the shedding of blood & murther of both sydes. wch wrought in them such a dislyke that they purposed & vowed revenge if euer tyme & opertunity serued which being deliuerd from the fathers to the children. & from one generation to another stood for a law to those which should be borne & nothing was wanting at anny tyme since to put it in Execution but strangers for they could make no difference of Nations being as it is likely neuer acqainted with anny before Magilanus nor neuer after till our comeing & therefore we being the first they tooke opertunity to performe their vow being determined neuer to giue credit to stangers out of this then did surely rise (as I thinke) this mischief against vs at this tyme be these Giants when we had so peaceably conversed wth all others wee mett with in all their land. at our departure wee named the Iland the Iland of blood in respect of vs & Magilanus

Concerning entering the Strait of Magellan and reverence to the queen

This Gratious Exercise ended with prayer to God for her Most Excellent Majesty. her Honorable councell and the Church & the comon weale of England. wth singing of Psalms. & giueinge thankes for Gods great & singular graces bestowed vpon vs from tyme to tyme: Wee departed from the Bloody Iland & Port Julian setting our course for the supposed streight wth three shipps onely. That is the Pellicane being admirall. The Elizabeth the vice admirall & the barke marigold wherin Edward Bright a shipp carpenter was newly placed Captaine. Wee thus ariuing at the northerly cape or headland of the aforesaid streight, wee struck sale & made som stay for a tyme tell we had performd among ourselues that dewty & seruice to God who had safely brought vs thither which we had so long desired & *honor to her Highnes fitting for Subjects of so gratious & seuenfold blessed princes praying for the continuance of her dayes to be as the dayes of Heauen as long as the Sonn & Moone indureth* those things thus accomplished wee joyfully entered the Streight w**th** hope of Good sucess. at our first Entrance we conjectured that from the cape whence wee departed to the oposite land on ethe other side against it being a grat Iland & high & seemeth to make the mouth of a streight that it was about 10 Leagues or there about by Estima**co** but afterwards we found the passage in som place 4 in other places 3 & 2 leagues broad & where it was narrowest it was as a large league. In passing alongst wee plainly discouered that the same terra australis left & sett downe to bee terra incognita before we cam there to bee no continent (& therefore no streight) but broken Ilands, & large Passages amongst them

Concerning exiting the Strait of Magellan and reverence to the queen

Now God in mercy at the last brought vs through this labrinth wee so long had intangled vs w**th** so may Extremityes & Iminent dangers to that w**ch** we so long desired that is to the Southerly cape of America entering into the South Sea. *Wherein as we Exceedingly rejoyced: so were wee resolued there to stay a while to do as wee had don to our good God & in deuty to her majesty in other places & to haue sett vp a monumt for her Highnes vpon the cape for a wittnes of our Passing that way & arival at that place wch monument I had engrauen in mettle for the same purpose the like of that wch you may see in the End of this Booke but it seemed not good to god wee should accomplish that tyme in this place as by the sequel will appeare if first I shall lay downe the descriptions of such things as are spoken of in this stoary of the Imagined streight of Magilan, together wth the Land off Land [sic] of so much of Brasilia the Riuer of Siluer the land of Giants the sowtherly land & this part of America to this South most cape entering into Mare del Zur that is into the South Sea* [Four Illustrations].

Concerning the discovery of the Drake Passage

Wherefore hee caused the Sonn by day & the moon & starrs by night to shine vpon vs. he changed the sterne lookes of the grisly mountains, to yeild a fauorable continuance the heauens to Laugh the seas to obey & the winds to cease yea euery place wee looked vpon to yeeld vs kind entertainment to refresh our weather beaten bodyes & liues discomforted in so greauous miserys & forlorne state. *which thing fell out most happily in his providence att the vttmost Iland of terra incognita. to the Southward of America whereat wee arriueing made both the seas to be one & the same sea & that there was no farther land beyond the heights of that Iland being to the Southward of the Æquinoctiall 55 & certaine minutes, to diuide them: but that the way lay open for shipping in that heigth without lett, or stay being the mane sea.*

Concerning erecting a monument to the queen at Cape Horn

In this Iland were growing wonderfull plenty of the small berry with vs named Currants, or as the comon sort call them small raisins. My selfe being landed did with my Boy trauill to the Souther most point of the Iland to the Sea one that Syde. Where I found that Iland to be Southerly three partes of a degree then anny of the Rest of the Ilands. Where haueing Sett vp an End a Stone of som biggnes & wth such Tooles as I hadd of Purpose euer about mee when I went one shoare. had engrauen her Majestyes name. her Kingdom. the year of Christ & the day of the moneth. I returned againe in som Reasonable tyme to our Company.

Concerning the misunderstandings of contemporary cartographers

Wee departeing hence & takeing our farewell from the Southern most part of the world knowne or as wee think to be knowne there. Wee alterd the name of those Southerly Ilands from terra incognita (for so it was before our comeing thither & so should haue remained still with our good wills) to terra nunc bene cognita/ That is broken Iland, which in coasting it againe one that syde in returneing to the Northward wee prooued to be true, & were thoroughly confirmed in the same. But before I proceed anny farther in our after trauiles I thinck it not impertinent heere to incert a few wordes to resolue that great Errour which hath been & is maintained of many touching the Streight of Magilanus. Videlicet that it is a frete, or streight indeed whose reasons being ouer throwne. their purpose faileth the Cheife wherof is that the Southerly Land is a Continent, but 2 continents made a streit. therfore it is and must bee a straight. I answer it was euer vncertain from the first discouery of that Passage by the Spanyards & could not be determined by Magilane himselfe that that Land was a continent but left it vnder the name of Terra incognita: & what others afore or since haue written or said afore of it are but guesses and imaginations seeing hee himselfe could determine nothing & that from his tyme There was no other tryall made of it vntell now wee haue by manifest Experience put it out of doubt to be no continent or maine Land but broken Ilands disseuered by many Passages & compassed about wth the sea on euery side. & therfore no streight. Their second Reason is as light as this, Namely their runneth such a current betweene the land of America and this terra Australis (caused say they) by the continent. that it carieth shipps headlongs through to the South Sea but admitteth no returne back againe. which could not bee if their was anny other vent or Passage for the sea to keepe its naturall course. To this the Answer is ready that it were the strangest thing in the world that the land reaching to the southward from the maine of America but 3 degrees at vttmost should bee a cause of so violent a current to fall into that Passage to worke such an Effect thoug the land were a mane land maine land [sic] and one body reaching so farre a thing in auditum: much less (seeing as hath been sayd) it is distincted Ilands with the sea between & among them, form one end to the other from the north to the sowth & from the East to the west conteining many as large Passages as that whereof they affirme these things. besides this if they Imagine some fourth & remote cause to be that a currant hath or setteth his course this way and falleing wth the Maine land ofe America, haueing on other vent is and must be inforced in this way. That is allso friuolous seing both the seas in 55 degrees. & vnder are both one. & therfore this reason is without reason, to proue a streight by that which is not, neither hath anny ground to be likely. To be short som ground themselues vpon the Spanyards report, & credit that certainly there is such a currant that way as sendeth most swiftly into the South Seas but suffereth no returne that this is false is euidently prooued by Experience as well from Magilan himselfe, as from ourselues, for at what tyme he passed that way being come into the south sea. one of his shipps stoale away & returned into their country the same way they came. the like thing chanced to ourselues. for as you may read before our vice admirall was enforced into the supposed streight againe wth the violence of the storme & so passed the whole way wee had com without anny such hinderance yea they had more sweet, & quiet returne back againe then wee had a comeing forward. Add vnto these the accidts wch fell to vs in our saleing towards the South sea in this Passage.

That often tymes wee were driuen back by the winds coming from the mountaines more in an houre. then wee could raise vp againe in manye. both which had not chanced if their had been so violent a currant as they dreame of but to make an end of this matter the reason why the Spanyards haue abused the word wth such an vntruth, and lye is that thereby they would prevent all other nations to giue the attempt to trauaile that way being (as they feigned) so impossible a thing to haue anny returne but must be forced to compass the whole world to their country againe & thus much for the cleareing of the doubt whither those things holden & falsely maintained touching this southermost land & the passag between it & the maine land of America to make it knowne to the world that their is neither Continent current nor streight as Idely it hath been surmised, & foolishly defended. Toucheing Mr Candish a Gentileman worthy of immortall fame for his Rare Enterprises in Trauailes, I say littel: onely this I obserue that he was able to say no more either to prooue that that [sic] terra australis is a continent or that the passag a streite: then Magilanus himselfe as by the line of his course in the mapp may appeare leadeing imediatly from the Passage by the South cape of america towards the line without anny touch of anny point further to the Southward to make anny proofe or tryall. neither could he make anny mention of anny current. And now to our purpose againe ffrom *which for a reasonable cause I haue thus farr degressed at our departure from this Iland before remembred this one thing wee obserued the night was but 2 houres Long & yet the sonn was not in their tropick by eight degrees wherof we gatherd that when the sonn should be in the Tropick they should haue no night at all neither would the sonne be out of their sight at anny tyme for certaine dayes. by reason of its easterly & westerly course as it falleth out in Russia & other the Northerly Countryes, when the sonn is in our Tropick as our Country men trauaillers in those partes can witnes & myself in my former trauailes haue seen*

Concerning Mucho Island and whether or not God had forsaken them

Wherein whether God mett vs Jumpe in the South Sea or noe wee ourselues might Easily Judg this Iland is most Rich in Gold & silver and it aboundeth in many good things neccessary for the maintainance of Gods good poople flouishing wth trees & fruit continually wanting nothing but a poople feareing God to injoye it to glorify his Name for such Excellt blessings bestowed vpon it I may compare it fittly to her Majestyes Iland named weight vpon the Coast of England which in respect of its situation is called Vectis. a doore barr to the Land one that Syde of our country for so this Iland layeth in like sort right against a most golden province of the world named Valdinia wch being possessed by one Prince would make the one & the other invincible but of this Valdinia you shall heare more in the second part of this nauigation about the world wch I will attempt to finish wth all convenient speed I may & thus I end this first part of this trauaile wth description or mapp of the Iland of Much & the forme & the monuments wch I made to haue ben sett vp vppon the Southermost Cape of America att the entering into the Sowth Sea but could not as hath been said

[Illustration]

The account ends with Drake reminding the queen of his attempt to erect a monument on her behalf upon exiting the Strait of Magellan.

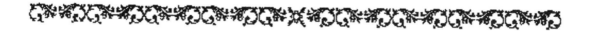

Appendix D

Comparison of Latitudes

In the following tables, I correlate the latitude data reported in *The World Encompassed* and "The Famous Voyage" against the true latitudes of Drake's identifiable landfalls for the purpose of checking the accuracy of Drake's measurements. I used Raymond Aker's latitude comparison as a starting point, incorporating his landfall identifications in instances where I agreed with them. Otherwise, I referred to modern charts and atlases.[1]

The "World Encompassed" Latitudes

Table D-1. Latitude Comparisons for "The World Encompassed" (*W.E.*)									
W.E. date	*W.E.* latitude position as reported	*W.E.* place name or description as reported	Place name from modern atlas	Latitude from modern atlas or chart	Diff. (min.)	Abs. diff. (min.)	Comments	Diff. for likely land reading	Diff. for likely sea reading
Dec. 25, 1577	32 deg. 30 min.	Cape Cantine	Cape Kantin, Morocco	32º 32′ N	-2	2	Sailed by		-2
Dec. 25, 1577	31 deg. 40 min.	Mogador	Mogodor I., Morocco	31º 30′ N	10	10	Landed	10	
Jan. 3, 1578	30 deg. [?] min.	Cape DeGuerre	Cape Rhir, Morocco	30º 38′ N			The account appears uncertain here		
Jan. 16, 1578	20 deg. 30 min.	Cape Blanck	Cape Blanc, Mauritania	20º 45′ N	-15	15	Trimmed the ships	-15	
Jan. 28, 1578	15 deg. 00 min.	Maio	Maio I., Cape Verdes	15º 15′ N	-15	15	Landed	-15	
Jan. 31, 1578	14 deg. 30 min.	Fogo	Fogo I., Cape Verdes	14º 55′ N	-25	25	Sailed by		-25
Apr. 5, 1578	31 deg. 30 min.	First landfall on Brazil					Not enough info to identify the landfall		
Apr. 14, 1578	35 deg.	Cape Saint Maria	Cape St. Maria, Uruguay	34º 40′ S	20	20	Sailed by		20
Apr. 27, 1578	36 deg. 20 min.	Shoal off south side of River Plate	Mendano Bank, Argentina?	36º 53′ S			Info is too ambiguous to definitively identify the landfall		
May 12, 1578	47 deg.	Cape Hope	Cabo Blanco, Argentina	47º 12′ S	-12	12	Landed	-12	
May 17, 1578	47 deg. 30 min.	Seal Bay	Puerto Deseado, Argentina	47º 45′ S	-15	15	Landed	-15	
June 14, 1578	50 deg. 20 min.	Bay 1 degree from the straits	Porto Santa Cruz, Argentina	50º 08′ S	12	12	Anchored		12
June 20, 1578	49 deg. 30 min.	Port St. Julian	Port St. Julian, Argentina	49º 17′ S	13	13	Landed	13	
Aug. 20, 1578	52 deg.	East entrance to strait	Strait of Magellan	52º 27′ S	-27	27	Sailed through		-27
Sept. 1578	53 deg. 15 min.	Middle of strait	Cape Froward, at the bend?	53º 56′ S?			Info is too ambiguous to definitively identify the landfall		

Continued on next page

1. Aker's latitude comparison is in *Report of Findings*, pp. 436–50. For the most part, I used the modern charts provided in Wagner's *Drake's Voyage* to identify Drake's circumnavigation landfalls and their associated true latitudes.

	**Table D-1. Latitude Comparisons for "The World Encompassed" (W.E.)**
W.E. date	W.E. latitude position as reported	W.E. place name or description as reported	Place name from modern atlas	Latitude from modern atlas or chart	Diff. (min.)	Abs. diff. (min.)	Comments	Diff. for likely land readin	Diff. for likely sea readin
Sept. 6, 1578	52 deg. 30 min.	West exit from strait	Strait of Magellan	52º 37′S	-7	7	Sailed through		-7
Sept. 1578	57 deg.	Farthest point south					No land in sight		
October 1578	55 deg.	Among the islands south of the strait					Not enough information to identify the landfall		
Oct. 28, 1578	56 deg.	Uttermost cape	Cape Horn, Chile	55º 59′ S	1	1	Landed. Could be considered contested	1	
Nov. 25, 1578	37 deg. or thereabout	Mucho Island	Isla Santa Maria, Chile	37º 00′ S			Drake landed but did not make a reading. Contested		
Nov. 30, 1578	32 deg. or thereabout and six leagues north of Valperizo	Philips Bay	The bay in the area of Port Papudo, Chile	32º 30′ S			Anchored but Drake does not appear to have landed. Contested		
Dec. 5, 1578	35 deg. 40 min.	Valperizo	Valparaíso, Chile	33º 02′ S			An overt transcription error. The crew landed here but Drake had just been severely wounded		
Dec. 19, 1578	29 deg. 30 min.	Bay not far S of Cyppo	Port Herradura, Chile	29º 57′ S	-27	27	Anchored briefly		-27
Dec. 20, 1578	27 deg. 55 min.	A bay north of Cyppo	Chasco Cove, Chile	27º 40′ S	15	15	Trimmed the ship	15	
Jan. 26, 1579	22 deg. 30 min.	Town of Mormorena	Cobya Bay, Chile	22º 30′ S			Landed. Contested		
Feb. 7, 1579	20 deg.	Arica	Arica, Peru (probably Tarapacá)	18º 29′ S			This was likely the position Drake reported for Tarapacá at about 20º S, the episode of which the W.E reports out of sequence with that for "Mormorena"		
Feb. 15, 1579	12 deg. 30 min.	Lima	Callao de Lima, Peru	12º 04′ S	26	26	They anchored at night and hurried off in the morning		26
Feb. 20, 1579	4 deg. 40 min.	Paita	Paita, Peru	05º 05′ S	-25	25	Sailed by		-25
Mar. 5, 1579	1 deg. S	Cape Francisco	Cape Francisco, Ecuador	00º 51′ N	9	9	Sailed by		9
Apr. 15, 1579	15 deg. 40 min.	Guatulco	Bahia de Santa Cruz, Mexico	15º 45′ N	-5	5	Landed	-5	
June 3, 1579	42 deg.	Farthest north before heading towards shore					No land in sight		
June 1579	48 deg.	Farthest N on Pacific coast	Queenhythe Bay, Wash	47º 40′ N			Anchored. Contested		
June 17, 1579	38 deg. 30 min.	Portus Nova Albion	Bodega Harbor	38º 19′ N			Landed. Contested		
Sept. 30, 1579	Islands lying about 8 deg. N	Island of Thieves	Babelthuap, Palau Is.?	07º 30′ N			Info is too ambiguous to definitively identify the landfall		

Continued on next page

Table D-1. Latitude Comparisons for "The World Encompassed" (*W.E.*)

W.E. date	W.E. latitude position as reported	W.E. place name or description as reported	Place name from modern atlas	Latitude from modern atlas or chart	Diff. (min.)	Abs. diff. (min.)	Comments	Diff. for likely land readin	Diff. for likely sea readin
Oct. 16, 1579	Four islands standing in 7 deg. 5 min.	Watering spot on the largest island?	The mouth of the Davao R., Davao Gulf?	07º 05′ N			Info is too ambiguous to definitively identify the landfall		
Oct. 25, 1579	3 deg. 40 min.	Talao	Sangi Island	03º 32′ N	8	8	Sailed by		8
October 1579	3 deg.	The middle island of a chain south of Talao	Siaoe Island of the Sangi chain?	02º 43′ N			Sailed by. The account appears confused here		
Nov. 1, 1579	1 deg. 30 min. N	Suaro	Maju Island	01º 20′ N	10	10	Sailed by		10
Nov. 4, 1579	27 min.	Ternate	Ternate, Moluccas	00º 48′ N	-21	21	Drake did not land		-21
Nov. 14, 1579	1 deg. 40 min. S	Crab Island	Lifamatola Island (near Mangola I.)	01º 50′ S			Trimmed the ship. Contested		
Jan. 9, 1580	2 deg. S less 3–4 min.	A shoal	Vesuvious Reef off Banggai grp?	02º 05′ S?			Info is too ambiguous to definitively identify the landfall		
Jan. 12, 1580	3 deg. 30 min.	A shoal	Near Manui I., Celebes?	03º 15′ S			Info is too ambiguous to definitively identify the landfall		
Jan. 14, 1580	4 deg. 6 min.	Coast of Celebes	Wowoni I., Celebes	04º 08′ S	-2	2	Landed. Could be considered contested	-2	
January 1580	5 deg.	Southernmost Cape of Celebes	Butung I., Celebes?	Approx. 5 degrees S			Info is too ambiguous to definitively identify the landfall		
Feb. 8, 1580	7 deg. 13 min.	Barativa	Damar I., Indonesia	07º 09′ S	4	4	Landed. Could be considered contested	4	
Feb. 12, 1580	8 deg. 4 min.	Green Island to the south	Moa I., Indonesia	08º 10′ S	-6	6	Sailed by		-6
Feb. 16, 1580	9 deg. 40 min.	Four or five big islands they passed through	They were in the Savu Sea off SW Timor				Info is too ambiguous to definitively identify the landfall		
Mar. 9, 1580	8 deg. 20 min.	High land to the north	They were off the S coast of Java	08º 12′ S?			Info is too ambiguous to definitively identify the landfall		
Mar. 11, 1580	7 deg. 30 min.	The middle of Java		07º 30′ S?			Info is too ambiguous to definitively identify the landfall		
May 21, 1580	31 deg. and half	In sight of west coast of Africa					Not enough info to identify landfall		
Count	50			44	25	25		11	14
Range					-27–26	1–27		-15–15	-27–26
Mean					-3	13		-2	-4
Std dev					16			11.5	19
Abs mean deviation						8		6	9

Summary – "The World Encompassed"

Of the fifty latitude readings provided in *The World Encompassed*, there are forty-four that can be potentially correlated with positions on modern charts and atlases. Of the forty-four, there are ten where the information in the account is too ambiguous to definitively identify the landfall (the "Island of Thieves" for example); there are six where the landfalls are still hotly contested despite the fact the information in the account does not appear ambiguous (Portus Nova Albion for example); there are three where the information associated with the latitude entry appears uncertain or confused

(Cape DeGuerre, Arica, Talao); and one where there appears to have been an overt transcription error (Valparaíso). Of the remaining twenty five, eleven appear to have been taken ashore and fourteen at sea. For those taken on land, the mean deviation between the reported position and the actual latitude is about minus two minutes within a range of minus fifteen minutes to plus fifteen minutes and the standard deviation is about eleven and a half minutes. For those taken at sea, the mean deviation is about minus four minutes within a range of about minus thirty to plus thirty minutes, and the standard deviation is about nineteen minutes.

By this statistical analysis, assuming the 38 degree 30 minute reading reported for the California landfall was taken on land, there is less than a 5 percent probability that the actual position lay outside of 38 degrees 9 minutes and 38 degrees 55 minutes (outside two standard deviations of the datum).

"The Famous Voyage" Latitudes

Table D-2. Latitude Comparisons for "The Famous Voyage" (*F.V.*)									
F.V. dates	*F.V.* position as reported	*F.V.* name or description as reported	Place name from modern atlas	Latitude from modern atlas or chart	Diff. (min.)	Abs. diff. (min.)	Comments		
April 5, 1578	33 degrees	First sighting on Brazil					Not enough info to identify the landfall		
April 1578	Six and thirty degrees	Mouth of River Plate	Middle of entrance?	35°–36° S	30	30	Seems more of an approximate geographic location than a navigational detail		
October 1578	55 degrees and a terce [third]	Among islands southward of the strait	Cape Horn	56° S	-40	40	Seems to be from the Fletcher account		
Nov. 29, 1578	38 deg. of latitude	Isle of Moucha	Isla Santa Maria, Chile	37° S	60	60	Could be Isla Mocha at 38° 22′		
June 3, 1579	42 degrees	Farthest N before heading toward land					No land in sight		
June 17, 1579	Within 38 degrees towards the line	Portus Nova Albion	Bodega Harbor	38° 19′ N	-19	19	Seems to be an approximation		
Oct. 13, 1579	Islands 8 degrees to the northward of the line	Island of Thieves	Babelthuap, Palau Islands	07° 30′ N	30	30	Applies to island group, not a landfall		
Count	7			5	5	5			
Range					-40–60	19–60			
Mean Deviation					12				
Std Dev					36				
Abs Mean Deviation						36			

Summary – "The Famous Voyage"

For the seven latitudes provided in Richard Hakluyt's "The Famous Voyage," 1589 edition, there are five that can be correlated with locations on a modern world atlas. For these five, the mean deviation between the reported position and the actual latitude is about twelve minutes within a range of minus forty to plus sixty minutes, and the standard deviation is about thirty-six minutes. Except in the case of Cape Horn, Hakluyt gave his positions to the nearest whole degree.

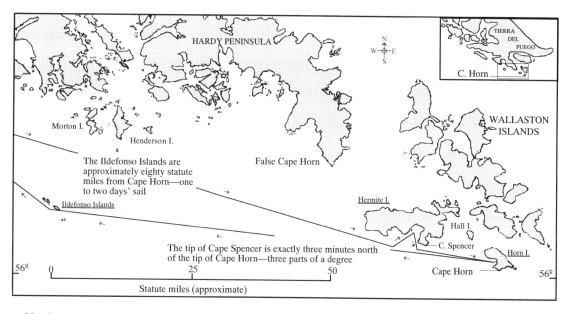

Map 31. Outline map showing Drake's route off the tip of South America and the "uttermost cape" (landfalls <u>underlined</u>).

The "Uttermost Cape"

As discussed in chapter 5, heated controversy concerning Drake's avowed discovery of the southernmost cape of South America (today's Cape Horn) erupted in his lifetime (see figures 35 and 36 and related discussion). Considering all the evidence in the contemporary accounts that support Drake's discovery of the Horn, it is remarkable that the controversy has persisted into the twentieth century.[2]

According to *The World Encompassed*, Drake discovered "near in 56. deg. . . . the southernmost cape of all the islands" lying off the tip of South America. Presuming Drake made this reading on land, and that the observation is accurate to within fifteen minutes, this can only be Cape Horn located at 55 degrees 59 minutes south latitude. The following corroborating evidence provides overwhelming support for a Cape Horn landfall. (1) *The World Encompassed*, Nuño da Silva's second relation, and the Fletcher narrative all report that approximately one to two days after departing the "uttermost cape" on a heading for the west coast of Chile, the English encountered some low islands, separated from the rest of the islands, that were virtual storehouses of birds and eggs—The Ildefonso Islands located northwest and within two days sail of Cape Horn correlate perfectly with this evidence. (2) The Fletcher narrative reports that the English found the "uttermost cape" to be "more southerly three parts of a degree [three minutes] than of the rest of the islands"—As first pointed out by Robert Power, the tip of Cape Horn is indeed exactly three minutes south of the tip of Cape Spencer.

Although the contemporary evidence shows that Drake landed at Cape Horn, it further supports the common sense notion that he initially anchored the *Golden Hind* at Hermite Island rather than Horn Island where there were many overt disadvantages. *The World Encompassed* states that they were "in and about" the islands at the very tip of South America, while the Fletcher narrative alludes to an excursion to "the island to the sea on that side."

---

2. Conflicting opinions include the following: Wagner (*Drake's Voyage*, pp. 84–86) and Morison (*Southern Voyages*, pp. 648–49) for Henderson Island at about 55 degrees 35 minutes south latitude; Power (*Geographical Magazine*, May 1980, p. 537) and Aker ("Francis Drake at Cape Horn") for Cape Horn; and Riesenberg (*Cape Horn*, pp. 81–83) for an island that allegedly sank into the sea in that area of today's Pactolus Bank at about 56 degrees 36 minutes south latitude.

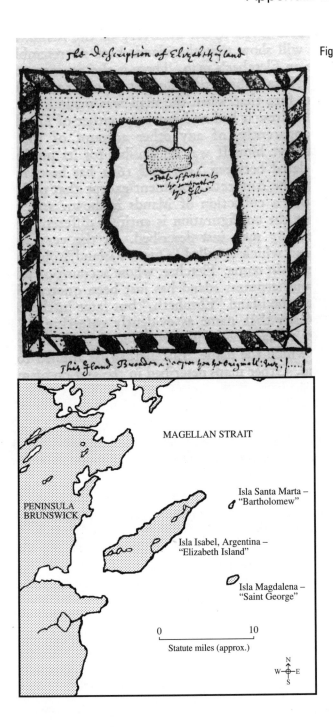

Fig. 174. Elizabeth Island (*above*) from the Francis Fletcher narrative, compared with today's Isla Isabel, Argentina (*below*), as depicted on a modern chart. Based on the location where the illustration falls in the Fletcher narrative, historians and Drake researchers have long assumed that it depicts the island Drake discovered on October 28, 1578, at the southernmost tip of South America. If we are to trust *The World Encompassed*, however, the sketch can only represent the Elizabeth Island Drake discovered and named for his queen on August 24 of that year, some twelve leagues inside the mouth of the Strait of Magellan, easily identifiable on the modern chart detail below as today's Isla Isabel, Argentina. Although *The World Encompassed* states that Drake applied the name "Elizabeth" to the group of islands at the tip of South America, it includes specific language that makes it clear that he did not apply the name to the single island farthest south. The proof of the matter lies in the "pool of fresh water" depicted on what the notation on the Drake illustration states is the south part of the island. As first pointed out by Robert Power in 1980, today's Isla Isabel has several such pools including one on the south side. A careful examination of the Fletcher narrative suggests that the error in placement was made when the narrative was transcribed by John Conyers in 1677: the text does not contain the usual introductory lead-in for this particular illustration. *By permission of the British Library, London.*

"Mucho Island"

Was Drake's "Mucho Island" today's Isla Mocha, Chile, or was it today's Isla Santa Maria about 100 miles to the north? According to *The World Encompassed*, in the height of "37 degrees or thereabouts," the English fell with an island that the Spanish named Mucho in respect to its large size and circuit. Spain's "Mucho Island," now called Isla Mocha, however, is located at 38 degrees 22 minutes, almost a full degree outside the nominal accuracy of Drake's latitude observations. Despite the fact that there is another island roughly the same size and circuit as Isla Mocha that indeed lies at 37 degrees, not one historian or Drake researcher has ever called Elizabethan geography into question; they have all simply assumed that the latitude reported in *The World Encompassed* is

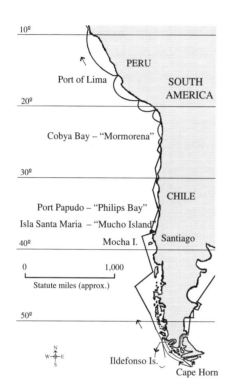

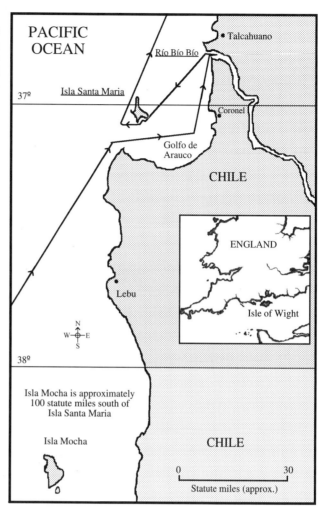

Map 32. Outline maps showing Drake's landfall at "Mucho Island" at about 37 degrees south latitude on the coast of Chile. The inset (*on right*) illustrates the similar geographic and strategic locations of Isla Santa Maria, Chile, and the Isle of Wight, England.

inaccurate. Consider the following corroborating evidence, however, strongly supporting an Isla Santa Maria landfall. (1) In his first interrogation by the Spanish in April 1579, Nuño da Silva states that he did not know the name of this island (Mucho) at the time of the visit. If da Silva did not know it, we can safely assume that the English did not either. (2) In his second interrogation by the Spanish in May 1579, da Silva described the island as "small and flat" which correlates exactly with Isla Santa Maria, but not Isla Mocha. (3) Sir Richard Hawkins, who visited both islands in 1594, described Isla Mocha as "a high mountainous hill" and Isla Santa Maria as "little and low, but fertile and well-peopled."[3] (4) From the deposition of one of the prisoners Drake took off the coast of Ecuador (San Juan de Antón), we know that just prior to his arrival at "Mucho Island," Drake reached the "River of Valdivia" and entered it for about half a league before he was turned back by the strong currents. This could easily have been the mouth of today's Río Bío Bío located just north of Isla Santa Maria. (5) In the Fletcher narrative, the author provides the following strategic information about Mucho Island: "I may compare it fittingly to Her Majesty's island named Wight, which in respect to its situation is called a door bar to the land on that side of the country, for so this island lays in like sort right against a most golden province of the world." Compare the situations of the Isle of Wight and Isla Santa Maria and you will appreciate the comparison; Isla Mocha, on the other hand, does not command the entrance to any bays, rivers or harbors.

3. Williamson, ed., *The Observations of Sir Richard Hawkins*, pp. 97 and 100.

Fig. 175. The "Mucho Island" illustration from the Francis Fletcher narrative. It was probably copied from a sketch that Drake or his cousin John drew on the spot in November 1578 while anchored off the lee side of today's Isla Santa Maria. *By permission of the British Library, London.*

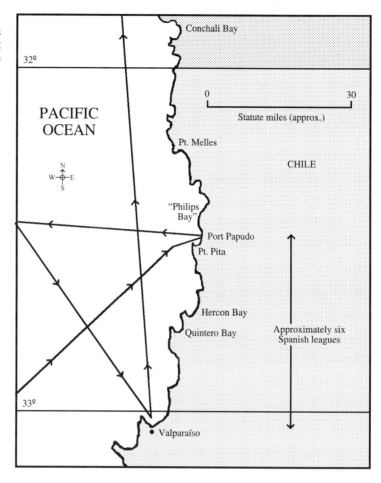

Map 33. Outline map showing Drake's "Philips Bay" landfall at about 32 degrees 30 minutes south latitude on the coast of Chile.

"Philips Bay"

According to *The World Encompassed*, the English spent five days at anchor in "Philips Bay" located "in 32 degrees or thereabout," using the ship's boat to scour the surroundings in a futile search for water and provisions. The account further indicates that after departing on December 4, 1578, it took the English a full day (until December fifth) to sail to Valparaíso though it was located "not far back to the southward." Notwithstanding this evidence and the Nuño da Silva account that places the anchorage six Spanish leagues (about twenty-five miles) north of Valparaíso, most authorities place Drake's Philips Bay anchorage at today's Quintero Bay (32º 45´) located just three Spanish leagues north of Valparaíso. They derive this conclusion from the contemporary Spanish account by Pedro Sarmiento de Gamboa (drawn from an interview with "Juan the Greek") that states that Drake "cast anchor at the port of Quintero, six leagues from the port of Santiago." The fact is, however, that the phrase "port of Quintero" could apply to any of several harbors located in the vicinity of today's Quintero Bay, and presumably to the one located about six Spanish leagues distant from Valparaíso. The weight of the evidence from all the accounts points to an anchorage at Port Papudo, located about six Spanish leagues distant from Valparaíso at about 32 degrees 30 minutes south latitude. Furthermore, in denominating a "bay" as opposed to a "port" or "harbor," Drake was likely referring to an inlet formed by two headlands, namely Point Melles (32º 15´ S) and Point Pita (32º 32´ S). Another point to consider here is that Drake, in stating that the bay was located at 32 degrees "or thereabout" seemed to be highlighting the fact that he showed up at the "appointed height wherein our ships were to be looked for, " rather than pinpointing the location of the anchorage (Drake, *The World Encompassed*, p. 49).

"Mormorena"

According to *The World Encompassed*, after anchoring near a small island (today's Constitution Island) off the north cape of the province of Mormorena on January 22, 1579, Drake sailed beyond the cape and up the coast of Chile for four days, passing by several Indian towns before casting anchor on January 26 "in 22 deg. 30 min." at Mormorena "another great town of the same people [native Indians]." Despite the fact that Nuño da Silva's logbook and second relation both suggest that Drake sailed well beyond today's Cape Mejillones by January 26, most authorities have placed the Mormorena landfall in Mejillones Bay at 23 degrees 6 minutes, the harbor formed by the cape itself on its north side. The weight of the evidence from all the accounts points to Cobya Bay that does indeed lie at about 22 degrees 30 minutes south latitude. From the de Gamboa account, the Indian town actually appears to have been named "Morro de Jorje" as opposed to "Mormorena."

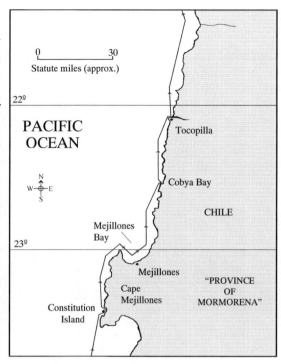

Map 34. Outline map showing the "Mormorena" landfall at 22 degrees 30 minutes south latitude on the coast of northern Chile.

The "Island of Thieves" and Mindanao

As with his California landfall, there has been much debate over the location of Drake's first landfall after the Pacific crossing. William Lessa performed an exhaustive evaluation of demographics and ethnological evidence, narrowed the possibilities down to the Islands of Yap and the Palau group, and declared in favor of the latter. Notwithstanding his dissertation on the "faults, flaws and defects" of *The World Encompassed*, Lessa ended up championing the validity of the navigational detail it provides for this landfall. Robert Power, after acknowledging that "usually the text of *The World Encompassed* can be accepted at face value for its accuracy," flatly rejected its navigational detail for the Island of Thieves in making his case for Yap.[4]

The World Encompassed reports that Drake's Island of Thieves was the first land sighted after the Pacific crossing; that the islands lay "about 8 degrees northward of the line"; that after departing the island, Drake sailed westward for thirteen days without sight of land; and that after sighting land, Drake coasted four islands "standing in 7 deg. 5 min. to the northward of the line, the largest called Mindanao," where they landed and took on water. This unambiguous navigational data points us to the northernmost islands of the Palau group, probably the north side of Babelthuap at about 7 degrees 40 minutes, with a Mindanao landfall at the mouth of the Davao River at about 7 degrees 5 minutes. As Lessa pointed out, the rather slow sailing time from the Palau Islands to Davao is consistent with the scant wind conditions that prevail in this region during the fall season. He also offers a good explanation for why the English reported seeing four islands in sailing along the shorelines of the Davao Gulf: they mistook the San Agustin Peninsula for an island (see inset on map 35).

---

4. Lessa's interesting work, published in 1975, is entitled, *Drake's Island of Thieves*. Power voiced his opinion in an unpublished monograph entitled "A Report on the Concept that Drake's Island of Thieves was Mindanao," 1971.

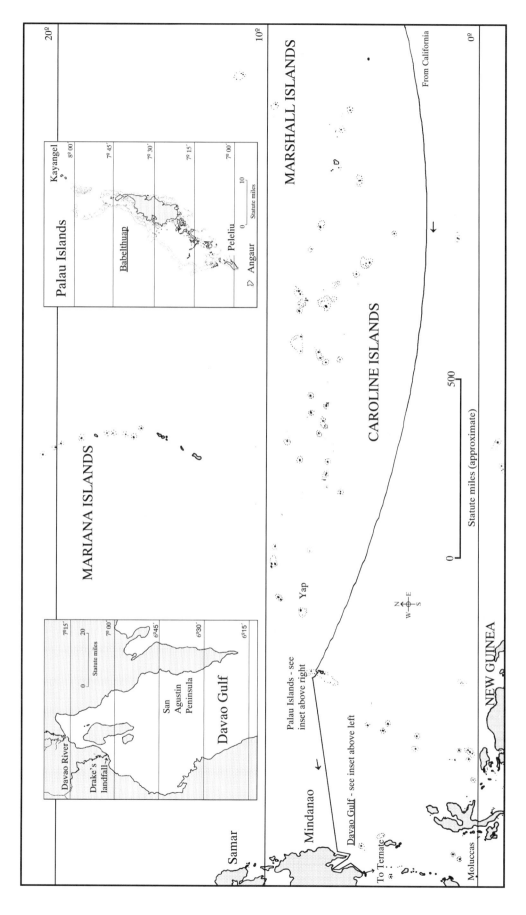

Map 35. Outline map showing Drake's route through Micronesia and his "Island of Thieves" (Babelthuap, Palau Islands) and Mindanao (Davao Gulf) landfalls (underlined). Redrawn and adapted from DRAKE'S ISLAND OF THIEVES by William Lessa, University Press of Hawaii, 1975.

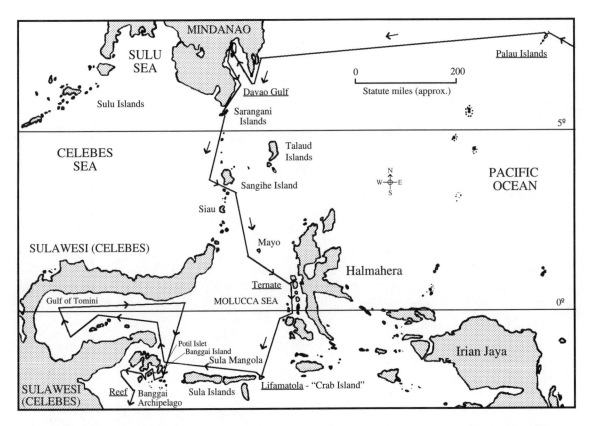

Map 36. Outline map showing Drake's route through Northern Indonesia and his "Crab Island" landfall (landfalls underlined).

"Crab Island"

There is only one locale in Indonesia that completely satisfies the detailed descriptive and navigational information that *The World Encompassed* provides for the small uninhabited island upon which the English built a fort, and spent three weeks in November 1579 trimming the *Golden Hind* —contrary to conventional wisdom, it is not in the Banggai Archipelago.

First of all, according to the account, Crab Island must lie not only near "1 deg. 40 min. toward the pole Antarctic," but also within five days' sail of Ternate in the season "before the coming of the breeze." In reaching Ternate, we know Drake's progress was slow in sailing south-southeast from Sangihe Island to Mayo Island: 170 miles in 6 days (about 30 miles per day). His progress was also slow sailing westerly from Palau to Davao: 500 miles in 13 days (about 40 miles per day); and for the short leg south-southeast from Mayo Island to Ternate: 60 miles in 2 days (about 30 miles per day). With five-days' sailing time it is not impossible for Drake to have reached the Banggai group located about 300 miles west-southwest of Ternate (60 miles per day), but given what is stated about the winds and the condition of the ship, it seems unlikely.

Secondly, according to *The World Encompassed,* after departing Crab Island, Drake did not sight the east coast of Celebes until he had sailed for four days on a westerly heading and become entangled in some islands (presumably the Banggai group). How could Drake have departed an island in the Banggai group and sailed west for four days without sight of the Celebes?

John Drake's second declaration also tends to refute the Banggai Archipelago theory. It states that "after leaving Ternate the English sailed among many islands to one they named Crab Island," implying their course was southerly through the Moluccas, not westerly into the Molucca Sea.

To agree with the accounts, the island should lack fresh water supplies, have a large island nearby to the west; and another island "somewhat to the south" where the English could secure water. It should also be covered with palm trees and densely populated with coconut crabs, fireflies, and large bats. Never having visited the locale, nor seen any descriptive information about it, I will venture to say that the small island located off the east end of Sula Mangola, about 170 miles south-southwest of Ternate, fits these descriptions to the letter. On all the maps I have seen, it appears unpopulated, presumably due to a lack of fresh water.

Lifamatola Island is located at 1 degree 50 minutes south latitude, within fifteen minutes of the position reported in *The World Encompassed* and within about four day's sail due west of the east coast of Celebes Island.

Most authorities have placed Crab Island at or near Banggai Island in the Banggai Archipelago which is also within fifteen minutes of the latitude reported in *The World Encompassed*. As Henry Wagner pointed out, there seems to be some solid evidence supporting a sojourn at Banggai in a Spanish report written in 1582 by one Francisco de Dueñas, who visited the Moluccas in the fall of 1581. Dueñas obtained his information on Drake's operations from the local Portuguese and islanders. The Dueñas account places Drake's encampment and "shipwreck" on "Coro Island" near "Zangay Island" (Banggai?) off the town of "Jape" on the coast of "Limbotan" in territory controlled by the "king of Bonga" (Banggai?). The account states that the English spent three months in the area, during which time they had a great deal of interaction with the native islanders.[5]

I regard the Dueñas account as untrustworthy in that the pertinent details appear to be garbled as would be expected for an account based on second or thirdhand information recorded two years after the fact. Furthermore, the impression from both *The World Encompassed* and John Drake's second declaration is that the English had little interaction with the islanders during their three-and-a-half-week interlude at Crab Island. Based on what is reported in *The World Encompassed*, the English would likely have interacted with the inhabitants of the Banggai Archipelago during the four-week period following their departure from Crab Island.

In any event, I stress that whether or not Drake's "Crab Island" was Lifamatola Island, Banggai Island (Wagner's choice), or Potil Islet off the northeast tip of Banggai Island (Aker's choice), Drake's land-based observation of latitude was accurate to within fifteen minutes.

An Unnamed Island and "Barativa"

The World Encompassed reports that on January 14, 1580, Drake cast anchor at an unnamed island in "4 deg. 6 min." on the south part of the Celebes where they spent the day "watering and wooding." This can only have been Wowoni Island located at about 4 degrees 8 minutes south.

The only question as to the location of Drake's two-day interlude at "Barativa," is which island it was in today's Barat Daya group: Wetar at 7 degrees 50 minutes south, Romang at 7 degrees 35 minutes south, or Damar at 7 degrees 9 minutes south. In reporting that the island "stands in 7 deg. 13 min. south the line," and that it was the largest of the last four islands of a five-island chain, *The World Encompassed* directs us to Damar.

---

5. Wagner, *Drake's Voyage*, pp. 172 and 183–86.

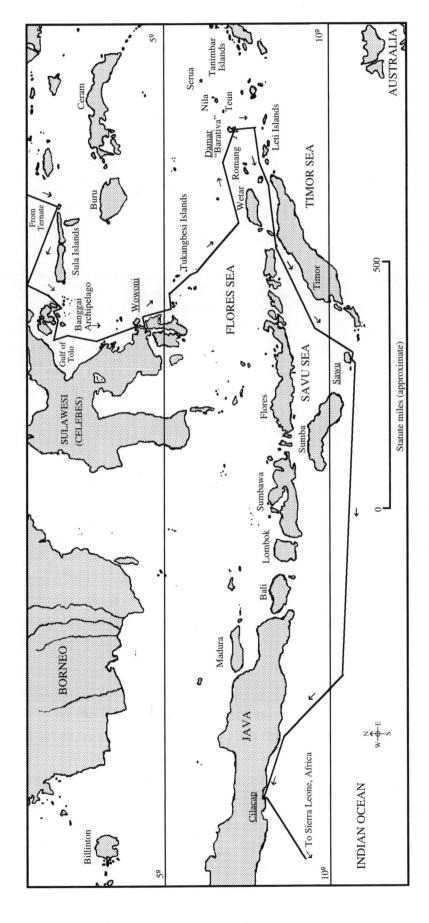

Map 37. Outline map showing Drake's route through Southern Indonesia and his landfall at a unnamed island (Wowoni Island) in the South Celebes and "Barativa" (Damar Island) (landfalls underlined).

Appendix E

The Pacific Northwest Landfall

According to *The World Encompassed*, in searching for the Northwest Passage, Drake took the *Golden Hind* northeast from 42 degrees north latitude on June 3, 1579 (June 13 by the modern calendar), to a bad bay at 48 degrees. From this bay, the English could see a coastline comprised of low plains, moderate hills and snow-capped peaks trending northwest as far as the eye could see. Forced by contrary winds to retreat south, Drake coasted down the shore, coming to anchor at 38 degrees 30 minutes on June 17 (June 27 by the modern calendar).

George Davidson and the Drake Navigators Guild, reflecting the view of most historians and scholars, elected to completely dismiss this contemporary evidence. They argue that Drake only made it as far north as 42 degrees or 43 degrees on the coast of today's Oregon, in an area where a mostly mountainous shoreline trends north-northeast. They cite the latitude evidence in Hakluyt's "The Famous Voyage" and contend that Drake, given prevailing winds and currents, could not possibly have reached 48 degrees in the fourteen days allotted in *The World Encompassed*.

Notwithstanding the views of the experts, I will demonstrate that the weight of the evidence is solidly in favor of Drake's reaching the coast of today's Washington State, where he very likely anchored on the lee side of Destruction Island in a bight located at 47 degrees 40 minutes: today's Queenhythe Bay.

Comparison with the Bruno de Hezeta Voyage of 1775

In addressing the question of how far north Drake ventured in his search for the Northwest Passage, it is useful to examine the records of some of the early navigators—in particular, those involved with the second Pacific Northwest Coast exploring expedition sent out from San Blas, Mexico, in the late 1770s by Viceroy Antonio María Bucareli y Ursúa. As it turns out, one of the vessels involved in this expedition, the *Santiago*, captained by Bruno de Hezeta, effectively duplicated the *Golden Hind's* route north and retreat south as reported in *The World Encompassed*, albeit in discontinuous segments, and for the last leg of the voyage, a little later in the season. I provided a brief account of the second Bucareli expedition in chapter 35. Here I summarize just the segments pertinent to the issue at hand.[1]

The "Golden Hind" northbound – According to *The World Encompassed*, on June 3, 1579, Drake took the two ships (*Golden Hind* and Tello's bark) past 42 degrees on a northward heading, averaging about twenty-nine leagues per day (1.3 degrees/day). By June 5, having traveled two degrees farther northward, the wind changed, presumably to the northwest, forcing them to "run in with the shore," which they "then first descried." They ultimately anchored in a "bad bay" at 48 degrees north latitude.

---

1. Galvin, ed., in *Journal of Explorations,* (1964) translated and published the diary of Father Miguel de la Campa, the Franciscan chaplain who accompanied Bruno de Hezeta aboard the frigate *Santiago*. The parallels between the northern voyages of the *Golden Hind* and the *Santiago* are extraordinary. The two square-riggers were nearly identical in size (about a hundred feet long) and draft (about thirteen feet fully laden). The ships left their respective Mexican ports (Guatulco at 15 degrees 45 minutes north and San Blas at 20 degrees 30 minutes north) on April 16 and March 16 respectively, with their captains both intent on sailing directly to a position about 60 degrees north on the then-uncharted Pacific Northwest Coast. Both ships were tailed by small consorts (Tello's bark/Bodega y Quadra's schooner) about forty feet in length. Both ships arrived off the California coast in the neighborhood of 42 degrees north latitude the first week of June, closer to land than expected. Hezeta sighted land with snow-capped hills and made a voluntary diversion to Trinidad Bay (41 degrees 5 minutes) for reprovisioning. After reaching 48 degrees and finding an uncomfortable roadstead at about 47 degrees 30 minutes, Hezeta made an excursion north to claim Nootka Sound for Spain. In his return down the coast from 48 degrees to 38 degrees, Hezeta coasted the shore, making observations of what he saw along the way. He appears to have passed Bodega Head in the wee hours of the morning at a time they were out of sight of land.

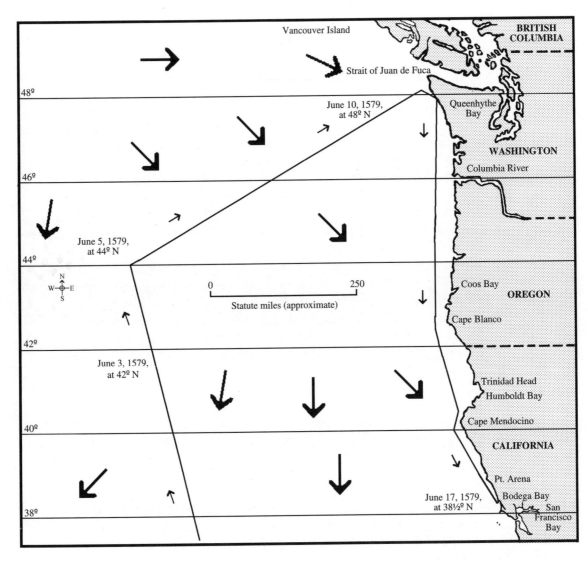

Map 38. Outline map showing Drake's route along the Pacific Northwest Coast, the "bad bay" landfall and the direction of the prevailing summer winds (*large arrows*). As first demonstrated by Kenneth Holmes, Drake's course north, as set forth in *The World Encompassed*, correlates quite well with the Pacific Northwest's prevailing summer wind patterns. Add ten days to the dates to convert them to the modern calendar system.

The "Golden Hind" southbound – According to *The World Encompassed* and "The Famous Voyage," leaving this dangerous roadstead at his first opportunity, Drake took the *Golden Hind* southward, and behind constant north to northwest winds, coasted along the shore without landing until June 17, at which time they fell upon a convenient harbor within 38 degrees towards the line. According to John Stow, Drake turned south on June 10. If so, it had taken the English about seven days to sail northeast from 42 degrees to 48 degrees, and another seven days to sail due south from 48 degrees to 38½ degrees.

The "Santiago" northbound – At noon, July 2, 1775, the two ships (*Santiago* and *Sonora*) were in 42 degrees 13 minutes north latitude, about ninety leagues from shore. By noon, July 4, having traveled two degrees farther northward with the wind west, they were at 44 degrees 12 minutes. On July 7, at 46 degrees 20 minutes, the wind veered to the northwest forcing them eastward. By noon, the ninth, they had sighted land in 47 degrees 36 minutes and hove to. As with Drake, it had taken them roughly seven days to sail northeast from 42 degrees to 47½ degrees.

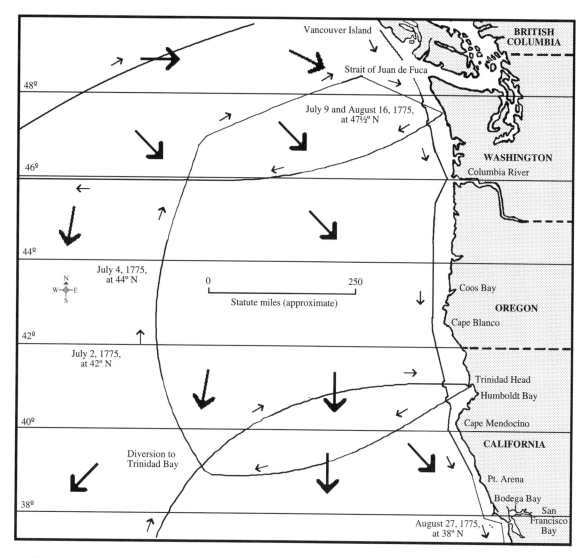

Map 39. Outline map showing the route of Bruno de Hezeta in the *Santiago* in the summer of 1775 and the prevailing summer winds (*large arrows*). In sailing from 42 degrees north latitude off the coast of California to 47½ degrees on the coast of Washington, then south to 38 degrees, it took Hezeta eighteen days to complete the route that *The World Encompassed* has Drake completing in fourteen days. Hezeta, however, was becalmed for three days on his way south. The track of the Santiago is derived from the contemporary accounts.

Juan Francisco de la Bodega y Quadra in the *Sonora* did not return south from the area of 48 degrees until well into the fall and made some explorations en route. Hezeta in the *Santiago* made his return in late August with the objective of covering the distance as quickly as possible while coasting along shore to examine the bays and harbors.

The "Santiago" southbound – At noon, August 16, the *Santiago* was at 47 degrees 14 minutes sailing south behind a northwest wind, but the wind shifted, halting further progress until ten o'clock when the wind shifted back to the northwest. Hezeta then sailed south until ten o'clock the night of the seventeenth and hove to until three in the morning. By noon on August 20, the Spanish were in 42 degrees 34 minutes, having covered about 5 degrees of latitude in about 4 days (about 1.25 degrees/day). Shortly after noon on August 20, the wind fell off and they were engulfed in fog. For the next three days, they were becalmed, and due to fog, could not take an observation again until noon on the twenty-fourth when they found themselves at 41 degrees 27 minutes (about 0.3

degree/day). About five o'clock that afternoon, the north wind rose, though very light until midnight, and they stood their course south until eight o'clock the evening of the next day. Weighing at four in the morning, they coasted south behind a good northwest wind, finding themselves at 37 degrees 56 minutes at noon on the twenty-seventh. Between the twenty-fourth and twenty-seventh, they had covered 3½ degrees of latitude in three days (1.2 degrees/day). Though it ultimately took Hezeta eleven days to cover the roughly 9½ degrees of latitude running south from 47½ degrees to 38 degrees, had he not been becalmed, he was on a pace to cover it in about eight days compared to Drake's seven, even with periods of adverse winds.

Evaluation – The navigational records correlate very well for the *Golden Hind* and the *Santiago*. Assuming Stow's June 10, 1579, departure date from the bad bay is valid, Drake would have had to make his way down the coast from 48 degrees to 38½ degrees at a slightly faster pace then did Hezeta when Hezeta was not becalmed (1.3 degrees/day vs. 1.2 degrees/day), but Drake would have had both stronger and more constant northerly winds in mid-June 1579 than did Hezeta in late August 1775.

Additional Correlations for the Retreat South

In preparing the following table, I provided data for just those segments of the respective voyages where I could accurately identify the positions and dates with a fair degree of certainty from the information at hand (see bibliography). In most cases, the commanders, like Drake, were taking pains to closely examine the shoreline while coasting south.

Table E-1. Sailing Rates for Square-Riggers Running South Along the Coast to the San Francisco Area							+
Vessel/captain	Start date/location	Latitude (approx.)	End date/location	Latitude (approx.)	Total days	Degrees traveled	Speed (deg./day)
Golden Hind/Drake	June 10, 1579, off Washington	47º 40′	June 17, 1579, off Bodega Head	38º 20′	7	9.3	1.3
San Agustin/Cermeño	Nov. 2, 1595, off Cape Mendocino	40º 30′	Nov. 4, 1595, off Point Reyes	38º	2	2.5	1.25
Santiago/Hezeta	Aug. 24, 1775	41º 30′	Aug. 27, 1775, off Point Reyes	38º	3	3.5	1.2
Sonora/Bodega y Quadra	Sept. 26, 1775	43º 15′	Oct. 1, 1775, off Tomales Point	38º 15′	5	6	1.2
Argonaut/Colnett	Sept. 11, 1790, at Havens Anchorage	38º 50′	Sept. 11, 1790, off Bodega Head	38º 20′	9.5 hours (0.4 days)	.5	1.25
Discovery/Vancouver	Oct. 14, 1793, off Cape Orford (Blanco)	43º	Oct. 17, 1793, off Point Arena	39º	3	4	1.3
Average not including Drake							1.25

Summary of Evidence in Favor of Drake Reaching 48 Degrees

- *The World Encompassed* reports the bad bay was located at 48 degrees. As demonstrated in appendix D, the latitudes that *The World Encompassed* reports for readings made at sea are consistently accurate to within about thirty minutes, and were very likely provided by Drake himself.

- *The World Encompassed* states the shoreline trended northwest at the point they gave up the search. From map 39 one can see that for the entire stretch of coastline between 40 degrees and 48 degrees, a pronounced northwest trend is found only in the vicinity of 48 degrees.

- *The World Encompassed* and "The Famous Voyage" both report that the vista in the area of the "bad bay" featured low plains, hills of moderate height, and snow-capped peaks. When approaching the coast just south of Destruction Island at about 47 degrees 40 minutes (Queenhythe Bay), the shoreline indeed trends northwest as far as the eye can see, with the land abreast "low and quite flat," backed first by wooded hills that rise to "moderate heights" and then by the "snow-peaks" of Washington's Olympus Range.

- *The World Encompassed* and "The Famous Voyage" both indicate that between June 5 when the English were at 44 degrees north latitude, and June 17 when they anchored at about 38 degrees, Drake took twelve days to cover a distance (six degrees) that, given the prevailing winds and currents, he could have easily covered in six days or less if sailing due south—there was clearly a diversion.

- *The World Encompassed* reports a quick escape from the "bad bay," not a prolonged stay— "in this place was no abiding for us."

- *The World Encompassed*, John Stow account, "The Famous Voyage," and the "Anonymous Narrative" collectively report or strongly imply that the English, desirous of escaping the cold weather, coasted south without landing, and caught just fleeting glimpses of the shorelines they were passing.

- John Stow reported that Drake reached 47 degrees and turned back on June 10, 1579.
- John Drake reported the English reached 48 degrees.
- The "Anonymous Narrative" reports that Drake reached 48 degrees.
- The French Drake Map, circa 1582–85, shows Drake reaching 48 degrees.
- The Dutch Drake Map, circa 1582–85, shows Drake reaching 48 degrees.
- The Silver Map, 1589, shows Drake reaching 48 degrees.
- The Hondius Broadside Map, circa 1589, showed Drake reaching 48 degrees until the track was changed circa 1596 to correlate with the text of Hakluyt's account.
- The Molyneux Globe, circa 1592, shows Drake reaching 48 degrees.

Rebutting the Oregon Landfall Theory

I believe that George Davidson and Raymond Aker, both of whom were infinitely more qualified to speak on the matter of navigation than this seasick-prone landlubber, strayed off course in the interpretation of the following rather ambiguous statement in *The World Encompassed*, key to the issue at hand (*emphasis added*): "The 5 day of June, we were forced by contrary winds to run in with the shore, which we *then first descried*, and to cast anchor in a bad bay . . ."

Davidson and Aker interpreted the phrase "then first descried" to mean that the English sighted the coast on June 5, 1579, in about 44 degrees north latitude, and immediately entered the "bad bay." Taken in the context of preceding and subsequent statements, however, *The World Encompassed* statement should literally be interpreted to read that on June 5 Drake was forced by contrary winds to run toward the land which they "then" (subsequently or consequently) "first descried" first saw signs of." In the discussion immediately preceding, Drake has successfully encouraged his men to push on past 44 degrees which they had already reached by June 5. In the discussion immediately following, they have reached a bad bay at 48 degrees north latitude.

Webster defines the word "descry" as follows: "to perceive with a special effort of the senses; to catch sight of obscure objects; to discern; to detect." There is every reason to believe that in using the term "descried," the account's author was indicating that the English merely saw signs of the coast as they came within 44 degrees—not that they actually sighted it. In this respect, consider the following warning that Bodega y Quadra relayed to the commanders of the Alejandro Malaspina and Dionisio Alcalá Galiano expedition before it departed San Blas, Mexico, in 1791 to explore the Pacific Northwest Coast:

> [As you continue your course northwesterly to]. . . 41º to 45º you will see rough waters and will see various signs indicating land nearby, such as plants, insects, and birds . . . Signs of the coast are found at a distance of 70 to 80 leagues [about 210 to 240 miles], increasing with plants and sea otters, and when nearby, ducks and the color of very rough water.[2]

Taking the case for a landfall at 42 degrees or 43 degrees, one is very hard pressed to explain how Drake occupied his time. With the date of the "bad bay" landfall moved up to June 5, this leaves him twelve full days to cover just 4 to 5 degrees of latitude in coasting the shore due south without landing to 38 degrees. Hezeta, at the pace he sailed in late August, would have made the distance in about one-third of this time (three to four days). The concept of Drake making a prolonged stay in the "bad bay" or taking a torturous path to closely examine the coast, is not consistent with the language in either the Drake or Hakluyt accounts.

In truth, there is no credible contemporary evidence to support the theory that Drake's Pacific Northwest landfall was at 42 to 43 degrees north latitude on the coast of Oregon. "The Famous Voyage," as with many other details of the voyage, provides no information on when or where Drake first sighted the coast after encountering adverse winds on June 5, 1579.

---

2. Cited by Cutter, *Malaspina & Galiano*, p. 10.

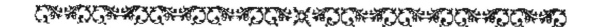

Selected Bibliography

Sixteenth-Century Accounts and Cartography of Drake's Voyages

Andrews, Kenneth R. *The Last Voyage of Drake and Hawkins*. London: Cambridge University Press, 1972. Contains numerous contemporary documents.

"Anonymous Narrative." Harleian Manuscript No. 280, Folio 23, British Museum, circa 1589. Reprinted in Vaux, ed., 1854; and Wagner, 1926. A contemporary account of the circumnavigation.

Antón, Don Juan de. Deposition in Panama, 1579. Translated in Nuttall, 1914; and Wagner, 1926.

Blunderville, Thomas. *M. Blunderville, His Exercises Containing Six Treatises*. London, 1594. Excerpts reprinted in Wagner, 1926.

Bravo, Benito Díaz. Deposition in Panama, 1579. Translated in Nuttall, 1914; and Wagner, 1926.

British Museum. *Sir Francis Drake's Voyage Round the World 1577-1580, Two Contemporary Maps*. London: British Museum, 1927. The Hondius Broadside and French Drake Map.

Bry, Theodore de. *Americae, pars VIII*. Frankfurt, 1599. Reiterates Hakluyt's "The Famous Voyage" in Latin with several illustrations.

_____. *Grande et Petit Voyages*. Frankfurt, 1590–1634. A series of volumes in Latin, some published by the son, Johann-Theodore de Bry.

Camden, William. *Annales Rerum Anglicarum, et Hibernicarum, regnante Elizabetha, ad Annum Salutis MDLXXXIX*. London, 1615. Excerpts reprinted in Wagner, 1926.

Christy, Miller. *The Silver Map of Drake's Voyage*. London: Henry Stevens, 1900.

Churchill, Awnsham. "Sir William Monson's Naval Tracts." *A Collection of Voyages and Travels*. Vol. 8 of 8 vols. London, 1704.

Cliffe, Edward. Edward Cliffe account. Written in England, circa 1589. First published by Hakluyt, 1600. Reprinted in Vaux, ed., 1854; Penzer, ed., 1926; and Hampden, ed., 1972. A contemporary account of the circumnavigation.

Colchero, Alonso Sánchez. Deposition in Panama, 1579. Translated in Nuttall, 1914.

Conyers, John. *See* Sir Francis Drake and Fletcher (Sloane Manuscript No. 61).

Cooke, John. John Cooke narrative. Written in England, circa 1589. First published in Vaux, ed., 1854. Reprinted in Penzer, 1926; Wagner, 1926; and Hampden, ed., 1972. A contemporary account of the circumnavigation.

Donno, Elizabeth S., ed. *An Elizabethan in 1582: The Diary of Richard Madox Fellow of All Souls*. London: Hakluyt Society, 1976.

Doughty, Thomas. "Documents relating to Thomas Doughty." Written in England, circa 1580. First published in Vaux, ed., 1854.

Drake, Sir Francis, and Francis Fletcher. "The First Part of the Second Voyage ..." (Francis Fletcher narrative). Written in England, circa 1591. Transcribed by John Conyers, London, 1677. Sloane Manuscript No. 61, British Library, London. Excerpts published by Vaux, ed., 1854. First published in full by Penzer, ed., 1926. Sometimes called the "Fletcher notes" or "Fletcher narrative." A draft contemporary account of the circumnavigation probably coauthored by Drake and Fletcher.

_____. *The World Encompassed*. Written at Plymouth, England, by Drake and Fletcher circa 1591–95. Edited and published by Sir Francis Drake, baronet, London: Nicholas Bourne, 1628. Reprinted 1635; also in 1653 as part of a collection of voyages, *Sir Francis Drake's Voyages*. Reprinted by Vaux, ed., 1854; and Penzer, ed., 1926. Reprinted in facsimile by Readex Microprint Corp., 1966.

Drake, Sir Francis, and Philip Nichols. *Sir Francis Drake Revived*. Written at Plymouth, England, by Drake and Nichols circa 1591–95. Edited and published by Sir Francis Drake, baronet, London: Nicholas Bourne, 1626. Reprinted by Hampden, ed., 1972.

Drake, Sir Francis, baronet. *See* Drake, Sir Francis.

Dudley, Robert. *Arcano del Mare*. Florence, Italy, 1647. An atlas of early seventeenth-century maps.

Dueñas, Francisco de. Report of a Spanish expedition to the Moluccas in 1582. Extracts translated in Wagner, 1926.

Fletcher, Francis. *See* Drake, Sir Francis.

Gamboa, Pedro Sarmiento de. "Relacion," 1579. Translated in Nuttall, 1914; and Wagner, 1926.

Hakluyt, Richard. *See also* D. B. Quinn.

_____ . *Divers Voyages*. London, 1582. Reprinted for the Hakluyt Society, London, 1850.

_____ . "A particular discourse concerning the great necessity and manifold commodities that are likely to grow to this realm of England by the western discoveries lately attempted, written in the year 1584." Called "The discourse on western planting."

_____ . *Peter Martyr's Decades*. Paris, 1587.

_____ . *The Principal Navigations, Voyages and Discoveries of the English Nation*. London, 1589. Contains "The Famous Voyage" and the Lopez Vaz discourse of the 1572 Nombre de Dios expedition. A facsimile of the 1589 edition was printed by Cambridge University Press for the Hakluyt Society, 1965, in 2 vols.

_____ . Same, 1598–1600 edition. 3 vols. Vol. 3, 1600, contains "The Famous Voyage"; "The Course"; the Nuño da Silva account; the Edward Cliffe account; and the Lopez Vaz discourse.

Hampden, John, ed. *Francis Drake, Privateer: Contemporary Narratives and Documents*. Alabama: University of Alabama Press, 1972. Reprints *Sir Francis Drake Revived*; *The World Encompassed*; the John Winter account; "The Third and Troublesome Voyage" of John Hawkins; and the John Cooke narrative.

Hawkins, Sir John. "A true declaration of the troublesome voyage of M. John Hawkins to the parties of Guinea and the West Indies, in the years of our Lord 1567 and 1568." London, 1569. Published by Hakluyt, 1589.

Hawkins, Sir Richard. *The Observations of Sir John Hawkins, Knight, in his Voyage into the South Sea*. London, 1622. Reprinted by Markham, 1878.

Jorje, Nicolás. Deposition in Panama, 1579. Translated in Nuttall, 1914; and Wagner, 1926.

Keeler, Mary F., ed. *Sir Francis Drake's West Indian Voyage, 1585–86*. London: Hakluyt Society, 1981. Contains numerous contemporary documents.

Kerr, Willis H. "The Treatment of Drake's Circumnavigation in 'Hakluyt's Voyages,' 1589." *Bibliographical Society of American Papers* 34 (fourth quarter, 1940).

Kraus, Hans P. *Sir Francis Drake, a Pictorial Biography*. Amsterdam: N. Israel, 1970. Contains a photographic reprint of "The Famous Voyage," 1589 ed.; the Hondius Broadside Map; the French Drake Map; and many other contemporary maps and documents.

Lanberd, Cornieles. Deposition in Panama, 1579. Translated in Nuttall, 1914; and Wagner, 1926.

Laughton, John K., ed. *The Spanish Armada, Anno 1588*. 2 vols. New York: Lennoz Hill, first published 1894, reprinted 1971. Contains numerous contemporary documents.

Madox, Richard. *See* Donno ed.

Markham, Sir Clements. *The Hawkins Voyages*. London: Hakluyt Society, 1878.

Martyr, Peter. *See also* Hakluyt, 1587.

_____ . *Decades*. Paris, 1587.

Montanus, Arnold. *Die Nieuwe en Onbekonde Weereld: Beschryvning van Americo en 1 + Zuid-Land. . . .* Amsterdam, 1671. Contains the Montanus illustration.

Morena, N. de. *See* Zárate y Salmerón.

Nebenzahl, Kenneth, ed. *Atlas of Columbus and the Great Discoveries*. New York: Rand McNally, 1990. Contains the Hondius Broadside Map; the 1569 Mercator world map; and the 1599 Hakluyt-Wright world map (Twelfth-Night Map).

Nichols, Philip. *See* Drake, Sir Francis.

Nuttall, Zelia. *New Light on Drake*. London: Hakluyt Society, New Series, No. 34, 1914. Includes translations of many Spanish documents dealing with Drake, such as depositions of Nuño da Silva, John Drake, Nicolás Jorje, and San Juan de Antón.

Oxenham, John. Deposition in Panama, 1579. Translated in Nuttall, 1914.

Penzer, N. M., ed. *The World Encompassed*. London: Argonaut Press, 1926. Contains *The World Encompassed*; Francis Fletcher's narrative; John Cooke's narrative; Thomas Doughty's oration; Nuño da Silva's second relation; Edward Cliffe's account; Pedro Sarmiento de Gamboa's report; Francisco de Zárate's letter; Juan de Antón's testimony; Drake's letter of safe conduct; and the charges against Drake of hanging captives.

Purchas, Samuel. *Purchas, His Pilgrims*. London: William Stansby, 1625.

Quinn, David B., and R. A. Skelton. *The Principal Navigations, Voyages and Discoveries of the English Nation*. 2 vols. Cambridge, England: Cambridge University Press, 1965.

Rengifo, Francisco Gómez. Deposition in Panama, 1579. Translated in Nuttall, 1914.

Silva, Nuño da. *See also* Hakluyt, 1600, and Penzer, ed., 1926.

_____ . Several accounts, declarations and logbook entries. Translations published in Nuttall, 1914 (complete set); and Wagner, 1926 (composite account).

Sloane Manuscript No. 61. *See* Drake, Sir Francis, and Francis Fletcher, circa 1591.

Stow, John. *The Chronicles of England, from Brute unto this Present Year of Christ, 1580*. Reprinted in London, 1592, with a short account entitled "Francis Drake, His Voyage Round About the World." Excerpts reprinted in Wagner, 1926.

Taylor, E. G. R. "Francis Drake and the Pacific: Two Fragments." *Pacific Historical Review* 1 (1932): 360–69. Contains excerpts from the Richard Madox diary.

Ulbaldino, Petruccio. "Commentario della impresa. . . ." An unpublished account of the Armada campaign prepared in 1589, ostensibly with Drake's input. The account is discussed at length in Corbett, *Drake and the Tudor Navy*, vol. 2, pp. 442–451.

Vargas, Gaspar de. Letter to Viceroy, April 13, 1579. Translated in Nuttall, 1914; and Wagner, 1926.

Vaux, W. S. W., ed. *The World Encompassed by Sir Francis Drake*. London: Hakluyt Society, 1854. Reprints *The World Encompassed*; "The Famous Voyage"; "The Course"; the "Anonymous Narrative"; Francis Fletcher's narrative (excerpts); John Cooke's narrative; Thomas Doughty's oration; Nuño da Silva's second relation; and Edward Cliffe's account.

Vaz, Lopez. Lopez Vaz discourse. Written circa 1580. First published in Hakluyt, 1600. Reprinted in Vaux, ed., 1854; Penzer, ed., 1926; and Hampden, ed., 1972. A contemporary account of the circumnavigation.

Wallis, Helen. *The Voyage of Sir Francis Drake Mapped in Silver and Gold*. Berkeley, California: Bancroft Library, 1979. Contains photographic plates of the following maps in the British Library: the Francis Fletcher map of the tip of South America; the Drake-Mellon Map (cropped); the French Drake Map (cropped); the Drake Silver Map; and the Hondius Broadside Map.

_____ . "The Cartography of Drake's Voyage." In Thrower, ed., *Sir Francis Drake and the Famous Voyage, 1577–1580*, pp. 121–63. *See* Thrower, ed., in the Biographical section of this bibliography.

Waters, David W., ed. *The True and Perfecte Newes of that Valient Knight Sir Francis Drake*. Hartford, England: Henry C. Taylor, 1955. Reprint of the 1587 account of Thomas Greepe.

Wernham, R. B., ed. *The Expedition of Sir John Norris and Sir Francis Drake to Spain and Portugal, 1589*. Aldershot, England: Temple Smith for the Navy Records Society, 1988. Contains numerous contemporary documents.

Winter, John. "Report to George and Sir William Winter, 1579." In Hampden, ed., 1972.

Zárate, Don Francisco de. Letter to the Viceroy, April 16, 1579. Translated in Nuttall, 1914; and Wagner, 1926.

Zárate y Salmerón, Father Gerónimo de. The N. de Morena account. *Documentos para la historia de Mexico*. Ser. 3, vol. 4, published circa 1626, a part of which was translated as "Relation of Events in California and New Mexico up to 1626," an excerpt from which was published by Lummis, ed., in *Land of Sunshine* 12 (Feb. 1900): 13.

Biographical and Tudor England

Allen, Michael. "Charles Fitzgeffrey's Commendatory Lamentation on the Death of Drake." In Thrower, ed., *Sir Francis Drake and the Famous Voyage, 1577–1580*, pp. 99–111.

Andrews, Kenneth R. *Drake's Voyages: A Reassessment of Their Place in Elizabethan Maritime Expansion.* New York: Charles Scribner, 1967.

_____ . "The Aims of Drake's Expedition of 1577–1580." *American Historical Review* 73, no. 3 (February 1968): 724–41.

_____ . "Drake and South America." In Thrower, ed., *Sir Francis Drake and the Famous Voyage, 1577–1580*, pp. 49–59.

Bancroft, Eleanor. "One Hundred Works Relating to Sir Francis Drake." In *The Plate of Brass.* San Francisco: California Historical Society special publication no. 13, 1937. Reprinted in special publication no. 25, 1953.

Bancroft, George. *History of the United States.* Boston: Charles C. Little, 1852.

Barrow, John. *The Life, Voyages, and Exploits of Admiral Sir Francis Drake, Knight.* London: John Murray, 1863.

Belleville, Bill. "A Dead Man's Tale: Plundering the Lore of the Spanish Main." *Destination Discovery*, the magazine of the Discovery Channel (January 1994): 19–25.

Benson, E. F. *Sir Francis Drake.* London: John Lane, 1927.

Bradford, Ernle. *The Wind Commands Me, A Life of Sir Francis Drake.* New York: Harcourt, Brace & World, 1965.

Cartography. A calendar with thirteen full-color reproductions of sixteenth-century maps by Abraham Ortelius. San Francisco: Cavallini, 1995.

Corbett, Julian S. *Sir Francis Drake.* London: Macmillan, 1890.

_____ . *Drake and the Tudor Navy.* 2 vols. London: Longmans Greene, 1898.

Cumming, Alex A. *Sir Francis Drake and The Golden Hind.* Norwich, England: Jarrold & Sons, 1975.

Cummins, John. *Francis Drake: the Lives of a Hero.* New York: St. Martin's Press, 1995.

Drake, Samuel G. "Passengers of the *Golden Hind.*" *New England Historical Genealogical Register* 1 (Apr. 1847): 127–31.

Draper, Benjamin P. "A Collection of Drake Bibliographic Items: 1569–1659." In Thrower, ed., *Sir Francis Drake and the Famous Voyage, 1577–1580*, pp. 173–206.

_____ . "Drake Bibliography, 1569–1979." Typescript. Manuscript copies containing close to six hundred entries are held by the British Library, London, and the Bancroft Library, Berkeley, California.

Eliott-Drake, Lady Elizabeth. *The Family and Heirs of Sir Francis Drake.* 2 vols. London: Smith, Elder, 1911.

Erickson, Carolly. *The First Elizabeth.* New York: Summit Books, 1983.

Fitzgeffrey, Charles. *Sir Francis Drake, His Honorable Life's Commendation, and His Tragic Death's Lamentation.* London, 1595. Reprinted by Michael Allen, in Thrower, ed.

Froude, James Anthony. *History of England from the Fall of Wolsey to the Defeat of the Armada.* Vol. 5. London, 1872–75.

_____ . *English Seamen in the Sixteenth Century.* New York, 1895.

Gerhard, Peter. *Pirates on the West Coast of New Spain, 1575–1742.* Glendale, California: Arthur H. Clarke, 1960.

Gill, Crispin. "Drake and Plymouth." In Thrower, ed., *Sir Francis Drake and the Famous Voyage, 1577–1580*, pp. 78–98.

Hough, Richard. *The Great Admirals.* New York: William Morrow, 1977.

Jewkes, W. T. "Sir Francis Drake Revived. From Letters to Legend." In Thrower, ed., *Sir Francis Drake and the Famous Voyage, 1577–1580*, pp. 112–20.

Lessa, William. *Drake's Island of Thieves.* Honolulu, Hawaii: University Press of Hawaii, 1975.

_____ . "Drake in the South Seas." In Thrower, ed., *Sir Francis Drake and the Famous Voyage, 1577–1580,* pp. 60–77.

Mainwaring, Sir Henry. *Seaman's Dictionary.* London, 1623. Excerpts reprinted in Aker, 1972.

Mason, A. E. W. *The Life of Francis Drake.* London: Hodder and Stoughton, 1941.

McDowell, Cleta Seward. "Builders of the *Golden Hind.*" *Pacific Historian* (Spring 1976): 46–48.

Menzies, Winfred M. *A Visit to Buckland Abbey.* Ross, California: Sir Francis Drake Association, Leaflet no. 2, October 14, 1941.

Morgan, Kenneth O., ed. *The Oxford Illustrated History of Britain.* New York: Oxford University Press, 1984.

Morison, Samuel E. *The Oxford History of the American People.* Vol. 1. New York: Oxford University Press, 1965.

_____ . *The European Discovery of America, The Northern Voyages, A.D. 1500–1600.* New York: Oxford University Press, 1971.

_____ . *The European Discovery of America, The Southern Voyages, A.D. 1492–1616.* New York: Oxford University Press, 1974.

Ortelius, Abraham. *Theatrum Orbis Terraum.* Antwerp, 1570. Fifteen maps.

Parry, John. "Drake and the World Encompassed." In Thrower, ed., *Sir Francis Drake and the Famous Voyage, 1577–1580,* pp. 1–11.

Quinn, David B. "Early Accounts of the Famous Voyage." In Thrower, ed., *Sir Francis Drake and the Famous Voyage: 1577–1580,* pp. 33–48.

_____ . *Set Fair for Roanoke.* Chapel Hill, North Carolina: University of North Carolina Press, 1985.

_____ . *The Roanoke Voyages: 1584–1590.* 2 vols. New York: Dover Publications, 1991. Originally published in 1955 by the Hakluyt Society, London.

Roche, T. W. E. *The Golden Hind.* New York: Praeger, 1973.

Rowse, A. L. "The Elizabethans and America." *American Heritage* (Apr. 1959).

Sanderlin, George. *The Sea Dragon.* New York: Harper Row, 1969.

Sugden, John. *Sir Francis Drake.* New York: Henry Holt, 1990.

Thomson, George Malcolm. *Sir Francis Drake.* New York: William Morrow, 1972.

Thrower, Norman J., ed. *Sir Francis Drake and the Famous Voyage, 1577–1580.* Berkeley, California: University of California Press, 1984.

Wagner, Henry. *Sir Francis Drake's Voyage Around the World.* San Francisco: John Howell, 1926.

Wallis, Helen. "The Cartography of Drake's Voyage." In Thrower, ed., *Sir Francis Drake and the Famous Voyage: 1577–1580,* pp. 121–63.

Waters, David A. *The Art of Navigation in Elizabethan and Early Stuart Times.* London: Hollis & Carter, 1958.

_____ . "Elizabethan Navigation." In Thrower, ed., *Sir Francis Drake and the Famous Voyage, 1577–1580,* pp. 12–32.

White, Robert. *A Witness for Eleanor Dare.* Lagunitas, California: Lexikos, 1991.

Williamson, James A. *Sir John Hawkins, the Time and the Man.* Oxford, 1927.

_____ . *The Age of Drake.* London: Adam & Charles Black, 1938.

_____ . *The Ocean in English History.* Oxford: Clarendon Press, 1941.

_____ . *Hawkins of Plymouth.* London, 1949.

_____ , ed. *The Observations of Sir Richard Hawkins.* New York: Argonaut Press, 1970.

Wilson, Derek. *The World Encompassed.* New York: Harper & Row, 1977.

Winchester, Simon. "Sir Francis Drake is Still Capable of Kicking up a Fuss." *Smithsonian* (January 1997): 82–91.

Wright, Helen, and Samuel Rappnet. *The Great Explorers.* New York: Harper & Brothers, 1957.

Regional History

Allen, Sidney P. "Battle of Bodega Head." *Public Utilities Fortnightly* (June 18, 1964): 37–40.

Anderson, Bern. *Surveyor of the Sea: The Life and Times of Captain George Vancouver*. Seattle: University of Washington Press, 1960.

Ascensión, Father Antonio de la. *The Voyage of Sebastián Vizcaíno. See* Wagner, 1929.

Bancroft, Hubert Howe. *The Works of Hubert Howe Bancroft*. Vol. 17: *History of the Northwest Coast*. Vol. 1. San Francisco: A. L. Bancroft, 1884.

_____ . *The Works of Hubert Howe Bancroft*. Vol. 28: *History of California*. Vols. 1–4. San Francisco: A. L Bancroft, 1884

_____ . *The New Pacific*. 3d edit. New York: Bancroft Company, 1915.

Baranov, Alexander. "October 14, 1808: Instructions from Alexander Baranov to his Assistant, Ivan A. Kuskov, Regarding the Dispatch of a Hunting Party to the Coast of Spanish California." In Dmytryshyn et al., eds., *The Russian American Colonies, 1798–1867*, vol. 3, pp. 165–74.

Barbour, M., R. Craig, F. Drysdale, and M. Ghiselin. *Coastal Ecology of Bodega Head*. Berkeley, California: University of California Press, 1973.

Barrett, Samuel A. "The Ethno-Geography of the Pomo and Neighboring Indians." *University of California Publications, American Archaeology and Ethnology* 6, no. 1 (1908).

_____ . "Pomo Buildings." Written in 1916. Reprinted in *Seven Early Accounts of the Pomo Indians and Their Culture* by Robert Heizer, 1975, pp. 37–63. *See* Heizer, ed., 1975.

Barrington, Daines, ed. *See also* Russell, ed., 1920.

_____ . *Miscellanies*. London, 1781. Contains Mourelle's "Voyage of the *Sonora*": the account of the Juan Francisco de la Bodega y Quadra's 1775 voyage.

Bean, Lowell John, and Dorothea Theodoratus. "Western Pomo and Northeastern Pomo." Chapter in Heizer, ed., *Handbook of North American Indians*. Vol. 8, *California*, pp. 289–305. Washington D.C.: Smithsonian Institute, 1978.

Beard, Yolanda. *The Wappo: A Report*. Banning, California: Malki Museum Press, 1979.

Beechey, Frederick W. *Narrative of a Voyage to the Pacific in 1825–1828*. 2 vols. London, 1831.

Bolanos, Francisco de. *Derrotero* [1603]. Preserved in the Biblioteca Nacional in Madrid. Excerpts translated by Wagner, 1929. Contemporary account of the Vizcaíno expedition.

Bolton, Herbert E. *Outpost of Empire*. New York: Alfred and Knopf, 1931.

_____ , ed. *Historical Memoirs of New California by Fray Francisco Palóu*. 4 vols. Berkeley, California: University of California Press, 1926.

_____ , ed. *Font's Complete Diary*. Berkeley: University of California Press, 1933.

Brown, Alan. "San Francisco Bay Costanoan." *International Journal of American Linguistics* 39 (3): 184–89.

Browning, Peter, ed. *The Discovery of San Francisco Bay, the Portola Expedition of 1769–1770*. Lafayette, California: Great West Books, 1992.

California Coastal Commission. *California Coastal Access Guide*. Berkeley, California: University of California Press, 1991.

California Department of Fish and Game. *Natural Resources of Bodega Harbor*. Sacramento, May 1975.

Callaghan, Catherine. *Bodega Miwok Dictionary*. Berkeley, California: University of California Press, 1970.

Campa, Father Miguel de la. *See* Galvin, ed., 1964.

Cañizares, José de. "The Report of José de Cañizares, First Sailing Master of the *San Carlos*, to Captain Ayala." *See* Galvin, ed., 1971.

Carter, Charles F., ed. "Duhaut-Cilly's Account of California in the Years 1827–1828." *California Historical Society Quarterly* 8 (1929): 239–43.

Cermeño, Sebastián Rodríguez. See Wagner, 1924 and 1929.

Chapman, Charles E. *History of California: The Spanish Period*. New York: McMillan, 1921.

Choris, Louis. *See* Mahr.

Colley, Charles C. "The Missionization of the Coast Miwok Indians of California." *California Historical Society Quarterly* 69 (1970): 143–63.

Collier, Mary E. T., and Sylvia B. Thalman, eds. *Interviews with Tom Smith and Maria Copa: Isabel Kelly's Ethnographic Notes on the Coast Miwok Indians.* San Rafael, California: Miwok Archaeological Preserve of Marin, 1991.

Cook, Sherburne F. *The Population of the California Indians, 1769–1970.* Berkeley, California: University of California Press, 1976.

_____. *The Conflict Between the California Indian and White Civilization.* Berkeley, California: University of California Press, 1976.

Conzett, Nancy. *The Battle of Bodega Bay.* Pamphlet. Bodega Bay, California: Rancho Bodega Historical Society, reprinted from *Ridge Review* 9, no. 4 (Oct. 1991).

Corney, Peter. Excerpts from his journal are in Sullivan ed., *The Russian Settlement in California.*

Cronise, Titus Fey. *The Natural Wealth of California.* San Francisco: H. H. Bancroft & Company, 1868.

Cutter, Donald C. *California in 1592, A Spanish Naval Visit.* Norman, Oklahoma: University of Oklahoma Press, 1990.

_____. *Malaspina and Galiano: Spanish Voyages to the Northwest Coast, 1791 and 1792.* Seattle: University of Washington Press, 1991.

Dana, Julian. *Sutter of California.* New York: Halcyon House, 1934.

Davidson, George C. *Examination of Some of the Early Voyages of Discovery and Exploration of the Northwest Coast of America, from 1539 to 1603.* Washington D.C., 1886.

Davis, James T. *Trade Routes and Economic Exchange Among the Indians of California.* Ramona, California: Ballena Press, 1974.

Davis, William H. *Seventy-Five Years in California.* San Francisco: John Howell, 1929.

Denis, Alberta J. *Spanish Alta California.* New York: Macmillan Company, 1927.

Dickinson, A. Bray. *Tomales Township: A History.* Tomales, California: Tomales History Center, 1993.

Dmytryshyn, Basil, E. A. P. Crownhart-Vaughan, and Thomas Vaughan, eds. *Colonial Russian America: Kyrill T. Khlebnikov's Reports, 1817–1832.* Portland: Oregon Historical Press, 1976.

_____. *The Russian American Colonies, 1798–1867.* Vol. 3. Portland: Oregon Historical Press, 1989.

Driver, Harold. "Wappo Ethnology." *University of California Publications in American Archaeology and Ethnology* (3): 179-220. Berkeley, California.

Duflot de Mofras, Eugène. *See also* Wilbur, ed., 1937.

_____. *Exploration of Oregon and California.* 2 vols. Paris, 1844.

Duhaut-Cilly, Auguste. *See* Carter.

Eastwood, Alice, ed. "Archibald Menzies Journal of the Vancouver Expedition." *California Historical Society Quarterly* 2 (Jan. 1924): 265–340.

Edwards, Clinton. "Wandering Toponyms: El Puerto de la Bodega and Bodega Bay." Berkeley, California: University of California Press. *Pacific Historical Review* 33, no. 3 (Aug. 1964): 253–72.

Eliza, Francisco de. *See* Wagner, 1931, "The Last Spanish Exploration."

Engelhardt, Fr. Zephyrin. *Missions and Missionaries of California: Upper California.* Vols. 2 and 3. San Francisco: James H. Barry, 1913.

Essig, E. O., A. Ogden, and C. J. Dufour. *Fort Ross: California Outpost of Russian Alaska, 1812–1841.* Kingston, Ontario: Limestone Press, 1991.

Fairley, Lincoln. *Mount Tamalpais: A History.* San Francisco: Scottwall Associates, 1987.

Farris, Glenn. "The Story of the Purchase of Fort Ross and Payment for Bodega Bay by the Russian Promyshlennik, Tarakanov." Typescript, 1994. California Department of Parks and Recreation.

_____. "The Bodega Miwok as seen by M. T. Tikhanov in 1818." Typescript. Paper presented at the California Indian Conference, UCLA, October 7, 1995. California Department of Parks and Recreation.

Fenega, Franklin. "Field Notes and Manuscript Report on Excavation of 4-Son-299, Bodega Head, California." Long Beach, California: California State University, n.d.

Fowler, Stephen and James Fowler. "Journal of Stephen and James Fowler: 1849–1864." 2 vols. BANC MSS C-F 144. Manuscript held by the Bancroft Library, Berkeley, California.

Fredrickson, David. "Archaeological Investigations Within Construction Site Area of Unit 1 of Pacific Gas & Electric Company's Atomic Park, Sonoma County, California." San Francisco: Pacific Gas & Electric Company, 1962.

_____ . "Manuscript Report on Excavations in Advance of PG&E Plant Construction, 4-Son-293, 294." San Francisco: Pacific Gas & Electric Company, 1962.

_____ . "Archaeological Resources and Proposed Maintenance Dredging and Improvements, Bodega Harbor Navigation Project." U.S. Army Corps of Engineers, San Francisco District, California, 1974.

Gaffney, Rose, et al. "Catalog of the Rose Gaffney Collection." Petaluma, California: Northwestern California Archaeological Society, n.d.

Galvin, John, ed. *Journal of Explorations Northward along the Coast from Monterey in the year 1775*, by Father Miguel de la Campa. San Francisco: John Howell Books, 1964.

_____ . *The First Spanish Entry into San Francisco Bay, 1775*. San Francisco: John Howell Books, 1971. A translation of the narratives of Father Vicente de Santa María and José de Cañizares.

Gibson, James R. *Imperial Russia in Frontier America*. New York: Oxford University Press, 1976.

_____ . *Otter Skins, Boston Ships, and China Goods*. Seattle: University of Washington Press, 1992.

Gilliam, Harold. *San Francisco Bay*. Garden City, New York: Doubleday, 1957.

_____ . *Island in Time*. San Francisco: Sierra Club, 1962.

Gleason, Duncan. *The Islands and Ports of California*. New York: Delvin-Adair, 1958.

Golovnin, Vasili M. *Around the World on the Kamchatka*. Translated with introductory remarks by Ella Lury Wiswell. Honolulu: University Press of Hawaii, 1979.

Goycoechea, Felipe de. *See* Wagner, 1931, "The Last Spanish Exploration."

Gudde, Erwin G. *California Place Names*. Berkeley, California: University of California Press, 1969.

Hanson H., J. Miller, and D. Peri. *Wild Oats in Eden*. Santa Rosa, California, 1962.

Hart, James D. *A Companion to California*. Berkeley, California: University of California Press, 1987.

Hedgpeth, Joel W. "Science at the Seashore." *Pacific Discovery* 7, no. 2 (Mar.–Apr. 1959): 16–21.

_____ . "Bodega Head—A Partisan View." *Bulletin Atomic Scientists* (Mar. 1965): 2–7.

_____ . "The Battle of Bodega Head." *Per/Se* (Stanford, California: Stanford University): (Fall 1966): 42–47.

_____ . Papers, 1958–78. Three boxes (two linear feet). BANC MSS 78/156c. Bancroft Library, Berkeley, California.

Heig, Adair. *History of Petaluma: A California River Town*. Petaluma: Scottwall Associates, 1982.

Heig, James, ed. *Pictorial History of Tiburon*. San Francisco: Scottwall Associates, 1984.

Heizer, Robert F. *See also* Stross and Heizer.

_____ , ed. *Seven Early Accounts of the Pomo Indians and their Culture*. Berkeley, California: Archaeological Research Facility, University of California, 1975.

_____ , ed. *Handbook of North American Indians—Vol. 8: California*. Washington, D.C.: Smithsonian Institution, 1978.

Hendry, George and J. N. Bowman. Part II, Sonoma, in *The Spanish and Mexican Adobe and Other Buildings in the Nine San Francisco Bay Counties, 1776 to about 1850*. Berkeley, California: University of California, 1940. In Bancroft Library holdings.

Herrara, Antonio de. *Descripción de las Indias Occidentalis*. 8 vols. Madrid, 1601. Translated by John Stevens, 1725.

Hudson, John W. "Pomo Wampum Makers, An Aboriginal Double Standard." *Overland Monthly* 30 (1897): 101–8. Reprinted in *Seven Early Accounts of the Pomo Indians and their Culture*, Robert Heizer, ed., 1975, pp. 9–20.

Hunt, R. D. *John Bidwell, Prince of California Pioneers*. Caldwell, Idaho: Caxton Printers, 1942.

Hutchinson, William Henry. *California*. Palo Alto, California: American West, 1969.

Imray, James. *Sailing Directions for the West Coast of North America between Panama and Queen Charlotte Islands.* London: James Imray & Sons, 1868.

———. *North Pacific Pilot: Part I. The West Coast of North America between Panama and Queen Charlotte Islands.* London: James Imray & Sons, 1881.

Ingles, Lloyd. *Mammals of the Pacific States.* Stanford, California: Stanford University Press, 1965.

Istomin, Alexei A. *The Indians at the Fort Ross Settlement.* Fort Ross, California: Fort Ross Interpretive Association, 1992.

Jameson, E. W., and H. J. Peeters. *California Mammals.* Berkeley, California: University of California Press, 1988.

Jones & Stokes Associates, Inc. "Environmental Impact Statement Bodega Bay Public Utility District Proposed Sewerage Facilities. Sacramento, 1972–74.

Kelly, Isabel. "Coast Miwok." Chapter in Heizer, ed. *Handbook of North American Indians.* Vol. 8, *California,* pp. 414–25. Washington D.C.: Smithsonian Institute, 1978.

———. *Interviews with Tom Smith and Maria Copa.* See Collier and Thalman, eds.

Khlebnikov, Kirill. *Colonial Russian America: Kirill T. Khlebnikov's Reports, 1817–1832.* Translated with an introduction and notes by Basil Dmytryshyn and E. A. P. Crownhart-Vaughan. Portland: Oregon Historical Press, 1976.

———. *The Khlebnikov Archive, Unpublished Journal (1800–1837) and Travel Notes (1820, 1822, 1824).* Translated with an introduction and notes by Leonid Shur and John Bisk. Fairbanks, Alaska: University of Alaska Press, 1990.

King, Thomas. "Son-320; An Unusual Site on Bodega Head." Archives of California Archaeology, Society for California Archaeology, Berkeley, California, 1966.

Koenig, J. B. "The Geologic Setting of Bodega Head." *Mineral Information Service,* July 1963. Sacramento: California Division of Mines and Geology.

Kostromitinov, Peter. "Observations on the Indians of Upper California." *See* Wrangell; Pierce, ed., 1980; and Stross and Heizer, eds.

Kotzebue, Otto von. *See also* Mahr. Excerpts from his journal are in Sullivan ed., *The Russian Settlement in California.*

———. *A New Voyage Around the World in the Years 1823, 24, 25, and 26.* Vol. 1. London, 1830.

Lane, Jana. *A History of the Territory That Was Called Rancho Bodega.* Bodega Bay, California: Rancho Bodega Historical Society, 1993.

Langsdorff, George von. *Voyages and Travels in Various Parts of the World.* Reprint. New York: Da Capo Press, 1968.

Larrison, Earl. *Mammals of the Northwest.* Seattle: Audubon Society, 1976.

Lewis, Oscar. *George Davidson, Pioneer West Coast Scientist.* Berkeley, California: University of California Press, 1954.

Lewis Publishing Company. *An Illustrated History of Sonoma County, California.* Chicago: Lewis Publishing, 1889.

Lightfoot K. G., T. A. Wake, and A. M. Schiff. *The Archaeology and Ethnohistory of Fort Ross, California.* Vol. 1. Berkeley, California: Archaeological Research Facility, University of California, 1991.

Lütke, Fedor. "Observations on California, Sept. 4–28, 1818": excerpts from the diary of his circumnavigation aboard the sloop *Kamchatka,* 1817–1819. Chapter in Dmytryshyn et al., eds., *The Russian American Colonies 1798–1867,* vol. 3, pp. 257–85.

Mahr, August. *The Visit of the "Rurik" to San Francisco in 1816.* Stanford, California: Stanford University Press, 1932.

Marine, Gene. "Outrage on Bodega Head." *The Nation* (June 22, 1963): 524–27.

Marliave, Elmer and Donald Tocher. "Geologic and Seismic Investigation of the Site for a Nuclear Electric Power Plant on Bodega Head." Unpublished. San Francisco: Pacific Gas & Electric Company, 1964.

Martinez y Zayas, Juan. *See* Wagner, 1931, "Last Spanish Exploration."

Mason, Jack. *Point Reyes, The Solemn Land.* Inverness, California: North Shore Books, 1970.

_____ . *Early Marin.* Petaluma, California: House of Printing, 1971.

_____ . *Earthquake Bay, A History of Tomales, California.* Inverness, California: North Shore Books, 1976.

Mason, Jack, and Thomas J. Barfield. *Last Stage for Bolinas.* Petaluma, California: North Shore Books, 1973.

Mathes, W. Michael. *Vizcaíno and Spanish Expansion into the Pacific Ocean, 1580–1630.* San Francisco: California Historical Society special publication no. 44, 1968.

Matiushkin, Fyodor. *See* Watrous, ed.

Matute, Juan Bautista. *See* Wagner, 1931, "Last Spanish Exploration."

McLendon, Sally, and Robert Oswalt. "Pomo: Introduction." Chapter in Heizer, ed. *Handbook of North American Indians.* Vol. 8, *California,* pp. 274–88. Washington D.C.: Smithsonian Institute, 1978.

Menzies, Archibald. *See* Eastwood, ed.

Merriam, C. Hart. "Indian Names in the Tamalpais Region." *California Out-of -Doors.* (April 1916): 118.

Milliken, Randall. *A Time of Little Choice: The Disintegration of Tribal Culture in the San Francisco Bay Area, 1769–1810.* Menlo Park, California: Ballena Press, 1995.

Mofras, Eugène Duflot de. *See* Duflot de Mofras.

Moraga, Gabriel. "Diario de su expedición al puerto de Bodega," San Francisco, October 1810. *Provincial State Papers,* 19, pp. 276–77. Manuscript in the Bancroft Library, Berkeley, California.

Mourelle, Francisco Antonio. *See* Barrington ed.; and Russell, ed.

Munro-Fraser, J. P. *History of Marin County.* San Francisco: Alley, Bowen, 1880.

_____ . *History of Sonoma County.* San Francisco: Alley, Bowen, 1880

National Park Service. *Land Use Survey, Proposed Point Reyes National Seashore.* Washington, D.C., U.S. Government Printing Office, Feb. 1961.

Newmark, N. M. "Report to the AEC Regulatory Staff: Seismic Effect on Bodega Bay Reactor," 1964. Bancroft Library.

Northern California Association to Preserve Bodega Head and Harbor, Inc. *The Battle of Bodega Bay.* Pamphlet. Berkeley, California, 1963.

_____ . *Earthquakes, the Atom, and Bodega Head.* Pamphlet. Berkeley, California, 1965.

O'Brien, Bickford, ed., and Diane Spencer-Hancock, writer. *Fort Ross: Indians-Russians-Americans.* Fort Ross, California: Fort Ross Interpretive Association, 1980.

Ogden, Adele. *See also* Essig et al.

_____ . *The California Sea Otter Trade, 1784–1848.* Berkeley, California: University of California Press, 1941.

Pacific Gas & Electric Company. *See also* Public Utility Commission.

_____ . *PG&E Progress.* San Francisco. The following issues reference the "Atomic Park" project: Aug. 1961, Apr. 1963, Aug. 1963, Oct. 1963, Nov. 1963, Apr. 1964, and Dec. 1964.

_____ . 1961 Annual Report.

_____ . PG&E News Bureau press release, June 28, 1961.

Palóu, Father Francisco. *See* Bolton, ed., 1926.

Pesonen, David. *See also* Northern California Association to Preserve Bodega Head and Harbor.

_____ . "A Visit to the Atomic Park." San Francisco: Sierra Club, 1962. Originally published in the *Sebastopol Times,* Sept. 27 and Oct. 4, 11, and 18, 1962.

Phelps, William Dane. Logbooks of the *Alert,* 1840–43. Originals in the Widner Memorial Library, Harvard University, Cambridge, Massachusetts.

_____ . "Solid Men of Boston in the Northwest." BANC MSS P-C 31. Manuscript. Original in the Bancroft Library, Berkeley, California.

Pierce, Richard A. *Russian America: A Biographical Dictionary.* Kingston, Ontario: Limestone Press, 1990.

_____ , ed. *Russian America Statistical and Ethnographic Information* by Ferdinand Petrovich von Wrangell. Kingston, Ontario: The Limestone Press, 1980.

_____, ed. *The Russian-American Company: Correspondence of the Governors Communications Sent: 1818.* Kingston, Ontario: Limestone Press, 1984.

_____, ed. *The Odyssey of a Russian Scientist: I. G. Voznesenskii in Alaska, California and Siberia, 1839–1849,* by A. I. Alekseev. Kingston, Ontario: Limestone Press, 1987.

Pierce, Richard, and Alton Donnelly. *History of the Russian-American Company,* by Peter A. Tikhmenev, 1861–63. Seattle and London: University of Washington Press, 1978.

Powers, Stephen. *Tribes of California.* Reprint. Berkeley, California: University of California Press, 1976.

Preston, R. N., ed. *Early California Atlas, Northern Edition.* Portland: Binford & Mort, 1974.

Price, A. G., ed. *The Explorations of Captain James Cook in the Pacific.* New York: Dover, 1971.

Public Utilities Commission, State of California. The files for the Pacific Gas & Electric Company Atomic Park permit application can be viewed at the PUC's San Francisco office under Application No. 43808 and Decision No. 64537. They are archived in Sacramento.

Quinn, Arthur. *Broken Shore: The Marin Peninsula.* Salt Lake City: Peregrine Smith, 1981.

Rancho Bodega Historical Society. *Around Bodega Harbor.* Pamphlet. Bodega Bay, California: Rancho Bodega Historical Society, August 23, 1996.

Resanov, Nikolai. "June 17, 1806: A confidential report to Minister of Commerce Nikolai P. Rumiantsev, concerning trade and other relations between Russian America, Spanish California and Hawaii." Chapter in Dmytryshyn et al., eds., *The Russian American Colonies, 1798–1867,* vol. 3, pp. 112–48.

Revere, Joseph W. *Naval Duty in California.* Oakland: Biobooks, 1947.

Richman, Irving B. *California Under Spain and Mexico, 1535–1847.* New York: Houghton Mifflin, 1911.

Russell, Thomas C. *See also* Barrington, ed.

_____, ed. *The Voyage of the Sonora in the Second Bucareli Expedition, 1775.* San Francisco, 1920.

Saint Amand, Pierre. *Geologic and Seismologic Study of Bodega Head.* Berkeley, California: Northern California Association to Preserve Bodega Head and Harbor, Inc., 1963.

Santa María, Father Vicente de. *See* Galvin, ed., 1971.

Schenk, Robert, ed. *Contributions of the Archaeology of Point Reyes National Seashore: A Compendium in Honor of Adan E. Treganza.* San Francisco: San Francisco State College, Treganza Museum Papers no. 6, December 15, 1970.

Schlocker, Julius, and M. G. Bonilla. *Engineering Geology of the Proposed Nuclear Power Plant Site on Bodega Head, Sonoma County, California.* U.S. Department of the Interior, Geological Survey, 1964.

Schurz, William Lytle. *The Manila Galleon.* New York: Dutton, 1939.

Shur, Leonid, ed. *The Khlebnikov Archive, Unpublished Journal (1800–1837) and Travel Notes (1820, 1822, 1824).* Fairbanks, Alaska: University of Alaska Press, 1990.

Slaymaker, Charles. *Cry for Olompali.* San Rafael, California: Miwok Archaeological Preserve of Marin, 1972.

_____. *The Material Culture of the Cotomkotca.* San Rafael, California: Miwok Archaeological Preserve of Marin, 1977.

Smilie, Robert S. *The Sonoma Mission.* Fresno, California: Valley Publishers, 1975.

Smith, Stephen. Papers, 1843–54. Four folders in one portfolio. Bancroft Library, University of California, Berkeley.

Sonoma County Planning and Harbor Commissions. *Bodega Bay: A Summary of Improvement Plans and Problems.* Santa Rosa, California, 1960.

Sonoma County Planning Commission. *Master Plan of Development for Bodega Bay.* Santa Rosa, California, 1956.

Stevens, John. *The General History of the Vast Continent and Islands of America.* 8 vols. London, 1725. Translation of Herrara's history of the West Indies, 1601.

Stross, Fred, and Robert Heizer, eds. "Ethnographic Observations on the Coast Miwok and Pomo by Contre-Admiral F. P. von Wrangell and P. Kostromitinov of the Russian Colony Ross, 1839." Berkeley, California: University of California Archaeological Research Facility, 1974.

Sullivan, Joseph, ed. *The Russian Settlement in California: Fort Ross,* by R. A. Thompson. Reprint. Oakland: Biobooks, 1951.

Sumerset Publishers. *Encyclopedia of California.* New York: Somerset Publishers, 1994.

Taylor, C. W., ed. *Eminent Californians.* 2d edit. Palo Alto, California: C. W. Taylor, 1956.

Thalman, Sylvia Barker. *See also* Collier and Thalman, eds.

_____ . *The People of the Good and Fair Bay.* San Rafael, California: Miwok Archaeological Preserve of Marin, 1991.

_____ . *The Coast Miwok of the Point Reyes Area.* Point Reyes, California: Point Reyes National Seashore Association, 1993.

Thompson, Robert A. *Historical and Descriptive Sketch of Sonoma County, California.* Philadelphia, 1877.

_____ . *The Russian Settlement in California Known as Fort Ross.* Santa Rosa, California: Sonoma Democrat Publishing Company, 1896. Reprinted by Sullivan, ed., 1951.

Thompson, Thomas H. & Co. *Historical Atlas Map of Sonoma County, California.* Oakland, California: Thos. H. Thompson, 1877.

Tibesar, Antonine. *Writings of Junipero Serra.* 2 vols. Washington D.C.: Academy of Franciscan History, 1955.

Tomlin, Kaye, and Stephen Watrous. *Outpost of an Empire.* Pamphlet. Fort Ross, California: Fort Ross Interpretive Association, 1993.

Treganza, Adan E. "The Indian Story." In *Land Use Survey, Proposed Point Reyes National Seashore,* by the National Park Service. Washington D.C.: U.S. Government Printing Office, 1961.

Trussel, Margaret E. "Settlement of the Bodega Bay Region." Graduate thesis. University of California, Long Beach, 1960.

Tuomey, Honoria. *History of Sonoma County.* 2 vols. San Francisco: S. J. Clarke, 1926.

U.S. Atomic Energy Commission. *Summary Analysis by the Division of Reactor Licensing in the Matter of Pacific Gas and Electric Company Bodega Head Nuclear Power Plant.* Washington, D.C., 1964.

Vancouver, Captain George. *See also* Anderson, 1960; and Wilbur, ed., 1953.

_____ . *Voyage of Discovery to the North Pacific Ocean, 1790–1795.* 3 vols. London, 1798.

Venegas, Miguel. *Noticias de la California.* Madrid, 1757. 3 vols. Translated in *A Natural and Civil History of California.* London, 1759.

Vizcaíno, Sebastián. *See* Wagner, 1929, "The Last Spanish Exploration."

Wagner, Henry R. "The Voyage of Sebastián Rodríguez Cermeño in 1595." *California Historical Society Quarterly* 3 (April 1924): 3–24.

_____ . *Spanish Voyages to the Northwest Coast of California in the Sixteenth Century.* San Francisco: California Historical Society, 1929. Includes the Cabrillo, Cermeño and Vizcaíno accounts.

_____ . "The Last Spanish Exploration of the Northwest Coast and the Attempt to Colonize Bodega Bay." *California Historical Society Quarterly* 10, no. 4 (1931): 313–45.

_____ . "George Davidson, Geographer of the Northwest Coast of America." *California Historical Society Quarterly* 11 (1931): 299–320.

_____ . *Cartography of the Northwest Coast of America to the Year 1800.* San Francisco: California Historical Society, 1933.

_____ . "Creation of Rights of Sovereignty Through Symbolic Acts." *Pacific Historical Review* 7 (1938).

Watrous, Stephen. *See also* Tomlin and Watrous.

_____ , ed. "A Journal of a Round-the-World Voyage on the Sloop *Kamchatka,* under the Command of Captain Golovnin," by Fyodor F. Matiushkin. Translated from the Russian text in the September–October 1996 newsletter of the Fort Ross Interpretive Association.

Weber, David J. *The Spanish Frontier in North America.* New Haven: Yale University Press, 1992.

Wellock, Thomas. "The Battle for Bodega Bay: the Sierra Club and Nuclear Power, 1958–1964." *California History* (Summer 1993): 193–214.

Wilbur, Marguerite Eyer, ed. *Duflot de Mofras' Travels on the Pacific Coast*. 2 vols. Santa Ana, California: Fine Arts Press, 1937.

_____ , ed. *Vancouver in California*. 2 vols. Los Angeles: Glen Dawson, 1953.

Wilcox, Del. *Voyagers to California*. Elk, California: Sea Rock Press, 1991.

Winsor, Justin, ed. *Narrative and Critical History of America*. 8 vols. Vol. 3, *English Explorations and Settlements in North America, 1497–1689*. Boston and New York: Houghton Mifflin, 1884–89.

Woodward, Arthur. *Indian Trade Goods*. Portland, Oregon: Oregon Archaeological Society Publication no. 2, 1965.

Wrangell, Ferdinand Petrovich von. *Russian America Statistical and Ethnographic Information* (1839). Translated with introductory notes by Richard Pierce. Kingston, Ontario: Limestone Press, 1980. *See also* Stross and Heizer, eds.

Early Maps and Charts of Bodega Bay

Bodega y Quadra, Juan de la. Plano del Puerto de la Bodega . . . 1775. Bancroft Library G4362.B64 1775 .B6 Case XB.

Bodega y Quadra, Juan de la. Plano del Cap. Bodega . . . 1775. Bancroft Library G4362.B64 1775 .B65 Case XB.

Bodega y Quadra, Juan de la. Plano del Capitan Bodega . . . 1775. Bancroft Library G4362.B64 1775 .B7 Copy 1, Case A, Copy 2, Case XB.

Colnett, Captain James. A sketch by compass of Port Sir Francis Drake, 1790. Public Record Office, London.

Matute, Juan Bautista. Pto. de la Bodega . . . 1793. Bancroft Library G4365.B63 1793 .P8 Case XB.

Martinez y Zayas, Juan. Plano del Puerto de la Bodega . . . 1793. Bancroft Library G4365.B63 1793 .M3 Copy 1, Case A, Copy 2, Case XB.

Golovnin, Vasili. Chart of Rumiantsev Bay . . . 1818. Contained in volume 2 of his journal.

Hall, Captain John. Sketch of Port Bodega . . . 1822. In appendix I of *California: A History of Upper and Lower California*, by Alexander Forbes. London: Smith Elder, 1839.

Belcher, Captain Edward. Bodega Bay . . . 1839. London: Hydrographic Office of the Admiralty, 1849. Bancroft Library G4362.B62P5 1839 .B4 Case C.

Duflot de Mofras, Eugéne. Carte detaile du mouillage de Fort Ross et du Port de la Bodega . . . 1841. Library of Congress.

Vioget, Jean J. Plano de terreno que pretende Don Estevan Smith . . . 1843. Bancroft Library Land Case Map D-74.

Teben'kov, Mikhail D. Karta zalivov Bodego . . . 1848. University of California, Berkeley Map Room G4362.C6 1848 .T4 Case D.

Bowers, A. B. Map of Sonoma County, California . . . 1863. Bancroft Library G4365.S75 1867 .B6 Case D.

United States Coast and Geodetic Survey. Bodega Bay, California . . . 1862. University of California, Berkeley Map Room G4362.B63P5 1862 .U5 Case XB.

Thos. H. Thompson & Co. Historical Atlas Map of Sonoma County, California . . . 1877. Oakland, California: Thompson & Co., 1877. University of California, Berkeley Map Room fG1528.S85 .T4 1877b.

The Anchorage Debate

Aker, Raymond. "The Cermeño Expedition at Drakes Bay, 1595." Typescript. Point Reyes, California: Drake Navigators Guild, 1965.

_____ . *Report of Findings Related to Identification of Sir Francis Drake's Encampment at Point Reyes National Seashore*. Point Reyes, California: Drake Navigators Guild, 1970.

_____ . *Sir Francis Drake at Drakes Bay: A Summation of Evidence Relating to the Identification of Sir Francis Drake's Encampment at Drakes Bay, California*. Point Reyes, California: Drake Navigators Guild, 1978.

_____ . "A Safe Harbor at Point Reyes." *Geographical Magazine* (May 1980): 544–46.

_____ . "Francis Drake at Cape Horn." *Sea History* (Winter 1996–97): 12–13. Publication of the National Maritime Historical Society, Peekskill, New York.

Aker, Raymond and Edward Von der Porten. *Discovering Portus Novae Albionis, Francis Drake's California Harbor*. Palo Alto, California: Drake Navigators Guild, 1979.

Aker, Raymond, M. P. Dillingham, and R. Parkinson. "Nova Albion Rediscovered." Typescript. Point Reyes, California: Drake Navigators Guild, 1956.

Allen, Robert W. "An Examination of the Botanical References in the Accounts Relating to Drake's Encampment at Nova Albion in 1579." Typescript. Point Reyes, California: Drake Navigators Guild, 1971.

_____ . "Identification of an 'Herbe Much Like Our Lettuce'." Typescript. Point Reyes, California: Drake Navigators Guild, 1971.

Allen, Robert W., and R. Parkinson. "Identification of the Nova Albion Conie." Typescript. Point Reyes, California: Drake Navigators Guild, 1971.

Asaro, Frank. *See* Michel and Asaro.

Bancroft, Hubert Howe. *The Works of Hubert Howe Bancroft*—Vol. 28: *History of California*. Vol. 1, 1542–1800. San Francisco: History Company, 1884.

Bancroft Library. *The Plate of Brass Reexamined*. Berkeley, California: Bancroft Library, 1977.

_____ . *The Plate of Brass Reexamined: A Supplemental Report*. Berkeley, California: Bancroft Library, 1979.

Barrett, Samuel A. "The Ethno-Geography of the Pomo and Neighboring Indians." *University of California Publications, American Archaeology and Ethnology* 6, no. 1 (1908).

Becker, Robert H. "An Historical Survey of Point Reyes." Chapter in *Land Use Survey, Proposed Point Reyes National Seashore*. Washington D.C.: U.S. Department of the Interior, National Park Service, 1961.

Berthoud, Edward. "Sir Francis Drake's Anchorage." *National Geographic* 6 (Dec. 29, 1894): 208–14.

Bishop, R. P. "Drake's Course in the North Pacific." *British Columbia Historical Quarterly* (Vancouver, B.C.) (July 1939).

Bodega y Quadra, Juan Francisco de la. "Viaje de 1775." BANC MSS P-B 26. Unpublished manuscript in Bancroft Library, Berkeley, California.

Bolton, Herbert E. "The Plate of Brass." In *Drake's Plate of Brass: Evidence of his Visit to California in 1579*. San Francisco: California Historical Society special publication no. 13, 1937. Reprinted in special publication no. 25, 1953.

Brereton, R. M. *Did Sir Francis Drake Land on Any Part of the Oregon Coast?* Portland, Oregon: J. M. Gill, 1907.

Brown, Alan K. "Did Drake Land Here?" *La Peninsula: Journal of the San Mateo County Historical Association* 10, no. 4 (Feb. 1960): 3–5.

Bryant, William Cullen. *See* Hale.

Burney, Captain James. *Chronological History of the Discoveries in the South Seas or Pacific Ocean, Part I*. London: Luke Hansard, 1803.

California Historical Society. Newsletter. *CHS October Events*: "Where Did Drake Land in 1579? Special Tour October 27, 1974." San Francisco: California Historical Society, October 1974.

Chickering, Allen L. "Some Notes with Regard to Drake's Plate of Brass." *California Historical Society Quarterly* 16, no. 3 (Sept. 1937): 275–81.

———."Further Notes on the Drake Plate." *California Historical Society Quarterly* 18, no. 3 (September 1939): 251–2.

———. "Drake's Plate of Brass: Its Past and Present Briefly Told." In *Drake's Plate of Brass Authenticated*. San Francisco: California Historical Society special publication no. 14, 1938. Reprinted in special publication no. 25, 1953.

———. "Drake in California." *California Historical Society Quarterly* 36 (1957): 21.

Chickering, Allen L., and Robert Heizer. Preface to *The Plate of Brass: Evidence of the Visit of Francis Drake to California in the Year 1579*. San Francisco: California Historical Society special publication no. 25, 1953.

Colnett, James. *Journal of Captain James Colnett Aboard the Argonaut from April 26, 1789 to November 3, 1791*. Reprinted by Judge F. W. Howay, ed. Toronto, Canada: Champlain Society, 1940.

———.*A Voyage to the South Atlantic and Round Cape Horn into the Pacific Ocean*. London, 1798. Reprinted by De Capo Press, New York, 1968.

Davidson, George. *Pacific Coast Pilot: California, Oregon and Washington*. Washington D.C.: U.S. Government Printing Office, 1869.

———. *Discoveries and Explorations on the Northwest Coast of America from 1539 to 1603*. Washington D.C.: U.S. Government Printing Office, 1887.

———. *Pacific Coast Pilot: California, Oregon and Washington*. Washington D.C.: U.S. Government Printing Office, 1889.

———. *Identification of Sir Francis Drake's Anchorage on the Coast of California in the Year 1579*. San Francisco: California Historical Society/Bacon & Company, 1890.

———. *Francis Drake on the Northwest Coast of America in the Year 1579: The "Golden Hind" Did Not Anchor in San Francisco Bay*. San Francisco: Geographical Society of the Pacific, 2d series, vol. 5, 1908.

Davies, Arthur. "Drake at San Francisco." *Geographical Magazine* (July 1980): 690–92.

———. "Drake and California." *Pacific Discovery* 34 (Nov. – Dec. 1981): 14–24.

Dillingham, Matthew P. *See also* Aker, Dillingham and Parkinson.

———. "A Review of the Findings of Dr. Adan E. Treganza Relative to the Site of Drake's Landing in California." Typescript. Point Reyes, California: Drake Navigators Guild, 1960.

Dines, Glen. *Drake in Marin: Fact and Fancy*. Pamphlet. San Rafael, California: Marin County Free Library, 1975.

Dixon, James. "Drake on the Pacific Coast." *Publications of the Historical Society of Southern California (Los Angeles)* 9 (1912): 86–96. Also, *The Overland Monthly* 63 (1914): 537–45.

Dobson, Richard. *See* Wiegand.

Doerr, Albert E. "Drake's California Harbor: Another Look at William Caldeira's Story." *Terrae Incognitae* 9 (1977): 49–59.

Ellison, Joseph W. "True or False?" *Saturday Evening Post* (Apr. 3, 1943): 32.

Elmendorf, William. *See* Heizer and Elmendorf.

Elsasser, Albert. "Francis Drake's Landing in California." *Pacific Discovery* 32 (July-Aug. 1979): 1–6.

Epperson, George. "Drake's Port of Nova Albion." In *Summary of an Exhibit of Artifacts Related to Sir Francis Drake and his Visit to Marin County in 1579*. Ross, California: Marin Cultural Center & Museum, 1994.

Farquhar, Francis P. "A Review of the Evidence." *California Historical Society Quarterly* 36 (1957): 22–30.

Felix, Charles, and Neil Malloch. "Documentation of Sir Francis Drake on the Farallones, 1579." Typescript. January 1983.

Fink, Colin, and E. P. Polushkin. "The Report on the Plate of Brass." In *Drake's Plate of Brass Authenticated*. San Francisco: California Historical Society special publication no. 14, 1938. Reprinted in special publication no. 25, 1953.

Forbes, Alexander. *California: A History of Upper and Lower California*. London: Smith Elder, 1839. Reprinted by John Henry Nash, San Francisco, 1937.

Gilliam, Harold. "Harbor of the *Golden Hind*." Chapter in *Island in Time: The Point Reyes Peninsula*. 2d edit. New York: Sierra Club/Charles Scribner's Sons, 1973.

Gilmore, James. "Five Iron Cannon May Solve the Mystery of Sir Francis Drake's Fair and Good Bay." *Santa Barbara Magazine*, (Winter 1981).

Greenhow, Robert. *The History of Oregon and California*. Boston: Charles C. Little and James Brown, 1845.

Hale, Edward Everett. "Dudley and the Arcano del Mare." *American Antiquarian Society's Proceedings*, Oct. 21, 1873. Also in William Cullen Bryant. *Popular History of the United States*. 4 vols. New York: Scribner, Armstrong, 1876. Also reprinted in Justin Winsor, ed., *Narrative and Critical History of America*. Vol. 3. New York, 1884–89.

Hanna, Warren. *Lost Harbor: The Controversy over Drake's California Anchorage*. Berkeley, California: University of California Press, 1979.

_____. "Legend of the Nicasios—The Men Drake Left Behind at Nova Albion." *California Historical Society Quarterly* 58 (Summer 1979): 154–165.

Hart, James. *See also* Bancroft Library.

_____. *The Plate of Brass*. Pamphlet. Berkeley, California: Bancroft Library, 1977.

Haselden, R. B. "Is the Drake Plate of Brass Genuine? *California Historical Society Quarterly* 16 (1937): 271–74.

Heizer, Robert F. *See also* Chickering and Heizer; and Meighan and Heizer.

_____. "Archaeological Evidence of Sebastián Rodríguez Cermeño's California Visit in 1595." *California Historical Society Quarterly* 20 (1941): 315–28.

_____. *Francis Drake and the California Indians, 1579*. Berkeley, California: University of California Press, 1947. Reprinted in *Elizabethan California*, by Robert F. Heizer, 1974.

_____. *Elizabethan California*. Ramona, California: Ballena Press, 1974.

Heizer, Robert F., and William W. Elmendorf. "Francis Drake's California Anchorage in the Light of the Language Spoken There." *Pacific Historical Review* 11 (1942): 213–17. Reprinted in Heizer, *Elizabethan California*, 1974.

Hildebrand, Joel H. "A Word Concerning Professor Colin G. Fink." In *Drake's Plate of Brass Authenticated*. San Francisco: California Historical Society special publication no. 14, 1938. Reprinted in special publication no. 25, 1953.

Hittell, John S. *History of the City of San Francisco*. San Francisco, 1878.

Hittell, Theodore H. *History of California*. Vol. 1. San Francisco: N. J. Stone, 1885.

Hoffman, Leonard, Jr. "The Fair and Good Bay." Typescript, draft, July 19, 1975. Gift to California Historical Society Library, San Francisco.

Holliday, J. S. "The Francis Drake Controversy." *California Historical Society Quarterly* 53, no. 2 (1974): 197–292.

Holmes, Kenneth L. "Francis Drake's Course in the Northern Pacific." Monmouth, Oregon: Gamma Theta Upsilon, Department of Social Science, Oregon College of Education. *The Geographic Bulletin* 17 (June 1979).

Howay, Judge F.W. *See* Colnett, 1940.

Hubbard, Gardiner. "Discoveries of America." *National Geographic* 5 (Apr. 7, 1893): 1–20.

Humboldt, Alexander von. *Essai Politique sur de Royaume de Nouvelle Espagne*. 2 vols. Paris, 1811. Translated by John Black as *Political Essay on the Kingdom of New Spain*. 4 vols. London, 1811.

Hunt, John C. "Sir Francis Drake Visits California." *American History Illustrated* (Dec. 1967): 47–53.

Jensen, M.W. *See* Viles and Jensen.

King, Thomas, and Ward Upson. "Protohistory of Limantour Sandspit: Archaeological Investigations at 4-Mrn-216 and 4-Mrn-298. In Robert Schenk, ed. *Contributions of the Archaeology of Point Reyes National Seashore: A Compendium in Honor of Adan E. Treganza.* San Francisco: San Francisco State College, Treganza Museum Papers no. 6, December 15, 1970.

Kroeber, A. L. *Handbook of the Indians of California.* 2d ed. Berkeley, California: California Book Co., 1953.

Lummis, Charles, ed. "Pioneers of the Far West—Fray Zárate y Salmerón's 'Relación,' concluded." *Land of the Sunshine* 7, no. 3 (Feb. 1900): 180, 184–85. The N. de Morena account.

Malloch, Neil. *See* Felix and Malloch.

McAdie, Alexander G. "Nova Albion–1579." *American Antiquarian Society Proceedings*, new series, vol. 28 (1918): 189–98.

Meighan, Clement W. "Report on the 1949 Excavations of Sixteenth-Century Sites of Drakes Bay." *University of California Archaeological Survey,* ms. no. 79, 1949.

_____ . "Excavations in Sixteenth-Century Shellmounds at Drakes Bay, Marin County." *Reports of the University of California Archaeological Survey,* no. 9, paper 9. Berkeley, California, 1950. Reprinted in California Historical Society special publication no. 25, 1953.

Meighan, Clement W., and Robert F. Heizer. "Archaeological Exploration of Sixteenth-Century Indian Mounds at Drakes Bay." *California Historical Society Quarterly* 31 (1952): 98–108.

Michel, Helen V. and Frank Asaro. "Chemical Study of the Plate of Brass," *Archaeometry* 21 (1979).

Morena, N. de. *See* Lummis.

Morison, Samuel Eliot. *The European Discovery of America: The Southern Voyages, A. D. 1492 – 1616.* New York: Oxford University Press, 1974.

Morrison, Harry B. "An Investigation of the Louise Welshons Buell Theory of the California Anchorage of Francis Drake, 1579." Typescript. Pinole, California, 1982.

National Park Service. *Land Use Survey, Proposed Point Reyes National Seashore.* Washington, D.C., U.S. Government Printing Office, Feb. 1961.

Neasham, V. Aubrey, and William E. Pritchard. *Drake's California Landing: The Evidence for Bolinas Lagoon.* Sacramento, California: Western Heritage, 1974.

Nichols, William. *Why a Sir Francis Drake Association in California.* Ross, California: Sir Francis Drake Association, leaflet no. 1, Aug. 25, 1922.

Nimitz, Chester. "Drake's Cove: A Navigational Approach to Identification." *Pacific Discovery* 11 (Mar. - Apr. 1958): 12–20.

Oko, Captain Adolph E. "Francis Drake and Nova Albion." *California Historical Society Quarterly* 43 (1964):135–50.

Parkerson, Margaret. *A Sir Francis Drake Bibliography for Marin County Civic Center Library.* Pamphlet. San Rafael, California: Marin County Civic Center Library, 1979.

Parkinson, Robert. *See* Aker, Dillingham and Parkinson; and Allen and Parkinson.

Pate, Robert, and Ellen Pate. "Drake's Treasure." *Western Treasures* (Jan. 1969): 14–20.

Pate, Robert, and Richard Dobson (Pirates Cove). *See* Wiegand.

Pike, Douglas G. "Historiography of the Drake Controversy." *California Historical Society Quarterly* 52 (1973): 128–30.

Polushkin, E. P. *See* Fink and Polushkin.

Power, Robert H. "Portus Novae Albionis Rediscovered?" *Pacific Discovery* 7 (May - June 1954): 10–12. San Francisco, California Academy of Sciences.

_____ . "Early Discoveries of San Francisco Bay." Unpublished paper prepared for the Senate Parks and Recreation Subcommittee. Nut Tree, California, 1968.

_____ . "A Report on the Concept that Drake's Island of Thieves was Mindanao." Monograph. Nut Tree, California: Robert Power, 1971.

_____ . "Drake's Landing in California, A Case for San Francisco Bay." *California Historical Society Quarterly* 52 (1973): 101–28.

_____ . *Francis Drake and San Francisco Bay: A beginning of British Empire.* Keepsake no. 6 in a series issued by Library Associates of the University Library. Davis, California: University of California, 1974.

_____ . "A Plate of Brass by Me . . . C. G. Francis Drake." *California History* 57 (1978): 172–85.

_____ . *A Study of Two Historic Maps.* Pamphlet. Nut Tree, California: Robert Power, 1978.

_____ . *Francis Drake, N. Morera and the Island of California.* Pamphlet. Nut Tree, California: Robert Power, 1979.

_____ . Book review for Warren Hanna's *Lost Harbor. California Historical Society Quarterly* 58 (Winter 1979/1980): 367.

_____ . "Cape Horn Discovered—and then to San Francisco Bay." *Geographical Magazine* (May 1980): 537–43.

Pritchard, William. *See* Neasham and Pritchard.

Quinn, David B. *Drake's Circumnavigation of the Globe: A Review.* Monograph: the fifteenth Harte Lecture, delivered in the University of Exeter on Nov. 14, 1980. Exeter, England: University of Exeter, 1981.

Reinstadt, Randall A. "Monterey's Mysterious Drake Plate." *Monterey Savings Locale* (issue number and date unknown). A copy of the article can be found in the Monterey Public Library's California Room clippings file under the heading "Historic Objects."

Riesenberg, Felix. *Cape Horn.* New York: Dodd Mead, 1939. Reprinted by Ox Bow Press, Woodbridge, Connecticut, 1995.

Robertson, John W. *Francis Drake and other Early Explorers Along the Pacific Coast.* San Francisco: Grabhorn Press, 1927.

_____ . *The Harbor of St. Francis.* San Francisco: Grabhorn Press, 1927.

Ruhge, Justin. *The Historic Cannons of Goleta.* Goleta, California: Justin Ruhge, 1982.

_____ . *Goleta, Pueblo de Las Islas.* Goleta, California: Justin Ruhge, 1984.

_____ . *Gunpowder and Canvas.* Goleta, California: Quantum Imaging, 1987.

_____ . *Drake in Central California.* Goleta, California: Quantum Imaging, 1990.

Sagen, Ethel Ames. *Sir Francis Drake's Secret Landfall.* Bakersfield, California: Ethel Sagen, 1978.

Shangraw, Clarence, and Edward Von der Porten. *The Drake and Cermeño Expeditions' Chinese Porcelains at Drakes Bay, California, 1579 and 1595.* Santa Rosa and Palo Alto, California: Santa Rosa Junior College and Drake Navigators Guild, 1981.

Sir Francis Drake Society. *News: Sir Francis Drake is Coming: Celebrating the Quatercentennial of Sir Francis Drake's Landing . . .* San Francisco: Sir Francis Drake Society, newsletter, circa June 1979.

Soule, Frank, J. H. Gihon, and J. Nisbet. *The Annals of San Francisco.* New York: D. Appleton, 1855.

Starr, Walter. "Evidence of Drake's Visit to California, 1579." *California Historical Society Quarterly* 36 (1957): 31–34.

_____ . "Drake Landed in San Francisco Bay in 1579: The Testimony of the Plate of Brass." *California Historical Society Quarterly* 41 (1962): 1–29.

Stillman, J. D. B. "Did Drake Discover San Francisco Bay?" *Overland Monthly* 1 (1868): 332–37.

_____ . *Seeking the Golden Fleece.* San Francisco: A. Roman & Co., 1877.

Thomas, Robert C. *Drake at Olompali.* San Francisco: A-Pala Press, 1979.

Treganza, Adan E. "The Examination of Indian Shellmounds in the Tomales and Drakes Bay Area with Reference to Sixteenth-Century Historical Contacts." Manuscript on file at the State of California Department of Parks and Recreation, History Section.

_____ . "The Examination of Indian Shellmounds Within San Francisco Bay With Reference to the Possible 1579 Landfall of Sir Francis Drake." Vacaville, California, 1957.

Tuthill, Franklin. *History of California.* San Francisco: H. H. Bancroft Co., 1866.

Twiss, Sir Travers. *The Oregon Territory: Its History and Discovery,* sometimes cited as *The Oregon Question Examined.* London, 1846.

Upson, Ward. *See* King and Upson.

Vallejo, Mariano. His centennial speech commemorating the founding of Mission Dolores is in the appendix of William Heath Davis, *Seventy-Five Years in California*, pp. 354–68. San Francisco, John Howell, 1929.

Vancouver, Captain George. *Voyage of Discovery to the North Pacific Ocean, 1790–1795*. 3 vols. London, 1798.

Varley, F. J. "Drake's Plate of Brass." *Geographical Review* 95: 159–60.

Verne, Jules. *Works of Jules Verne*. Vol. 15, *Exploration of the World, Famous Travels and Travelers*. New York: Vincent Park, 1911.

Viles, D. M., and M. W. Jensen, Jr. "The Northern Mystery." In *Vanguard*, a publication of Portland State University, Portland, Oregon, Nov. 16, 1971.

Villiers, Alan. "Sir Francis Drake." *National Geographic* (Feb. 1975): 216–53.

Von der Porten, Edward P. *See also* Aker and Von der Porten; and Shangraw and Von der Porten.

_____ . "Our First New England." Annapolis, Maryland: *U.S. Naval Institute Proceedings* (Dec. 1960).

_____ . "Drakes Bay Shellmound Archaeology, 1951–1962." 2 vols. Typescript. Point Reyes, California: Drake Navigators Guild, 1963.

_____ . "Drake-Cermeño: An Analysis of Artifacts." Typescript. Point Reyes, California: Drake Navigators Guild, 1965.

_____ . "The Porcelains and Terra Cotta of Drakes Bay." Typescript. Point Reyes, California: Drake Navigators Guild, 1968.

_____ . "Drake's First Landfall." *Pacific Discovery* 28 (Jan. - Feb. 1975): 28–30.

_____ . "The Drake Puzzle Solved." *Pacific Discovery* 37 (July - Sept. 1984): 22–26.

Wagner, Henry R. "The Voyage to California of Sebastián Rodríguez Cermeño in 1595." *California Historical Society Quarterly* 3 (1924): 3–24.

_____ . *Sir Francis Drake's Voyage Around the World*. San Francisco: John Howell, 1926.

_____ . *Drake on the Pacific Coast*. Los Angeles: Zamorano Club, 1970.

Watson, Douglas. "Drake and California." In *Drake's Plate of Brass: Evidence of His Visit to California in 1579*. San Francisco: California Historical Society special publication no. 13, 1937. Reprinted in special publication no. 25, 1953.

Wiegand, Mary. "Pirate's Cove." *True West* 15, no. 2 (Nov. - Dec. 1967): 6–9, and 48. The article covers the Pirates Cove/Mallagh's Landing anchorage theory advanced by Robert Pate and Richard Dobson.

Williams, Lawrence A. "Digging for History at Drakes Bay." *Pacific Discovery* 6 (July - Aug. 1953): 10–17.

Winsor, Justin. *See* Hale.

Clippings for most of the newspaper articles cited below can be found at the Marin Civic Center Library in San Rafael, California.

Newspaper Articles – San Francisco Papers

Dwinelle, John W. Review of William Bryant's "Popular History of the United States." San Francisco Evening Bulletin, Oct. 5, 1878, p. 1.

Staff writer. "Drake's Ghost Titters: Finding of His Plaque Stirs Up Debate." San Francisco Examiner (*SFE*), Apr. 17, 1937.

Staff writer. "Drake May Have Hit San Francisco Bay: Scientist Re-Examines Old Map." *San Francisco Chronicle* (*SFC*), Apr. 29, 1954, p. 1.

Horstmeyer, Harold. "Did Drake Sail Into Golden Gate? Power Claims Map Proves He Did." *SFE*, June 15, 1959.

Draper, Benjamin. "Celebrating Drake in California." *San Francisco Sunday Examiner and Chronicle (SFSEC)*, July 1, 1973, *This World*.

Staff writer. "In Quest of Drake's Landfall." *SFC*, Sept. 17, 1973.

Staff writer. "Oregon: A New Theory of Drake's Voyage." *SFC*, Sept. 23, 1973.

Staff writer. "Bolinas Find May be Drake's Fort." *SFC*, Jan. 11, 1974.

Brewer, Jim, and Jerry Burns. "Drake's Mystery: the Site of 'Fort' Buried in Mud." *SFC*, Jan. 12, 1974.

Staff writer. "An Exciting Drake's Fort Discovery in Marin." *SFC*, Aug. 4, 1974, p. 3.

Grieg, Michael. "Drake Plaque Called Hoax [by Morison]." *SFC*, Oct. 11, 1974, p. 1.

Staff writer. "Drake's Plaque Doubted." *SFSEC*, Oct. 20, 1974, *This World*.

Champion, Dale. "New Clue to Drake Landing: an Extraordinary Find in Marin." *SFC*, Nov. 9, 1974, p. 3.

Power, Robert. "Only Sir Francis Knows For Sure." *SFSEC*, June 27, 1976, *This World*, p. 1.

Hearst, Will. "A Self-Taught Archaeologist Digs into Controversy." *SFE*, Aug. 8, 1976.

Irving, Carl. "The Drake Plate a Hoax, UC Says." *SFE*, July 27, 1977, p. 1. Related articles: "A Historic Landing in Search of a Site"; "Drake left his Name Behind"; and "The Adventure of the *Golden Hind*," p. 8.

Harris, Michael. "Drake's Plate is a Fake UC Reveals." *SFC*, July 28, 1977, p. 1.

Irving, Carl. "He Was Here 400 Years Ago." *SFSEC*." Dec. 11, 1977, p. A15.

Staff writer. "Proof of Drake Landing Well into Bay Claimed." *SFC*, Apr. 12, 1978, p. 1.

Irving, Carl. "Drake Buffs Miss the Boat with State Historical Commission." *SFE*, Oct. 24, 1978.

Nolan, Dick. "Drakes Bay is His Place." *SFE*, Feb. 16, 1979, p. 37.

Irving, Carl. "Issue for Compromise: Where Drake Rested 400 Years Ago." *SFE*, Mar. 1, 1979, p. 9.

Staff writer. "Francis Drake Still at Sea in Marin County." *SFC*, Mar. 3, 1979, p. 2.

Verran, Roger. "Francis Drake's Silent Witness." *SFSEC*, Mar. 18, 1979, *This World*, pp. 24–30.

Gilliam, Harold. "The Mystery of Sir Francis' Bay Visit." *SFSEC*, Mar. 25, 1979, *This World*, p. 39.

Irving, Carl. "Correct or Not, Drake Gets Plaque." *SFE*, May 8, 1979, p. 3.

Hogan, William. "Quadricentennial." Book review for Hanna's *Lost Harbor*. *SFC*, June 8, 1979, p. 63.

Northwood, Bill. "Drake Drops Anchor in Oakland." *SFSEC*. June 10, 1979, p. 40.

Champion, Dale. "The Big Dig for Drake's Lost Fort." *SFC*, June 15, 1979, p. 4.

Irving, Carl. "A Giant Landmark to Honor Drake." *SFSEC*, June 17, 1979, p. A4.

Staff writer. "Sir Francis: Difficult to Live With, But He Had a Sense of Humor." *SFSEC*, June 17, 1979.

Doss, Margot Patterson. "Following Drake by 400 Years." *SFSEC*, June 17, 1979, p. Pun. 6.

Hamilton, Mildred. "Sir Francis Drake." *SFSEC*, July 1, 1979, p. Sce. 6.

Gilliam, Harold. "Another Clue in the 400-Year Mystery." *SFSEC*, July 1, 1979, *This World*, p. 42.

Clucas, Lowell. "The Land Called 'New Albion'." *SFSEC*, July 21, 1979, p. Pun. 6.

Clucas, Lowell. "Drake's 400th and Captain Cook." *SFSEC*, July 21, 1979, p. 34.

Creib, Ralph. "Pieces of China May Solve Drake Puzzle." *SFC*, May 5, 1980, p. 3.

Newhall, Scott. "Where Drake Landed—Exactly." *SFC*, May 12, 1980, p. 6.

Bender, Frederick. "Where Drake Landed" *SFC*, May 19, 1980, p. 56, letter to the editor.

Staff writer. "A New *Golden Hind* Ends World Trip." *SFC*, June 4, 1980, p. 16.

Irving, Carl. "Did Sir Francis Drake Really Land in Marin." *SFE*, May 9, 1983.

Irving, Carl. "Find Could Prove Francis Drake Didn't Land Here." *SFSEC*, Oct. 9, 1983, p. A1.

Irving, Carl. "A Hard Look/Proof that Golden Hind Docked at Drake's Bay." *SFE*, July 6, 1984, p. A1.

Pimsleur, J. "Mysterious Letter: Buried Drake Treasure Claimed to Be in Marin." *SFC*, Aug. 10, 1984, p. 28.

Taylor, Michael. "*Golden Hind* Discovers Oakland's Port Festival." *SFC*, Sept. 26, 1987, p. A2.

Todd, John. "Drake's Landing Discovered, Maybe." *SFSEC*, Jan. 22, 1989, p. B1.

Petit, Charles. "Another Volley Fired in Drake's Landing Debate." *SFC*, May 28, 1993, p. B3.

Newspaper Articles – "San Rafael Independent-Journal"

Staff writer. "Marin County in 1579 and First English Religious Service Held in America," Jan. 7, 1915.

Staff writer. "Drake Service Set for June 25," May, 1916 (photo of "big cross").

Staff writer. "Drakes Bay Still Entitled to Name," Oct. 30, 1919 (the Bolinas breast plate).

Staff writer. "Sir Francis Drake to Be Honored," June 16, 1921.

Staff writer. "English Lady Gives Five Pounds to Sir Francis Drake Monument," July 27, 1930.

Staff writer. "Sir Francis Drake Society Hears of Loss of Relics," Mar. 13, 1938 (a fire at Buckland Abbey).

Staff writer. "Bay Drake Plate Found Authentic: Seven Month Study by Columbia Savant Proves Relic Genuine," Dec. 1938.

Staff writer. "Drake's Plaque on Exhibit at Fair," Aug. 31, 1939.

Staff writer. "Drake's Plaque to be Exhibited at Manufacturers' Dinner," Sept. 10, 1946.

Livermore, Mrs. Norman B. "How Drake's Landing Spot was Acquired and a Suggestion for a Museum to Mark Historic Site," Feb. 13, 1948.

Staff writer. "Francis Drake, on Famed Journey Around the World, Came to Marin in 1579, Claimed Land for Britain," Oct. 16, 1948, p. M1.

Donnelly, Florence. "Monument to be Dedicated to Drake," Aug. 13, 1949, p. 1 (photo of 3,600-pound granite cross erected at Drakes Bay).

Staff writer. "County Accepts Memorial," Aug. 22, 1949.

Barthelmess, Mary. "Marin Indians Were Not Hiawatha Types," Nov. 11, 1950, p. M9.

Bagshore, Fred. "Did Drake Land in Marin? Good Evidence Found," Mar. 29, 1952 (Nicasio Indian Legend).

Takeshita, Wat. "Public Display Commemorates Drake's Landing in Marin," June 20, 1953.

Staff writer. "Marin Historian Pooh-Poohs Findings of Drake Guild," Sept. 25, 1954.

Dickinson, Bray. "Drake's Landing Place in California: Tomales Man Analyzes Evidence," Part 1, Oct. 2; Part 2, Oct. 9; Part 3, Oct. 16; Part 4, Oct. 22; and Part 5, Oct. 29, 1954.

Dickinson, Bray. "Historian Questions Authenticity of Drake's Plate of Brass," Dec. 11, 1954, letter to editor.

Cain, Mary. "English Visitor Sees Drakes Bay," Dec. 10, 1955, p. M3.

Cain-Minton, Mary. "Drake's California Landing Site Believed Pinpointed," June 16, 1956, p. M7 (Drake Navigators Guild's discovery announcement).

Staff writer. "Monterey Writer Slams Marin Claim," July 6, 1956.

Dickinson, Bray. "Drake Landing Site Claims Argued," July 28, 1956.

Staff writer. "Archaeologist at Site Where Halberd Found," Dec. 1, 1958, p. 10.

Power, Robert H. "Drakes Estero or Golden Gate? Robert Power Presents Theory That *Golden Hind* Entered Bay," Aug. 8, 1959, p. M1.

Draper, Benjamin. "A Marinite Visits Drake's Home Base: Mill Valley Man Reports on Stay at Plymouth During British Trip," May 19, 1962, p. M6.

Staff writer. "Oko's Article on Drake is Sent Out," Aug. 27, 1964, p. 34.

Staff writer. "Did a Corny Sir Francis Ever Nap in Shade of a Marin Tree," Dec. 9, 1965, p. 28.

Farrara, Frank. "Ship Relic Found at Drakes Bay," Apr. 19, 1966, p. 18.

Staff writer. "Visitor's Bureau Works on Worthy Drake Project," June 15, 1966, p. 34 (plans for statue and museum).

Staff writer. "Pirate's Cove as Drake's Landing Site," Feb. 23, 1967, p. 1.

Yates, John. "The Sandspit Was Named Limantour," Mar. 31, 1973, p. M11.

Spiro, Leon. "Thinks 'Plate of Brass' is Merely Hoax," Jan. 3, 1974, letter to editor.

Greer, Jeff. "Researchers Believe Drake Built Bolinas Lagoon Fort," Jan. 10, 1974.

Staff writer. "Brass Mortar Found in Church May Have Belonged to Drake," Jan. 10, 1974, p. 21.

Staff writer. "Drake's Flagship Sails Again," Jan. 16, 1974 (*Golden Hind II*, picture).

Dines, Glen. "Thinks Greenbrae Site More Likely Landing," Jan. 19, 1974, letter to editor.

Spiro, Leon. "Seek Drake Artifacts Farther North," Jan. 28, 1974, letter to editor.

Spiro, Leon. "Drake's Plate is Claimed a Hoax," Mar. 8, 1974, letter to editor.

Drimmel, J. R. "More Mysteries About Plate of Brass," Mar. 16, 1974, letter to editor.

Greer, Jeff. "Diggers Seek Drake Clues at Bolinas Lagoon Site," July 23, 1974.

Staff writer. "New Clues on Drake's Fort," July 26, 1974, p. 27.

Greer, Jeff. "New Evidence on Drake's Landing," Aug. 3, 1974, p. 1.

Leydecker, Mary. "Competing Drake Fort 'Sites' Inspected," Oct. 28, 1974, p. 1.

Leydecker, Mary. "Drake Coin at Olompali?" Nov. 8, 1974, p. 1.

Staff writer. "The Drake Question, and One Answer," Apr. 5, 1975, p. M1.

Gardiner, Dorothy. "Draper Explores Drake after Scholarly Voyage," Oct. 18, 1975, p. M5.

Staff writer. "Queen Invited to Marin," May 6, 1977.

De Brow, Rick. "Sir Francis Drake Gave Vision of an Empire to Elizabeth I," May 14, 1977, p. 30.

Greer, Jeff. "Bolinas, Drake's Bay Elsewhere?" May 14, 1977, p. 30.

Staff writer. "Drake's Plate Tagged a Forgery," July 27, 1977.

Staff writer. "New Group Making Plans for Drake Celebration," Nov. 3, 1977.

Staff writer. "Drake Began Voyage 400 Years Ago Today," Dec. 13, 1977, p. 17.

Locati, Jeffrey. "Sir Francis Drake's Exotic Welcome in Marin, 1579," May 6, 1978, p. 7.

United Press Service. "Proposed Drake Stamp," Apr. 22, 1978, p. 8.

Smith, Ed. "British Inspect Drake Site," Sept. 13, 1978, p. 1.

Smith, Ed. "A New Twist in Drake Tale," Sept. 25, 1978, p. 1. Bolinas Plate.

Smith, Ed. "Commission Will Decide Drake's Landing Spot," Oct. 12, 1978, p. 1, third section.

Smith, Ed. "Stinking Fog Hits Shipload of Drake Buffs," Oct. 23, 1978, p. 1.

Staff writer. "Frustrating Hunt for Drake's Safe Harbor," Oct. 26, 1978, p. 50.

Mouate, Lucia. "A Seafaring Adventure to Commemorate," Jan. 12, 1979, p. 1.

Smith, Ed. "Drake Fete is Planned for Summer," Jan. 18, 1979, p. 1.

Staff writer. "Marinites are Named to Drake Board," Jan. 31, 1979.

Staff writer. "No Plaque for Drake in Marin," Mar. 3, 1979, p. 1.

Nixon, Stuart. "Argument Continues to Rage Over Where Drake Landed," Mar. 10, 1979, p. 5.

Smith, Ed. "Two Drake Plates Planned," May 12, 1979, p. 3.

Staff writer. "Drake's Plate Spawns Satirical Berkeley Skit," Apr. 6, 1979.

Staff writer. "Drake Players on Film," Apr. 16, 1979, p. 1.

Smith, Ed. "Drake's Plate Probably Fake Scholars Say," May 4, 1979, p. 1.

Staff writer. "A Marin Visitor Who Made History," June 5, 1979, p. 14.

Staff writer. "Drake's Visit May Remain a Mystery," June 5, 1979, p. 1.

Larsen, Rebecca. "Drake and the Indians: an Ominous Encounter," June 7, 1979, p. 1.

Smith, Ed. "Rerun of Drake's Landing," June 8, 1979, p. 1.

Angle, Pat. "Drake's Voyage—Another Epic in Exciting Times," June 9, 1979, p. 1.

Smith, Ed. "Drake Challenges Spain," June 11, 1979, p. 1.

Horowitz, Donna. "Drake Hated the Spanish," June 12, 1979, p. 1.

Staff writer. "Where did Drake Land? Three Possible Marin Sites," June 12, 1979, p. 12.

Leydecker, Mary. "Drake: Explorer Held Religious Services," June 13, 1979, p. 1.

Smith, Ed. "Ceremonies for Drake Plaque Set," June 14, 1979.

Greer, Jeff. "New Drake Salvo," June 14, 1979, p. 1.

Whittington, Mark. "Actors More Interested in Drake Than Debate," June 15, 1979, p. 1.

Sias, Spencer. "Drake Chronicler Didn't Like Fog," June 15, 1979, p. 10.

Bernstein, Ed. "Drake's Ghostly Visit," June 16, 1979, p. 1.

Young, Barbara. "A Visit with an Oddly Reluctant Drake Heir," June 16, 1979, p. 40.

Angle, Pat. " 'Drake' Gives View of Landing Site," June 16, 1979, p. 40.

Voris, Linda. "Drake Opened New Vistas to Trade, Travel," June 17, 1979, p. 45.

Smith, Ed. "Wind Whips Drake' Landing," June 18, 1979, p. 1.

Staff writer. "Drake Honored with Guild Plaque," July 30, 1979, p. 3 (photo of Raymond Aker).

Miller, Jack. "Americans Can Find Their Historical Roots in Plymouth," Jan. 12, 1980, p. 33.

Greer, Jeff. "Pottery Fragments Provide Piece in Drake Puzzle," May 5, 1980, p. 1.

Staff writer. "Chinese Puzzle Over Drake's Visit," May 29, 1980, p. 46.

Staff writer. "A New *Golden Hind* Ends World Trip," June 4, 1980.

Keown, Don. "Drake's Lost Harbor: is it at Bolinas Beach?" June 10, 1980, p. 15.

Keown, Don. "Did Drake Land at Bolinas?" June 11, 1980.

Staff writer. "Drake Post Card Delayed," Sept. 10, 1980.

Greer, Jeff. "Goleta Guns Aim at Drake Theory," Mar. 31, 1983, p. A1.

Greer, Jeff. "More Tests Set on 'Drake' Guns," Oct. 10, 1983.

Ashley, Beth. Lifestyle column: "The Mystery of Drake's Legend," June 24, 1984.

Greer, Jeff. "Did Drake Land at Agate Beach?" July 6, 1984, p. 1.

Staff writer. "Drake Slept Here; No, it Was There," July 9, 1984.

Greer, Jeff. "$15 Million Treasure Buried in Marin?" Aug. 11, 1984, p. A1.

Chui, Glennda. "Drake's Legacy: Tests May End Brass-Plate War," Nov. 24, 1986.

Akre, Brian. "New Theory Proposes Drake's Bay Really Belongs in Oregon," Feb. 2, 1987, p. 1.

Porter, Patricia. "Funds for Drake: Explorer's Home to Be Renovated," Nov. 8, 1987, p. A1.

Epperson, George. "Drake Treasures" in *Reader's Forum*, Sept. 8, 1987.

Greer, Jeff. "First-Day Stamp for Marin," Nov. 7, 1989, p. 1.

Horowitz, Donna. "New Test May Prove Drake Left That Plate," July 18, 1991, p. A1.

Horowitz, Donna. "New Test: Drake Plate a Fake," July 21, 1991, p. 1.

Neill, Alex. "County to Consider Plaque for What May Be Drake's Landing," June 24, 1992.

Horowitz, Donna. "Drake Finally Gets His Due: Supervisors Declare Explorer Landed Off Point Reyes," June 16, 1994.

Horowitz, Donna. "Debate Rages Over Drake's Landing Site," Aug. 24, 1994.

Newspaper Articles – Other Papers

Dana, Julian. "The Drake Anchor." *Pony Express Courier*, Placerville, June 1937.

Staff writers. "Discovery Here of Drake Scroll May Change Pacific History—Seemingly Irrefutable Evidence That English Navigator Predated Vizcaíno; Drake Story is too Big to Keep Secret; Check Authenticity of Lead Tablet Found in Ancient Bottle; Study of Drake Plate Suggested by Historian; Monterey Drake Scroll Checks with Known Historical Facts; Could Be Fake, But Clever One." *Monterey Peninsula Herald*, Dec. 7, 1949, p. 1.

Costello, Jimmy. "Myron Oliver Says Bottle Real Antique; Tells of Removal of 400 Year Old Scroll. *Monterey Peninsula Herald*, Dec. 12, 1949.

Staff writer. "Rare Old Teacup Found in Search for New Relics of Drake's Visit." *Monterey Peninsula Herald*, Feb. 25, 1950.

Staff writer. "Anchor to Mark Drake's Landing." *The Baywood Press*, Mar. 25, 1954, p. 7.

Staff writer. "Robert H. Power." *Redwood City Tribune*, June 18, 1959.

Sagen, Ethel. "Kern Researcher Says Drake Landed at San Luis Obispo Bay." *Bakersfield Californian*, Feb. 24, 1968.

Yarish, Alice. "The Landing of Francis Drake." *Pacific Sun*, Nov. 5, 1969, p. D3.

Lyhne, Bob. "San Francisco Bay Discovery: the Case for Drake." *Redwood City Tribune*, Apr. 3, 1971, *Peninsula Living*, p. 8.

Theiler, Don. "*Elizabethan California*: Book on Drake's Landing Theory is Timely." *Perra Linda News*, Mar. 19, 1975, p. 16.

Staff writer. "Remains of Sir Francis Drake's Ship Are Sought in London." *New York Times*, Oct. 30, 1977.

Staff writer. "Drake Exhibit at Civic Center Library." *San Rafael News Pointer*, July 18, 1979, p. 1.

Sandrock, Fred. Book Review: Warren Hannah, *Lost Harbor*. *Twin Cities Times*, July 26, 1979.

Breithaupt, Brad. "First Day Issue for San Rafael." *San Rafael News Pointer*, Nov. 24, 1980.

Sullivan, Steve. "Did Sir Francis Drake Repair His Ship at Goleta Slough?" *Santa Barbara News-Press*, Mar. 21, 1982.

Power, Robert. "Those Cannons Weren't Drake's." *Santa Barbara News-Press*, June 20, 1982.

Gilmore, Jim. "Those Goleta Beach Cannon." *Santa Barbara News-Press*, Aug. 8, 1982.

Sullivan, Steve. "Goleta Cannons Not Left by Drake, Claims History Buff." *Santa Barbara News-Press*, Nov. 2, 1982.

Frost, Frank. "Goleta Cannons Remain a Mystery." *Santa Barbara News-Press*, May 28, 1983.

Murphy, William. "Porcelain Indicates *Golden Hind* Anchored at Point Reyes." *Los Angeles Times*, June 15, 1984.

Reutinger, Joan. "Drake's Secret Centers in Bolinas." *Coastal Post*, July 16, 1984.

Reutinger, Joan. "Drake's Buried Treasure on West Marin Coast." *Coastal Post*, July 30, 1984.

Reutinger, Joan. "The Biggest Treasure in All History Is Out There," *Coastal Post*, Aug. 20, 1984.

Merchant, Jennifer. "New Claim that Drake Really Landed at Bolinas." *Point Reyes Light*, Dec. 1, 1988, p. 1.

Zimmerman, Joy. "The Great Drake Debate." *Pacific Sun*, Dec. 9, 1988, p. 1.

Staff writer. "Foundation to Hold Funds for Sir Francis Drake Statue." *Twin Cities Times*, Jan. 4, 1989.

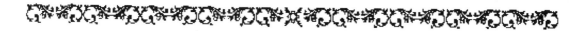

Index

W

Y

Z

Fig. 176 The author completing his photographing at Bodega Head. Brian Kelleher, a self-employed, free-lance consultant in the environmental management field, resides in Cupertino, California. He conducts the majority of his work for the California Superior Court, assisting special masters, mediators and private parties resolve complicated lawsuits associated with contaminated commercial and industrial properties. The author grew up in Medfield, Massachusetts, obtained his B.A. (Environmental Biology) at Lake Forest College, Lake Forest, Illinois (1974), and his M.S. (Environmental Engineering) at the University of Lowell, Lowell, Massachusetts (1981). This is his first book. *Photograph by Teri Fung, November 1996.*

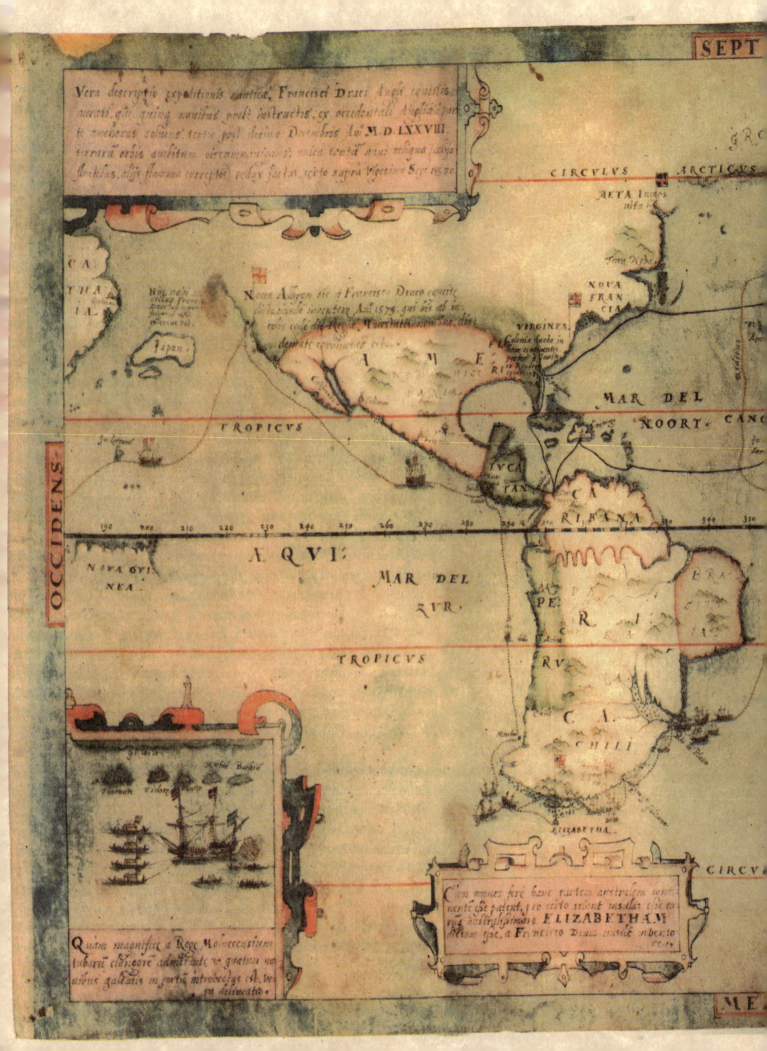

Vera descriptio expeditionis nautica, Francisci Draci Angli, equitis aurati, qui quinq nautibus probe instructis, ex occidentali Angliæ parte ancheras soluens, tertio post decimo Decembris Añ M.D.LXXVIII. terrarū orbis ambitum circumnauigans, unica tentā auaus reliqua satis soluebus, alijs flumina correptis, redux factus, ipso te supra vigesimo Sept 1580.

CIRCVLVS ARCTICVS

META Incognita

Tera Noba

NOVA FRANCIA

CA THA IA

Japan

VIRGINEA

Noua Albyon sic a Francisco Draco equite illustrissimo inventore Añ 1579. qui eis ab incolis cede die Regiæ Maiestatis nomine, diademate coronatus est.

A M E R I C A

MAR DEL NOORT. CANC.

TROPICVS

IVCATAN

OCCIDENS

190 200 210 220 230 240 250 260 270 280 290 300 310 320 330 340 350

CA RIBANA

NOVA GVINEA

A Q V I

MAR DEL ZVR

PE RV

BRA

R A

TROPICVS

RV

C A CHILI

ELIZABETHA

CIRCVL

Quam magnifice a Rege Moluccensium tubarū cldr oore admiratus et quatuor nauibus galeatis in portu introductus est, te in delineatio.

Cum omnes fere hanc partem australem continente ese putent, pro certo sciant insulas ese terras australissimam ELIZABETHAM dictam ipse, a Francisco Draco eques inberto eius.

ME